T0185958

Studies in Universal Logic

This series is devoted to the universal approach to logic and the development of a general theory of logics. It covers topics such as global set-ups for fundamental theorems of logic and frameworks for the study of logics, in particular logical matrices, Kripke structures, combination of logics, categorical logic, abstract proof theory, consequence operators, and algebraic logic. It includes also books with historical and philosophical discussions about the nature and scope of logic. Three types of books will appear in the series: graduate textbooks, research monographs, and volumes with contributed papers. All works are peer-reviewed to meet the highest standards of scientific literature.

More information about this series at http://www.springer.com/series/7391

Andrzej Indrzejczak

Sequents and Trees

An Introduction to the Theory
and Applications of Propositional Sequent
Calculi

 Birkhäuser

Andrzej Indrzejczak
Institute of Philosophy
University of Łódź
Łódź, Poland

ISSN 2297-0282 ISSN 2297-0290 (electronic)
Studies in Universal Logic
ISBN 978-3-030-57147-4 ISBN 978-3-030-57145-0 (eBook)
https://doi.org/10.1007/978-3-030-57145-0

Mathematics Subject Classification: 03F03, 03F05, 03F07, 03F52, 03B05, 03B35, 03B45, 03B50, 03B20, 03B53, 03B70

This book is published under the imprint Birkhäuser, www.birkhauser-science.com by the registered company Springer Nature Switzerland AG
The registered company address is: Gewerbestrasse 11, 6330 Cham, Switzerland

Preface

There are a lot of excellent works in proof theory using various forms of Sequent Calculi (SC) as the basic formal systems.[1] This book is of a different character. It is not a systematic exposition of proof theory using sequent calculus but it is a methodological study of sequent calculi as such. Of course it is not possible to present sequent calculi and their applications and to avoid a presentation of important issues in proof theory. However, our aim is to focus on the tools not on the results; the latter will serve as examples illustrating proof techniques developed in the framework of sequent calculi. In my opinion sequent calculi are truly amazing kind of formal systems and they deserve a separate study. In what follows we will present several variants of rules and calculi, describe their merits and focus on their applications. In particular, we will present in detail several methods of proving the most important results like the cut-elimination theorem, completeness, decidability, interpolation.

The book is elementary and self-contained. All technical details will be introduced both formally and with informal explanations. Proofs will be presented in detail but with a lot of gaps left as exercises for the careful reader. As our aim is to show how to use sequent calculi we must underline here that it is essential to do most of the exercises to gain a working knowledge of the methods presented. There is a tendency in contemporary works to present very general results. This is theoretically valuable but for a novice it may be hard to follow. In this book we opt for a different route; we rather present several case studies which are much simpler to grasp. After careful study of such special cases, the general and uniform treatment of broad classes of similar problems may be developed more easily.

We focus on proofs. Not so much on proofs in SC since their construction is rather simple and any reader who has worked with tableau systems should have no problem with proof search in SC. We rather focus on proofs of the most profound results concerning SC. We also want to show how easily

---

[1]For example: Takeuti [261] Troelstra and Schwichtenberg [264], Buss [43], Negri and von Plato [185, 186], Bimbo [31] to mention just a few.

SC may be used as a general and extremely elegant framework for doing general metalogic, usually presented in the setting of axiom systems.

In particular, we are going to focus on different methods of proving the cut-elimination (or admissibility) theorem which is one of the most important results of proof theory established in the framework of SC. References to Cut-Elimination Theorem (CET) are ubiquitous in the logical literature. One may find them in so diverse fields as the foundations of mathematics, automated deduction, logical programming or philosophical logic. Despite the widespread applications of this result, it is formulated in the specific framework of sequent calculi and related systems like tableaux. Roughly speaking, the cut-elimination theorem shows that every application of this important rule called cut may be eliminated from any proof in SC for some logic or theory. Although the rule of cut is in a sense essential for any SC, the possibility of its elimination yields many important consequences like consistency, decidability, interpolation, for many logics and theories.

SC was developed in its modern form by Gerhard Gentzen [95] for intuitionistic and classical logic in 1934, although the notion of a sequent was introduced earlier by Paul Hertz [114]. In fact, Gentzen was primarily concerned with Natural Deduction (ND) and SC was devised as a technical tool for investigations on the structure of proofs in ND. Gentzen's idea was not only to provide rules of ND but also to show that every ND-proof may be transformed into normal form (with no detours). However, he was able to show it directly only for intuitionistic but not for classical logic, so he resigned from its publication (see von Plato [197]). Instead he introduced SC for both logics, proved that every proof in SC may be transformed into a cut-free proof, and finally that this implies the existence of ND normal proofs for every thesis of intuitionistic and classical logic.

Although SC was treated originally as a technical tool for investigations on the properties of ND-proofs, soon it became of interest on its own for researchers in proof theory like Ketonen [148], Curry [55, 56], Kleene [149], to mention just a few names important for the earliest stage of development.

Since then many variants of SC were devised which are suitable for dealing with various non-classical logics and formal theories. In addition, many generalised versions of SC were provided like display calculi [25], hypersequent calculi [10], many-sided SC [221] and the like. SC was also influential in the field of investigation on automated proof search leading to the invention of early forms of tableau calculi. The development of SC and its methodology was particularly important for research on mathematical theories and proofs [186], proof theory of non-classical logics [31], and investigations on the philosophy of meaning (of logical constants) [229]. Summing up, two things seem to be indisputable, that SC is a standard tool of modern proof theory, and that the cut-elimination theorem is treated as the most important result in the field.

One may wonder why this result is so important that already Gentzen used the name 'Hauptsatz' (the main theorem) for it. In fact, the rule of cut is essential for any SC; Gentzen needed it for showing the equivalence of SC with axiomatic formulations of intuitionistic and classical logic (namely for the simulation of applications of Modus Ponens). This is not all—depending on the interpretation of sequents, the cut rule may be seen as expressing the transitivity of the consequence relation induced by SC, or as encoding the process of using lemmata in proof construction, or even as expressing the principle of bivalence (see section 1.8) On the other hand, it is welcome not to have cut as a primitive rule of SC. Gentzen was well aware of the importance of cut elimination and used it not only for showing the existence of normal ND-proofs, but also for showing consistency and decidability of propositional intuitionistic and classical logic. The list of important consequences of cut elimination may be enlarged with many technical (e.g. automated deduction, interpolation) and philosophical (e.g. analytic proofs, proof-theoretical semantics) results. We will describe some of these consequences but the main part of the book is concerned with an analysis of different proofs of cut elimination as these methods are of interest on its own and provide beautiful examples of the application of formal methods.

The Gentzen's original proof is brilliant yet quite complicated. Since then many other proof methods for CET were proposed. One can divide them roughly into indirect (semantic) and direct (constructive) proofs. The former show how to obtain cut-free proofs just from the beginning, whereas the latter are based on syntactical transformations of proofs either of local (e.g. Dragalin [64], Schütte [232]) or of global character (e.g. Curry [56], Buss [43]). Some of the methods were suitably abstracted and generalised in order to apply them in the framework of generalised variants of SC, and for many non-classical logics.

Although we focus on the presentation of a variety of proofs of the cut-elimination theorem, this book contains also a number of proofs of other remarkable results. In particular, we provide numerous examples showing that the application of SC yields simpler or more direct or elegant proofs of well-known results like Kalmar's or Henkin's proofs of completeness. We believe that a reader, even if well acquainted with SC and its applications, will find also some relatively new things. In particular, connections between invertibility and permutability of rules, confluence, proof search and decidability of systems are described in detail. Some of the proofs may seem difficult to follow. In order to make their understanding easier a lot of details, usually omitted, are explicitly stated. Following Bertrand Russell's (or perhaps Albert Einstein's) advice, I was trying to put things as simple as possible but not simpler.

In chapter 1, after a brief introduction to classical propositional logic, we present the simple sequent calculus K and its features like the invertibility of all rules. Sequents in K are built from sets of formulae which simplify some proofs considerably and make it particularly well-behaved for beginners. We will show several proofs of completeness for K, a simple proof of admissibility of cut rule, as well as a proof of decidability. The chapter ends with considerations of strong completeness and cut elimination in the presence of additional axiomatic sequents.

Chapter 2 is concerned with Gentzen's original system LK which significantly differs from K. Sequents of LK are built from finite lists of formulae, many rules have different shapes and are not invertible and, in addition to rules for the connectives, there are special structural rules. This chapter contains a presentation of many proofs of eliminability or admissibility of cut and their analysis. Moreover, questions of the generation of equivalent rules and permutability of rules are discussed.

Chapter 3 introduces calculus G3 which is similar to K but allows for a more subtle analysis of proof structure. Several admissibility results for structural rules are presented and some more proofs of cut admissibility of different sorts. Moreover, this chapter contains a discussion of different interpretations of sequents as well as a presentation of types of sequent calculi and sequent rules. Additionally some constructive proofs of the interpolation theorem are presented.

The last two chapters show how SC may be applied to non-classical logics. They are more concise in comparison with the three chapters devoted to classical propositional logic and strongly dependent on the results presented earlier. The focus is on the presentation of a variety of approaches and a variety of techniques for resolving problems. We try to be comprehensive, but of course it was unavoidable to refrain from presenting many interesting and important results. In many places we decided to present in more detail things which are not widely known and sometimes not easily accessible but interesting in the Author's opinion. On the other hand, many important but well-documented issues are only mentioned with references to accessible sources.

In chapter 4 we focus on a few modal logics which are extensions of classical logic. It is shown how standard techniques may be extended to logics which require rules of a different kind from those of classical logic. Since for some modal logics standard solutions do not work, we must introduce generalised sequent calculi. This is illustrated on the basis of **S5**, one of the most important modal logics.

In chapter 5 we are concerned with non-classical logics which are weaker than classical logic. We discuss applications of SC to intuitionistic, substructural (including relevant logics) and many-valued logics. In each case the problems connected with the extension of standard solutions are discussed and some generalised sequent calculi are introduced that are well suited for the particular logics under consideration. In both chapters we focus on the cut-elimination theorem but other issues like proof search and decidability are also discussed.

Let me finish with a comment concerning the restriction of this volume to propositional logics. It was mainly connected with the aim of presenting essential methods and techniques of SC without being involved in discussing too many details concerning richer languages and semantic structures. However, a continuation of this volume is in progress which will be devoted to the applications of sequent calculi to quantified logics, including some non-classical ones, to logics in extended languages, and to some important applications to mathematical theories. In a similar way this sequel is planned as a series of detailed case studies showing the scope of applications, their limitations and ways of overcoming problems.

Parts of this book are based on my earlier publications, in particular:

Sections 1.6–1.10, 2.1–2.2, 2.4–2.5, 3.1–3.3 are partially based on my book (in Polish) 'Sequent Calculi in Classical Logic' published in 2013 by the University of Lodz Publishers.

Section 3.5 is partially based on [133].

Parts of sect 4.5 are based on [138] and [139].

Parts of sect 4.7 are based on Bednarska and Indrzejczak [24].

Łódź, Poland   Andrzej Indrzejczak
2019

Acknowledgments

I would like to express my gratitude to my colleagues from the Department of Logic of the University of Łódź for fruitful cooperation and advices. I'm also greatly indebted to Andrzej Pietruszczak, Michał Zawidzki, Nils Kürbis and Yaroslav Petrukhin for cooperation and valuable comments, and to Jean-Yves Beziau for his encouragement to writing this book. As for the technical side I am greatly indebted to Sam Buss for the possibility of using bussproofs.sty—this makes life with trees easier. Last but not least I would like to thank my family for support and patience.

The work on the book was supported by the National Science Centre, Poland (grant number: DEC-2017/25/B/HS1/01268).

Łódź, Poland   Andrzej Indrzejczak
2019

Contents

Chapter 1

Analytic Sequent Calculus for CPL

The sequent calculus was invented by Gerhard Gentzen in the 1930s and since then many variants of this system were proposed. In what follows we will be using a generic term 'sequent calculus' (SC) and specific names for some concrete variants. We start our considerations not with the Gentzen's original system LK but with the system K which is essentially due to Oiva Ketonen. K differs from Gentzen's original calculus in many respects, concerning the rules and even the notion of a sequent. However, the system K is better suited as a starting point for showing the most important techniques and results developed for sequent calculi, as we are not disturbed with too many technical details. The presentation of Gentzen's calculus LK is postponed to chapter 2.

In this chapter we start with semantical and axiomatic presentations of classical propositional logic **CPL**. In section 1.2 we introduce the notion of a sequent and sequent rules of the calculus K. Proofs and derivations in K (as well as in any other SC) will be characterised and illustrated with a few examples in section 1.3. The problem of additional, derivable and admissible, rules is discussed in section 1.4, where also a notion of invertibility of rules is presented. Section 1.5 contains a simple proof that K adequately characterises **CPL**, whereas the next section discusses the decision procedures at length. In section 1.7 a different kind of proof of completeness is presented which is more general than that of section 1.5. In brief, sections 1.2–1.7 contain a presentation of essential features and techniques of the analytic form of SC.

Section 1.8 introduces the cut rule, one of the most important rules in the framework of SC, and we prove that this rule is admissible in K. Cut itself destroys the analytic character of K but enables additional important applications of SC. As an illustration of advantages of having cut, a purely syntactical proof of equivalence

© Springer Nature Switzerland AG 2021
A. Indrzejczak, *Sequents and Trees*, Studies in Universal Logic,
https://doi.org/10.1007/978-3-030-57145-0_1

of K with an axiomatic characterisation of **CPL** is demonstrated in section 1.9. The next section shows that in the presence of cut we can also apply standard Henkin-style techniques for proving the completeness of K. However, such a proof is not constructive and not analytic in the sense explained in section 1.7. Finally in section 1.11 we take up a problem of taking control of the applications of cut in proofs. In particular, we present a slightly modified completeness proof with some restrictions on the applications of cut which restore analyticity. Moreover, we show that in the presence of (restricted) cut we can strengthen the completeness result of section 1.7 to cover theories and to prove the interpolation theorem. This section contains also a brief discussion of some other advantages of cut connected to the complexity of proofs.

1.1   The Classical Propositional Calculus

Let us recall some basic facts concerning **CPL** since most of the material of this volume works on sequent formalizations of this logic. The reader may skip this section and take a look at it when necessary to understand notation and conventions that will be used extensively in the text.

1.1.1   Language

In what follows we will use a rather standard form of a propositional language for classical logic. Let $\mathbf{L_{CPL}}$ denote an abstract algebra of formulae:

$$\langle FOR, \neg, \wedge, \vee, \rightarrow \rangle \tag{1.1}$$

with a denumerable set of propositional symbols, also called the atomic formulae (or just atoms).

$$PROP = \{\, p, q, r, \ldots, p_1, q_1, \ldots \} \subseteq FOR \tag{1.2}$$

Operations of this algebra correspond to the well-known functors (connectives) of *negation, conjunction, disjunction* and *implication*. The signs φ, ψ, $\chi, \ldots$ will be used to denote any formulae in $\mathbf{L_{CPL}}$ and in each language considered further. $\Gamma, \Delta, \Pi, \Sigma, \ldots$ will be used to denote sets (or multisets or lists in later chapters) of formulae. For unary functors we will use prefix notation, and for binary functors — infix notation; in particular the rules for generating the set FOR of all formulae of **CPL** look as follows:

- if $\varphi \in FOR$, then $\neg\varphi \in FOR$;

- if $\varphi \in FOR$ and $\psi \in FOR$, then $(\varphi \odot \psi) \in FOR$, where $\odot \in \{\wedge, \vee, \rightarrow\}$

Incidentally, we will use also propositional constants: $\bot, \top$ and binary equivalence connective $\leftrightarrow$, defined in a standard way:

Definition 1.1. ($\bot, \top, \leftrightarrow$).

$$\bot := p \wedge \neg p; \quad \top := \neg\bot$$

$$\varphi \leftrightarrow \psi := (\varphi \rightarrow \psi) \wedge (\psi \rightarrow \varphi)$$

To limit the number of necessary parentheses, we employ a convention concerning the strength of argument's binding. For binary functors we assume that $\wedge$ binds tighter than $\vee$, and $\vee$ binds tighter than $\rightarrow$[1]. Negation, and all unary functors in general, are assumed to bind their arguments tighter than binary functors. Additionally we omit outer parentheses for any formula and inner parentheses in case of many occurrences of associative operations like conjunction or disjunction (we admit in this way n-ary conjunctions and disjunctions for $n > 2$). Thus:

$$p \vee \neg q \wedge r \rightarrow p \wedge \neg s \vee q \wedge s \wedge t$$

is a legal abbreviation of the official:

$$((p \vee (\neg q \wedge r)) \rightarrow ((p \wedge \neg s) \vee (q \wedge (s \wedge t))))$$

Let us define some special types of formulae:

Definition 1.2. (Normal forms).

- *An atom or its negation is called a literal (positive or negative).*
- *An n-ary conjunction of literals is an elementary conjunction.*
- *An n-ary disjunction of literals is a clause.*
- *Horn clause is a clause with at most one positive literal (atom).*
- *A formula is in conjunctive normal form (CNF) iff it is a conjunction of clauses.*
- *A formula is in disjunctive normal form (DNF) iff it is a disjunction of elementary conjunctions.*

Remark 1.1. Notice that in case of clauses, elementary conjunctions and normal forms we admit as a special case one-element (reduced) conjunctions and disjunctions. For example, p is at the same time a clause (a Horn clause in particular), an elementary conjunction and a formula in CNF or DNF; $p \vee q$ is a clause (but not a Horn clause), but also a formula in CNF (a one-element conjunction) and in DNF (two-element disjunction of one-element elementary conjunctions). Notice also that if $\bot$ and $\top$ are in use they are also treated as literals.

A set of atoms occurring in a formula or a set of formulae will be denoted by $PROP(\varphi)$ or $PROP(\Gamma)$. A set of subformulae of φ or Γ will be denoted by $SF(\varphi)$ ($SF(\Gamma)$) and the complexity of a formula as $c(\varphi)$. These notions are defined as follows:

---

[1] In case we use the additional functor $\leftrightarrow$, we assume that $\rightarrow$ binds tighter than $\leftrightarrow$.

Definition 1.3. (Subformulae, complexity).

- *for $\varphi \in PROP$, $SF(\varphi) = \{\varphi\}$ and $c(\varphi) = 0$;*
- *for $\varphi := \neg\psi$, $SF(\varphi) = \{\varphi\} \cup SF(\psi)$ and $c(\varphi) = c(\psi) + 1$;*
- *for $\varphi := \psi \odot \chi$, where $\odot \in \{\wedge, \vee, \rightarrow\}$, $SF(\varphi) = \{\varphi\} \cup SF(\psi) \cup SF(\chi)$ and $c(\varphi) = c(\psi) + c(\chi) + 1$;*
- *$SF(\Gamma) = \bigcup\{SF(\varphi) : \varphi \in \Gamma\}$.*

ψ and χ are immediate subformulae of φ, and φ is (the only) improper subformula of itself.

1.1.2 Semantics

We will interpret the language of **CPL** by means of valuations, these being mappings of the form: $V : PROP \longrightarrow \{1, 0\}$, where 1 is interpreted as truth and 0 as falsity. Every valuation uniquely determines an interpretational structure (a model $\mathfrak{M}$) for the whole language. It is specified by means of a satisfaction relation $\vDash$ in the following way:

$$
\begin{array}{lll}
\mathfrak{M} \vDash \varphi & \text{iff} & V(\varphi) = 1, \text{ for any } \varphi \in PROP \\
\mathfrak{M} \vDash \neg\varphi & \text{iff} & \mathfrak{M} \nvDash \varphi \\
\mathfrak{M} \vDash \varphi \wedge \psi & \text{iff} & \mathfrak{M} \vDash \varphi \text{ and } \mathfrak{M} \vDash \psi \\
\mathfrak{M} \vDash \varphi \vee \psi & \text{iff} & \mathfrak{M} \vDash \varphi \text{ or } \mathfrak{M} \vDash \psi \\
\mathfrak{M} \vDash \varphi \rightarrow \psi & \text{iff} & \mathfrak{M} \nvDash \varphi \text{ or } \mathfrak{M} \vDash \psi
\end{array}
$$

Hence $\mathfrak{M} \vDash \varphi$ means that a formula φ is true (satisfied) in the model $\mathfrak{M}$, or simply that $\mathfrak{M}$ is a model of φ. $\mathfrak{M} \nvDash \varphi$ means that a formula φ is false (falsified) in this model. $\mathfrak{M} \vDash \Gamma$ means that $\mathfrak{M} \vDash \psi$ for every $\psi \in \Gamma$. $\mathfrak{M} \nvDash \Gamma$ means that at least one formula in Γ is false in this model. Below we display the most important semantical notions:

Definition 1.4.   - *φ (or Γ) is satisfiable iff there exists a model in which it is satisfied (it has a model);*

- *φ (Γ) is falsifiable iff there exists a model in which it is false;*
- *φ (Γ) is unsatisfiable iff no model satisfies it;*
- *φ is valid (tautology) ($\vDash \varphi$) iff $\mathfrak{M} \vDash \varphi$, for every $\mathfrak{M}$;*
- *φ logically follows from Γ, or is a logical consequence of Γ ($\Gamma \vDash \varphi$) iff $\mathfrak{M} \vDash \Gamma$ implies $\mathfrak{M} \vDash \varphi$, for every $\mathfrak{M}$.*

$\nvDash \varphi$ denotes an invalid formula and $\Gamma \nvDash \varphi$ means that φ does not follow from Γ.

Let us notice the following:

Claim 1.1.   - *$\nvDash \varphi$ iff φ is falsifiable iff $\neg\varphi$ is satisfiable.*

- $\Gamma \models \varphi$ *iff* $\Gamma \cup \{\neg\varphi\}$ *is unsatisfiable.*

Claim 1.2. *1. For any clause φ, $\models \varphi$ iff it contains some atom and its negation.*

2. For any φ in CNF, $\models \varphi$ iff $\models \psi$, for any clause ψ which is an element of φ.

1.1.3 Axiomatization

The earliest, and still the most popular, syntactic style of defining logics was axiomatic, especially in the form provided by Hilbert. An axiomatic system consists of two sets: (schemata of) axioms and rules. In what follows such systems, after Hilbert will be called H-systems (although the name 'Frege systems' is also popular). We do not aim to focus on axiomatic formalizations and their properties, but it is handy to specify some concrete system for further references. Any axiomatic (or Hilbert) formalisation of the logic **L** will be denoted as H-**L**. In particular our chosen H-**CPL** consists of a single rule called MP (Modus Ponens) which allows us to deduce (a conclusion) ψ from (the premises) $\varphi \to \psi$ and φ, and the following (schemata of) axioms:

$$
\begin{array}{ll}
1 & \varphi \to (\psi \to \varphi) \\
2 & (\varphi \to (\psi \to \chi)) \to ((\varphi \to \psi) \to (\varphi \to \chi)) \\
3 & \varphi \wedge \psi \to \varphi; \quad \text{and} \quad \varphi \wedge \psi \to \psi \\
4 & \varphi \to (\psi \to \varphi \wedge \psi) \\
5 & \varphi \to \varphi \vee \psi; \quad \text{and} \quad \psi \to \varphi \vee \psi \\
6 & (\varphi \to \chi) \to ((\psi \to \chi) \to (\varphi \vee \psi \to \chi)) \\
7 & (\neg\varphi \to \neg\psi) \to (\psi \to \varphi)
\end{array}
$$

Remark 1.2. It is an invariant version of an axiomatic system in which instead of a finite set of axioms in the object language we are using a set of schemata of axioms in the metalanguage. As a result the set of axioms is infinite, but we do not need a rule of substitution (for propositional symbols) as primitive. In what follows, for simplicity both schemata and their concrete instances in the object language will be called axioms.

The key notion of the provability (deducibility, derivability) relation ($\vdash$), and the related notions of a thesis (of **CPL**) and consistency, may be defined in the following way:

Definition 1.5. • $\Gamma \vdash \varphi$ *iff there is a finite sequence of formulae (a proof) in which the last element is φ and every element is either:*

1. *an instance of an axiom, or*

2. *an element of Γ(assumption), or*

3. *a result of application of MP to earlier elements in the sequence.*

- $\vdash \varphi$ *iff* $\varnothing \vdash \varphi (\varphi$ *is a thesis)*

- Γ *is inconsistent iff* $\Gamma \vdash \bot$, *otherwise it is consistent.*

Remark 1.3. Usually a symbol $\vdash$ occurs with indices denoting a kind of a system or a logic, like $\vdash_H, \vdash_{H-L}, \vdash_L^S$; for simplicity we apply these decorations only when there is a risk of confusion, i.e. when we compare different relations (logics, systems).

Note the following properties of $\vdash$:

Lemma 1.1.   *1. if $\varphi \in \Gamma$, then $\Gamma \vdash \varphi$*

2. if $\Gamma \vdash \varphi$ and $\Gamma \subseteq \Delta$, then $\Delta \vdash \varphi$

3. $\vdash \varphi$ iff $\Gamma \vdash \varphi$, for any set Γ

4. if $\Gamma \vdash \varphi$ and $\Delta, \varphi \vdash \psi$, then $\Gamma, \Delta \vdash \psi$

5. $\Gamma \vdash \varphi$ iff $\Delta \vdash \varphi$, for some finite $\Delta \subseteq \Gamma$

6. $\Gamma, \varphi \vdash \psi$ iff $\Gamma \vdash \varphi \rightarrow \psi$

7. if $\Gamma, \neg\varphi \vdash \varphi$, then $\Gamma \vdash \varphi$

8. if $\Gamma, \varphi \vdash \neg\varphi$, then $\Gamma \vdash \neg\varphi$

9. $\Gamma, \neg\varphi \vdash \bot$ iff $\Gamma \vdash \varphi$

10. if $\Gamma, \varphi \vdash \psi$ and $\Delta, \neg\varphi \vdash \psi$, then $\Gamma, \Delta \vdash \psi$

11. if $\Gamma, \varphi \vdash \psi$ and $\Delta, \varphi \vdash \neg\psi$, then $\Gamma, \Delta \vdash \neg\varphi$

12. if $\Gamma \vdash \varphi$ and $\Delta \vdash \neg\varphi$, then $\Gamma, \Delta \vdash \psi$, for any ψ

13. $\Gamma, \varphi, \psi \vdash \chi$ iff $\Gamma, \varphi \wedge \psi \vdash \chi$

14. if $\Gamma \vdash \varphi \vee \psi$ and $\Delta, \varphi \vdash \chi$ and $\Pi, \psi \vdash \chi$, then $\Gamma, \Delta, \Pi \vdash \chi$

15. if $\vdash \varphi \leftrightarrow \psi$, then $\vdash \chi \leftrightarrow \chi[\varphi//\psi]$, where $\chi[\varphi//\psi]$ means that at least one occurrence of φ in χ is replaced by ψ.

We omit the (mostly easy) proofs of these properties; more involved (inductive) proofs of (one-half of) the point 6 and 15 are displayed in the Appendix. Note that the last point is the extensionality principle, that the left-right implication of 6 is the deduction theorem (DT), and that its converse expresses the closure of $\vdash$ under MP which may be equivalently stated:

1. $\varphi, \varphi \rightarrow \psi \vdash \psi$ (inferential MP)

2. if $\Gamma \vdash \varphi$ and $\Delta \vdash \varphi \rightarrow \psi$, then $\Gamma, \Delta \vdash \psi$ (deductive MP)

The system presented above is an adequate formalisation of **CPL**. We recall that the link between H-systems (and syntactic formalizations in general) and suitable classes of models is obtained via soundness and completeness theorems of the form:

- (Soundness) if $\Gamma \vdash \varphi$, then $\Gamma \models \varphi$

- (Completeness) if $\Gamma \models \varphi$, then $\Gamma \vdash \varphi$

The last one is often formulated equivalently:

- if Γ is consistent, then Γ is satisfiable.

If the first theorem holds, then the system is sound with respect to a suitable class of models, if the second holds, then it is (strongly) complete. It is adequate if it is both sound and complete. Note that if Γ is empty or finite, weak completeness holds, otherwise we have a strong form (i.e. admitting infinite Γ).

Standard proofs of completeness apply the well-known construction of a model due to Henkin (based on Lindenbaum's earlier result). Essentially, it consists in showing that there is a unique infinite model that falsifies every formula unprovable in **CPL**. This method of proof is therefore not constructive. For questions of decidability and automated theorem proving it is more important that there are constructive methods of proving completeness. They show how to find for any unprovable formula some finite falsifying model.

CPL is a decidable logic, in the sense that there is an effective procedure which, for every $\varphi \in \mathbf{L_{CPL}}$, resolves whether it is a tautology of this logic or not. This is a special case of the decision problem generally called the *validity problem* for a logic **L**. Very often the decidability of some logic **L** is posed as the so called *satisfiability problem* (or shortly, sat-problem) for **L**: given φ in the language of this logic, decide whether φ is satisfiable (in this logic). Clearly, these instances of the decision problem are complementary since, for every logic considered in this volume, it holds that $\models \varphi$ iff $\neg\varphi$ is not satisfiable. Hence, for any logic, the validity problem is decidable iff the sat-problem is decidable.

The notion of an effective procedure, or algorithm, used in the characterisation of decidability needs some explanation. There are plenty of formal explications of it, in terms of Turing machines, Markov algorithms, recursive functions, etc., to mention just the oldest and the most popular. Advanced investigations on these mathematical models of effectiveness form the core of computability theory. In this book we do not need any formal treatment of these matters, however. Informally, it must be a method which is mechanical (it works without any need for ingenuity), fair (it does what it is assumed to do for every input) and terminating (it works in finite time). On the other hand, in all mathematical models of effective procedure no real bounds are put on the time of performance or amount of memory needed to store the data. Hence effective is not the same as efficient. Investigations on practical (time and space) requirements of algorithms, and the classification of decidable problems belong to complexity theory.

1.2 Sequent Calculus

There are many sequent calculi with different rules and features. However, any SC may be roughly described as a collection of (schemata) of rules of the form:

$$S_1, ..., S_n \ / \ S_{n+1}, \ n \geq 0,$$

where any S_i is a sequent. In case $n = 0$ we call S_{n+1} an axiomatic sequent; otherwise it is a conclusion of a rule and $S_1, ..., S_n$ are premisses. As we will see, what is common (and fundamental) to different versions of SC is the notion of a sequent.

1.2.1   Sequents and their Interpretation

Sequents are ordered pairs of the form $\Gamma \Rightarrow \Delta$ (or $\varphi_1, ..., \varphi_k \Rightarrow \psi_1, ..., \psi_n$, with $k \geq 0, n \geq 0$). Γ is called the antecedent and Δ the succedent of a sequent with Γ, Δ being finite sets of formulae[2].

In order to understand the meaning of a sequent it is good to appeal to Gentzen's original interpretation by changing sequents into formulae, thus:

- any sequent $\varphi_1, ..., \varphi_i \Rightarrow \psi_1, ..., \psi_k$ with $i > 0, k > 0$ may be interpreted as a formula $\varphi_1 \wedge ... \wedge \varphi_i \rightarrow \psi_1 \vee ... \vee \psi_k$;

- in case $i = 1$ or $k = 1$ we have a reduced (one-element) conjunction or disjunction; the latter case is the common treatment of intuitionistic sequents;

- the empty antecedent is interpreted as $\top$, the empty succedent as $\bot$.

In what follows we will use $I(S)$ for a formula that results from Gentzen's transformation of a sequent S. Note that in this interpretation a sequent with empty antecedent and succedent denotes simply $\bot$, since $\top \rightarrow \bot \leftrightarrow \neg\top \vee \bot \leftrightarrow \bot \vee \bot \leftrightarrow \bot$. One may provide many other (syntactical) interpretations of sequents (cf. Paoli [193]). Of particular importance is the interpretation of $\Rightarrow$ as a kind of consequence relation induced by SC (see Scott [234], Shoesmith and Smiley [238]). In the basic form it reads:

$$\Gamma \vdash^s \Delta \text{ iff for some finite } \Gamma' \subseteq \Gamma \text{ and } \Delta' \subseteq \Delta \text{ there is a proof of } \Gamma' \Rightarrow \Delta' \ .$$

Of course in order for this characterisation to be meaningful, we must have a notion of a calculus and a proof of a sequent in it. This will be given below.

We can characterise sequents also semantically in terms of truth and validity. Let $\mathfrak{M}$ be a classical propositional model (extended valuation), we say that:

- S is true (satisfied) in $\mathfrak{M}$, or $\mathfrak{M}$ is a model of S ($\mathfrak{M} \vDash S$) iff at least one formula in the antecedent is false, or at least one in the succedent is true;

- S is falsified in $\mathfrak{M}$ ($\mathfrak{M} \nvDash S$) iff all formulae in the antecedent are true and all in the succedent false;

- S is valid ($\vDash S$) iff it is true in every interpretation.

Exercise 1.1. *Prove that S is true in $\mathfrak{M}$ iff $I(S)$ is true in $\mathfrak{M}$.*

---

[2]Later we will introduce sequents built from different data structures like multisets or finite lists.

1.2.2 System K

The first sequent calculus introduced in this book is not the original system of Gentzen. We start with a simplified version called K after Aiva Ketonen who first proposed a similar variant of SC in [148]. Our system K for **CPL** consists of the following logical rules:

$$(\neg\Rightarrow) \ \frac{\Gamma\Rightarrow\Delta,\varphi}{\neg\varphi,\Gamma\Rightarrow\Delta} \qquad\qquad (\Rightarrow\neg) \ \frac{\varphi,\Gamma\Rightarrow\Delta}{\Gamma\Rightarrow\Delta,\neg\varphi}$$

$$(\wedge\Rightarrow) \ \frac{\varphi,\psi,\Gamma\Rightarrow\Delta}{\varphi\wedge\psi,\Gamma\Rightarrow\Delta} \qquad\qquad (\Rightarrow\wedge) \ \frac{\Gamma\Rightarrow\Delta,\varphi \quad \Gamma\Rightarrow\Delta,\psi}{\Gamma\Rightarrow\Delta,\varphi\wedge\psi}$$

$$(\Rightarrow\vee) \ \frac{\Gamma\Rightarrow\Delta,\varphi,\psi}{\Gamma\Rightarrow\Delta,\varphi\vee\psi} \qquad\qquad (\vee\Rightarrow) \ \frac{\varphi,\Gamma\Rightarrow\Delta \quad \psi,\Gamma\Rightarrow\Delta}{\varphi\vee\psi,\Gamma\Rightarrow\Delta}$$

$$(\Rightarrow\to) \ \frac{\varphi,\Gamma\Rightarrow\Delta,\psi}{\Gamma\Rightarrow\Delta,\varphi\to\psi} \qquad\qquad (\to\Rightarrow) \ \frac{\Gamma\Rightarrow\Delta,\varphi \quad \psi,\Gamma\Rightarrow\Delta}{\varphi\to\psi,\Gamma\Rightarrow\Delta}$$

There is only one type of axiomatic sequents of the form: $\Gamma \Rightarrow \Delta$ where $\Gamma\cap\Delta$ is nonempty.

One should notice a remarkable symmetry involved in the system. Every logical constant is characterised by means of a pair of rules introducing a formula with this constant either to the antecedent or to the succedent. Moreover, no other occurrences of any constant are displayed.

It is useful to introduce some terminology for the formulae that are displayed in the schemata of rules:

- A formula introduced by the application of a logical rule is the *principal formula* of this rule application.

- Formulae used for the derivation of the principal formula are *side formulae* of this rule application.

- All other elements of Γ and Δ are *parametric formulae* or the *context* of this rule application.

- The principal and side formulae are the *active formulae* of this rule application.

By convention we take the formula which occurs on both sides of any axiomatic sequent as an active formula. It is immediate that all rules satisfy the principle of distributivity: every formula in any instance of a rule is either a parametric or a principal or a side formula. There is also an important connection between formulae in premisses and conclusions. The former are immediate ancestors of the latter in the following sense: the side formulae are the immediate ancestors of the principal formula and every parametric formula in premisses is an immediate ancestor of the same parametric formula in the conclusion. A converse of this relation is that of immediate descendant. As for parameters a relation of congruence holds between them in the sense that every parameter in the conclusion is congruent to itself and to all ancestors in all premisses.

Example 1.1. $$\frac{p \wedge q, q \to r \Rightarrow \neg q, p \to r \qquad p \wedge q, q \to r \Rightarrow \neg q, \neg s}{p \wedge q, q \to r \Rightarrow \neg q, (p \to r) \wedge \neg s}$$

provides an illustration of the application of $(\Rightarrow \wedge)$ *with* $p \to r$ *and* $\neg s$ *as side formulae,* $(p \to r) \wedge \neg s$—*the principal formula, and* $p \wedge q, q \to r, \neg q$—*parametric formulae. The identification of ancestors and congruent formulae is obvious.*

One may notice two important features of the rules of K: the subformula property and validity-preservation.

Subformula property: If some formula occurs in a premiss, it occurs in the conclusion as a parameter or as a subformula of the principal formula. We can say that SC has this property if all its (primitive) rules have it.

This is an extremely important property which for K may be verified by inspection. A particular consequence of this feature is that each premiss is always less complex than the respective conclusion, in the sense of having one occurrence of some connective deleted. We will see (in sections 1.3 and 1.6) that this feature has a strong impact on proof search in K (and other SC with the subformula property). At this place we note some interesting consequence of the subformula property which is more theoretical, namely *separability*. We say that SC has a property of separation for a constant $\star$ iff every provable S with no occurrence of $\star$ has a proof with no application of rules for $\star$.

It is easy to prove:

Claim 1.3. *If SC has the subformula property, then it has a property of separation for all constants.*

PROOF: Assume that $\star$ does not occur in S. If a rule introducing $\star$ would be applied in the proof of S, then by the subformula property this introduced occurrence of $\star$ would be present in S—contradiction.   □

The second important property is semantic:

Validity-preservation: If all premisses are valid, then the conclusion is valid.

We prove that it holds for K.

Lemma 1.2. (Validity-preservation). *All rules of K are validity-preserving in* **CPL**.

PROOF: We provide one example for $(\to\Rightarrow)$.

Assume that $\models \Gamma \Rightarrow \Delta, \varphi$ and $\models \psi, \Gamma \Rightarrow \Delta$ but $\not\models \varphi \to \psi, \Gamma \Rightarrow \Delta$. So in some $\mathfrak{M}$ both $\varphi \to \psi$ and all elements of Γ are true but all elements of Δ are false. If $\mathfrak{M} \models \varphi \to \psi$, then $\mathfrak{M} \not\models \varphi$ or $\mathfrak{M} \models \psi$. Both cases lead to contradiction since either the left or the right premiss must be falsified by $\mathfrak{M}$.   □

Exercise 1.2. *Prove the remaining 7 cases.*

In fact, for all rules of K one may show also a weaker semantical feature which we call normality or truth-preservation. Let us say that a rule is **normal** whenever, if all premisses are true in some $\mathfrak{M}$, then the conclusion is true in the same $\mathfrak{M}$.

Exercise 1.3. *1. Prove that all rules of K are normal.*

2. Show that every normal rule is also validity-preserving.

1.3 Proofs and Derivations

Proofs in SC are usually built as trees of sequents (i.e. with nodes labelled with sequents) with axioms as leaves and the proven sequent as a root. Formally this notion may be defined inductively as follows:

1. An axiomatic sequent S is a *proof of a sequent S*.

2. If $\mathcal{D}$ is a proof of a sequent S, then $\dfrac{\mathcal{D}}{S'}$ is a proof of a sequent S', provided that S is an instance of the premiss and S' an instance of the conclusion of some one-premiss rule.

3. If $\mathcal{D}$ is a proof of a sequent S and $\mathcal{D}'$ is a proof of a sequent S', then $\dfrac{\mathcal{D}\ \mathcal{D}'}{S''}$ is a proof of a sequent S'', provided that S and S' are instances of the premisses and S'' an instance of the conclusion of some two-premiss rule.

We can relax the definition of a proof in K (or any SC) by admitting any sequents (not only axiomatic ones) as leaves of a tree. In this case we say that there is a *derivation* of S from $S_1, ..., S_n$. So a derivation of S from $S_1, ..., S_n$ is a tree where some leaves are decorated with sequents that are not axioms but belong to the list $S_1, ..., S_n$. If $n = 0$ we have a proof of S as a special case of a derivation. In semantical terms we will say that S **logically follows** from $S_1, ..., S_n$ ($S_1, ..., S_n \models S$) iff S is satisfied in every model in which all $S_1, ..., S_n$ are satisfied. In what follows we will be often using the term derivation as covering proofs as a special case.

Note that the notion of immediate ancestor (descendant) may be naturally extended on proofs as the transitive closures of the relation of immediate ancestry (descendancy). In a similar way we may generalise the notion of congruency between parameters from the level of rules to the level of derivations.

We will use the symbol $\vdash S$ if there is a proof of S in K (or any other SC) and in case of a derivation we will write $S_1, ..., S_n \vdash S$. Although this is a different sense of derivability than that considered for axiomatic systems or induced by $\Rightarrow$ for SC (see $\vdash^s$ in subsection 1.2.1), the application of the same sign $\vdash$ should not lead to misunderstanding. Note however that it is possible to define different relations of logical consequence on the basis of SC. In particular, we may introduce a consequence relation between sequents and this is just what we have done. From now on this meaning of $\vdash$ will be the basic one in the framework of SC. In case of a tree $\mathcal{D}$ that is a proof (or a derivation) of S we will write $\mathcal{D} \vdash S$ or $S_1, ..., S_n \vdash_{\mathcal{D}} S$.

Another handy concept is that of a *subtree* of a tree of sequents. A single sequent is its only subtree. If S has a proof (a derivation) $\mathcal{D}$ of the form $\dfrac{\mathcal{D}'}{S}$, then $\mathcal{D}$ itself is its (improper) subtree as well as $\mathcal{D}'$ and all its subtrees. Note that the subtrees of a tree of sequents are in natural one-one correspondence with all sequents in this tree in the sense that every subtree of a proof (derivation) is also a proof (derivation) of its root sequent. In particular, a subtree of a derivation may be a proof of its root sequent.

Remark 1.4. We have stressed that the new usage of $\vdash$ is a bit different since no object on the left or right side of $\vdash$ is a formula. However, as we pointed out, this ambiguity should not lead to misunderstandings and seems to be a better solution than the addition of a new sign. Anyway, we should provide some explanation concerning this special application of $\vdash$, since sequents are not the items we meet in standard axiomatic or natural deduction systems, where we are proving theses or demonstrate derivability from assumptions. Also in the semantic framework we rather speak about tautologies or valid formulae and about relations of consequence or entailment. How to relate SC to this more common approach? It is straightforward in case of theses: we can say that φ is a thesis if $\vdash \Rightarrow \varphi$.

What about consequence relations in Tarski's sense, i.e. holding between formulae? One possible answer was suggested in section 1.2.1 where $\Rightarrow$ was identified with some such relation $\vdash^s$ (Scott's relation). Leaving aside the problem that on the right side of $\vdash^s$ we have a set of formulae, not a single formula as in the standard (Tarski-style) approach, this is a reasonable option[3]. If we want to have something more resembling a Tarski-style relation we can say that φ is derivable from Γ ($\Gamma \vdash^t \varphi$) iff $\vdash \Gamma' \Rightarrow \varphi$ for some finite $\Gamma' \subseteq \Gamma$. We have used $\vdash^t$ ('t' for 'truth') to denote this kind of a Tarski-style consequence relation leaving $\vdash$ for the derivability relation between sequents.

It is certainly not the only way one can define a Tarski-style consequence relation on the basis of K, or any other SC. Avron [9] provides another one: $\Gamma \vdash^v \varphi$ iff $\Rightarrow \psi_1, ..., \Rightarrow \psi_k \vdash \Rightarrow \varphi$, for $\{\psi_1, ..., \psi_k\} \subseteq \Gamma$. In fact, in case of K, $\vdash^t$ and $\vdash^v$ ('v' for validity) coincide, which will be shown in subsection 1.9.2. For the time being we may notice only the obvious consequence of our definitions: $\vdash^t \varphi$ iff $\vdash^v \varphi$ iff $\vdash \Rightarrow \varphi$.

1.3.1   Constructing Proofs

In order to obtain a mastery in using SC we must try to construct proofs. But how to do it? Let us analyse an example.

Example 1.2. *A proof of Frege's syllogism:*

---

[3]In the appendix we will consider the problem of generalised theory of consequence relations defined between sets of formulae.

$$(\to\Rightarrow) \dfrac{\dfrac{p \Rightarrow r, q, p \qquad q, p \Rightarrow r, q}{p \to q, p \Rightarrow r, q}}{\dfrac{}{}}$$

$$\dfrac{p \to q, p \Rightarrow r, p \qquad \dfrac{p \Rightarrow r, q, p \quad q, p \Rightarrow r, q}{p \to q, p \Rightarrow r, q} \qquad \dfrac{r, p \to q, p \Rightarrow r}{q \to r, p \to q, p \Rightarrow r} (\to\Rightarrow)}{\dfrac{p \to (q \to r), p \to q, p \Rightarrow r}{\dfrac{p \to (q \to r), p \to q \Rightarrow p \to r}{\dfrac{p \to (q \to r) \Rightarrow (p \to q) \to (p \to r)}{\Rightarrow (p \to (q \to r)) \to ((p \to q) \to (p \to r))} (\Rightarrow\to)} (\Rightarrow\to)} (\Rightarrow\to)}$$

A tree displays a ready proof but the important issue is how to find it. As with many things here too, there is a great difference between a process (here of proving) and a result (here a proof). The great advantage of SC (K in particular) over axiomatic systems is the possibility of easy proof search. One may start not with the axioms but with a sequent to be proved and proceed in a root-first manner, i.e. by applying all possible rules upside-down, from conclusion to premises. We will illustrate this with a proof of a thesis $(p \to q \vee r) \to (p \to q) \vee (p \to r)$. If we start a root-first proof search we first obtain:

$$\dfrac{p \to q \vee r \Rightarrow (p \to q) \vee (p \to r)}{\Rightarrow (p \to q \vee r) \to (p \to q) \vee (p \to r)} \ (\Rightarrow\to)$$

Now we have a choice: either to apply $(\to\Rightarrow)$ or $(\Rightarrow \vee)$. In general, we should first apply a nonbranching rule which leads to $p \to q \vee r \Rightarrow p \to q, p \to r$ and three choices of what to do next. Again we prefer the nonbranching rule $(\Rightarrow\to)$ which after two applications gives $p \to q \vee r, p \Rightarrow q, r$, and nothing can save us from the application of $(\to\Rightarrow)$. We get:

$$\dfrac{\dfrac{\dfrac{\dfrac{p \Rightarrow q, r, p \qquad q \vee r, p \Rightarrow q, r}{p \to q \vee r, p \Rightarrow q, r} \ (\to\Rightarrow)}{p \to q \vee r, p \Rightarrow q, p \to r} \ (\Rightarrow\to)}{p \to q \vee r \Rightarrow p \to q, p \to r} \ (\Rightarrow\to)}{\dfrac{p \to q \vee r \Rightarrow (p \to q) \vee (p \to r)}{\Rightarrow (p \to q \vee r) \to (p \to q) \vee (p \to r)} \ (\Rightarrow\to)} \ (\Rightarrow \vee)$$

The leftmost sequent is an axiom so we must only apply $(\vee \Rightarrow)$ to the right branch and obtain a proof:

$$\dfrac{\dfrac{\dfrac{p \Rightarrow q, r, p \qquad \dfrac{q, p \Rightarrow q, r \qquad r, p \Rightarrow q, r}{q \vee r, p \Rightarrow q, r} \ (\vee \Rightarrow)}{p \to q \vee r, p \Rightarrow q, r} \ (\to\Rightarrow)}{p \to q \vee r, p \Rightarrow q, p \to r} \ (\Rightarrow\to)}{\dfrac{p \to q \vee r \Rightarrow p \to q, p \to r}{\dfrac{p \to q \vee r \Rightarrow (p \to q) \vee (p \to r)}{\Rightarrow (p \to q \vee r) \to (p \to q) \vee (p \to r)} \ (\Rightarrow\to)} \ (\Rightarrow \vee)} \ (\Rightarrow\to)$$

Why do we not use branching rules first? Do we obtain a different result? No – if something is provable it may be provable in many ways so we also will find a proof but more complicated one. Let us suppose that we apply the strategy of always first applying a rule to the leftmost complex formula. Then we obtain the following proof of the same thesis:

$$
\cfrac{
  \cfrac{
    (\Rightarrow \lor)\ \cfrac{
      (\Rightarrow \to)\ \cfrac{p \Rightarrow q, p \to r, p}{\Rightarrow p \to q, p \to r, p}
    }{\Rightarrow (p \to q) \lor (p \to r), p}
  }{}
  \qquad
  \cfrac{
    (\Rightarrow \lor)\ \cfrac{
      (\Rightarrow \to)\ \cfrac{q, p \Rightarrow q, p \to r}{q \Rightarrow p \to q, p \to r}
    }{q \Rightarrow (p \to q) \lor (p \to r)}
  }{}
  \qquad
  \cfrac{
    (\Rightarrow \lor)\ \cfrac{
      (\Rightarrow \to)\ \cfrac{
        (\Rightarrow \to)\ \cfrac{r, p \Rightarrow q, r}{r, p \Rightarrow q, p \to r}
      }{r \Rightarrow p \to q, p \to r}
    }{r \Rightarrow (p \to q) \lor (p \to r)}
  }{}
}{
  \cfrac{
    \cfrac{p \to q \lor r \Rightarrow (p \to q) \lor (p \to r)}{}
  }{\Rightarrow (p \to q \lor r) \to (p \to q) \lor (p \to r)}\ (\Rightarrow \to)
}
$$

$$(\lor \Rightarrow) \qquad (\lor \Rightarrow) \qquad q \lor r \Rightarrow (p \to q) \lor (p \to r)$$

Not much bigger than the preceding one (13 sequents versus 9) but we may easily find cases where the difference is really significant. Of course in this way we can build proofs not only for theses but for any provable sequents.

Exercise 1.4. *Construct proofs for the following sequents:*

$$\Rightarrow (p \to q) \lor (\neg p \to \neg q)$$
$$\Rightarrow \neg (p \land q) \to (q \to \neg p)$$
$$\neg p \to q \land r, q \to (r \to s) \Rightarrow p \lor s$$
$$p \to q \lor r, q \to s \Rightarrow \neg (r \to s), p \to s$$

1.3.2   Disproofs

In fact we can use K not only for proving what is provable. The same method may be applied for testing if a sequent is provable and showing its unprovability as well as its provability. Let us consider a formula $(p \land q \to r \land s) \to (p \to r) \land (q \to s)$. It is not a thesis and after first two moves we obtain:

$$
\cfrac{
  p \land q \to r \land s \Rightarrow p \to r \qquad p \land q \to r \land s \Rightarrow q \to s
}{
  \cfrac{p \land q \to r \land s \Rightarrow (p \to r) \land (q \to s)}{\Rightarrow (p \land q \to r \land s) \to (p \to r) \land (q \to s)}\ (\Rightarrow \to)
}\ (\Rightarrow \land)
$$

Now, how to proceed in case of branching? We can apply one of the two basic strategies: depth-first or breath-first. In the former approach we always choose one branch (e.g. the leftmost one) and continue until it ends with an axiomatic or atomic sequent, and then we return to the next unfinished branch. In the latter case we apply successively the rules to all unfinished branches one by one, proceeding, say, from the leftmost to the rightmost branch. Let us try a depth-first strategy. Always selecting the leftmost branch we continue until we get:

$$
\cfrac{
  (\Rightarrow \land)\ \cfrac{
    (\to \Rightarrow)\ \cfrac{
      (\Rightarrow \to)\ \cfrac{
        (\Rightarrow \land)\ \cfrac{
          \cfrac{p \Rightarrow r, p \qquad p \Rightarrow r, q}{p \Rightarrow r, p \land q} \qquad r \land s, p \Rightarrow r
        }{p \land q \to r \land s, p \Rightarrow r}
      }{p \land q \to r \land s \Rightarrow p \to r}
    }{}
  }{}
  \qquad p \land q \to r \land s \Rightarrow q \to s
}{
  \cfrac{p \land q \to r \land s \Rightarrow (p \to r) \land (q \to s)}{\Rightarrow (p \land q \to r \land s) \to (p \to r) \land (q \to s)}\ (\Rightarrow \to)
}
$$

Note that the second top sequent is not axiomatic but atomic. It means that if we are concerned only in testing if the input formula is a thesis, we are done. We do not need to continue the proof search with $r \wedge s, p \Rightarrow r$, then with $p \wedge q \to r \wedge s \Rightarrow q \to s$, since one branch beginning with a nonaxiomatic and atomic sequent is enough to show that there is no proof. Moreover, we can easily find a model falsifying our sequent. Take any one satisfying p but with false r and q (the value of s is inessential), then an easy calculation shows that $p \wedge q \to r \wedge s$ is satisfied but $(p \to r) \wedge (q \to s)$ is not. In general, to falsify a root sequent we define a model for some nonaxiomatic atomic sequent which satisfies all atomic formulae in the antecedent, and falsifies all in the succedent.

It seems that a depth-first strategy of constructing a tree is better for proof search since in case of unprovable sequents, we can finish the job earlier, if we are lucky (e.g. there may be only one nonaxiomatic branch and it is just the rightmost one). However, for some reasons we may still be interested in constructing a tree where all rules were applied; let us call it a completed tree. In case of our example a completed tree may look like that (other ones are possible):

$$
(\Rightarrow \wedge)\,\dfrac{\dfrac{q \Rightarrow s, p \qquad q \Rightarrow s, q}{q \Rightarrow s, p \wedge q} \qquad \dfrac{r, s, q \Rightarrow s}{r \wedge s, q \Rightarrow s}\,(\wedge \Rightarrow)}{\dfrac{\dfrac{p \wedge q \to r \wedge s, q \Rightarrow s}{p \wedge q \to r \wedge s \Rightarrow q \to s}\,(\Rightarrow \to)}{}}\,(\to \Rightarrow)
$$

$$
\dfrac{\mathcal{D} \qquad \dfrac{p \wedge q \to r \wedge s, q \Rightarrow s}{\dfrac{p \wedge q \to r \wedge s \Rightarrow q \to s}{p \wedge q \to r \wedge s \Rightarrow (p \to r) \wedge (q \to s)}\,(\Rightarrow \wedge)}}{\Rightarrow (p \wedge q \to r \wedge s) \to (p \to r) \wedge (q \to s)}\,(\Rightarrow \to)
$$

where $\mathcal{D}$ is a proof tree for the left conjunct provided above.

We may note that on the right side of this completed tree there is another atomic and nonaxiomatic sequent which also falsifies the root sequent. The above tree may be treated thus as providing us with the information that there are only two different partial valuations which generate falsifying models for this unprovable sequent. How do we know that there are only these two? We will get back to this problem later but the reader is invited to construct another tree for this sequent this time applying first $(\to \Rightarrow)$ to $p \wedge q \to r \wedge s$ in the second move. The tree will be different but interestingly enough the only nonaxiomatic sequents will be the same.

Exercise 1.5. *Construct completed trees for the following sequents (try different strategies of proof search):*

$\Rightarrow (p \to q) \to (\neg p \to \neg q)$
$\Rightarrow \neg(p \wedge q) \to \neg(p \vee q)$
$p \to q, q \vee r \Rightarrow \neg p \vee (r \wedge q)$
$p \vee q \to (r \to s) \Rightarrow p \to r, \neg q \to s$

Remark 1.5. Readers familiar with tableau methods may at this point notice some similarities with K. Indeed one may treat tableaux as a simplified form of sequent calculus, at least in the case of classical logic. In Hintikka-style tableaux one is indirectly testing if a formula φ (or an argument $\Gamma \,/\, \varphi$) is valid by starting with $\neg \varphi$ (with a set $\Gamma \cup \{\neg \varphi\}$) and developing a down-growing tree. We successively apply rules to finite sets of formulae (nodes of a tableau) until we get a proof (every leaf-set containing contradictory formulae), if φ is valid (φ follows from Γ). One could easily obtain all tableau-rules of this kind from rules of K by changing sequents $\psi_1, ..., \psi_i \Rightarrow \varphi_1, ..., \varphi_j$ into sets $\psi_1, ..., \psi_i, \neg\varphi_1, ..., \neg\varphi_j$ and turning all rules upside-down. Thus, e.g. from ($\Rightarrow \vee$) we obtain a tableau rule which allows to deduce $\Gamma, \neg\varphi, \neg\psi$ from $\Gamma, \neg(\varphi \vee \psi)$.

Notice that we can read all the trees displayed in this subsection not only in terms of the unprovability of the root sequent but also in terms of its derivability from nonaxiomatic sequents. It is because, formally, they are derivations in the sense defined at the beginning of this section. Thus, e.g. the first tree shows that $\Rightarrow (p \wedge q \to r \wedge s) \to (p \to r) \wedge (q \to s)$ is derivable from $p \wedge q \to r \wedge s \Rightarrow p \to r$ and $p \wedge q \to r \wedge s \Rightarrow q \to s$, whereas a completed tree (the last one) yields the information that the same sequent is derivable also from the atomic sequents $p, r \Rightarrow q$ and $q \Rightarrow p, s$. However, in these examples (including the ones specified in the exercise), the construction of derivations is not a primary task but rather a byproduct of (failed) proof search. We can of course also treat derivations directly and start with a task of the form: show that $S_1, ..., S_n \vdash S$, but note that in general, in this case a proof search maybe not so easy to perform. What is more important, even if S really follows from $S_1, ..., S_n$ it may be not possible to obtain a tree demonstrating the derivability of S from $S_1, ..., S_n$. In other ways K as stated is not complete with respect to derivability, although it is complete with respect to provability which will be shown in section 1.5. We postpone a discussion of the problem of completeness with respect to derivability to section 1.11.

1.3.3   Analyticity

We have called K an 'analytic' system in the title of the chapter and after studying the examples of proof search in the preceding subsection we may explain in what sense we are using this term[4]. Roughly we will say that SC is analytic in the sense that the root-first proof search is bounded, i.e. that there is always a finite number of choices of what to do next. It is obvious that K is analytic in this sense due to the subformula property of all its rules. Although this feature is primarily attributed to rules we may also say that a proof (and a disproof) satisfies the subformula property if all formulae occurring in it are subformulae of the root sequent. Finally we may say that a system has the subformula property if all proofs in the system have this feature. It is obvious that if all rules of the system have the subformula property, then all proofs have it and consequently, the calculus has it. So why not say that a system has the subformula property instead of saying that it is analytic? In K it is the same, but in general we must keep these qualifications distinct. Notice, e.g. that a proof in some SC may have the subformula property in the sense specified above even if not all rules applied in this system have the subformula property like in K. It is sufficient that all formulae occurring in the premisses of

---

[4]It is important since the term is ambiguous even in proof theory—see, e.g. remarks in Poggiolesi [198].

some rule application are subformulae of the root sequent, not necessarily of the conclusion of this rule application. Also a system may be analytic even if not all proofs have the subformula property, e.g. some closure conditions may be specified that restrict the number of formulae which may be used in proof construction even if they are not subformulae of the root sequent. So in general, a system may be analytic even if it does not satisfy a subformula property. In other words the subformula property (of rules) is a sufficient (but not necessary) condition for the analyticity of SC.

In what follows we will illustrate these phenomena (see in particular section 1.11) but first we derive important consequences of the subformula property. The decidability of **CPL** will be the most important one, but this requires special attention and will be presented in section 1.6. This section will be finished with a result which is very important but extremely easy to prove.

Theorem 1.1. *K is consistent*

PROOF: It is a straightforward consequence of the subformula property of all rules. Every proof must be cumulative in the sense that nothing is lost when starting from axioms and going down the tree. Hence $\Rightarrow$ (which express $\perp$—see section 1.2.1) cannot be provable by any rule. □

1.4 Additional Rules

The notion of additional rules for K, or any other SC, is a bit ambiguous. We may think about rules for additional connectives enriching the basic language, or about introduction of some rules serving for special purposes. Let us start with the first issue.

1.4.1 Rules for Other Connectives

In section 1.1.1, we introduced as definitional shortcuts two propositional constants $\top, \perp$ and one binary connective $\leftrightarrow$. In particular, $\perp$ is very often used as a primitive logical constant in proof-theoretic investigations (see e.g. Negri and von Plato [185]). $\perp$ may be characterised by means of the additional axiomatic sequent $\perp, \Gamma \Rightarrow \Delta$.

One can easily verify the correctness of this axiom syntactically on the basis of the fact that $\perp$ allows for a definition of $\neg$ in the following way: $\neg\varphi := \varphi \rightarrow \perp$. Both rules for $\neg$ are easily derivable by means of the rules for $\rightarrow$ and the axiom for $\perp$. We will demonstrate the derivation of $(\Rightarrow \neg)$:

$$(\Rightarrow W) \frac{\varphi, \Gamma \Rightarrow \Delta}{\varphi, \Gamma \Rightarrow \Delta, \perp}$$
$$(\Rightarrow \rightarrow) \frac{}{\Gamma \Rightarrow \Delta, \varphi \rightarrow \perp}$$

Note that in the second step we have used an additional rule of weakening ($\Rightarrow W$) allowing for the addition of any formula to the succedent; its correctness will be justified in the next paragraph.

Exercise 1.6. *Provide a demonstration of* ($\neg \Rightarrow$) *on the basis of the axiom for* $\bot$ *and* ($\rightarrow\Rightarrow$).

Characterise $\top$; *remember that it may be defined as* $\neg\bot$.

Note that we may also easily prove an axiomatic sequent for $\bot$ if we define it as $\neg(\varphi \rightarrow \varphi)$. Here is a proof:

$$(\neg \Rightarrow) \dfrac{(\Rightarrow\rightarrow) \dfrac{\varphi, \Gamma \Rightarrow \Delta, \varphi}{\Gamma \Rightarrow \Delta, \varphi \rightarrow \varphi}}{\neg(\varphi \rightarrow \varphi), \Gamma \Rightarrow \Delta}$$

One may introduce suitable rules also for other truth-functional connectives. For example for $\leftrightarrow$ we need the following pair of rules:

$$(\Rightarrow\leftrightarrow) \ \dfrac{\psi, \Gamma \Rightarrow \Delta, \varphi \quad \varphi, \Gamma \Rightarrow \Delta, \psi}{\Gamma \Rightarrow \Delta, \varphi \leftrightarrow \psi} \qquad (\leftrightarrow\Rightarrow) \ \dfrac{\varphi, \psi, \Gamma \Rightarrow \Delta \quad \Gamma \Rightarrow \Delta, \varphi, \psi}{\varphi \leftrightarrow \psi, \Gamma \Rightarrow \Delta}$$

We consider three more interesting binary connectives:

1. Strong (exclusive) disjunction $\veebar$ ('exor') is true iff both arguments have different values; hence it may be defined as negation of $\leftrightarrow$.

2. Scheffer's stroke $\uparrow$ is false iff both arguments are true; hence it may be defined as negation of $\wedge$ ('nand').

3. Peirce's arrow $\downarrow$ is true iff both arguments are false; hence it may be defined as negation of $\vee$ ('neither ... nor').

In particular, the last two are very important since each of them may be used as the only primitive constant of **CPL** allowing for the definition of all other connectives. Suitable rules for the last one are the following:

$$(\Rightarrow\downarrow) \ \dfrac{\Gamma \Rightarrow \Delta, \varphi, \psi}{\varphi \downarrow \psi, \Gamma \Rightarrow \Delta} \qquad (\downarrow\Rightarrow) \ \dfrac{\varphi, \Gamma \Rightarrow \Delta \quad \psi, \Gamma \Rightarrow \Delta}{\Gamma \Rightarrow \Delta, \varphi \downarrow \psi}$$

Their shape should be of no surprise if we realise that $\downarrow$ is just a negated $\vee$.

Exercise 1.7. *Define a pair of rules for* $\veebar$ *and* $\uparrow$ *on the basis of the information that the former is a negation of* $\leftrightarrow$ *and the latter a negation of* $\wedge$.

There are exactly 16 binary extensional two-valued connectives. Find truth-functional characterizations of the remaining ones and try to define pairs of rules for each of them.

In the context of rules characterising other connectives we may consider more general questions: Can we define (logical) constants by means of rules? What conditions should such rules satisfy? The answer to the first question is quite obvious. For example, we can systematically build normal rules for any connective on the basis of its truth table and our definition of satisfiability of a sequent (see Beziau [29]). Each row of the truth table induces a single normal rule with one premiss for each side formula. We just put side formulae (in the premisses) and the principal formula (in the conclusion) in the antecedent if they are false, and in the succedent if they are true. Thus for every binary connective we obtain four two-premiss rules corresponding to every row of a truth table. For example, truth table for $\wedge$ gives us a rule ($\Rightarrow \wedge$) but instead of a single ($\wedge \Rightarrow$) we obtain three different two-premiss rules with the principal formula in the antecedent of the conclusion of the form:

$$(2\wedge \Rightarrow) \quad \frac{\psi, \Gamma \Rightarrow \Delta \quad \varphi, \Gamma \Rightarrow \Delta}{\varphi \wedge \psi, \Gamma \Rightarrow \Delta} \qquad \frac{\psi, \Gamma \Rightarrow \Delta \quad \Gamma \Rightarrow \Delta, \varphi}{\varphi \wedge \psi, \Gamma \Rightarrow \Delta} \qquad \frac{\Gamma \Rightarrow \Delta, \psi \quad \varphi, \Gamma \Rightarrow \Delta}{\varphi \wedge \psi, \Gamma \Rightarrow \Delta}$$

In a similar way we can produce three rules for ($\Rightarrow \vee$) and ($\Rightarrow \rightarrow$). Such rules are called full by Avron and Lev [14] who provide a syntactical way for their construction. The upshot of such a mechanical production of rules is obvious—we obtain four rules instead of two, and if we apply such a procedure for example to ternary connectives (like 'if then else') we will obtain eight three-premiss rules. But we prefer to have a pair of rules and this may be treated as the first condition which rules should obey. In fact, one may link this requirement with the strict reading of the content of truth tables on the basis of the dual interpretation of sequents. Now, ($\wedge \Rightarrow$) is intact and for ($\Rightarrow \wedge$) we have only one rule but with three premisses corresponding to three cases when it is false:

$$(3 \Rightarrow \wedge) \quad \frac{\varphi, \Gamma \Rightarrow \Delta, \psi \quad \psi, \Gamma \Rightarrow \Delta, \varphi \quad \Gamma \Rightarrow \Delta, \varphi, \psi}{\Gamma \Rightarrow \Delta, \varphi \wedge \psi}$$

and similarly for ($\vee \Rightarrow$) and ($\rightarrow \Rightarrow$). Although this solution may be seen as economic and perfectly matching the semantic conditions, it is rather redundant and we prefer the solution with two premisses (with incomplete semantic information). However, in section 4.8 we will see that the choice of these three-premiss rules may have some advantages.

Exercise 1.8. *Define for disjunction and implication the remaining 2− (six full rules) and 3−style (only two) rules.*

What other conditions are required for well-behaved rules? We will focus more closely on these matters in chapter 3 and conclude this subsection with some remarks on the approach which is now often called proof-theoretical semantics.

In fact, Gentzen may be treated as a father of this approach although he tried to provide meaning postulates for logical constants in terms of natural deduction (introduction) rules. In the wider perspective proof-theoretical semantics is closely related to the semantical program of inferentialism and other anti-realist approaches to meaning. We do not attempt to provide a detailed treatment of such an approach and its contemporary variants; a concise survey may be found in Schroeder-Heister [229] but some remarks are in order. Popper [200, 201] seems to have been the first who tried to provide definitional rules for connectives in terms of sequent calculus, however his approach was not successful (see Schroeder-Heister [226]). Prior [206] has shown that we cannot use arbitrary rules as meaning postulates since it may trivialise the system. His natural deduction rules for a constant called 'tonk' may be expressed in SC as follows:

$$\Gamma \Rightarrow \Delta, \varphi \ / \ \Gamma \Rightarrow \Delta, \varphi \ tonk \ \psi$$
$$\Gamma \Rightarrow \Delta, \varphi \ tonk \ \psi \ / \ \Gamma \Rightarrow \Delta, \psi$$

A reader may easily 'prove' $\varphi \Rightarrow \psi$ using these rules and this is not surprising. Prior's rules do not satisfy even the most general conditions which should be satisfied by definitions. Already Leśniewski [165] formulated two elementary conditions for correct definitions: eliminability and conservativity. The latter may be informally stated as follows: the definition of a new term is conservative iff it does not change the truth values of sentences not containing this term. In the framework of SC and definitional rules (see Poggiolesi [198]) this property may be stated in the following way:

Definition 1.6. *Let $L\star$ denote the language L with added $\star$ and $SC\star$ denote SC with added rules for $\star$, then $SC\star$ conservatively extends SC iff for every S in L, if $SC\star \vdash S$, then $SC \vdash S$.*

Thus the new rules are conservative iff they allow for proving new theorems only in the enriched language. It is obvious that if SC has the subformula property, then it has also the property of conservativity.

Claim 1.4. (Conservative extensions). *If $SC\star$ has subformula property, then it is a conservative extension of SC.*

Some advocates of proof-theoretical semantics (or more generally of inferentialism), like e.g. Dummett [66], claim that conservativity is a necessary condition for correct definitional rules. A debate on necessary and sufficient conditions for such rules is still open; at present various proposals concerning a notion of harmony are discussed (see e.g. Schroeder-Heister [230]). Anyway, one may easily check that Prior's rules are not conservative.

1.4.2 Derivability and Admissibility

Although K is complete for **CPL** (which will be shown in the next section), for several purposes we may need to introduce additional rules. Such an example of an additional rule called weakening and its application already occurred in the previous paragraph. The notion of a proof and its generalisation to the notion of a derivation allows for formal definitions of important notions concerning such secondary rules in any SC.

- A rule $\dfrac{S_1, ..., S_n}{S}$ is derivable in SC iff there is a derivation of S from $S_1, ..., S_n$.

- A rule $\dfrac{S_1, ..., S_n}{S}$ is admissible in SC iff if there are proofs of sequents $S_1, ..., S_n$, then $\vdash S$.

It is rather obvious that:

Lemma 1.3. *Every derivable rule in some SC is also admissible in it.*

PROOF: Consider an instance $S_1, ..., S_n/S$ of a derivable rule r, then we have a derivation $\mathcal{D}$ of S with some leaves being not axioms but instances of $S_i, i \leq n$. Assume that all $S_i, i \leq n$ have proofs, then we may append these proofs to the respective leaves of $\mathcal{D}$. This way we get a proof of S. □

Note that the converse statement does not hold in general. One should also notice that proofs of derivability of rules are rather simple whereas proofs of admissibility are usually involved and need special techniques. We will illustrate the point in many places of this book (see in particular section 1.8 on admissibility of the cut rule, or several such proofs of the admissibility of structural rules in later chapters).

A recursive definition of proofs allows for proofs by induction on proof trees (in particular proofs of the admissibility of some rules)[5]. One of the most useful measures is that of the height of a proof (or derivation):

- let $\mathcal{D}$ be a proof of an axiom, then $h\mathcal{D} = 0$;

- let $\mathcal{D}$ be a proof of S with S deduced from S' and $\mathcal{D}'$ be a proof of S', then $h\mathcal{D} = h\mathcal{D}' + 1$;

- let $\mathcal{D}$ be a proof of S with S deduced from S' and S'' and with $\mathcal{D}'$ and $\mathcal{D}''$ being their proofs, then $h\mathcal{D} = Max(h\mathcal{D}', h\mathcal{D}'') + 1$.

Sometimes we will write $\vdash_n S$ if a proof of S is of height at most n. In case $\mathcal{D}$ is a proof of S of height at most n we will write $\mathcal{D} \vdash_n S$.

Another handy measure of a proof is its size denoted as $\|\mathcal{D}\|$. The definition is like for the height with the only difference in the last clause which reads:

- let $\mathcal{D}$ be a proof of S with S deduced from S' and S'' and with $\mathcal{D}'$ and $\mathcal{D}''$ being their proofs, then $\|\mathcal{D}\| = \|\mathcal{D}'\| + \|\mathcal{D}''\| + 1$.

As an example of the application of the above notions we will prove by (strong, or complete) induction the admissibility of the rules of weakening of every provable sequent. Incidentally, we will provide the result in a stronger form of height-preserving admissibility. It means that if the premiss has a proof of some height, then the conclusion has a proof which is not longer. The advantage of having such rules was already shown in the previous paragraph. Both rules have the form:

---

[5]A reader who is not accustomed to such proofs should perhaps first read the Appendix where necessary information on inductive proofs is provided.

$$(W \Rightarrow) \quad \frac{\Gamma \Rightarrow \Delta}{\varphi, \Gamma \Rightarrow \Delta} \qquad\qquad (\Rightarrow W) \quad \frac{\Gamma \Rightarrow \Delta}{\Gamma \Rightarrow \Delta, \varphi}$$

In contrast to all primitive rules of K, both weakening rules are structural, i.e. they do not display any logical constants. Such rules only allow for the manipulation of sequents in proofs. The name derives from the fact that a provable sequent with fewer formulae is more informative than its extension either in the antecedent or in the succedent. In what follows we will use W as a short name for both rules or their multiple application.

Lemma 1.4. (Height-preserving admissibility of W for K). *If $\vdash_n \Gamma \Rightarrow \Delta$, then $\vdash_n \Gamma' \Rightarrow \Delta'$ for $\Gamma \subseteq \Gamma', \Delta \subseteq \Delta'$*

PROOF: By induction on the height of the premiss. In the basis, if $\Gamma \Rightarrow \Delta$ is axiomatic, then the addition of any formula on the left or right of it is unable to destroy its status. In other words, $\vdash_0 \Gamma \Rightarrow \Delta$ implies $\vdash_0 \Gamma' \Rightarrow \Delta'$. In the inductive step we must consider all the 8 cases of rules as applied to $\Gamma \Rightarrow \Delta$. We take the case of $(\Rightarrow \wedge)$ and leave the remaining ones as exercises. Now our $\Gamma \Rightarrow \Delta := \Gamma \Rightarrow \Pi, \varphi \wedge \psi$ and the premisses are $\Gamma \Rightarrow \Pi, \varphi$ and $\Gamma \Rightarrow \Pi, \psi$, respectively, both with proofs having height $< n$. By the induction hypothesis both $\Gamma' \Rightarrow \Pi', \varphi$ and $\Gamma' \Rightarrow \Pi', \psi$, where $\Pi' = \Delta' - \{\varphi \wedge \psi\}$, are provable with proofs of height $< n$, hence $\Gamma' \Rightarrow \Delta'$ is provable with the height n.   □

Note that we applied strong induction (see Appendix) since the proof of one of the premisses may have the height $< n - 1$. Although in case of strong induction a separate proof of the basis is not necessary we will usually provide it for better readability.

Admissibility of Weakening in K is in a sense a consequence of some important features of rules which were analysed first by Curry [56]. We collectively call them:

1.4.2.1   Context independence of rules:

Validity-preservation of rules is intact by:

1. deletion of the same parameters in premisses and conclusion;

2. addition of the same parameters to premisses and conclusion[6];

3. interchange of some parameter for a different one in all occurrences.

Note that the third point is just a consequence of the first and the second but these two are independent and it is possible to have rules which satisfy the first but not the second (see chapter 4 for examples).

Exercise 1.9.   *1. Complete a proof of lemma 1.4*

   2. Check that rules of K are context-independent.

---

[6]This is called *purity* of rules by Avron [9] and closure under rule expansion by Beziau [29].

3. *Show that every derivable rule is normal.*

4. *Show that every admissible rule is validity-preserving.*

As an example of the application of W we can prove that the usual $(\Rightarrow \wedge)$ of K is derivable in SC with $(3 \Rightarrow \wedge)$, introduced in the previous paragraph, as a primitive rule:

$$(W \Rightarrow) \quad \dfrac{\Gamma \Rightarrow \Delta, \varphi}{\psi, \Gamma \Rightarrow \Delta, \varphi} \qquad \dfrac{\Gamma \Rightarrow \Delta, \psi}{\varphi, \Gamma \Rightarrow \Delta, \psi} \qquad \dfrac{\Gamma \Rightarrow \Delta, \varphi}{\Gamma \Rightarrow \Delta, \varphi, \psi} (\Rightarrow W)$$
$$(3 \Rightarrow \wedge) \quad \overline{\qquad\qquad\qquad\qquad \Gamma \Rightarrow \Delta, \varphi \wedge \psi \qquad\qquad\qquad\qquad}$$

Exercise 1.10. *Prove derivability of* $(\vee \Rightarrow)$ *and* $(\rightarrow \Rightarrow)$ *by means of their* $3-variants$ *and W.*

Prove that each of the $2-$ *(or full) rules is derivable in K from only a single premiss by means of W and ordinary* $(\wedge \Rightarrow), (\Rightarrow \vee), (\Rightarrow \rightarrow)$.

To prove the derivability of $3-$ variants in K, or the standard rules $(\wedge \Rightarrow)$, $(\Rightarrow \vee), (\Rightarrow \rightarrow)$ by means of $2-$variants, we need another rule called cut which will be introduced in section 1.8.

1.4.3 Invertibility of Rules

There is one more specific and very important feature of all rules of K. Let us say for any rule $S_1, ..., S_n / S$, that the inversions of this rule are rules: $S / S_1, ..., S / S_n$. In K all the rules are invertible in the sense that if the conclusion is valid, then premiss(es) are valid as well. In other words, we can say that inversions of the original rules are additional validity-preserving rules. We prove below this result as it is essential for the proof of the completeness theorem in the next section.

Lemma 1.5. (Invertibility). *All rules of K are invertible.*

PROOF: There are 8 cases to consider; we illustrate the proof with two, leaving the remaining ones to the reader.

The case of $(\wedge \Rightarrow)$: Assume that $\models \varphi \wedge \psi, \Gamma \Rightarrow \Delta$ but $\not\models \varphi, \psi, \Gamma \Rightarrow \Delta$. Then there is an $\mathfrak{M}$ which falsifies $\varphi, \psi, \Gamma \Rightarrow \Delta$ but it also falsifies $\varphi \wedge \psi, \Gamma \Rightarrow \Delta$ since $\mathfrak{M} \models \varphi \wedge \psi$, if $\mathfrak{M} \models \varphi$ and $\mathfrak{M} \models \psi$ – a contradiction.

The case of $(\rightarrow \Rightarrow)$: Assume that $\models \varphi \rightarrow \psi, \Gamma \Rightarrow \Delta$ but neither $\models \Gamma \Rightarrow \Delta, \varphi$ nor $\models \psi, \Gamma \Rightarrow \Delta$. In the former case, in some model $\mathfrak{M} \models \Gamma$ and $\mathfrak{M} \not\models \Delta, \varphi$. But then $\mathfrak{M} \models \varphi \rightarrow \psi$ which yields $\mathfrak{M} \not\models \varphi \rightarrow \psi, \Gamma \Rightarrow \Delta$ – a contradiction. Analogously for the second premiss $\psi, \Gamma \Rightarrow \Delta$. □

Exercise 1.11. *Provide proofs of the remaining 6 cases.*

Again, we are concerned with the stronger notion of validity-preservation, since this is what we need for the completeness proof in the next section. However, one may easily demonstrate that any rule which is the inversion of a primitive rule of K is also normal.

Exercise 1.12. *Show normality of the inversions of all primitive rules of K.*

We can show that all rules and their inversions are normal also in a syntactical way if we refer to the informal reading of sequents due to Gentzen. We can prove the following 6 equivalences corresponding to our rules (we omit the cases of $(\wedge \Rightarrow)$ and $(\Rightarrow \vee)$ since they are trivial). In all the schemata below γ corresponds to (the conjunction of) parameters from the antecent, δ to (the disjunction of) parameters from the succedent and φ, ψ denote active formulae of respective rules.

Lemma 1.6. *The following hold:*

1. $\vdash \Rightarrow \gamma \rightarrow \delta \vee \varphi \leftrightarrow \neg\varphi \wedge \gamma \rightarrow \delta$

2. $\vdash \Rightarrow \varphi \wedge \gamma \rightarrow \delta \leftrightarrow \gamma \rightarrow \delta \vee \neg\varphi$

3. $\vdash \Rightarrow (\gamma \rightarrow \delta \vee \varphi) \wedge (\gamma \rightarrow \delta \vee \psi) \leftrightarrow \gamma \rightarrow \delta \vee \varphi \wedge \psi$

4. $\vdash \Rightarrow (\varphi \wedge \gamma \rightarrow \delta) \wedge (\psi \wedge \gamma \rightarrow \delta) \leftrightarrow (\varphi \vee \psi) \wedge \gamma \rightarrow \delta$

5. $\vdash \Rightarrow \varphi \wedge \gamma \rightarrow \delta \vee \psi \leftrightarrow \gamma \rightarrow \delta \vee (\varphi \rightarrow \psi)$

6. $\vdash \Rightarrow (\gamma \rightarrow \delta \vee \varphi) \wedge (\psi \wedge \gamma \rightarrow \delta) \leftrightarrow (\varphi \rightarrow \psi) \wedge \gamma \rightarrow \delta$

PROOF: We provide proofs of 5 and 6. As for 5 we obtain a proof by $(\Rightarrow\leftrightarrow)$ from the two proofs:

$$(\Rightarrow \wedge) \;\; \cfrac{\cfrac{\gamma, \varphi \Rightarrow \delta, \psi, \varphi \quad \gamma, \varphi \Rightarrow \delta, \psi, \gamma}{\gamma, \varphi \Rightarrow \delta, \psi, \varphi \wedge \gamma} \quad \cfrac{\cfrac{\delta, \gamma, \varphi \Rightarrow \delta, \psi \quad \psi, \gamma, \varphi \Rightarrow \delta, \psi}{\delta \vee \psi, \gamma, \varphi \Rightarrow \delta, \psi} (\vee \Rightarrow)}{}}{\cfrac{\cfrac{\varphi \wedge \gamma \rightarrow \delta \vee \psi, \gamma, \varphi \Rightarrow \delta, \psi}{\varphi \wedge \gamma \rightarrow \delta \vee \psi, \gamma \Rightarrow \delta, \varphi \rightarrow \psi} (\Rightarrow \rightarrow)}{\cfrac{\varphi \wedge \gamma \rightarrow \delta \vee \psi, \gamma \Rightarrow \delta \vee (\varphi \rightarrow \psi)}{\varphi \wedge \gamma \rightarrow \delta \vee \psi \Rightarrow \gamma \rightarrow \delta \vee (\varphi \rightarrow \psi)} (\Rightarrow \rightarrow)} (\Rightarrow \vee)} (\rightarrow \rightarrow)$$

and

$$\cfrac{\cfrac{\varphi, \gamma \Rightarrow \delta, \psi, \gamma \quad \cfrac{\delta, \varphi, \gamma \Rightarrow \delta, \psi \quad \cfrac{\cfrac{\varphi, \gamma \Rightarrow \delta, \psi, \varphi \quad \psi, \varphi, \gamma \Rightarrow \delta, \psi}{\varphi \rightarrow \psi, \varphi, \gamma \Rightarrow \delta, \psi} (\rightarrow \Rightarrow)}{\delta \vee (\varphi \rightarrow \psi), \varphi, \gamma \Rightarrow \delta, \psi} (\vee \Rightarrow)}{\gamma \rightarrow \delta \vee (\varphi \rightarrow \psi), \varphi, \gamma \Rightarrow \delta, \psi} (\rightarrow \Rightarrow)}{\gamma \rightarrow \delta \vee (\varphi \rightarrow \psi), \varphi \wedge \gamma \Rightarrow \delta, \psi} (\wedge \Rightarrow)}{\cfrac{\gamma \rightarrow \delta \vee (\varphi \rightarrow \psi), \varphi \wedge \gamma \Rightarrow \delta \vee \psi}{\gamma \rightarrow \delta \vee (\varphi \rightarrow \psi) \Rightarrow \varphi \wedge \gamma \rightarrow \delta \vee \psi} (\Rightarrow \vee)} (\Rightarrow \rightarrow)$$

In a similar way we divide the work for 6. To prove $(\varphi \rightarrow \psi) \wedge \gamma \rightarrow \delta \Rightarrow (\gamma \rightarrow \delta \vee \varphi) \wedge (\psi \wedge \gamma \rightarrow \delta)$ we provide two proofs:

$$
\begin{array}{c}
(\Rightarrow\rightarrow) \dfrac{\varphi,\gamma \Rightarrow \delta,\varphi,\psi}{\gamma \Rightarrow \delta,\varphi,\varphi\rightarrow\psi} \qquad \gamma \Rightarrow \delta,\varphi,\gamma \\
(\Rightarrow\wedge) \dfrac{}{} \\
(\rightarrow\Rightarrow) \dfrac{\gamma \Rightarrow \delta,\varphi,(\varphi\rightarrow\psi)\wedge\gamma \qquad\qquad \delta,\gamma \Rightarrow \delta,\varphi}{} \\
(\Rightarrow\vee) \dfrac{(\varphi\rightarrow\psi)\wedge\gamma\rightarrow\delta,\gamma \Rightarrow \delta,\varphi}{} \\
(\Rightarrow\rightarrow) \dfrac{(\varphi\rightarrow\psi)\wedge\gamma\rightarrow\delta,\gamma \Rightarrow \delta\vee\varphi}{(\varphi\rightarrow\psi)\wedge\gamma\rightarrow\delta \Rightarrow \gamma\rightarrow\delta\vee\varphi}
\end{array}
$$

and

$$
\begin{array}{c}
(\Rightarrow\rightarrow) \dfrac{\varphi,\psi,\gamma \Rightarrow \delta,\psi}{\psi,\gamma \Rightarrow \delta,\varphi\rightarrow\psi} \qquad \psi,\gamma \Rightarrow \delta,\gamma \\
(\Rightarrow\wedge) \dfrac{}{} \\
(\rightarrow\Rightarrow) \dfrac{\psi,\gamma \Rightarrow \delta,(\varphi\rightarrow\psi)\wedge\gamma \qquad\qquad \delta,\psi,\gamma \Rightarrow \delta}{} \\
(\wedge\Rightarrow) \dfrac{(\varphi\rightarrow\psi)\wedge\gamma\rightarrow\delta,\psi,\gamma \Rightarrow \delta}{} \\
(\Rightarrow\rightarrow) \dfrac{(\varphi\rightarrow\psi)\wedge\gamma\rightarrow\delta,\psi\wedge\gamma \Rightarrow \delta}{(\varphi\rightarrow\psi)\wedge\gamma\rightarrow\delta \Rightarrow \psi\wedge\gamma\rightarrow\delta}
\end{array}
$$

and apply $(\Rightarrow\wedge)$.

To prove $(\gamma\rightarrow\delta\vee\varphi)\wedge(\psi\wedge\gamma\rightarrow\delta) \Rightarrow (\varphi\rightarrow\psi)\wedge\gamma\rightarrow\delta$ we provide a proof:

$$
\begin{array}{c}
\qquad\qquad \dfrac{\delta,\psi\wedge\gamma\rightarrow\delta,\varphi\rightarrow\psi,\gamma \Rightarrow \delta \qquad \mathcal{D}}{} (\vee\Rightarrow) \\
\dfrac{\psi\wedge\gamma\rightarrow\delta,\varphi\rightarrow\psi,\gamma \Rightarrow \delta,\gamma \qquad \delta\vee\varphi,\psi\wedge\gamma\rightarrow\delta,\varphi\rightarrow\psi,\gamma \Rightarrow \delta}{\gamma\rightarrow\delta\vee\varphi,\psi\wedge\gamma\rightarrow\delta,\varphi\rightarrow\psi,\gamma \Rightarrow \delta} (\rightarrow\Rightarrow) \\
\dfrac{}{\gamma\rightarrow\delta\vee\varphi,\psi\wedge\gamma\rightarrow\delta,(\varphi\rightarrow\psi)\wedge\gamma \Rightarrow \delta} (\wedge\Rightarrow) \\
\dfrac{}{\gamma\rightarrow\delta\vee\varphi,\psi\wedge\gamma\rightarrow\delta \Rightarrow (\varphi\rightarrow\psi)\wedge\gamma\rightarrow\delta} (\Rightarrow\rightarrow) \\
\dfrac{}{(\gamma\rightarrow\delta\vee\varphi)\wedge(\psi\wedge\gamma\rightarrow\delta) \Rightarrow (\varphi\rightarrow\psi)\wedge\gamma\rightarrow\delta} (\wedge\Rightarrow)
\end{array}
$$

where $\mathcal{D}$ is:

$$
\begin{array}{c}
\qquad\qquad \dfrac{\psi,\varphi,\gamma \Rightarrow \delta,\psi \qquad \psi,\varphi,\gamma \Rightarrow \delta,\gamma}{\psi,\varphi,\gamma \Rightarrow \delta,\psi\wedge\gamma} (\Rightarrow\wedge) \\
(\rightarrow\Rightarrow) \dfrac{\varphi,\gamma \Rightarrow \delta,\psi\wedge\gamma,\varphi}{} \\
(\rightarrow\Rightarrow) \dfrac{\varphi,\varphi\rightarrow\psi,\gamma \Rightarrow \delta,\psi\wedge\gamma \qquad\qquad\qquad \delta,\varphi,\varphi\rightarrow\psi,\gamma \Rightarrow \delta}{\varphi,\psi\wedge\gamma\rightarrow\delta,\varphi\rightarrow\psi,\gamma \Rightarrow \delta}
\end{array}
$$

$\square$

Exercise 1.13. *Prove the remaining sequents.*

1.5 The Adequacy of K

First of all we must check that K is sound, i.e. that everything we can prove in it is correct. On the basis of validity-preservation lemma (lemma 1.2), one easily obtains for K:

Lemma 1.7. (Soundness). *If* $\vdash S$*, then* $\models S$.

PROOF: By induction on the height of a proof of S. The basis is trivial since every axiomatic sequent must be valid. For the inductive step, we assume that the claim holds for every sequent having a proof of the height $< n$, then by lemma 1.2 (validity-preservation) it holds for S having a proof of the height n.   □

One may demonstrate the completeness of K in many ways (see e.g. section 1.7.). We will present here one of the simplest proofs based on an application of the invertibility lemma (i.e. lemma 1.5.) which is due to Buss [43]. First we prove some preliminary result. Let us call K' a version of K with all axioms restricted to those with atomic formulae as active. We must show that such a restriction does not weaken the calculus, i.e.:

Lemma 1.8. (Atomic K). *If S is provable in K, then it is provable in K'*

PROOF: It is enough to show in K' that for any formula φ it is provable that $\Gamma, \varphi \Rightarrow \varphi, \Delta$. We do it by induction on the complexity of φ. The basis is trivial since if φ is atomic we have an axiom of K'. So one must check 4 cases depending on the shape of φ. Let it be a conjunction, then we may provide the following proof schema:

$$(\wedge \Rightarrow) \frac{\dfrac{\Gamma, \varphi, \psi \Rightarrow \varphi, \Delta}{\Gamma, \varphi \wedge \psi \Rightarrow \varphi, \Delta} (\wedge \Rightarrow) \quad \dfrac{\Gamma, \varphi, \psi \Rightarrow \psi, \Delta}{\Gamma, \varphi \wedge \psi \Rightarrow \psi, \Delta} (\wedge \Rightarrow)}{\Gamma, \varphi \wedge \psi \Rightarrow \varphi \wedge \psi, \Delta} (\Rightarrow \wedge)$$

where all leaves are provable by the induction hypothesis.   □

Exercise 1.14. *Provide proofs of the remaining 3 cases.*

Since K' is trivially included in K, we have that both systems are equivalent, i.e. both yield the same provable sequents. We will provide a completeness proof for K'. Moreover, this proof allows for the estimation of the size of a proof. In fact, the following proof may be established also directly for K but lemma 1.8 is of interest on its own right and will be needed later as well.

Theorem 1.2. (Completeness). *If $\models S$ and there are n occurrences of logical constants in S, then $\vdash S$ with a proof $\mathcal{D}$ such that $\|\mathcal{D}\| < 2^n$.*

PROOF: By induction on n.

The basis: if $n = 0$, then S must be axiomatic in order to be valid, hence it has a proof of the size 0.

The induction hypothesis states that the claim holds for any valid sequent with occurrences of constants $< n$. We show that it holds for any valid sequent with n occurrences of constants. Compound formulae in S are of 4 forms and we must consider them as elements of the antecedent or the succedent. Let us take a conjunction in both positions.

$\varphi \wedge \psi$ in the antecedent. So $S := \varphi \wedge \psi, \Gamma \Rightarrow \Delta$. By invertibility we have $\models \varphi, \psi, \Gamma \Rightarrow \Delta$. The latter has $n-1$ occurrences of constants so by the induction hypothesis $\vdash \varphi, \psi, \Gamma \Rightarrow \Delta$ and $\|\mathcal{D}\| < 2^{n-1}$. Adding to $\mathcal{D}$, $\varphi \wedge \psi, \Gamma' \Rightarrow \Delta$ as the root deduced by $(\wedge \Rightarrow)$ we get a proof $\mathcal{D}'$ of this sequent such that $\|\mathcal{D}'\| < 2^{n-1} + 1 \leq 2^n$.

$\varphi \wedge \psi$ in the succedent. So $S := \Gamma \Rightarrow \Delta, \varphi \wedge \psi$. By invertibility we have $\models \Gamma \Rightarrow \Delta, \varphi$ and $\models \Gamma \Rightarrow \Delta, \psi$. Each sequent has $< n$ occurrences of constants so by the induction hypothesis it has a proof $\|\mathcal{D}\| < 2^{n-1}$. We add to both proofs $\Gamma \Rightarrow \Delta, \varphi \wedge \psi$ as the root deduced by $(\Rightarrow \wedge)$ and get a proof $\mathcal{D}'$ of this sequent such that $\|\mathcal{D}'\| < 2^n$, since for any $i, j < 2^{n-1}$, $i + j + 1 < 2^n$. $\qquad\square$

Exercise 1.15. *Provide proofs of the remaining 3 cases.*

1.6 Decidability

The proof of the completeness theorem is very simple but in some sense not very informative. It is not constructive in the sense that it gives no hints how to obtain a proof of any valid sequent. However, in section 1.3 we have shown informally that the very construction of the rules of K, namely their subformula property, allows for simple root-first proof search. In fact, such a method, when applied in a systematic way, gives not only a constructive proof of completeness, but also a proof of decidability. We examine the latter issue first.

1.6.1 Proof Trees

We have noted that root-first proof search may be performed in different ways but, since sequents are finite, the number of choices is bounded at every step. In section 1.3. we have suggested that such a procedure always leads to the expected result, i.e. gives a proof for provable sequent, and a falsifying model otherwise. Now we will show why this is so. First we define precisely the notion of a proof tree which was used informally in section 1.3.

Definition 1.7. (A proof tree for a sequent S).

1. *A one-node tree which consists of S is a proof tree for S.*

2. *If S' is nonatomic leaf in a proof tree for S, then a tree obtained by appending a sequent T above S' is a proof tree for S, provided that T is an instance of the premiss, and S' is an instance of the conclusion of some of the one-premiss rules.*

3. *If S' is nonatomic leaf in a proof tree for S, then a tree obtained by appending sequents T and T' above S' is a proof tree for S, provided that T is an instance of the left premiss, T' is an instance of the right premiss, and S' is an instance of the conclusion of some of the two-premiss rules.*

4. *Nothing more is a proof tree for S.*

This definition is different from the definition of a proof or derivation stated in section 1.3 since it starts from the root, not from the leaves, in accordance with the actual proof search. Of course not every proof tree is a proof but we have admitted such trees with nonaxiomatic leaves $S_1, ..., S_n$ as representing a derivation of S from $S_1, ..., S_n$. Despite the differences, both definitions are inductive definitions of particular kinds of trees. The present one is more suitable for our purpose of showing decidability since it allows for the application of reverse induction proofs (from the root to the leaves)[7]. The relation between both notions may be stated as follows:

Claim 1.5. $\mathcal{D}$ *is a derivation of S from the set X of sequents iff $\mathcal{D}$ is a proof tree for S in which X is a set of nonaxiomatic leaves.*

PROOF: By comparison of induction clauses—in both cases the same rules regulate the generation of a tree.   □

   Particularly important are trees which cannot be extended:

Definition 1.8. (Completed proof tree). *Any proof tree with only atomic sequents as leaves is completed.*

   For proof trees we can prove:

Lemma 1.9. (Termination). *Any proof tree may be extended to a completed proof tree.*

PROOF: Take any proof tree which is not yet completed. Take the leftmost branch with a nonatomic leaf and apply a suitable rule to some complex formula. Since every sequent is finite and all rules satisfy the subformula property, every application of a rule diminishes the number of complex formulae. Consequently, always selecting the leftmost nonatomic sequent, after a finite time we must finish with an atomic leaf. We move to the right, to the next nonatomic leaf, and repeat the procedure. Since our trees are finitely branching, this procedure must terminate. In particular, if we start with a sequent of complexity n, i.e. having n occurrences of connectives, we obtain a tree having at most $2^n + 1$ nodes and branches of height at most n.   □

Exercise 1.16. *Provide a proof of the estimation of the size of a completed proof tree by induction on n.*

   A simple consequence of this lemma is:

Lemma 1.10. *Every completed proof tree for S is either a proof of S or a derivation of S from a nonempty set of atomic nonaxiomatic sequents.*

---

[7]One may find a rigorous treatment of both kinds of inductive sets based on trees and, respectively, two variants of inductive proofs in Segerberg [235]. Here, for the sake of readability, we decided to explain things in a less formal way, but it is possible to restate all the results in a strict form.

1.6.2 Confluency

But this is not enough for demonstrating decidability. Now we must consider if every completed tree for S which is not a proof but a derivation, guarantees that S is unprovable. In general this is not always so (see a proof of decidability for LK in chapter 2), but fortunately for K it holds: every completed tree decides the problem of provability of its root sequent. In other words K is confluent. We define this property generally for any SC as follows:

Definition 1.9. *SC is confluent iff for a provable sequent S, every completed proof tree for S yields a proof of S.*

This means that in a confluent SC it does not matter in which order the application of rules are taken. The trees may be different but the result will be always the same—a proof of a provable sequent. By contraposition, if we construct a completed proof tree with at least one nonaxiomatic leaf, the root sequent is unprovable. Confluent systems behave nicely from the standpoint of automated deduction because they overload the memory of the program less and require simpler algorithms. In particular, we are not forced to backtrack to earlier stages if we made 'wrong' choices, since there are no wrong choices. Our choices may have a significant impact on the size of a proof tree but not on the result.

Of course confluency is not necessary for proving decidability but it makes it particularly easy (again check decidability proofs in chapters 2, 4 and 5). Hence before proving decidability we will show:

Lemma 1.11. (Confluency). *K is confluent.*

PROOF: Assume that despite the provability of S we have constructed a completed proof tree which is not a proof of S. Take a branch with some atomic and nonaxiomatic leaf S'. Any model falsifying S' must falsify also any other sequent on this branch since all rules are invertible. Hence S is also falsified which—by soundness – leads to contradiction. □

The above proof shows the close connection between confluency and invertibility. In fact we can prove some more general result. Let us consider any logic with an adequate formalisation in terms of some SC, then we have:

Theorem 1.3. *An adequate SC for L is confluent iff all its rules are invertible.*

PROOF: Assume that SC is confluent but that some rule r is not invertible. Let us consider a proof of S where r was applied. Consider the first (from the root) instance $S_1, ..., S_n/S'$ of this rule on some branch. By soundness $\models S$, also all rules used on this branch below S' are invertible, so $\models S'$ and by completeness $\vdash S'$. By confluency any proof tree for S' yields a proof of S', in particular, a tree where r is applied first. Since every possible extension of $\dfrac{S_1, ..., S_n}{S'}$ yields a proof of S', then each S_i is provable and—by soundness—valid, but this means that r is also invertible, contrary to our assumption.

Now assume that all rules are invertible but SC is not confluent. Hence for some provable S there exists a completed proof tree with at least one atomic leaf S' which is nonaxiomatic and falsifiable. But, because of invertibility, it makes also S invalid which is impossible.   □

In fact we could show the confluency of K by referring not to invertibility but to permutability of rules. One may check that for any two rules their order of application may be reversed. This phenomenon will be examined in the next chapter in the more demanding framework of the calculus LK. For now we will illustrate this with the following example:

$$(\wedge \Rightarrow) \ \dfrac{\dfrac{\varphi, \psi, \Gamma \Rightarrow \Delta, \chi}{\varphi \wedge \psi, \Gamma \Rightarrow \Delta, \chi} \qquad \dfrac{\varphi, \psi, \Gamma \Rightarrow \Delta, \phi}{\varphi \wedge \psi, \Gamma \Rightarrow \Delta, \phi} \ (\wedge \Rightarrow)}{\varphi \wedge \psi, \Gamma \Rightarrow \Delta, \chi \wedge \phi} \ (\Rightarrow \wedge)$$

may be changed into:

$$\dfrac{\dfrac{\varphi, \psi, \Gamma \Rightarrow \Delta, \chi \qquad \varphi, \psi, \Gamma \Rightarrow \Delta, \phi}{\varphi, \psi, \Gamma \Rightarrow \Delta, \chi \wedge \phi} \ (\Rightarrow \wedge)}{\varphi \wedge \psi, \Gamma \Rightarrow \Delta, \chi \wedge \phi} \ (\wedge \Rightarrow)$$

Since we can perform such permutations for all rules of K it is evident that the order of the applications of rules in the root-first proof search cannot change the result.

Exercise 1.17. *Check permutability of at least 4 from 64 possible combinations of two rules.*

1.6.3 Decidability

Now we are in the position to strengthen lemma 1.10.

Theorem 1.4. (Decidability). *For any S in finite time we can either provide a proof, or find a countermodel.*

PROOF: It is enough to apply the procedure from a proof of lemma 1.9 which produces a finite proof tree for any S. If it is not a proof, then at least one leaf S' is atomic and nonaxiomatic. By invertibility, any model falsifying S' falsifies also S.   □

We finish this section with slightly more formal description of our decision procedure for K in pseudocode. Axiomatic sequents (not necessarily atomic) will be labelled as closed and such branches are not extended further.

Input: a sequent S

Output: a completed proof tree for S (giving either a proof of S or a denial of its provability).

0. **Start** Take S as the root of a proof tree.
1. **Select** the leftmost leaf S' which is not closed.
2. **If** S' is not axiomatic, **then**
    2.1. **If** S' is atomic, **then**
        stop: $\nvdash S$
        else apply a rule to selected complex formula **goto** 1.
    else close S'.
3. **If** S' is the rightmost leaf, **then**
    stop: $\vdash S$
    else goto 1.

That the procedure is fair and terminating should be obvious from previous results. In this book we are not going to investigate in detail issues of automatization or implementation but it is worthwhile to comment on a few features of this procedure:

1. a DEPTH-FIRST strategy;

2. quick branch-closure test;

3. bounded indeterminism;

4. no strategy of preference.

There are different strategies of building (or searching through) trees. The most popular are breadth-first and depth-first strategies. In the former type we would append a new sequent to each leaf in one stage, say from left to right instead of extending just one branch until the end. The examples of breadth-first algorithms for SC may be found in Gallier [93], Buss [43], or Kleene [150]; they are more appropriate for undecidable logics.

The choice of depth-first strategy may lead to earlier termination of search in case of unprovable sequents since we extend one branch until the end, and if the leaf is nonaxiomatic we do not consider any remaining branches. Also the fact that after every rule application we test the leftmost leaf and go to the right if it is axiomatic, reduces the time and size of the tree. On the other hand in automated search it is better to extend every branch to atomic leaf and eventually check if it is axiomatic or not, since such test are less memory-consuming.

Bounded indeterminism means that in case of a nonatomic leaf we select which complex formula is attacked first. Such a weak form of indeterminism may be avoided by introducing some order of selecting complex formulas, e.g. from the left to right. Clearly in case of automatization, it is better to use sequents built from lists of formulas not sets (see e.g. Gallier [93]). On the other hand, we could apply some strategies of optimization, for example, the obvious one of always choosing first those formulae which do not lead to branching. We did not apply any strategy of preference in order to avoid a complication in describing the algorithm.

Exercise 1.18. *1. Show that the procedure described above must terminate.*

2. *Modify the algorithm introducing a strategy of preference for nonbranching rules.*

3. *Put axiom checking at the end of extending a branch or a whole tree.*

4. *(For computer science students) Rewrite the procedure in your favourite programming language and implement.*

1.7   An Analytic Proof of Completeness

The automated procedure of proof search presented in the preceding section for the needs of a decidability proof may be also applied for proving the completeness of K. Such a proof is constructive in the sense that for every nonprovable sequent we receive a recipe for the construction of a falsifying model. In contrast to the proof based on the invertibility of rules, this approach is more general and can be applied to SC of different kinds and for different logics.

1.7.1   Hintikka's Tuples

The key notion used in such a kind of completeness proofs is that of a downward saturated set, originally introduced by Hintikka. Here we specify this notion for sequents, hence it will be better to define it not for sets (like it is commonly defined in the context of tableau methods), but for ordered pairs of sets of formulae. It can be done in at least two ways:

a. directly, by providing a definition of the downward saturated pair of sets;

b. indirectly (e.g. Goré [103]) by introducing first the notion of closure for rules.

We start with the first method:

Definition 1.10. (Γ, Δ) *is downward saturated iff it satisfies the following conditions:*

1. if $\neg\varphi \in \Gamma$, then $\varphi \in \Delta$

2. if $\neg\varphi \in \Delta$, then $\varphi \in \Gamma$

3. if $\varphi \wedge \psi \in \Gamma$, then $\varphi \in \Gamma$ and $\psi \in \Gamma$

4. if $\varphi \wedge \psi \in \Delta$, then $\varphi \in \Delta$ or $\psi \in \Delta$

5. if $\varphi \vee \psi \in \Gamma$, then $\varphi \in \Gamma$ or $\psi \in \Gamma$

6. if $\varphi \vee \psi \in \Delta$, then $\varphi \in \Delta$ and $\psi \in \Delta$

7. if $\varphi \rightarrow \psi \in \Gamma$, then $\varphi \in \Delta$ or $\psi \in \Gamma$

8. if $\varphi \rightarrow \psi \in \Delta$, then $\varphi \in \Gamma$ and $\psi \in \Delta$

Note that there is an obvious relationship between these conditions and the rules of K (applied root-first). It is also clear that any pair of sets of atomic formulae (even empty) is saturated. The close relationship of saturation conditions to rules enables an alternative solution (see e.g. Goré [103]) based on the notion of the closure of a sequent under (the application of) a rule, which may be defined in the following way:

Definition 1.11. $\Gamma \Rightarrow \Delta$ *is closed under the application of:*

a) a one-premiss rule, if whenever the principal formula of this rule (application) belongs to this sequent, then also side formulae belong to it;

b) a two-premiss rule, if whenever the principal formula of this rule (application) belongs to this sequent, then the side formulae from at least one premiss belong to it.

One may check by inspection that $\Gamma \Rightarrow \Delta$ is closed under every rule iff the pair (Γ, Δ) is downward saturated. Let us say that a pair (Γ, Δ) is consistent iff $\Gamma \cap \Delta = \varnothing$; otherwise it is inconsistent. We will call every saturated pair which is a consistent Hintikka tuple. The key lemma says:

Lemma 1.12. (Truth lemma). *If (Γ, Δ) is a Hintikka tuple (i.e. saturated and consistent), then there is a model $\mathfrak{M}$, such that:*

- *if $\varphi \in \Gamma$, then $\mathfrak{M} \vDash \varphi$;*
- *if $\varphi \in \Delta$, then $\mathfrak{M} \nvDash \varphi$.*

PROOF: We define a model assuming that for any p, $V(p) = 1$ iff $p \in \Gamma$ and prove by induction on the complexity of formulae that it satisfies the conditions. The basis holds by the definition of V and consistency. For the inductive step let us consider $\varphi := \neg \psi$:

If $\neg \psi \in \Gamma$, then—by condition 1 of the definition—$\psi \in \Delta$. By the induction hypothesis $\mathfrak{M} \nvDash \psi$ but then $\mathfrak{M} \vDash \neg \psi$. The case of $\neg \psi \in \Delta$ is proved symmetrically. $\square$

Exercise 1.19. *Prove the remaining cases.*

1.7.2 Completeness

Now we may show the crucial lemma:

Lemma 1.13. (Completed tree). *For any $\Gamma \Rightarrow \Delta$, either $\vdash \Gamma \Rightarrow \Delta$ or there is a Hintikka tuple (Π, Σ) of finite sets, such that $\Pi \cup \Sigma \subseteq SF(\Gamma \cup \Delta)$;*

Recall that $SF(\Gamma \cup \Delta)$ denotes the set of all subformulae of formulae from $\Gamma \cup \Delta$.

PROOF: is simple since we can use the procedure of proof search defined in the previous section. We know that for each sequent we obtain either a proof or not a proof. We must examine the latter. If $\nvdash \Gamma \Rightarrow \Delta$, then in a completed tree there is at least one branch with an atomic and nonaxiomatic leaf $\Gamma_n \Rightarrow \Delta_n$. Going from the root to the leaf we obtain a finite list of sequents: $\Gamma \Rightarrow \Delta, ..., \Gamma_n \Rightarrow \Delta_n$.

Let us define: $\Pi = \bigcup \Gamma_i$, $\Sigma = \bigcup \Delta_i$, where $i \leq n$ (i.e. $\Gamma_i \Rightarrow \Delta_i$ is a sequent occurring somewhere on this branch). It is sufficient to show that this pair is a Hintikka tuple. Now, consistency is satisfied by definition (the branch is open) so we must show that (Π, Σ) is saturated. Saturation follows from the fact that the conditions defining saturation are just the converses of the rules applied in the branch and we know that all of them were applied during the application of the proof search procedure. $\square$

This strategy of proving the existence of Hintikka tuples for nonprovable sequents is called a tree strategy by Hodges [118]. The characteristic feature of it is that we consider all possibilities of building a Hintikka tuple and forget about consistency until the last step. This works well in connection with some proof search algorithms which allows for the systematic construction of the whole tree of possible choices. We can alternatively apply a direct strategy. The difference is that we start not with any sequent but with an unprovable sequent and we build a Hintikka tuple from it piecemeal. Now it is essential that we must check the unprovability of our construction at every step. Such a strategy is applied e.g. by Fitting [81] and by Goré [103], but to tableau system and apparently without using the notion of Hintikka set (by application of the notion of closure of a set under (the application of) a rule).

This time we just define a procedure of saturation for any unprovable sequent instead of a procedure for proof search for any sequent. It is a finite chain of sequents each one extending the previous element by the addition of suitable formulae to the antecedent or the succedent. What we must show is that every sequent in the list is unprovable.

Now we must show that for any nonprovable sequent there exists its saturated and consistent extension:

Lemma 1.14. (Saturation). *If $\nvdash \Gamma \Rightarrow \Delta$, then there is a saturated pair (Π, Σ) of finite sets, such that:*

(a) $\Pi \cup \Sigma \subseteq SF(\Gamma \cup \Delta)$;

(b) $\nvdash \Pi \Rightarrow \Sigma$

Note that the unprovability of $\Pi \Rightarrow \Sigma$ implies that (Π, Σ) is consistent.

PROOF: Again we build a finite list of sequents: $\Gamma \Rightarrow \Delta, ..., \Gamma_n \Rightarrow \Delta_n = \Pi \Rightarrow \Sigma$ but now with the property that for each $i < n$, $\Gamma_i \subseteq \Gamma_{i+1}$ and $\Delta_i \subseteq \Delta_{i+1}$. Moreover, at each stage we must check if the result is unprovable.

Let $\Gamma_i \Rightarrow \Delta_i$ be not saturated, then some of the conditions 1–8 of the definition of Hintikka tuple do not hold. As an example we consider the cases of $\wedge$—we proceed as follows:

Assume that $\varphi \wedge \psi \in \Gamma_i$ but either $\varphi \notin \Gamma_i$ or $\psi \notin \Gamma_i$. Then add the lacking formula to Γ_i. The resulting sequent $\Gamma_{i+1} \Rightarrow \Delta_{i+1} = \varphi, \psi, \Gamma_i \Rightarrow \Delta_i$ cannot be provable, otherwise by $(\wedge\Rightarrow)$ $\Gamma_i \Rightarrow \Delta_i$ would be provable as well.

Assume that $\varphi \wedge \psi \in \Delta_i$ but neither $\varphi \in \Delta_i$ nor $\psi \in \Delta_i$. Then add to Δ_i, one of the lacking formula, namely this one which yields an unprovable sequent. At least one of them must be unprovable because if both $\vdash \Gamma_i \Rightarrow \Delta_i, \varphi$ and $\vdash \Gamma_i \Rightarrow \Delta_i, \psi$, then by $(\Rightarrow\wedge)$ also $\Gamma_i \Rightarrow \Delta_i$ would be provable.

By the subformula property this procedure must terminate. The last pair is a Hintikka tuple. □

Exercise 1.20. *Demonstrate the remaining cases.*

Notice one important difference with the preceding proof—here the sequents are saturated, not the branches. One may obtain the result directly on the level of the calculus by applying what is called Kleene's trick. It consists in replacing all rules with variants that have their principal formula present also in the premisses. Thus, e.g. the modified rules for conjunction look like this:

$$(\wedge\Rightarrow) \quad \frac{\varphi, \psi, \varphi \wedge \psi, \Gamma \Rightarrow \Delta}{\varphi \wedge \psi, \Gamma \Rightarrow \Delta} \qquad (\Rightarrow\wedge) \quad \frac{\Gamma \Rightarrow \Delta, \varphi \wedge \psi, \varphi \quad \Gamma \Rightarrow \Delta, \varphi \wedge \psi, \psi}{\Gamma \Rightarrow \Delta, \varphi \wedge \psi}$$

Such rules are derivable in K due to the fact that we define sequents as built from sets. The usual rules are easily derivable from these rules by W. So this refined calculus—let us call it KK—is equivalent to K. It is worth noting that in KK we have all subformulae of a proved sequent already present in axioms, and the application of rules is in fact a process of elimination of the principal formulae from the premisses.

Notice also that Goré's proof is based on the invertibility of rules but this is in fact not necessary for a proof to go through.

Both strategies of proof lead to the demonstration of:

Theorem 1.5. (Completeness:) *If* $\models \Gamma \Rightarrow \Delta$, *then* $\vdash \Gamma \Rightarrow \Delta$.

PROOF:
1. $\nvdash \Gamma \Rightarrow \Delta$ (assumption),
2. there exists saturated and consistent (Π, Σ) (Saturation lemma),
3. there exists a model falsifying (Π, Σ) (Truth lemma),
4. there exists a model falsifying (Γ, Δ) since $\Gamma \subseteq \Pi$ and $\Delta \subseteq \Sigma$,
5. $\nvDash \Gamma \Rightarrow \Delta$. □

We can easily prove also the converse of truth lemma.

Lemma 1.15. *If* $\Gamma \Rightarrow \Delta$ *has a falsifying model, then it may be extended to a Hintikka tuple* (Π, Σ).

PROOF: Let $\mathfrak{M}$ be a falsifying model for $\Gamma \Rightarrow \Delta$. Assume that $\Pi = \{\varphi : \mathfrak{M} \models \varphi\}$, and $\Sigma = \{\varphi : \mathfrak{M} \nvDash \varphi\}$. It is obvious that $\Pi \cap \Sigma = \varnothing$, and $\Gamma \subseteq \Pi$ and $\Delta \subseteq \Sigma$.

By comparison of the conditions of saturation with the definition of satisfaction $\vDash$ we see that (Π, Σ) must satisfy these conditions, therefore it is saturated and consistent.   □

Hence we get:

Theorem 1.6. $\Gamma \Rightarrow \Delta$ *is falsifiable iff it can be extended to a Hintikka tuple.*

1.8   The Cut Rule

Now we introduce one of the most famous rules of any SC, which is called cut. It is often included in the set of primitive rules of SC (e.g. in the original system of Gentzen described in the next chapter). Here we have chosen to introduce it as an admissible rule. In fact, in the approaches with primitive cut, one later tends to demonstrate that it is eliminable, which has the same effect. We will explain later why this rule is important, and in particular, why it is good not to have it as a primitive rule. The most popular version of this rule is the following:

$$(Cut) \quad \frac{\Gamma \Rightarrow \Delta, \varphi \quad \varphi, \Pi \Rightarrow \Sigma}{\Gamma, \Pi \Rightarrow \Delta, \Sigma}$$

φ in the schema will be called cut-formula of this application of the rule. In contrast to other two-premiss rules of K it is not necessary to have the same sets of parameters in both premisses. In the conclusion we have the unions of parameters from premisses. We may introduce (and prove admissible) also a variant of cut with the same sets of parameters but the proof of admissibility is a bit more complicated in this case.

An examination of the shape of this rule should make obvious why we do not want to have it as a primitive (or have a proof of its elimination). Cut rule does not satisfy the subformula property; an arbitrary formula φ is present in the premisses but not in the conclusion. Thus if we have SC with primitive cut which cannot be eliminable, a proof search cannot be carried out reasonably in a root-first manner, since we do not know in advance if at some stage cut should be used with some arbitrary φ introduced. Also decidability cannot be proved in the way as we did, even if the logic in question is decidable.

On the other hand, the cut rule may be very useful—we will illustrate some of its technical advantages in the forthcoming sections. Here we only want to point out some possible informal readings of this rule which show its importance. If we interpret $\Rightarrow$ as a kind of *consequence relation*, then cut expresses its transitivity which is an essential feature of every such relation. Also, if Γ, Δ, Π are empty (and Σ a singleton), then we can see cut as a kind of (generalised) modus ponens. This case may be also understood as expressing the application of lemmata in a proof; φ is the result that has already been proven and we just use it to shorten a proof we are currently constructing.

Cut may also be interpreted in a semantical manner. If we consider a version of cut with the same sets of parameters in both premises, and we read this rule in a root-first manner, it can be treated as expressing the principle of bivalence. Specifically, let us say that the conclusion $\Gamma \Rightarrow \Delta$ is falsified, i.e. all formulae in Γ are true and all in Δ false in some model. Then, any φ is either true or false in this model, which means that either $\varphi, \Gamma \Rightarrow \Delta$ or $\Gamma \Rightarrow \Delta, \varphi$ is falsified. The same reading applies if we interpret $\Gamma \Rightarrow \Delta$ dually, as verified, then both premises are verified.

1.8.1 Admissibility versus Eliminability

Going back to technical matters, a novice in the field may wonder why the theorem on the elimination of cut is often formulated as a theorem on the admissibility of cut. Essentially it is the same result since admissibility is the converse of eliminability in the following sense: let SC be a calculus with cut as primitive rule and SC' its cut-free version, then cut is eliminable in SC iff cut is admissible in SC'. However, there are some differences. First, the theorem is formulated in different terms:

Eliminability: If $\vdash_{SC} S$, then $\vdash_{SC'} S$

Admissibility: If $\vdash_{SC'} \Gamma \Rightarrow \Delta, \varphi$ and $\vdash_{SC'} \varphi, \Pi \Rightarrow \Sigma$, then $\vdash_{SC'} \Gamma, \Pi \Rightarrow \Delta, \Sigma$

Moreover, there are some differences in the general strategy. In particular, proofs of admissibility do not have to take care of the multiplicity of cut applications in the proofs of premises of cut. In general this simplifies the proof but on the other hand, (some) proofs of cut elimination allow for estimation of complexity bounds of the resulting (transformed) proofs. In fact, the structure of Gentzen's proof shows that it is a proof of admissibility embedded into a proof of elimination—see section 2.4.

Proofs of cut admissibility may be divided in the first instance into direct and indirect (= semantical). In the latter case we get admissibility of cut by first proving directly the completeness of a cut-free calculus, as we did in section 1.5. Now we can prove the admissibility of both variants of cut very simply on the basis of the adequacy of K and the fact that both rules are validity-preserving. It is a consequence of the general result concerning the admissibility of any validity-preserving rule in any SC that has an adequate semantics.

Lemma 1.16. *If SC has an adequate semantics and a rule* (r) *is validity-preserving in this semantics, then* (r) *is admissible in SC.*

PROOF: Assume that all premises of (r) are provable in SC. By the soundness lemma, they are all valid. By validity-preservation of (r), the conclusion is also valid. Hence, by completeness, it is also provable in SC. $\qquad\square$

In the light of this lemma we only have to show that cut is validity-preserving. It is straightforward and we leave it to the reader.

Exercise 1.21. *Prove that cut is validity-preserving.*

1.8.2   Admissibility of Cut

However, much more interesting and valuable are direct, syntactical proofs of this result. They are based on performing some transformations of proofs involving cuts (or proofs of its premises in case of proving admissibility) into cut-free proofs (or proofs of the conclusion). Basically, we must consider all the ways in which the premisses of cut were proved in K and show how to prove the conclusion in K. We successively transform a proof $\mathcal{D}$ into a proof $\mathcal{D}'$ such that all applications of cut are replaced with combinations of other rules using two kinds of steps:

(a) direct elimination of an application of cut;

(b) reduction steps which transform an application of cut into other applications of cut of "simpler" character.

What "simpler" means is stated precisely in terms of parameters used for inductive proofs. Here we need two parameters hence we provide a proof by means of a double induction: (1) on the complexity of the cut-formula c and (2) on the sum of heights of proofs of both premisses h. It is usually organised in such a way that the first parameter is main, the second induction is subsidiary. It is as if for each value of c we perform an induction on h, so the second induction is performed twice: both inside the basis of the first induction and inside the induction step. Often it is stated as induction on the weight of cut $w = \langle c, h \rangle$ where there is a lexicographical order of c and h. So the general schema is:

> I. Induction on the complexity of the cut-formula
> 1. Basis: cut on atomic cut-formulae is admissible; proven by:
> > II. Induction on the height of the left (right) premiss of cut
> > 1.1. Basis: cut with at least one premiss of height 0 is admissible
> > 1.2. Inductive step: if cut with the height $< n$ is admissible, then cut with the height n is admissible too.
> > Conclusion of 1.1, 1.2: cut on atomic cut-formulae is admissible.
> 2. Inductive step: if cut on cut-formulae of length $< n$ is admissible, then cut on cut-formula of complexity n is admissible; proven by:
> > II. Induction on the height of the left (right) premiss of cut
> > 2.1. Basis: cut with the left premiss of height 0 is admissible
> > 2.2. Inductive step: if cut with the left height $< n$ is admissible, then cut with the height n is admissible too.
> > Conclusion of 2.1, 2.2: cut on every non-atomic formula is admissible.
> Conclusion of 1 and 2: cut on every formula is admissible.

Summarising a bit we have to show that there is a proof of the conclusion of any application of cut using no cuts at all or only cuts where either the cut-formula is of lower complexity or of the same complexity but the sum of the heights of the premisses is decreased. In practice to avoid inessential repetitions, we divide the cases according to the height of a proof of the one premiss, and a character of the cut-formula, into three exhaustive classes:

1. at least one premiss is axiomatic;

2. the cut-formula is principal in both premisses;

3. the cut-formula is not principal in at least one premiss.

It may easily be verified that these cover all the possibilities. Indeed, if the first does not hold, then no premiss of cut is axiomatic and both are of height > 0. In the latter case either in both premisses the cut-formula is principal or not, i.e. at least one occurrence is not principal.

Theorem 1.7. (Admissibility of cut in K). *If* $\vdash \Gamma \Rightarrow \Delta, \varphi$ *and* $\vdash \varphi, \Pi \Rightarrow \Sigma$, *then* $\vdash \Gamma, \Pi \Rightarrow \Delta, \Sigma$

PROOF:

Case 1. Let us consider the left premiss of the application of cut being axiomatic. There are two possibilities:

(i) the cut-formula is active (i.e. it is on both sides of the left premiss);

(ii) the cut-formula is parametric.

Subcase (i). Let $\Gamma := \varphi, \Gamma'$, we have the following:

$$(Cut) \frac{\varphi, \Gamma' \Rightarrow \Delta, \varphi \qquad \varphi, \Pi \Rightarrow \Sigma}{\varphi, \Gamma', \Pi \Rightarrow \Delta, \Sigma}$$

In this case we get the conclusion from the right premiss by the application of W.

In subcase (ii) we have:

$$(Cut) \frac{\psi, \Gamma' \Rightarrow \Delta', \psi, \varphi \qquad \varphi, \Pi \Rightarrow \Sigma}{\psi, \Gamma', \Pi \Rightarrow \Delta', \Sigma, \psi}$$

where $\Gamma := \psi, \Gamma'$ and $\Delta := \Delta', \psi$. Then the conclusion is already provable as an axiom.

If the right premiss is axiomatic the situation is analogous and we also get the conclusion as provable in K (i.e. without application of cut). Note that the above cases cover the bases of both inductions. In other cases one cannot in general eliminate the application of cut in one step. Instead we must perform some reduction steps of different sorts. There are two main kinds of reduction:

1. complexity reduction—we replace cuts on compound formulae with cuts on their subformulae;

2. height reduction—we push up applications of cut in a proof.

They correspond to inductive steps of induction on the complexity and on the height, respectively. Intuitively, performing both types of steps we must finish with the case of at least one premiss being axiomatic.

Case 2. The cut-formula is principal in both premisses. We have 4 cases to consider.

Subcase 2.1: the cut-formula is a negation:

$$(Cut) \cfrac{(\Rightarrow \neg)\cfrac{\varphi, \Gamma \Rightarrow \Delta}{\Gamma \Rightarrow \Delta, \neg\varphi} \quad \cfrac{\Pi \Rightarrow \Sigma, \varphi}{\neg\varphi, \Pi \Rightarrow \Sigma}(\neg \Rightarrow)}{\Gamma, \Pi \Rightarrow \Delta, \Sigma}$$

We replace it with:

$$(Cut)\cfrac{\Pi \Rightarrow \Sigma, \varphi \quad \varphi, \Gamma \Rightarrow \Delta}{\Gamma, \Pi \Rightarrow \Delta, \Sigma}$$

Now the original proof with the application of cut is not replaced with a proof of the conclusion in K without cut. However, in the new proof cut is performed on a formula of lesser complexity, hence it is eliminable by the first induction hypothesis.

Subcase 2.2: the cut-formula is a disjunction:

$$\cfrac{(\Rightarrow \vee)\cfrac{\Gamma \Rightarrow \Delta, \varphi, \psi}{\Gamma \Rightarrow \Delta, \varphi \vee \psi} \quad \cfrac{\varphi, \Pi \Rightarrow \Sigma \quad \psi, \Pi \Rightarrow \Sigma}{\varphi \vee \psi, \Pi \Rightarrow \Sigma}(\vee \Rightarrow)}{\Gamma, \Pi \Rightarrow \Delta, \Sigma}(Cut)$$

is replaced with:

$$(Cut)\cfrac{(Cut)\cfrac{\Gamma \Rightarrow \Delta, \varphi, \psi \quad \psi, \Pi \Rightarrow \Sigma}{\Gamma, \Pi \Rightarrow \Delta, \Sigma, \varphi} \quad \varphi, \Pi \Rightarrow \Sigma}{\Gamma, \Pi \Rightarrow \Delta, \Sigma}$$

In this case we even introduced two cuts instead of one but still both are eliminable by the induction hypothesis since both are performed on formulae of lesser complexity.

Exercise 1.22. *Prove subcases 2.3. (conjunction) and 2.4. (implication).*

Case 3. The cut-formula is not principal in at least one premiss. Let it be the right premiss. There are 8 subcases according to which rule is applied to the right premiss. We illustrate the procedure with two of them, one with the application of a one-premiss rule and the other with the application of a two-premiss rule.

Subcase 3.1: The right premiss is deduced by $(\Rightarrow \vee)$:

$$\cfrac{\Gamma \Rightarrow \Delta, \varphi \quad \cfrac{\varphi, \Pi \Rightarrow \Sigma', \psi, \chi}{\varphi, \Pi \Rightarrow \Sigma', \psi \vee \chi}(\Rightarrow \vee)}{\Gamma, \Pi \Rightarrow \Delta, \Sigma', \psi \vee \chi}(Cut)$$

is replaced with:

$$\frac{\dfrac{\Gamma \Rightarrow \Delta, \varphi \qquad \varphi, \Pi \Rightarrow \Sigma', \psi, \chi}{\Gamma, \Pi \Rightarrow \Delta, \Sigma', \psi, \chi} \ (Cut)}{\Gamma, \Pi \Rightarrow \Delta, \Sigma', \psi \vee \chi} \ (\Rightarrow \vee)$$

Now we did not reduce the complexity of the cut-formula, but we made a permutation of the application of rules after which the new cut is of the lesser height, and hence it is eliminable by the induction hypothesis of the subsidiary induction on the sum of the heights of the premisses.

Subcase 3.2: The right premiss deduced by $(\rightarrow \Rightarrow)$:

$$\frac{\Gamma \Rightarrow \Delta, \varphi \qquad \dfrac{\varphi, \Pi' \Rightarrow \Sigma, \psi \qquad \chi, \varphi, \Pi' \Rightarrow \Sigma}{\psi \rightarrow \chi, \varphi, \Pi' \Rightarrow \Sigma} \ (\rightarrow \Rightarrow)}{\psi \rightarrow \chi, \Gamma, \Pi' \Rightarrow \Delta, \Sigma} \ (Cut)$$

it is replaced with:

$$\frac{(Cut) \ \dfrac{\Gamma \Rightarrow \Delta, \varphi \qquad \varphi, \Pi' \Rightarrow \Sigma, \psi}{\Gamma, \Pi' \Rightarrow \Delta, \Sigma, \psi} \qquad \dfrac{\Gamma \Rightarrow \Delta, \varphi \qquad \chi, \varphi, \Pi' \Rightarrow \Sigma}{\chi, \Gamma, \Pi' \Rightarrow \Delta, \Sigma} \ (Cut)}{\psi \rightarrow \chi, \Gamma, \Pi' \Rightarrow \Delta, \Sigma} \ (\rightarrow \Rightarrow)$$

Now we have introduced two cuts instead of one but both of lower height hence eliminable by the induction hypothesis. $\qquad \square$

Exercise 1.23. *Prove the remaining 6 subcases for the right premiss and then all 8 subcases for the left premiss.*

Note that we can apply admissible rules (like W) but we do not need to consider them as introducing one of the premisses of cut. We should also observe that a realisation of case 3., i.e. height reduction steps, is in fact a simple consequence of the context independency of all rules, described in subsection 1.4.2.

Remark 1.6. In fact, we can change the order of inductions and even simplify the general structure of the proof. Namely, if we take the height as the main induction parameter we do not need to prove the basis by means of subsidiary induction on the complexity of the cut-formula. This is necessary only once, for proving the induction step. So the general schema will be: 1 basis of induction on h, 2 inductive step (by subsidiary induction on c). A reader should check that such changes in the organisation of the structure of the induction are still in accordance with the division of cases, in the sense that inductive hypotheses are correctly placed for the proof to go through. We presented the above version because it is closer to the organisation of Gentzen's original proof (see section 2.4). Note, however, that in the case of Gentzen's proof, the complexity of the cut-formula must be the main parameter, since instead of the height of the proof he used another parameter which is sensitive to the reduction of the complexity of cut-formula. With height the situation is different; neither the reduction of the complexity can affect the height, nor can the reduction of height change the complexity. Hence both parameters are independent in this sense and as such the re-organisation of the structure of the induction is possible.

1.9   Applications of Cut

Having cut at our disposal significantly strengthens K both in the sense of practical applications and theoretical utility. Concerning practice, one may find shorter proofs and derivations of the same results, as we noted in the preceding section when comparing applications of cut to applications of lemmata. But the search for these shorter proof trees may be not so easy as in cut-free SC; we will say more about this in section 1.11. We can also demonstrate derivability of several useful rules. As an example, we show that $(3 \Rightarrow \wedge)$ introduced in paragraph 1.4.1 is derivable in K with cut:

$$(Cut) \cfrac{\cfrac{\Gamma \Rightarrow \Delta, \varphi, \psi \qquad \psi, \Gamma \Rightarrow \Delta, \varphi}{\Gamma \Rightarrow \Delta, \varphi} \qquad \cfrac{\Gamma \Rightarrow \Delta, \varphi, \psi \qquad \varphi, \Gamma \Rightarrow \Delta, \psi}{\Gamma \Rightarrow \Delta, \psi} (Cut)}{\Gamma \Rightarrow \Delta, \varphi \wedge \psi} (\Rightarrow \wedge)$$

Exercise 1.24. *Prove derivability of* $(3\vee \Rightarrow)$ *and* $(3 \rightarrow\Rightarrow)$ *in K using cut.*

Prove that each of $(\wedge \Rightarrow), (\Rightarrow \vee), (\Rightarrow\rightarrow)$ *is derivable from full* $2-variants$ *with cut.*

Now we will show a few theoretical applications of cut.

1.9.1   Equivalence with Axiomatic Formulation of CPL

As an immediate consequence of the admissibility of cut we can demonstrate the correctness of K in a purely syntactic way, by showing that provability in K equals provability in the system H introduced in subsection 1.1.3. It was the way of Gentzen who did not provide any semantical proofs of adequacy for his SC.

Theorem 1.8. (Equivalence of H and K). $\vdash_H \varphi$ *iff* $\vdash_K \Rightarrow \varphi$

We have added subscripts to $\vdash$ since we talk about two different consequence relations in one section. We prove first:

Lemma 1.17. *If* $\vdash_H \varphi$, *then* $\vdash_K \Rightarrow \varphi$

PROOF: of this lemma is not very complicated. It requires a construction of K-proofs for all schemata of axioms and showing that every application of MP in any H-proof may be simulated in K. In section 1.3 we provided a proof of an instance of axiom 2; it is sufficient to replace in it every occurrence of p with φ, q with ψ and r with χ and we get a general schema of a proof of every instance of this axiom. Additionally we demonstrate a schema of a proof for axiom 6:

$$(\rightarrow\Rightarrow) \cfrac{\cfrac{\cfrac{\cfrac{\varphi, \psi \rightarrow \chi \Rightarrow \chi, \varphi \qquad \chi, \varphi, \psi \rightarrow \chi \Rightarrow \chi}{\varphi, \psi \rightarrow \chi, \varphi \rightarrow \chi \Rightarrow \chi} (\vee \Rightarrow) \cfrac{\psi, \varphi \rightarrow \chi \Rightarrow \chi, \psi \qquad \chi, \psi, \varphi \rightarrow \chi \Rightarrow \chi}{\psi, \psi \rightarrow \chi, \varphi \rightarrow \chi \Rightarrow \chi} (\rightarrow\Rightarrow)}{\varphi \vee \psi, \psi \rightarrow \chi, \varphi \rightarrow \chi \Rightarrow \chi} (\Rightarrow\rightarrow)}{\psi \rightarrow \chi, \varphi \rightarrow \chi \Rightarrow \varphi \vee \psi \rightarrow \chi} (\Rightarrow\rightarrow)}{\varphi \rightarrow \chi \Rightarrow (\psi \rightarrow \chi) \rightarrow (\varphi \vee \psi \rightarrow \chi)} (\Rightarrow\rightarrow)}{\Rightarrow (\varphi \rightarrow \chi) \rightarrow ((\psi \rightarrow \chi) \rightarrow (\varphi \vee \psi \rightarrow \chi))}$$

As for the simulation of MP, in K this is obtained in the following way:

$$
\cfrac{
\begin{array}{c}
\mathcal{D}_2 \\
\Rightarrow \varphi
\end{array}
\quad
\cfrac{
\begin{array}{c}
\mathcal{D}_1 \\
\Rightarrow \varphi \to \psi
\end{array}
\quad
\cfrac{\varphi \Rightarrow \varphi \qquad \psi \Rightarrow \psi}{\varphi \to \psi, \varphi \Rightarrow \psi}(\to\Rightarrow)
}{\varphi \Rightarrow \psi}(Cut)
}{\Rightarrow \psi}(Cut)
$$

where $\mathcal{D}_1$ and $\mathcal{D}_2$ are simulations (in K) of H-proofs of the two theses that are the premisses of the application of MP. Now it is clear why we need to prove the admissibility of cut first. □

Exercise 1.25. *Prove the remaining axiom schemata in K.*

To prove that $\vdash_K \Rightarrow \varphi$ implies $\vdash_H \varphi$ Gentzen has shown that his translation (see section 1.2) of every primitive rule of K yields a rule which is derivable in H. We omit Gentzen's original proof, as it is a rather involved exercise in constructing proofs in an axiomatic system and we are not concerned with this issue in this work. Instead we provide a much simpler proof due to Kleene [149] which yields the same result. It is based on the interpretation of $\Rightarrow$ as a kind of a consequence relation which was also briefly introduced in section 1.2. Here the specific consequence relation we are interested in is the derivability relation $\vdash_H$. To avoid problems with multiplicity of formulae in the succedent (note that $\vdash_H$ admits at most one formula in this position) we apply a simple trick of putting the denials of all succedent formulae in the antecedent and using $\perp$ instead of an empty set to the right of $\vdash_H$. So we will prove the following apparently more general result.

Lemma 1.18. *If $\vdash_K \Gamma \Rightarrow \Delta$, then $\Gamma, \neg\Delta \vdash_H \perp$*

PROOF: By induction on the height of a proof in K. It is sufficient to use properties of $\vdash$ from lemma 1.1. The proof of the basis is trivial since we have $\varphi \vdash_H \varphi$, and therefore $\varphi, \neg\varphi \vdash_H \perp$ by point 9.

For the induction hypothesis we assume that the lemma holds for any proof of the height $< n$ and show that it holds also for n by consideration of all possible applications of rules. We present one example:

$(\Rightarrow \wedge)$ was applied to obtain $\vdash_n \Gamma \Rightarrow \Delta', \varphi\wedge\psi$. Both premisses have proofs of lesser height so they fall under the induction hypothesis and we have $\Gamma, \neg\Delta', \neg\varphi \vdash_H \perp$ and $\Gamma, \neg\Delta', \neg\psi \vdash_H \perp$. By lemma 1.1, point 9, we have $\Gamma, \neg\Delta' \vdash_H \varphi$ and $\Gamma, \neg\Delta' \vdash_H \psi$. By axiom 4 and point 6 of lemma 1.1 we obtain $\varphi, \psi \vdash_H \varphi \wedge \psi$ which, after double application of point 4, yields $\Gamma, \neg\Delta' \vdash_H \varphi \wedge \psi$, therefore (by point 9) $\Gamma, \neg\Delta', \neg(\varphi \wedge \psi) \vdash_H \perp$. □

Exercise 1.26. *Prove the remaining cases using lemma 1.1 and axioms.*

From lemma 1.18 with empty Γ and $\Delta := \{\varphi\}$ we obtain, by lemma 1.1, the converse of lemma 1.17, and this yields a proof of theorem 1.8.

1.9.2 Equivalence of Tarskian Consequence Relations

In Remark 1.4 we have claimed that two different notions of Tarski-style consequence relations are equivalent. Now we prove this claim. Let us recall both definitions:

- $\Gamma \vdash^t \varphi$ iff $\vdash \Gamma' \Rightarrow \varphi$ for some finite $\Gamma' \subseteq \Gamma$;

- $\Gamma \vdash^v \varphi$ iff $\Rightarrow \psi_1, ..., \Rightarrow \psi_k \vdash \Rightarrow \varphi$, for $\{\psi_1, ..., \psi_k\} \subseteq \Gamma$.

  $\vdash$ is of course a derivability relation between sequents.

Claim 1.6. $\Gamma \vdash^t \varphi$ *iff* $\Gamma \vdash^v \varphi$.

PROOF: In both directions we assume that we deal with the same finite subset of Γ, i.e. that $\Gamma' = \{\psi_1, ..., \psi_k\}$. This is not a restriction since if these finite sets are different then we can always in case of $\vdash^t$ add the lacking formulae to the antecedent Γ' and in case of $\vdash^v$ add vacuously $\Rightarrow \phi$ for any $\phi \in \Gamma'$ not being one of ψ_i.

$\Longrightarrow$: If we assume that $\vdash \psi_1, ..., \psi_k \Rightarrow \varphi$ and that $\vdash \Rightarrow \psi_i$ for $1 \leq i \leq k$, then by k applications of cut we obtain $\vdash \Rightarrow \varphi$.

$\Longleftarrow$: Let $\mathcal{D}$ be a derivation of $\Rightarrow \varphi$ from $\Rightarrow \psi_1, ..., \Rightarrow \psi_k$. Change it into a proof tree $\mathcal{D}'$ of $\psi_1, ..., \psi_k \Rightarrow \varphi$ from $\psi_1, ..., \psi_k \Rightarrow \psi_1, \quad ... \quad , \psi_1, ..., \psi_k \Rightarrow \psi_k$ by adding $\psi_1, ..., \psi_k$ to all antecedents. Since all rules of K are context-independent and each $\psi_1, ..., \psi_k \Rightarrow \psi_i$ is axiomatic, $\mathcal{D}'$ is a proof of $\psi_1, ..., \psi_k \Rightarrow \varphi$. $\qquad \square$

1.10 Completeness Again

It should be obvious by now that having cut at our disposal makes it possible to use SC (K in particular) for any task which is ordinarily performed by means of axiomatic systems with MP as a primitive rule. Moreover, it seems that in many cases the well-known results in the setting of SC may be formulated in a simpler and more elegant way than for Hilbert systems. To illustrate this remark we will show some standard results reformulated suitably for K.

In this section we focus on the Lindenbaum/Henkin-style proof of the completeness theorem. There are a lot of variants of this kind of proof designed for Hilbert systems. However all of them are based on the idea that any consistent set may be extended in such a way that it is maximal in some sense. Such maximal consistent sets allow for direct construction of a model in which the input consistent set is satisfied. Since unprovability is interdefinable with consistency then this model is used for the falsification of unprovable sequents. There is a notable difference between this type of proof and the Hintikka-style proof presented in section 1.7. In the latter we have shown that for each unprovable sequent we can construct a specific and finite countermodel. Now we will construct one infinite model falsifying all unprovable sequents. Below we provide some variants of this kind of proof but at first we will show how to generalise some notions introduced earlier for Hilbert systems to apply also to SC.

1.10.1 Consistency

Definition 1.12. (Consequence, Consistency).

1. $\Gamma \vdash^s \Delta$ iff $\vdash \Gamma' \Rightarrow \Delta'$ for some finite $\Gamma' \subseteq \Gamma$ and $\Delta' \subseteq \Delta$.

2. $\Gamma \nvdash^s \Delta$ iff $\nvdash \Gamma' \Rightarrow \Delta'$ for any finite $\Gamma' \subseteq \Gamma$ and $\Delta' \subseteq \Delta$.

3. (Γ, Δ) is consistent iff $\Gamma \nvdash^s \Delta$.

4. (Γ, Δ) is inconsistent iff $\Gamma \vdash^s \Delta$.

These definitions are based on Scott's interpretation of $\Rightarrow$ as $\vdash^s$ (here in K for **CPL** but they work for other systems and logics as well) and extended to the infinite case. Let us comment on the rationale for such a form of defining (in)consistency. It is in fact a very natural generalisation of the notion of the (in)consistency of a set defined by means of a provability relation in H-system. Let us recall that, according to the standard definition, Γ is inconsistent iff $\Gamma' \vdash_H \bot$ for some finite $\Gamma' \subseteq \Gamma$, but on the basis of the equivalence of $\vdash_H$ and $\vdash_K$ for **CPL** (theorem 1.8) we have $\vdash_K \Gamma' \Rightarrow$ which means that $(\Gamma, \varnothing)$ is inconsistent. Let us note that if (Γ, Δ) is inconsistent, then we can add to Δ any formula by $(\Rightarrow W)$ which is in accordance with another standard characterisation of inconsistency of Γ as $Cn(\Gamma) = FOR$ $(Cn(\Gamma) = \{\varphi : \Gamma \vdash \varphi\})$.

A simple consequence is:

Lemma 1.19. *If (Γ, Δ) is consistent, then:*

- $\Gamma \cap \Delta = \varnothing$

- (Γ', Δ') *is consistent for any $\Gamma' \subseteq \Gamma$ and $\Delta' \subseteq \Delta$*

PROOF: In the first case if $\Gamma \cap \Delta \neq \varnothing$, then $\vdash \varphi \Rightarrow \varphi$ for some $\varphi \in \Gamma \cap \Delta$, hence there is finite $\Gamma' \subseteq \Gamma$, $\Delta' \subseteq \Delta$ such that $\vdash \Gamma' \Rightarrow \Delta'$—contradiction.

Similarly in the second case, if some (Γ', Δ') were inconsistent, then $\vdash \Gamma' \Rightarrow \Delta'$ which yields $\vdash \Gamma \Rightarrow \Delta$ and contradiction. $\square$

Lemma 1.20. *If (Γ, Δ) is consistent, then:*

1. if $\varphi \in \Gamma$, then $\varphi \notin \Delta$

2. if $\varphi \in \Delta$, then $\varphi \notin \Gamma$

3. if $\neg\varphi \in \Gamma$, then $\varphi \notin \Gamma$

4. if $\neg\varphi \in \Delta$, then $\varphi \notin \Delta$

5. if $\varphi \wedge \psi \in \Gamma$, then $\varphi \notin \Delta$ and $\psi \notin \Delta$

6. if $\varphi \wedge \psi \in \Delta$, then $\varphi \notin \Gamma$ or $\psi \notin \Gamma$

7. if $\varphi \vee \psi \in \Gamma$, then $\varphi \notin \Delta$ or $\psi \notin \Delta$

8. if $\varphi \vee \psi \in \Delta$, then $\varphi \notin \Gamma$ and $\psi \notin \Gamma$

9. if $\varphi \to \psi \in \Gamma$, then $\varphi \notin \Gamma$ or $\psi \notin \Delta$

10. if $\varphi \to \psi \in \Delta$, then $\varphi \notin \Delta$ and $\psi \notin \Gamma$

PROOF: The first two cases are direct consequences of the previous lemma. We prove cases 9 and 10.

For 9. Assume that $\varphi \to \psi \in \Gamma$ but $\varphi \in \Gamma$ and $\psi \in \Delta$. However $\vdash \varphi \to \psi, \varphi \Rightarrow \psi$, so we have $\Gamma \vdash \Delta$ which contradicts the assumption of the lemma.

For 10. Assume that $\varphi \in \Delta$ or $\psi \in \Gamma$ but $\varphi \to \psi \in \Delta$. Let us note that both $\Rightarrow \varphi \to \psi, \varphi$ and $\psi \Rightarrow \varphi \to \psi$ are provable, so in each case we have $\Gamma \vdash \Delta$—a contradiction. □

Exercise 1.27. *Prove the remaining cases.*

We have underlined that the application of the Lindenbaum–Henkin method for SC requires the application of cut. Here is the first such situation:

Lemma 1.21. *If (Γ, Δ) is consistent, then $(\Gamma \cup \{\varphi\}, \Delta)$ is consistent or $(\Gamma, \Delta \cup \{\varphi\})$ is consistent for any φ.*

PROOF: Assume that both pairs of sets are inconsistent, then $\vdash \Gamma', \varphi \Rightarrow \Delta'$ and $\vdash \Gamma'' \Rightarrow \Delta'', \varphi$, for some finite $\Gamma' \cup \Gamma'' \subseteq \Gamma$ and $\Delta' \cup \Delta'' \subseteq \Delta$. By the application of (Cut) we obtain $\vdash \Gamma', \Gamma'' \Rightarrow \Delta', \Delta''$, hence (Γ, Δ) is not consistent. □

1.10.2 Maximality

In the setting of Hilbert systems we are using the notion of maximal (consistent) set. In a similar way as we did with Hintikka sets (exchanging them into Hintikka tuples in section 1.7) we also generalise the notion of a maximal set, changing it into maximal tuples of sets of formulae:

Definition 1.13. (Maximality). *(Γ, Δ) is maximal iff $\Gamma \cup \Delta = FOR$.*

Some obvious consequences of this definition are displayed below. One may compare them with the first four conditions from lemma 1.20 characterising consistent pairs.

Lemma 1.22. *If (Γ, Δ) is consistent and maximal, then:*

1. if $\varphi \notin \Gamma$, then $\varphi \in \Delta$

2. if $\varphi \notin \Delta$, then $\varphi \in \Gamma$

3. if $\neg\varphi \notin \Gamma$, then $\varphi \in \Gamma$

4. if $\neg\varphi \notin \Delta$, then $\varphi \in \Delta$

PROOF: Case 3. If both $\neg\varphi \notin \Gamma$ and $\varphi \notin \Gamma$, then, by maximality, $\{\neg\varphi, \varphi\} \subseteq \Delta$. But $\vdash \Rightarrow \neg\varphi, \varphi$ which contradicts the consistency of (Γ, Δ). □

Exercise 1.28. *Prove the remaining cases.*

The next lemma is a counterpart of the standard Lindenbaum lemma but stated for pairs of sets.

Lemma 1.23. (Lindenbaum). *If (Γ, Δ) is consistent, then there exists consistent and maximal pair (Γ', Δ') such that $\Gamma \subseteq \Gamma'$ and $\Delta \subseteq \Delta'$.*

PROOF: Since FOR is denumerable we may order this set and add successive formulae one by one either to Γ or to Δ. More precisely, let $\varphi_1, \varphi_2, ...$ be an infinite list of all formulae. We define an infinite list of sequents $\Gamma_0 \Rightarrow \Delta_0$, $\Gamma_2 \Rightarrow \Delta_2$, ... such that $\Gamma_0 \Rightarrow \Delta_0 := \Gamma \Rightarrow \Delta$, and for each $i \geq 0$:

$$\Gamma_{i+1} \Rightarrow \Delta_{i+1} := \begin{cases} \Gamma_i \Rightarrow \Delta_i, \varphi_{i+1} & \text{if } (\Gamma_i, \Delta_i \cup \{\varphi_{i+1}\}) \text{ is consistent} \\ \Gamma_i, \varphi_{i+1} \Rightarrow \Delta_i & \text{otherwise} \end{cases}$$

By construction and lemma 1.21 we know that at each stage at least one pair of sequents is consistent, hence every subsequent pair of sequents is consistent and all must be consistent.

Let $\Gamma' = \bigcup \Gamma_i$, and $\Delta' = \bigcup \Delta_i$, for $i < \omega$. It is clear that $\Gamma' \cup \Delta' = FOR$, hence (Γ', Δ') is maximal. Assume that (Γ', Δ') is inconsistent. Since every proof is finite, there must be some finite pair (Π, Σ) such that $\Pi \subseteq \Gamma', \Sigma \subseteq \Delta'$ and $\vdash \Pi \Rightarrow \Sigma$. By finiteness of (Π, Σ) there must be a stage i in the construction of our infinite list such that $\Pi \subseteq \Gamma_i$ and $\Sigma \subseteq \Delta_i$ which means that $\vdash \Gamma_i \Rightarrow \Delta_i$ and this pair is inconsistent, contrary to our assumption. By definition we also have that $\Gamma \subseteq \Gamma'$ and $\Delta \subseteq \Delta'$. Therefore (Γ, Δ) can be extended to maximal and consistent pair. □

1.10.3 Completeness

Lemma 1.24. (Truth lemma). *If (Γ, Δ) is maximal, then there is a model $\mathfrak{M}$ such that: for any φ, $\varphi \in \Gamma$ iff $\mathfrak{M} \vDash \varphi$.*

PROOF: We define a model $\mathfrak{M}$ taking as true all atomic formulae from $PROP(\Gamma)$ and prove that it satisfies the claim by induction on the complexity of φ. The basis follows by definition.

a. Let $\varphi := \neg\psi$:

If $\neg\psi \in \Gamma$, then, by lemma 1.20, point 3, $\psi \notin \Gamma$, and by lemma 1.22, point 1, $\psi \in \Delta$. By the induction hypothesis $\mathfrak{M} \nvDash \psi$, so $\mathfrak{M} \vDash \neg\psi$.

Similarly if $\mathfrak{M} \vDash \neg\psi$ which means that $\mathfrak{M} \nvDash \psi$, then by the induction hypothesis $\psi \in \Delta$. By lemma 1.20, point 4, $\neg\psi \notin \Delta$, hence, by lemma 1.22, point 2, $\neg\psi \in \Gamma$.

b. Let $\varphi := \psi \wedge \chi$:

If $\varphi \in \Gamma$, then $\psi \notin \Delta$ and $\chi \notin \Delta$ (lemma 1.20, point 5). Hence, by lemma 1.22, point 2, $\psi \in \Gamma$ and $\chi \in \Gamma$. By the induction hypothesis $\mathfrak{M} \vDash \psi$ and $\mathfrak{M} \vDash \chi$, so $\mathfrak{M} \vDash \varphi$.

If $\varphi \notin \Gamma$, then by lemma 1.22, point 1, $\varphi \in \Delta$, so by lemmata 1.20 and 1.22, $\psi \notin \Gamma$ or $\chi \notin \Gamma$ ($\psi \in \Delta$ or $\chi \in \Delta$). By the induction hypothesis $\mathfrak{M} \nvDash \psi$ or $\mathfrak{M} \nvDash \chi$; in any case we have $\mathfrak{M} \nvDash \varphi$. □

Exercise 1.29. *Prove the remaining 2 cases.*

Theorem 1.9. (Completeness). *If* $\vDash \Gamma \Rightarrow \Delta$, *then* $\vdash \Gamma \Rightarrow \Delta$.

PROOF: Assume that $\nvdash \Gamma \Rightarrow \Delta$, hence (Γ, Δ) is consistent. By lemma 1.23 there is a maximal consistent extension (Γ', Δ') of it, and by lemma 1.24, there is a model such that all elements of Γ' (and so of Γ) are satisfied and all elements of Δ' (hence Δ) are falsified in it. But this means that it is not true that $\vDash \Gamma \Rightarrow \Delta$. □

1.10.4   Some Variant Proofs

Henkin-like proofs of completeness may be carried out also in a different way. We sketch here some variations. First of all instead of proving separately some features of consistent (lemma 1.20) and maximal pairs (lemma 1.22), we can directly prove a lemma about important features of maximal consistent pairs:

Lemma 1.25. *If* (Γ, Δ) *is consistent and maximal, then it has the following properties:*

1. $\neg \varphi \in \Gamma$ iff $\varphi \in \Delta$

2. $\neg \varphi \in \Delta$ iff $\varphi \in \Gamma$

3. $\varphi \wedge \psi \in \Gamma$ iff $\varphi \in \Gamma$ and $\psi \in \Gamma$

4. $\varphi \wedge \psi \in \Delta$ iff $\varphi \in \Delta$ or $\psi \in \Delta$

5. $\varphi \vee \psi \in \Gamma$ iff $\varphi \in \Gamma$ or $\psi \in \Gamma$

6. $\varphi \vee \psi \in \Delta$ iff $\varphi \in \Delta$ and $\psi \in \Delta$

7. $\varphi \rightarrow \psi \in \Gamma$ iff $\varphi \in \Delta$ or $\psi \in \Gamma$

8. $\varphi \rightarrow \psi \in \Delta$ iff $\varphi \in \Gamma$ and $\psi \in \Delta$

PROOF: We consider cases 1 and 7.

If $\neg \varphi \in \Gamma$, then, by consistency of (Γ, Δ), $\varphi \notin \Gamma$ since $\vdash \varphi, \neg \varphi \Rightarrow$ which yields $\Gamma \vdash \Delta$. By maximality $\varphi \in \Delta$ since $\varphi \in \Gamma \cup \Delta$. The converse follows similarly.

If $\varphi \rightarrow \psi \in \Gamma$, then clearly $\varphi \notin \Gamma$ or $\psi \notin \Delta$; otherwise $\Gamma \vdash \Delta$ since $\vdash \varphi \rightarrow \psi, \varphi \Rightarrow \psi$. Hence by maximality $\varphi \in \Delta$ or $\psi \in \Gamma$. In the other direction assume that $\varphi \in \Delta$ or $\psi \in \Gamma$, and that $\varphi \rightarrow \psi \notin \Gamma$. Hence $\varphi \rightarrow \psi \in \Delta$ by maximality. In both cases we obtain that $\Gamma \vdash \Delta$ because in the first case $\vdash \Rightarrow \varphi \rightarrow \psi, \varphi$, and in the second $\vdash \psi \Rightarrow \varphi \rightarrow \psi$. □

Exercise 1.30. *Prove the remaining cases.*

The above lemma makes the proof of the truth lemma even easier since the properties naturally correspond to truth clauses. For example the proof for conjunction goes as follows:

$\varphi \wedge \psi \in \Gamma$ iff (by lemma 1.25) $\varphi \in \Gamma$ and $\psi \in \Gamma$ iff (by the induction hypothesis) $\mathfrak{M} \vDash \varphi$ and $\mathfrak{M} \vDash \psi$ iff $\mathfrak{M} \vDash \varphi \wedge \psi$.

Exercise 1.31. *It is also possible to define directly the notion of a maximal and consistent pair as a pair satisfying the 8 conditions of lemma 1.25. In this approach we do not need to prove them but the proof of lemma 1.23 is different. We do not need to change the construction but now we must prove that the resulting infinite pair satisfies all 8 conditions. The reader is encouraged to do it.*

Remark 1.7. The approach realised above (as well as in section 1.7) deviates from more standard solutions by treating both sides of a sequent as equally important for the construction of suitable models. However, for those who are accustomed to the standard Lindenbaum–Henkin proof for Hilbert systems we point out that one may dispense with the notion of a consistent (and maximal or downward saturated) pair of sets and apply the more traditional solution based on the use of sets. The main difference with the present approach is that such constructions are based rather on the antecedents of unprovable sequents. For example Maehara [169] defines the notion of a complete consistent system directly in terms of unprovable sequents in the following way: Π is a Complete Consistent System iff for all finite $\Gamma \subseteq \Pi$ and $\Delta \cap \Pi = \varnothing$ the sequent $\Gamma \Rightarrow \Delta$ is not provable. Using this notion we can provide a counterpart of Lindenbaum's lemma for unprovable sequents, i.e. we can prove that if $\nvdash \Gamma \Rightarrow \Delta$, then there exists a complete consistent system Π such that $\Gamma \subseteq \Pi$ and $\Delta \cap \Pi = \varnothing$. The proof goes by maximalisation, and the construction of a model is standard. Beziau [29] provides an even more general construction working for some class of SC and based on the more general notion of a relatively maximal consistent set due to Asser [6]. However in the case of classical logic this notion coincides with the notion of a maximal consistent set. Another general construction of Consistency Properties introduced by Smullyann [245] and developed by Fitting [85] may be applied to SC. One such application is provided by Boolos, Jeffrey and Burgess [38].

1.11 Restricted Cut

We have shown that cut may be useful, especially when we are interested in the transfer of results obtained for Hilbert systems to SC setting. But the list of profits is longer and we demonstrate some of them in this section and in the next chapters. One advantage of using cut in proofs is connected with the problem of the size of proofs. We have mentioned in section 1.8 that cut may simulate the applications of lemmata in proofs. This way we can use already proved results to obtain quickly the new ones instead of proving everything in isolation. It was shown that clever applications of cut may decrease proofs even exponentially; one may find examples (formulated in the setting of tableau systems) in Boolos [37] and Fitting [85]. Investigations on the relative complexity of different proof systems, carried out in terms of a simulation of proofs in one system by proofs in another, showed that SC with cut behaves much better than SC without cut. Loosely speaking, Hilbert systems, natural deduction and SC with cut may simulate their proofs with polynomial increase of the size of proofs only, whereas SC without cut provides exponential increase for many classes of proofs[8].

However, one cannot overlook that there is also a price to be paid for having cut — the subformula property and all its nice consequences exploited so far are lost. Uncontrolled applications of cut in a root-first proof search introduce full indeterminism in the sense that we do not have a bound space of choices—any formula φ may be used as a cut-formula. That is why we are interested in showing that cut may be treated as an admissible rule not a primitive one, or that we can eliminate it if it is introduced as primitive.

On the other hand, we have mentioned in subsection 1.3.3 that the subformula property of all rules is not necessary for analyticity of SC. In particular, we can think about some restricted applications of cut in proof search. We have mentioned above the computational benefits of the applications of cut. If it is possible to combine them with reasonable restrictions on indeterminism introduced by cut, then we can get an interesting and practically more efficient alternative to cut-free SC. In fact, resolution systems are based on this idea since resolution (at least in the propositional setting—see Avron [9]) is nothing more than (analytically restricted) cut, applied to atomic sequents (i.e. clauses).

Hence, even in case of logics which, like **CPL**, have a cut-free SC, admissibility of restricted forms of cut may lead to some unexpected advantages. Moreover, as we will see in chapter 4, for some non-classical logics one cannot get rid of cut but may be able to restrict its applications to some analytic version. In this section we will take up the question of taming cut.

Let us call an application of cut analytic iff its cut-formula is a subformula of the root sequent. If we restrict all applications of cut in this way in proofs/derivations, then we retain the subformula property for the system. We can go even further and introduce an analytical version of the rule of cut which satisfies the subformula property by addition of a side condition that the cut-formula must be a subformula of the conclusion sequent. As we will see we can strengthen even this side condition by demanding that the cut-formula is an atomic subformula of the conclusion sequent. Such solutions are stronger since we retain the subformula property of rules not only of proofs. However, the former notion is more general,

---

[8]We restrict considerations of these issues only to very general remarks. There are many subtle questions connected with how exactly we define the size of a proof or whether we are dealing with trees or direct acyclic graphs as representing proofs. A readable introduction to this field may be found in Urquhart [267].

and consequently more useful in proof search, so it is easier to prove results about it. Let K+AC denote K with cut so restricted. Note that restricting applications of cut to analytic ones only allow for root-first proof search and proving decidability by means of K+AC. However, we also show that atomically restricted analytic cut has some special, interesting applications.

1.11.1 Completeness with Analytic Cut

First we will show that one can combine some features of the Lindenbaum/Henkin proof with the constructive approach of Hintikka to obtain another adequacy result. Basically, the proof shows that for any consistent pair (Γ, Δ) we can build a relatively maximal pair from elements of $SF(\Gamma \cup \Delta)$ (lemma 1.26) and show that it is downward saturated (lemma 1.27).

The next definition shows how to change Lindenbaum's construction to work for finite sets.

Definition 1.14. (Relative maximality). *A pair* (Γ, Δ) *is relatively maximal in the set* $\Pi \subseteq FOR$ *iff for any* $\varphi \in \Pi, \varphi \in \Gamma$ *or* $\varphi \in \Delta$ *(i.e.* $\Gamma \cup \Delta = \Pi$*).*

This definition is in fact a generalisation of the definition of maximality since Π may be also FOR, but for us it is more interesting if Π is some finite and suitably defined subset of FOR. At first we consider a case of $SF(\Gamma \cup \Delta)$.

Here we have a counterpart of Lindenbaum's lemma.

Lemma 1.26. *If* (Γ, Δ) *is consistent, then there exists a pair* (Γ_n, Δ_n) *such that* $\Gamma \subseteq \Gamma_n$ *and* $\Delta \subseteq \Delta_n$, *and it is consistent and relatively maximal in* $SF(\Gamma \cup \Delta)$.

PROOF: Analogous to the proof of Lindenbaum lemma.

Assume that (Γ, Δ) is consistent. We make a finite list of all formulae from $SF(\Gamma \cup \Delta)$: $\varphi_1, \varphi_2, \ldots, \varphi_n$. Next we define a list of sequents $\Gamma_0 \Rightarrow \Delta_0, \ldots, \Gamma_n \Rightarrow \Delta_n$ such that $\Gamma_0 \Rightarrow \Delta_0 := \Gamma \Rightarrow \Delta$, and for any $n \geq i \geq 0$: $\Gamma_{i+1} \Rightarrow \Delta_{i+1} := \Gamma_i \Rightarrow \Delta_i, \varphi_{i+1}$ if $(\Gamma_i, \Delta_i \cup \{\varphi_{i+1}\})$ is consistent, otherwise $\Gamma_{i+1} \Rightarrow \Delta_{i+1} := \Gamma_i, \varphi_{i+1} \Rightarrow \Delta_i$.

The relative maximality of (Γ_n, Δ_n) follows from the fact that every element of $SF(\Gamma \cup \Delta)$ was considered at some stage of the construction.

By construction and lemma 1.21 it follows that each element of the list, and in particular (Γ_n, Δ_n), is consistent. Note however that all applications of cut required for showing its consistency were analytic. □

It is sufficient to prove that the so constructed finite pair (Γ_n, Δ_n) is saturated. This guarantees the existence of a model falsifying also $\Gamma \Rightarrow \Delta$, and yields a completeness proof for K+AC.

Lemma 1.27. (Γ_n, Δ_n) *is downward saturated.*

PROOF: To show that $\Gamma_n \Rightarrow \Delta_n$ is saturated we may refer to its consistency and to suitable conditions from lemma 1.20. For example:

a) Assume that $\varphi \wedge \psi \in \Gamma_n$, then, by lemma 1.20, condition 5—we have that $\varphi \notin \Delta_n$ and $\psi \notin \Delta_n$, but both φ and ψ, being subformulae of $\Gamma \cup \Delta$, must belong to $\Gamma_n \Rightarrow \Delta_n$. Hence both belong to Γ_n.   □

Exercise 1.32. *Prove the remaining cases.*

Then completeness of K+AC follows exactly as for K (see theorem 1.5), i.e. by lemma 1.12 which shows how to build a falsifying model for $\Gamma \Rightarrow \Delta$ from Hintikka tuple.

For the sake of variety we will show how to demonstrate the weak completeness of K+AC by means of Kalmar's method (see e.g. Mendelson [175]). In fact, we need for this proof only the most strongly restricted version of cut, namely where the cut-formulae are atomic subformulae of the conclusion sequent. Moreover, formulation of this proof strategy in the setting of SC is much simpler than for Hilbert systems. In particular, it does not require proving additional theses.

Lemma 1.28. *Let $PROP(\varphi) = \Gamma \cup \Delta$ such that for some $\mathfrak{M}$ we have $\mathfrak{M} \vDash \Gamma$ and $\mathfrak{M} \nvDash \vee \Delta$, then:*

1. if $\mathfrak{M} \vDash \varphi$, then $\vdash \Gamma \Rightarrow \Delta, \varphi$;

2. if $\mathfrak{M} \nvDash \varphi$, then $\vdash \varphi, \Gamma \Rightarrow \Delta$.

PROOF: By induction on the complexity of φ. The basis is trivial since φ is just an atom and in both cases we have $\vdash \varphi \Rightarrow \varphi$.
    Let $\varphi = \psi \wedge \chi$. If $\mathfrak{M} \vDash \varphi$, then $\mathfrak{M} \vDash \psi$ and $\mathfrak{M} \vDash \chi$. By the induction hypothesis $\vdash \Gamma' \Rightarrow \Delta', \psi$ and $\vdash \Gamma'' \Rightarrow \Delta'', \chi$ for $\Gamma' \cup \Delta' = PROP(\psi)$ and $\Gamma'' \cup \Delta'' = PROP(\chi)$. By applications of W and $(\Rightarrow \wedge)$ we have $\vdash \Gamma \Rightarrow \Delta, \varphi$. Similarly if $\mathfrak{M} \nvDash \varphi$, then $\mathfrak{M} \nvDash \psi$ or $\mathfrak{M} \nvDash \chi$. In the former case, by the induction hypothesis we have $\vdash \psi, \Gamma' \Rightarrow \Delta'$ and by W and $(\wedge \Rightarrow)$ we obtain $\vdash \varphi, \Gamma \Rightarrow \Delta$. The second case is analogous.   □

Exercise 1.33. *Prove the lemma for the remaining cases.*

A direct consequence of this lemma is the following:

Claim 1.7. *If $\vDash \varphi$, then $\vdash \Gamma \Rightarrow \Delta, \varphi$ for all partitions of $PROP(\varphi)$ into disjoint sets Γ, Δ.*

Now we can prove:

Theorem 1.10. *If $\vDash \varphi$, then $\vdash \Rightarrow \varphi$.*

PROOF: By induction on the number of elements of $PROP(\varphi)$. For every n we have just 2^n partitions corresponding to every possible valuation of $PROP(\varphi)$. We systematically apply cut on these atoms decreasing their number until we get $\Gamma \cup \Delta = \varnothing$. Instead of a rigorous inductive proof we simply illustrate the case of $n = 2$:

$$(Cut) \, \frac{\dfrac{\Rightarrow \varphi, p, q \qquad q \Rightarrow \varphi, p}{\Rightarrow \varphi, p} \qquad \dfrac{p \Rightarrow \varphi, q \qquad q, p \Rightarrow \varphi}{p \Rightarrow \varphi} \, (Cut)}{\Rightarrow \varphi} \, (Cut)$$

$\square$

Exercise 1.34. *Provide inductive proof of this theorem.*

1.11.2 Constructive Proof of Analytic Cut Admissibility

In the preceding subsection we treated the case of completeness of K+AC semantically but it is possible to do it also syntactically. We can constructively prove that in K with cut we can replace all applications of cut with analytic ones. A proof is basically the same as in section 1.8, i.e. by induction on the complexity of the cut-formula and the height of the proofs of the premises. The only difference is that we do not want to eliminate all cuts; we admit analytic cuts.

Theorem 1.11. *A proof in K+Cut may be replaced with a proof in K+AC*

PROOF: In the basis a cut-formula is a propositional symbol and we perform a subinduction on the height. In case of an axiom an application of a cut is eliminable as in the proof from section 1.8, and the same transformations are applied in case of height reduction (i.e. if at least one cut-formula is not principal).

The case where the cut-formula is principal in both premises is treated exactly as in the proof of theorem 1.7. Cuts on smaller formulae are either acceptable if analytic or eliminable by the induction hypothesis.

The only difference is that the last rule introducing a premiss of cut may be another cut, which by assumption is analytic. Let us consider such a case with ψ being an analytic and φ a non-analytic cut-formula:

$$(Cut) \, \frac{\dfrac{\Gamma \Rightarrow \Delta, \varphi, \psi \qquad \psi, \Pi \Rightarrow \Sigma}{\Gamma, \Pi \Rightarrow \Delta, \Sigma, \varphi} \, (Cut) \qquad \varphi, \Lambda \Rightarrow \Theta}{\Gamma, \Pi, \Lambda \Rightarrow \Delta, \Sigma, \Theta}$$

We permute both applications of cut:

$$(Cut) \, \frac{\dfrac{\Gamma \Rightarrow \Delta, \psi, \varphi \qquad \varphi, \Lambda \Rightarrow \Theta}{\Gamma, \Lambda \Rightarrow \Delta, \Theta, \psi} \, (Cut) \qquad \psi, \Pi \Rightarrow \Sigma}{\Gamma, \Pi, \Lambda \Rightarrow \Delta, \Sigma, \Theta}$$

Now, the non-analytic cut is eliminable by the induction hypothesis (the height is reduced). In case φ is in the right premiss of the input schema, the reduction is similar. In case φ is in both premises of the first cut, the transformation is slightly more complex since we must apply cut on φ twice, on both premises of the cut on ψ. $\square$

Exercise 1.35. *The reader is invited to demonstrate the cases mentioned above.*

1.11.3   Proof Search

Since the restriction of cut to analytic applications makes the proof search finite it provides also a possibility of defining an alternative decision procedure for **CPL**. However, it opens a room for specific solutions which may optimise proof search and restrict the size of proof trees. Let us mention briefly two possible strategies which may be borrowed from other approaches[9].

Mondadori [184] (see in particular DeAgostini [1]) introduced a system KE which is a variant of Smullyan's tableaux which has cut (called the PB rule there) as the only branching rule (i.e. two-premiss rule in SC framework). All other branching rules are replaced with non-branching ones but requiring additional (minor) premiss. For example, having $\varphi \vee \psi$ we additionally need $\neg\varphi$ (or $\neg\psi$) to derive ψ (or φ) on the branch, instead of adding φ and ψ on independent extensions of this branch. KE allows for exponentially shorter proofs which is not surprising in the light of remarks from the beginning of this section. Moreover, cut in KE may be restricted not only to analytic applications but even to some selected ones. Constructive completeness proof for KE shows that cut is only necessary when we have some disjunction, implication or negated conjunction on the branch, but with no additional premiss allowing for the application of a suitable rule. In the setting of K+AC this suggests the strategy of selecting as cut-formulae only immediate subformulae of disjunctions and implications occurring in the antecedent, or conjunctions from the succedent.

Quine [208] introduced quite efficient decision method for **CPL** called resolution by him. Although semantical, in fact it is strictly connected with resolution introduced later by Robinson [219]. Also the method of binary decision diagrams, popular in automated deduction, is essentially based on the same idea. The method works by always selecting one propositional symbol and replacing it with $\top$ and $\bot$, respectively. This is the resolution move. Several rewriting rules based on the truth-functional definitions of connectives allow for the systematic reduction of input formula to $\top$ or $\bot$. Of course we can select another propositional formula and repeat the resolution move if we stop at some branch with no definitive solution. This method may suggest another strategy for the application of analytic cuts in proof search. Restrict the selection of cut-formulae only to propositional symbols occurring in the input sequent. From Kalmar's completeness proof recorded above we know that this is sufficient. As for the strategy it seems that it is better to choose an atom which has more occurrences than others.

We provided only informal hints concerning possible candidates for cut-formulae. One should remember that they are taken from systems considerably different from SC. If we would like to simulate suitable procedures from [1] or [208] in an exact way we should show how to simulate other rules and the effect may be not wholly satisfying. For example, if we want to provide a formal analogue of KE in SC framework, we must replace the two-premiss rules with suitable pairs of one-premiss rules. For example instead of $(\vee \Rightarrow)$ we must introduce:

---

[9]They are discussed in more detail in Indrzejczak [130] in the framework of natural deduction.

$$\frac{\varphi, \varphi \vee \psi, \Gamma \Rightarrow \Delta, \psi}{\varphi \vee \psi, \Gamma \Rightarrow \Delta, \psi} \qquad \frac{\psi, \varphi \vee \psi, \Gamma \Rightarrow \Delta, \varphi}{\varphi \vee \psi, \Gamma \Rightarrow \Delta, \varphi}$$

And similarly for $(\rightarrow\Rightarrow)$ and $(\Rightarrow \wedge)$. Note that rules like:

$$\frac{\varphi, \Gamma \Rightarrow \Delta}{\varphi \vee \psi, \Gamma \Rightarrow \Delta, \psi} \qquad \frac{\psi, \Gamma \Rightarrow \Delta}{\varphi \vee \psi, \Gamma \Rightarrow \Delta, \varphi}$$

although more similar to ordinary SC rules, do not work. It is necessary to apply what is called Kleene's trick, i.e. to rewrite the principal formula and (minor) premiss from the conclusion since they may be used many times during the proof search process. Similar remarks apply to the effect of simulating Quine rewriting rules. Hence, from the standpoint of SC architecture it is rather pointless. Anyway, both approaches may be useful as the sources for a strategy of selecting cut-formulae in proof search. We conclude with the remark that it is possible to construct also other kinds of SC with primitive analytic cut which are adequate for **CPL**. One such system, due to Smullyan, will be described in section 3.6

Exercise 1.36. *Define suitable one-premiss rules being analogues to the other KE rules.*

Try to develop procedures of proof search for K+AC on the basis of KE or Quine's strategy.

1.11.4 Strong Completeness

The techniques presented so far in this section may seem of modest use since they do not yield essentially new results for **CPL**. Now we will show that something essentially new may be obtained on this basis. In general, another application of cut is connected with the possibility of obtaining a stronger version of completeness for SC; not only for provable sequents but also for the derivability relation between sequents. This way we open the door for proving adequacy results for theories formulated as collections of sequents added to some basic SC[10]. We have already mentioned by the end of subsection 1.3.2 that K (without cut) is not strong enough for such a result. To do this we need cut but fortunately, as we will see, the applications of cut may be limited in a satisfactory way also in this case.

First, we provide a simple proof based on theorem 1.5 and the syntactical interpretation I of sequents as formulae, described in subsection 1.2.1. Let $S_1, ..., S_n \vdash S$ denote a derivability relation in K+Cut. We will show:

Theorem 1.12. (Strong Adequacy). $S_1, ..., S_n \vdash S$ *iff* $S_1, ..., S_n \models S$

where $\models$ between sequents is understood as in section 1.3, i.e. that all models of $S_1, ..., S_n$ are models of S.

---

[10] This topic will be treated in more detail in the second volume where SC for First-Order Logic will be investigated.

PROOF: Soundness is not difficult to prove and it follows from the fact that all rules of K (including cut) are not only validity-preserving but also normal, i.e. hereditary with respect to models (see subsection 1.2.2).

Exercise 1.37. *Prove that if $S_1, ..., S_n \vdash S$, then $S_1, ..., S_n \models S$*

As for the completeness, we assume that $S_1, ..., S_n \models S$ with $S := \Gamma \Rightarrow \Delta$. It follows that $\models I(S_1), ..., I(S_n), \Gamma \Rightarrow \Delta$, for otherwise there is a model $\mathfrak{M}$ such that $\mathfrak{M} \models \Gamma$ and $\mathfrak{M} \models I(S_i)$ for every $i \leqslant n$ but $\mathfrak{M} \nvDash \psi$ for every $\psi \in \Delta$. Such a model however validates every S_i and falsifies S which contradicts our assumption. It follows by (weak) completeness that $\vdash I(S_1), ..., I(S_n), \Gamma \Rightarrow \Delta$. Now it is easy to prove $S_i \vdash \Rightarrow I(S_i)$ by successive applications of $(\wedge \Rightarrow), (\Rightarrow \vee)$ and $(\Rightarrow \rightarrow)$ to each S_i. Finally, n applications of cut to $I(S_1), ..., I(S_n), \Gamma \Rightarrow \Delta$ and every $\Rightarrow I(S_i)$ yields the result. $\square$

As we may notice all applications of cut were already somewhat restricted in this proof but we can sharpen the result in a way which was first stated in the more specific form by Takeuti [261] and then independently, and under different names, proved in general form by Girard [99], Avron [9] and Buss [43]. Below we present two proofs: the first syntactical (although still dependent on weak completeness) and the second semantical, both following Avron's presentations.

Let $S_1, ..., S_n \vdash^C S$ denote a derivability relation in K+Cut but with all applications of cut restricted to cut-formulae occurring as formulae in $S_1, ..., S_n$. Such cuts, and proofs containing only such cuts, may be called *anchored* after Buss [43]. We will show:

Theorem 1.13. (Strong Completeness). *If $S_1, ..., S_n \models S$, then $S_1, ..., S_n \vdash^C S$*

PROOF 1: By induction on n. The basis is just the weak completeness theorem. As the induction hypothesis, we assume that the result holds for any derivation with $n - 1$ nonaxiomatic sequents used as leaves. Now consider $S_n := \varphi_1, ..., \varphi_k \Rightarrow \psi_1, ..., \psi_l$ and assume that $S_1, ..., S_n \models S$. By (unrestricted) strong completeness proved in Theorem 1.12 we have $S_1, ..., S_n \vdash S$ but we must prove that $S_1, ..., S_n \vdash^C S$. First note that for each $i \leq k, l$ we have $\vdash^C \varphi_1, ..., \varphi_k \Rightarrow \psi_1, ..., \psi_l, \varphi_i$ and $\vdash^C \psi_i, \varphi_1, ..., \varphi_k \Rightarrow \psi_1, ..., \psi_l$ since they are just instances of axioms. Due to the context independence of all rules we can transform a derivation $\mathcal{D}$ of $S_1, ..., S_n \vdash S$ into $k + l$ derivations with added $\varphi_i, i \leq k$ to the succedent of every sequent or $\psi_i \leq l$ to the antecedent. Hence we get k derivations $\mathcal{D}'_i$ of $S'_1, ..., S'_n \vdash \Gamma \Rightarrow \Delta, \varphi_i$ and l derivations $\mathcal{D}''_i$ of $S''_1, ..., S''_n \vdash \psi_i, \Gamma \Rightarrow \Delta$ where each sequent has added φ_i in the succedent or ψ_i in the antecedent. In particular, every leaf decorated with S_n in $\mathcal{D}$ obtains in the resulting derivations an axiomatic form $\varphi_1, ..., \varphi_k \Rightarrow \psi_1, ..., \psi_l, \varphi_i$ or $\psi_i, \varphi_1, ..., \varphi_k \Rightarrow \psi_1, ..., \psi_l$. Hence by the inductive hypothesis we have that $S'_1, ..., S'_{n-1} \vdash^C \Gamma \Rightarrow \Delta, \varphi_i$ and $S''_1, ..., S''_{n-1} \vdash^C \psi_i, \Gamma \Rightarrow \Delta$ for each $i \leq k, l$. Addition of $S_j, j \leq n - 1$ over each leaf of the form S'_j or S''_j change these $k + l$ derivations into derivations of $\Gamma \Rightarrow \Delta, \varphi_i$ or $\psi_i, \Gamma \Rightarrow \Delta$ from $S_1, ..., S_{n-1}$; new steps are justified by applications of W. Finally we apply cut to S_n and every root sequent of the form $\Gamma \Rightarrow \Delta, \varphi_i$ and $\psi_i, \Gamma \Rightarrow \Delta$. We can do this for instance in the following order:

$$
\cfrac{
  \cfrac{\mathcal{D}'_2}{\Gamma \Rightarrow \Delta, \varphi_2} \quad
  \cfrac{
    \cfrac{\mathcal{D}'_1}{\Gamma \Rightarrow \Delta, \varphi_1} \quad \varphi_1, \ldots, \varphi_k \Rightarrow \psi_1, \ldots, \psi_l
  }{\varphi_2, \ldots, \varphi_k, \Gamma \Rightarrow \Delta, \psi_1, \ldots, \psi_l}
}{\varphi_3, \ldots, \varphi_k, \Gamma \Rightarrow \Delta, \psi_1, \ldots, \psi_l}
$$

$$
\cfrac{
  \cfrac{\mathcal{D}'_k}{\Gamma \Rightarrow \Delta, \varphi_k} \quad
  \cfrac{\vdots}{\varphi_k, \Gamma \Rightarrow \Delta, \psi_1, \ldots, \psi_l}
}{\Gamma \Rightarrow \Delta, \psi_1, \ldots, \psi_l}
$$

Then we repeat this process to each sequent $\psi_i, \Gamma \Rightarrow \Delta$ until we obtain a derivation of $S_1, ..., S_n \vdash^C S$. $\qquad\square$

PROOF 2: In order to prove (strong) completeness semantically we first slightly modify a definition of downward saturation and demonstrate an analogon of lemma 1.14.

Let Θ denote the set of all formulae occurring in $S_1, ..., S_n$ and $S := \Gamma \Rightarrow \Delta$. Then in addition to usual conditions for a saturated pair (Π, Σ) we require:

(0) If $\varphi \in \Theta$, then $\varphi \in \Pi \cup \Sigma$.

We can prove the following

Lemma 1.29. (Saturation). *If $S_1, ..., S_n \not\vdash^C \Gamma \Rightarrow \Delta$, then there is a saturated pair (Π, Σ) of finite sets, such that:*

(a) $\Pi \cup \Sigma \subseteq SF(\Gamma \cup \Delta \cup \Theta)$;

(b) $S_1, ..., S_n \not\vdash^C \Pi \Rightarrow \Sigma$

PROOF: We combine the proofs of lemma 1.26 and lemma 1.14. Let $\Theta = \{\varphi_1, \varphi_2, ..., \varphi_n\}$ We first systematically add elements of Θ either to Γ or to Δ according to the recipe described in the proof of lemma 1.26. After n steps we obtain a consistent (not necessarily saturated) pair (Γ', Δ') where all elements of Θ occur either in Γ' or in Δ'. Notice that $S_1, ..., S_n \not\vdash^C \Gamma' \Rightarrow \Delta'$ since for each $\varphi \in \Theta$ it is not possible that both $S_1, ..., S_n \vdash^C \Gamma \Rightarrow \Delta, \varphi$ and $S_1, ..., S_n \vdash^C \varphi, \Gamma \Rightarrow \Delta$. Otherwise, by (Cut) on φ (which is admitted by $\vdash^C$) we obtain $S_1, ..., S_n \vdash^C \Gamma \Rightarrow \Delta$—a contradiction.

Then we continue with (Γ', Δ') as in the proof of lemma 1.14 until we obtain (Π, Σ). By construction it is guaranteed that (a) and (b) of the lemma hold. $\qquad\square$

The construction of a model and suitable truth lemma is proved in an analogous way as in section 1.7. However now we must additionally show that the model in question satisfies also every $S_i, i \leq n$.

PROOF: For every such $\Gamma_i \Rightarrow \Delta_i$ we have $\Gamma_i \cup \Delta_i \subseteq \Pi \cup \Sigma$ due to saturation. If $\Gamma_i \subseteq \Pi$ and $\Delta_i \subseteq \Sigma$, then $S_1, ..., S_n \vdash^C \Pi \Rightarrow \Sigma$ trivially by W on S_i. Hence there must be some $\varphi \in \Gamma_i$ such that $\varphi \in \Sigma$ or $\varphi \in \Delta_i$ such that $\varphi \in \Pi$. In the first case $\mathfrak{M} \nvDash \varphi$, hence $\mathfrak{M}$ is a model of $\Gamma_i \Rightarrow \Delta_i$ since there is a false formula in the antecedent. In the second case $\mathfrak{M} \vDash \varphi$ hence $\mathfrak{M}$ is a model of $\Gamma_i \Rightarrow \Delta_i$ since there is a true formula in the succedent. $\qquad\square$

From the above result strong completeness follows for $\vdash$ (since $\vdash^C \subseteq \vdash$).

It is worth noticing that from the strong completeness theorem we obtain the following result.

Theorem 1.14. (Generalised Cut Admissibility). *If* $S_1, ..., S_n \vdash \Gamma \Rightarrow \Delta, \varphi$ *and* $S_1, ..., S_n \vdash \varphi, \Pi \Rightarrow \Sigma$, *then* $S_1, ..., S_n \vdash^C \Gamma, \Pi \Rightarrow \Delta, \Sigma$.

Exercise 1.38. *Derive this result from theorem 1.13.*

Remark 1.8. In Buss [43] such a result is described as an elimination of free cuts (as opposed to anchored cuts). Avron calls this result simply strong cut elimination but we do not follow his custom since such a name is by many authors used for a different result proved first by Dragalin [64] and then extended by many authors. Roughly speaking this is a demonstration that any sequence of suitable transformations performed on an input proof terminates with a cut-free proof. We will postpone a presentation of this result to the second volume. That the above result is a generalisation of the ordinary theorem follows from the fact that if $n = 0$ we obtain ordinary cut admissibility, i.e. no cuts are required for the proof of $\Gamma, \Pi \Rightarrow \Delta, \Sigma$.

1.11.5   Interpolation Theorem

As the last application of analytic cut we provide a simple proof of Craig's interpolation theorem for **CPL**. Here is a semantic formulation of Craig's theorem:

Theorem 1.15. (Interpolation). *If* $\vDash \varphi \to \psi$ *and* $PROP(\varphi) \cap PROP(\psi) \neq \varnothing$, *then for some* χ:

- $PROP(\chi) \subseteq PROP(\varphi) \cap PROP(\psi)$
- $\vDash \varphi \to \chi$
- $\vDash \chi \to \psi$

χ is called an interpolant. Note that the following holds:

Lemma 1.30. *If* $\vDash \varphi \to \psi$ *and* $PROP(\varphi) \cap PROP(\psi) = \varnothing$, *then* φ *is unsatisfiable or* ψ *is a tautology.*

PROOF: Assume to the contrary, that φ is satisfiable and ψ is not a tautology. So there is a model $\mathfrak{M}_1 \vDash \varphi$ based on V_1 and $\mathfrak{M}_2 \nvDash \psi$ based on V_2. Now define $\mathfrak{M}_3$ by postulating that its valuation $V_3(p) = V_1(p)$ for all $p \in PROP(\varphi)$, otherwise

$V_3(p) = V_2(p)$. Since $PROP(\varphi) \cap PROP(\psi) = \varnothing$ it follows that $\mathfrak{M}_3 \vDash \varphi$ and $\mathfrak{M}_3 \nvDash \psi$ but this contradicts the hypothesis that $\vDash \varphi \rightarrow \psi$. □

Clearly in case φ is unsatisfiable or ψ is a tautology they have an interpolant even if they have no common atoms; it is $\bot$ in the first case and $\top$ in the second. Note that we cannot avoid addition of these constants by using $\neg p \wedge p$ and $\neg p \vee p$, respectively, for arbitrary p since the condition concerning common atoms may fail whereas in case of $\bot$ and $\top$ it is trivially satisfied.

The proof we will provide is similar to the semantical proof of Chagrov and Zakharyashev [48] based on the application of the tableau method. It is nonconstructive in the sense that no recipe for obtaining an interpolant for concrete example is provided; in chapter 2 we provide a constructive proof. However, this proof is relatively simple and yields another application of analytic cut. In fact, for the needs of the proof we introduce two symmetrical forms of cut of the shape:

$$(R\text{-}cut) \ \frac{\Gamma \Rightarrow \Delta, \varphi \quad \Gamma \Rightarrow \Delta, \neg\varphi}{\Gamma \Rightarrow \Delta} \qquad (L\text{-}cut) \ \frac{\varphi, \Gamma \Rightarrow \Delta \quad \neg\varphi, \Gamma \Rightarrow \Delta}{\Gamma \Rightarrow \Delta}$$

Both are easily interderivable with ordinary cut by means of negation rules. Note also that in both we have the same sets of parametric formulae in both premisses, so W is also needed to show that ordinary cut is derivable. In what follows we will extend the notion of interpolation for sequents in an obvious way. We will say that $\Gamma \Rightarrow \Delta$ has an interpolant iff $I(\Gamma \Rightarrow \Delta)$ has.

We will prove the contrapositive of Craig's theorem, i.e. we prove that if there is no interpolant for $\varphi \rightarrow \psi$, then it is not valid.

To this aim let us consider a proof tree for $\varphi \Rightarrow \psi$ but constructed only by means of analytic symmetric cut applications. Let $\chi_1, ..., \chi_k$ be an arbitrary enumeration all elements of $SF(\{\varphi\})$ and $\gamma_1, ..., \gamma_n$ of all elements of $SF(\{\psi\})$. We build a tree in a root-first manner, starting with $\varphi \Rightarrow \psi$ and at each stage branching with symmetric cut on some χ_i or γ_j. In the former case we add χ_i to the antecedent of the left, and $\neg\chi_i$ to the antecedent of the right premiss of $(L\text{-}cut)$, whereas in the latter we add γ_j and $\neg\gamma_j$ to the succedents of premisses of $(R\text{-}cut)$ made on the active sequent. Hence we obtain a tree of height $k+n+1$ and size 2^{k+n} where for each leaf $\Gamma \Rightarrow \Delta$, Γ is maximal in $SF(\{\varphi\})$ and Δ in $SF(\{\psi\})$ in the sense that for all χ_i either it or its negation is in Γ and similarly for every γ_j and Δ. For such a construction it holds:

Claim 1.8. *For each* $m \leq k+n$, *if* $\Gamma_m \Rightarrow \Delta_m$ *has no interpolant, then at least one of the premisses, i.e.* $\chi_i, \Gamma_m \Rightarrow \Delta_m$ *and* $\neg\chi_i, \Gamma_m \Rightarrow \Delta_m$, *or* $\Gamma_m \Rightarrow \Delta_m, \gamma_j$ *and* $\Gamma_m \Rightarrow \Delta_m, \neg\gamma_j$, *has no interpolant.*

PROOF: Assume on the contrary that both premisses have interpolants. So there are α and β such that $PROP(\{\alpha\}) \subseteq PROP(\Gamma_m \cap \Delta_m)$ and similarly for β. Moreover, in the first case: $\vdash \chi_i, \Gamma_m \Rightarrow \alpha$, $\vdash \alpha \Rightarrow \Delta_m$, $\vdash \neg\chi_i, \Gamma_m \Rightarrow \beta$ and

$\vdash \beta \Rightarrow \Delta_m$. In the second case $\vdash \Gamma_m \Rightarrow \alpha$, $\vdash \alpha \Rightarrow \Delta_m, \gamma_j$, $\vdash \Gamma_m \Rightarrow \beta$ and $\vdash \beta \Rightarrow \Delta_m, \neg\gamma_j$.

Now, in the first case:

$$(\vee \Rightarrow) \ \frac{\alpha \Rightarrow \Delta_m \qquad \beta \Rightarrow \Delta_m}{\alpha \vee \beta \Rightarrow \Delta_m}$$

and

$$
\begin{array}{c}
(\Rightarrow W) \ \dfrac{\chi_i, \Gamma_m \Rightarrow \alpha}{\chi_i, \Gamma_m \Rightarrow \alpha, \beta} \qquad \dfrac{\neg\chi_i, \Gamma_m \Rightarrow \beta}{\neg\chi_i, \Gamma_m \Rightarrow \alpha, \beta} \ (\Rightarrow W) \\[2mm]
(Cut) \ \dfrac{\rule{0pt}{0pt}}{(\Rightarrow \vee) \ \dfrac{\Gamma_m \Rightarrow \alpha, \beta}{\Gamma_m \Rightarrow \alpha \vee \beta}}
\end{array}
$$

Hence $\alpha \vee \beta$ is an interpolant for $\Gamma_m \Rightarrow \Delta_m$, contrary to our assumption. $\qquad\square$

Exercise 1.39. *Show that in the second case $\alpha \wedge \beta$ is an interpolant.*

The claim implies that, since, by assumption, $\varphi \Rightarrow \psi$ has no interpolant, then at least one of the leaves, say $\Gamma \Rightarrow \Delta$, has no interpolant either. We will prove now for any $\Gamma \Rightarrow \Delta$ being such a leaf:

Lemma 1.31. *There is a model $\mathfrak{M}$ such that:*

- *for every $\chi \in SF(\varphi)$, $\chi \in \Gamma$ iff $\mathfrak{M} \vDash \chi$*
- *for every $\chi \in SF(\psi)$, $\chi \in \Delta$ iff $\mathfrak{M} \nvDash \chi$.*

PROOF: Note that $\Gamma \cap \Delta = \varnothing$; otherwise any formula which occurs on both sides is trivially an interpolant for $\Gamma \Rightarrow \Delta$. Also there are no contradictory formulae in Γ and Δ since then $\bot$ would be an interpolant. Hence the set of atomic formulae in Γ and Δ is disjoint, and the set of literals in both Γ and Δ does not contain any contradictory pair, so we can define a model by postulating $V(p) = 1$ iff $p \in \Gamma$ or $\neg p \in \Delta$.

The basis: Take any $p \in SF(\varphi)$. If $p \in \Gamma$, then $p \in \Gamma$ or $\neg p \in \Delta$ and $\mathfrak{M} \vDash p$. If $\mathfrak{M} \vDash p$, then $p \in \Gamma$ or $\neg p \in \Delta$, but since $p \in SF(\varphi)$ it follows that $p \in \Gamma$.

Take any $p \in SF(\psi)$ and assume that $p \in \Delta$ and $\mathfrak{M} \vDash p$. So $p \in \Gamma$ or $\neg p \in \Delta$ but, as we noticed above, both are impossible, hence $\mathfrak{M} \nvDash p$. If $\mathfrak{M} \nvDash p$ but $p \notin \Delta$, then $\neg p \in \Delta$. Hence $p \in \Gamma$ or $\neg p \in \Delta$ which implies $\mathfrak{M} \vDash p$.

For inductive step we check as an example $\chi := \alpha \wedge \beta$.

Assume that $\chi \in SF(\varphi)$ and $\chi \in \Gamma$ but $\mathfrak{M} \nvDash \chi$. So $\mathfrak{M} \nvDash \alpha$ or $\mathfrak{M} \nvDash \beta$. By the induction hypothesis $\alpha \notin \Gamma$ or $\beta \notin \Gamma$ which implies by maximality that $\neg\alpha \in \Gamma$ or $\neg\beta \in \Gamma$. But both $\{\alpha \wedge \beta, \neg\alpha\}$ and $\{\alpha \wedge \beta, \neg\beta\}$ are inconsistent hence $\bot$ is an interpolant for $\Gamma \Rightarrow \Delta$. If for the same $\chi \in SF(\varphi)$, $\mathfrak{M} \vDash \chi$ but $\chi \notin \Gamma$, then $\neg\chi \in \Gamma$. Since $\mathfrak{M} \vDash \alpha$ and $\mathfrak{M} \vDash \beta$ we have $\alpha \in \Gamma$ and $\beta \in \Gamma$ by the induction hypothesis. But $\{\neg(\alpha \wedge \beta), \alpha, \beta\}$ is inconsistent, hence again $\bot$ is an interpolant.

Now assume that $\chi \in SF(\psi)$ and $\chi \in \Delta$ but $\mathfrak{M} \vDash \chi$. So $\mathfrak{M} \vDash \alpha$ and $\mathfrak{M} \nvDash \beta$. By the induction hypothesis $\alpha \notin \Delta$ and $\beta \notin \Delta$ which implies by maximality that $\neg\alpha \in \Delta$ and $\neg\beta \in \Delta$. But $(\alpha \wedge \beta) \vee \neg\alpha \vee \neg\beta$ is a tautology hence $\top$ is an interpolant for $\Gamma \Rightarrow \Delta$. If for the same $\chi \in SF(\psi)$ we have $\mathfrak{M} \nvDash \chi$ and $\chi \notin \Delta$, then $\neg\chi \in \Delta$ follows by maximality. $\mathfrak{M} \nvDash \alpha$ or $\mathfrak{M} \nvDash \beta$ so we have $\alpha \in \Delta$ or $\beta \in \Delta$ by the induction hypothesis. But both $\vDash \neg(\alpha \wedge \beta) \vee \alpha$ and $\vDash \neg(\alpha \wedge \beta) \vee \beta$, hence again $\top$ is an interpolant. $\qquad\square$

Exercise 1.40. *Check the cases where χ is a negation, disjunction and implication.*

Craig's interpolation theorem follows directly from these two results. For, if we construct such a tree for $\varphi \Rightarrow \psi$ that has no interpolant, then at least one leaf $\Gamma \Rightarrow \Delta$ has no interpolant either but has a falsifying model. But $\varphi \in \Gamma$ and $\psi \in \Delta$, so $\varphi \to \psi$ is not a tautology.

Remark 1.9. The reader may wonder why we did not simply apply cut instead of these two variants, and why we take care of the strict separation of the subformulae of φ and ψ in the course of the proof. We will show that these apparently superficial complications were essential, analysing again the example from the proof. First if we just apply cut, in the first case we could derive only $\alpha \vee \beta \Rightarrow \Delta_m, \chi_i$ (or in the second case $\gamma_j, \Gamma_m \Rightarrow \alpha \wedge \beta$), so we do not have an interpolant for $\Gamma_m \Rightarrow \Delta_m$. Similarly, if we do not care about the separation of subformulae, and, for example, we use symmetric cut with some $\gamma_j \in SF(\psi)$ but added to Γ_m not to Δ_m, we fail with the claim if $PROP(\gamma_j) \cap PROP(\Gamma_m) = \varnothing$. Let $PROP(\alpha) \cap PROP(\Gamma_m) = \varnothing$ but included in $PROP(\gamma_j)$ and similarly for $PROP(\beta)$. In such a case of course it is still true that α and β are interpolants for premises but $\alpha \vee \beta$ is not an interpolant for $\Gamma_m \Rightarrow \Delta_m$.

Chapter 2

Gentzen's Sequent Calculus LK

This chapter introduces the original sequent calculus of Gentzen, called LK (der Logistische Kalkül). There are considerable differences with the calculus K from chapter 1. A sequent is defined as an ordered pair of finite (including empty) lists of formulae, and the set of primitive rules is significantly different. LK is in many respects harder to deal with than K, however the subtleties of its construction are important for further development. After introducing the calculus in section 2.1 we investigate some applications of cut concerning the equivalence of some forms of sequents and sequent rules in section 2.2. Next some invertibility results for LK are established in section 2.3. Then we will focus on different strategies of proving the cut-elimination/admissibility theorem, dividing them into local and global proofs. The former are based on small transformations of particular steps of a proof, similarly as it was done in the proof from chapter 1. The latter are based on transformations of the whole proofs of the premises of cut. We start in section 2.4 with local proofs, in particular we present in detail the original proof of Gentzen. Three other local proofs are also described in this section; all of them allow for avoiding some difficulties of Gentzen's proof. In section 2.5 we present two global proofs due to Curry and Buss. Concerning proof search, LK seems to be a less friendly tool than K, however we can still provide a decidability result although on the basis of a different strategy. This will be discussed in section 2.6. Section 2.7 contains a discussion of Kleene's result concerning permutability of rules, which is essentially a generalisation of (some steps) of cut-elimination procedure present in local proofs.

2.1 The System LK

The first version of SC for classical (and intuitionistic) logic was introduced by Gentzen [95] under the name LK (der Logistische Kalkül), in contrast to his natural deduction system NK (der Natürliche Kalkül). Whereas the latter was provided

© Springer Nature Switzerland AG 2021

A. Indrzejczak, *Sequents and Trees*, Studies in Universal Logic,
https://doi.org/10.1007/978-3-030-57145-0_2

as a formal representation of 'natural reasoning' made on formulae, the former used artificial items called sequents, axioms instead of assumptions, and his role was mainly (at least in this paper) technical. Below we will present the restricted part of LK limited to the formalisation of **CPL**. The basic difference between Gentzen's LK and K is connected with the very notion of a sequent. We continue the use of $\Gamma, \Delta, \Theta, \Lambda, \Xi, \Pi, \Sigma$ for antecedents and succedents of sequents but we must remember that in this chapter they denote not sets but finite, possibly empty, lists of formulae. Also in the contexts like $\Gamma \Rightarrow \Delta, \Pi$, we have a concatenation of lists Δ, Π (in this order) not a union of two sets.

2.1.1　Rules

Further differences are in the number and shape of the rules. It is a characteristic feature of LK that apart from logical rules we have a set of structural rules which specify some general operations on elements of sequents without displaying any formulae of a specific kind. In fact, some such rules were already introduced in the preceding chapter, namely the rules of weakening and cut. These rules, and some other, now belong to the set of primitive rules of the system. It may be said that the structural rules form a theory of $\Rightarrow$, whereas the logical rules define a specific logic. LK for **CPL** consists of the following rules:

Structural rules

(AX)　$\varphi \Rightarrow \varphi$

(Cut)　$\dfrac{\Gamma \Rightarrow \Delta, \varphi \quad \varphi, \Pi \Rightarrow \Sigma}{\Gamma, \Pi \Rightarrow \Delta, \Sigma}$

$(W\Rightarrow)$　$\dfrac{\Gamma \Rightarrow \Delta}{\varphi, \Gamma \Rightarrow \Delta}$　　　　　　$(\Rightarrow W)$　$\dfrac{\Gamma \Rightarrow \Delta}{\Gamma \Rightarrow \Delta, \varphi}$

$(C\Rightarrow)$　$\dfrac{\varphi, \varphi, \Gamma \Rightarrow \Delta}{\varphi, \Gamma \Rightarrow \Delta}$　　　　$(\Rightarrow C)$　$\dfrac{\Gamma \Rightarrow \Delta, \varphi, \varphi}{\Gamma \Rightarrow \Delta, \varphi}$

$(P\Rightarrow)$　$\dfrac{\Pi, \varphi, \psi, \Gamma \Rightarrow \Delta}{\Pi, \psi, \varphi, \Gamma \Rightarrow \Delta}$　　　$(\Rightarrow P)$　$\dfrac{\Gamma \Rightarrow \Delta, \psi, \varphi, \Pi}{\Gamma \Rightarrow \Delta, \varphi, \psi, \Pi}$

Logical rules

$(\neg\Rightarrow)$　$\dfrac{\Gamma \Rightarrow \Delta, \varphi}{\neg\varphi, \Gamma \Rightarrow \Delta}$　　　　$(\Rightarrow\neg)$　$\dfrac{\varphi, \Gamma \Rightarrow \Delta}{\Gamma \Rightarrow \Delta, \neg\varphi}$

$(\wedge\Rightarrow)$　$\dfrac{\varphi, \Gamma \Rightarrow \Delta}{\varphi \wedge \psi, \Gamma \Rightarrow \Delta}$　　　$(\wedge\Rightarrow)$　$\dfrac{\psi, \Gamma \Rightarrow \Delta}{\varphi \wedge \psi, \Gamma \Rightarrow \Delta}$

$(\Rightarrow\wedge)$　$\dfrac{\Gamma \Rightarrow \Delta, \varphi \quad \Gamma \Rightarrow \Delta, \psi}{\Gamma \Rightarrow \Delta, \varphi \wedge \psi}$　　$(\vee\Rightarrow)$　$\dfrac{\varphi, \Gamma \Rightarrow \Delta \quad \psi, \Gamma \Rightarrow \Delta}{\varphi \vee \psi, \Gamma \Rightarrow \Delta}$

$$(\Rightarrow\vee)\ \frac{\Gamma\Rightarrow\Delta,\varphi}{\Gamma\Rightarrow\Delta,\varphi\vee\psi} \qquad\qquad (\Rightarrow\vee)\ \frac{\Gamma\Rightarrow\Delta,\psi}{\Gamma\Rightarrow\Delta,\varphi\vee\psi}$$

$$(\rightarrow\Rightarrow)\ \frac{\Gamma\Rightarrow\Delta,\varphi\quad\psi,\Pi\Rightarrow\Sigma}{\varphi\rightarrow\psi,\Gamma,\Pi\Rightarrow\Delta,\Sigma} \qquad\qquad (\Rightarrow\rightarrow)\ \frac{\varphi,\Gamma\Rightarrow\Delta,\psi}{\Gamma\Rightarrow\Delta,\varphi\rightarrow\psi}$$

The names of structural rules mean, respectively: W—weakening, C—contraction, and P—permutation. In what follows if we refer to structural rules without specifying which part of a sequent is dealt with we will either use their full names or, especially in proof figures, names (W), (C), (P), or even simpler W, C, P.

Note also that some of the logical rules are different than suitable rules in K. $(\wedge\Rightarrow)$ and $(\Rightarrow\vee)$ now display in the premiss only one component of the principal formula, and that is why we need a pair of such rules. The two-premiss rule $(\rightarrow\Rightarrow)$ has possibly different lists of parameters in premisses which are concatenated in the conclusion.

The example of the application of $(\rightarrow\Rightarrow)$:

$$\frac{p\wedge q,q\rightarrow r\Rightarrow\neg q,p\rightarrow r,\neg r\quad p\vee q,q\rightarrow r,q\wedge r\Rightarrow\neg(p\vee r),\neg s}{\neg r\rightarrow p\vee q,p\wedge q,q\rightarrow r,q\rightarrow r,q\wedge r\Rightarrow\neg q,p\rightarrow r,\neg(p\vee r),\neg s}$$

We may say that $(\rightarrow\Rightarrow)$ is an example of a context-free rule, in contrast to other two-premiss rules which are context-sharing. In chapter 5 we will see that these differences may lead to important consequences. However it is useful to introduce a special terminology right now. Let us compare the rules for $\wedge,\vee$, and for $\rightarrow$. The differences between them led many logicians (see in particular Girard [98]) to introduce a general distinction between:

- multiplicative (context-free, intensional, internal) rules;

- additive (context-sharing, extensional, combining) rules.

For example, multiplicative rules (M-rules) for $\wedge$ look like that:

$$(\Rightarrow\wedge)\ \frac{\Gamma\Rightarrow\Delta,\varphi\quad\Pi\Rightarrow\Sigma,\psi}{\Gamma,\Pi\Rightarrow\Delta,\Sigma,\varphi\wedge\psi} \qquad\qquad (\wedge\Rightarrow)\ \frac{\varphi,\psi,\Gamma\Rightarrow\Delta}{\varphi\wedge\psi,\Gamma\Rightarrow\Delta}$$

whereas the rules for $\wedge$ in LK provide an example of additive rules (A-rules). In particular, cut may be also multiplicative or additive:

$$(M\text{-}cut)\ \frac{\Gamma\Rightarrow\Delta,\varphi\quad\varphi,\Pi\Rightarrow\Sigma}{\Gamma,\Pi\Rightarrow\Delta,\Sigma} \qquad\qquad (A\text{-}cut)\ \frac{\Gamma\Rightarrow\Delta,\varphi\quad\varphi,\Gamma\Rightarrow\Delta}{\Gamma\Rightarrow\Delta}$$

Note that we can also use some combined form of cut where the same parameters in both premisses are treated as in A-cut, whereas different premisses are combined as in M-cut (see Herbelin [110]); it looks like that:

$$\frac{\Gamma,\Lambda\Rightarrow\Delta,\Theta,\varphi\quad\varphi,\Gamma,\Pi\Rightarrow\Delta,\Sigma}{\Gamma,\Lambda,\Pi\Rightarrow\Delta,\Theta,\Sigma}$$

In what follows we will be using simply the name 'cut' either for M-cut or for both forms if the difference between them is not important.

One can easily show the equivalence of all M- and A-rules by means of C and W (see section 5.2). However, as we shall see, their properties and behaviour in proofs is different and sometimes it matters which ones are taken as primitive.

2.1.2   Proofs

All terminologies concerning proofs, proof trees, elements of sequents, etc. remain unchanged. In addition to the application of height as a parameter of inductive proofs we may introduce the notion of the length of a proof (proof tree). It is defined similarly as height but only applications of logical rules are counted (note that in K it is just the same measure).

Example 2.1. *Once again we present a proof of Frege's syllogism, for comparison with a proof in K.*

$$
\cfrac{
p \Rightarrow p \qquad
\cfrac{
\cfrac{
(\rightarrow\Rightarrow)\cfrac{
\cfrac{p \Rightarrow p \qquad q \Rightarrow q}{p \rightarrow q, p \Rightarrow q} \qquad r \Rightarrow r
}{q \rightarrow r, p \rightarrow q, p \Rightarrow r}(\rightarrow\Rightarrow)
}{p \rightarrow (q \rightarrow r), p, p \rightarrow q, p \Rightarrow r}(\rightarrow\Rightarrow)
}{
\cfrac{
\cfrac{
\cfrac{
\cfrac{
\cfrac{
\cfrac{
}{p, p \rightarrow (q \rightarrow r), p \rightarrow q, p \Rightarrow r}(P\Rightarrow)
}{p, p \rightarrow (q \rightarrow r), p, p \rightarrow q \Rightarrow r}(P\Rightarrow)
}{p, p, p \rightarrow (q \rightarrow r), p \rightarrow q \Rightarrow r}(P\Rightarrow)
}{p, p \rightarrow (q \rightarrow r), p \rightarrow q \Rightarrow r}(C\Rightarrow)
}{p \rightarrow (q \rightarrow r), p \rightarrow q \Rightarrow p \rightarrow r}(\Rightarrow\rightarrow)
}{p \rightarrow q, p \rightarrow (q \rightarrow r) \Rightarrow p \rightarrow r}(P\Rightarrow)
}
}{p \rightarrow (q \rightarrow r) \Rightarrow (p \rightarrow q) \rightarrow (p \rightarrow r)}(\Rightarrow\rightarrow)
}{\Rightarrow (p \rightarrow (q \rightarrow r)) \rightarrow ((p \rightarrow q) \rightarrow (p \rightarrow r))}(\Rightarrow\rightarrow)
$$

One may easily notice the relatively large number of applications of structural rules. The height of this proof is 11 whereas its length is only 6. In what follows we will usually omit such "obvious" steps in proof figures and use double line for pointing out such condensations.

Notice that although cut is not necessarily needed as a primitive rule of LK (we will prove this formally in sections 2.4 and 2.5) we cannot get rid of the other structural rules. Since the axioms are in simple form (no context), one cannot prove admissibility of weakening for LK, and this in turn is necessary as a preprocessing step for the unification of contexts in the premisses of context-sharing rules like $(\Rightarrow \wedge)$ and $(\vee \Rightarrow)$. Permutation is indispensable since we are dealing with lists and the active formulae must be placed in concrete, leftmost or rightmost, position[1]. Also contraction must be used in the proofs of many theorems.

---

[1]However one may redefine the rules in such a way that the result of permutation is already absorbed—see Gallier [93] and remark 2.4 in section 2.6.

Example 2.2. *Here we display two examples of proofs of characteristic theses of* **CPL***: LEM and Peirce law, essentially based on the application of C:*

$$
\begin{array}{c}
(\Rightarrow \neg) \dfrac{p \Rightarrow p}{\Rightarrow p, \neg p} \\
(\Rightarrow \vee) \dfrac{}{\Rightarrow p, p \vee \neg p} \\
(\Rightarrow P) \dfrac{}{\Rightarrow p \vee \neg p, p} \\
(\Rightarrow \vee) \dfrac{}{\Rightarrow p \vee \neg p, p \vee \neg p} \\
(\Rightarrow C) \dfrac{}{\Rightarrow p \vee \neg p}
\end{array}
$$

$$
\begin{array}{c}
(\Rightarrow W) \dfrac{p \Rightarrow p}{p \Rightarrow p, q} \\
(\Rightarrow\rightarrow) \dfrac{}{\Rightarrow p, p \rightarrow q} \qquad p \Rightarrow p \\
(\rightarrow\Rightarrow) \dfrac{(p \rightarrow q) \rightarrow p \Rightarrow p, p}{} \\
(\Rightarrow C) \dfrac{(p \rightarrow q) \rightarrow p \Rightarrow p}{} \\
(\Rightarrow\rightarrow) \dfrac{}{\Rightarrow ((p \rightarrow q) \rightarrow p) \rightarrow p}
\end{array}
$$

In the first example ($\Rightarrow \vee$) with only one side formula was applied; in the second ($\rightarrow\Rightarrow$) with independent contexts. The reader may easily check (by exhausting all possibilities of backward proof search) that without cut we cannot provide contraction-free proofs for them. Moreover, we must apply A-cut if we want to avoid contraction in proofs of the above sequents.

Example 2.3. *A contraction-free proof of the Peirce law in LK with A-cut:*

$$
\begin{array}{c}
(\Rightarrow W) \dfrac{p \Rightarrow p}{p \Rightarrow p, q} \\
(\Rightarrow\rightarrow) \dfrac{}{\Rightarrow p, p \rightarrow q} \qquad p \Rightarrow p \\
(\rightarrow\Rightarrow) \dfrac{(p \rightarrow q) \rightarrow p \Rightarrow p, p}{} \\
(\Rightarrow\rightarrow) \dfrac{\Rightarrow p, ((p \rightarrow q) \rightarrow p) \rightarrow p}{} \qquad \dfrac{p \Rightarrow p}{(p \rightarrow q) \rightarrow p, p \Rightarrow p} \, (W \Rightarrow) \\
(\Rightarrow P) \dfrac{\Rightarrow ((p \rightarrow q) \rightarrow p) \rightarrow p, p}{} \qquad \dfrac{}{p \Rightarrow ((p \rightarrow q) \rightarrow p) \rightarrow p} \, (\Rightarrow\rightarrow) \\
(A\text{-}Cut) \dfrac{}{\Rightarrow ((p \rightarrow q) \rightarrow p) \rightarrow p}
\end{array}
$$

Notice two things. The A-cut applied in the proof is analytic and we have used Quine's strategy of taking atomic cut-formulae in proof search (see remarks in subsection 1.11.3). In the original LK with M-cut such a contraction-free proof is not possible. Moreover, if we introduce A-cut as a derivable rule of LK we can see that contraction is necessary for proving this.

Exercise 2.1. *Construct a contraction-free proof of LEM in LK with A-cut.*

The fact that contraction is practically indispensable in LK is of importance also for proof search in LK and for devising decision procedures. We will point out the difficulties and possible solutions in section 2.6.

2.2   Applications of Cut

In the remaining parts of this chapter we will show how to eliminate cut in LK and then consider some applications of cut elimination. However, first we will illustrate some further advantages of having it in addition to those which were discussed in chapter 1. Let us recall that one important application was shown in section 1.9—a proof of the equivalence of K with Hilbert's system for **CPL**. This may be done also for LK and is left to the reader.

Exercise 2.2. *Prove the equivalence of LK with H (be sensitive to differences between LK and K)*

In section 1.11 we discussed also additional profits of having cut. Here we will consider some more advantages connected with the provability of equivalents and the generation of extra rules.

2.2.1   Equivalent Sequents

Below we formally prove some intuitively obvious relations between formulae in antecedents and succedents of any (provable) sequent.

Lemma 2.1. *For any sequent* $\varphi_1, ..., \varphi_i \Rightarrow \psi_1, ..., \psi_k, (i, k > 0)$ *the following forms are equivalent:*

1. $\vdash \varphi_1, ..., \varphi_i \Rightarrow \psi_1, ..., \psi_k$

2. $\vdash \varphi_1 \wedge .... \wedge \varphi_i \Rightarrow \psi_1 \vee ... \vee \psi_k$

3. $\vdash \varphi_1, ..., \varphi_i, \neg\psi_1, ..., \neg\psi_k \Rightarrow$

4. $\vdash \Rightarrow \neg\varphi_1, ..., \neg\varphi_i, \psi_1, ..., \psi_k$

PROOF:

1. $\Longrightarrow$ 2. We perform the following deduction:

$$
\frac{
\frac{
\frac{
\frac{\varphi_1, ..., \varphi_i \Rightarrow \psi_1, ..., \psi_k}{\varphi_1 \wedge \varphi_2, \varphi_2, ..., \varphi_i \Rightarrow \psi_1, ..., \psi_k} (\wedge \Rightarrow)
}{\varphi_2, \varphi_1 \wedge \varphi_2, \varphi_3, ..., \varphi_i \Rightarrow \psi_1, ..., \psi_k} (P \Rightarrow)
}{\varphi_1 \wedge \varphi_2, \varphi_1 \wedge \varphi_2, \varphi_3, ..., \varphi_i \Rightarrow \psi_1, ..., \psi_k} (\wedge \Rightarrow)
}{\varphi_1 \wedge \varphi_2, \varphi_3, ..., \varphi_i \Rightarrow \psi_1, ..., \psi_k} (C \Rightarrow)
$$

and continue until we get $\varphi_1 \wedge, ..., \wedge \varphi_i \Rightarrow \psi_1, ..., \psi_k$. Next we provide an analogous deduction by means of $(\Rightarrow \vee), (\Rightarrow P)$ and $(\Rightarrow C)$ on the succedent until we obtain 2.

2. $\Longrightarrow$ 1. First note that for any $k \geq 2$ it holds $\vdash \psi_1 \vee ... \vee \psi_k \Rightarrow \psi_1, ..., \psi_k$, here is the beginning of the proof:

$$(\Rightarrow W) \cfrac{(\vee \Rightarrow) \cfrac{(\Rightarrow W) \cfrac{\cfrac{\psi_1 \Rightarrow \psi_1}{\psi_1 \Rightarrow \psi_1, \psi_2} \qquad \cfrac{\psi_2 \Rightarrow \psi_2}{\psi_2 \Rightarrow \psi_1, \psi_2}(\Rightarrow W),(\Rightarrow P)}{\psi_1 \vee \psi_2 \Rightarrow \psi_1, \psi_2}}{\psi_1 \vee \psi_2 \Rightarrow \psi_1, \psi_2, \psi_3} \qquad \cfrac{\psi_3 \Rightarrow \psi_3}{\psi_3 \Rightarrow \psi_1, \psi_2, \psi_3}(\Rightarrow W),(\Rightarrow P)}{\psi_1 \vee \psi_2 \vee \psi_3 \Rightarrow \psi_1, \psi_2, \psi_3}$$

In a similar way we prove for $i \geq 2$ that $\vdash \varphi_1, ..., \varphi_i \Rightarrow \varphi_1 \wedge .... \wedge \varphi_i$. Applying (Cut) twice to 2. and to the obtained sequents we get 1.

  1. $\implies$ 3. From 1. we get $\vdash \neg\psi_1, ..., \neg\psi_k, \varphi_1, ..., \varphi_i, \Rightarrow$ by k applications of $(\neg \Rightarrow)$ which yields 3. by permutation.

  3. $\implies$ 1. First note that for every $i \leq k$ by $(\Rightarrow \neg)$ we get $\vdash \Rightarrow \psi_i, \neg\psi_i$. From 3. by permutation we get $\vdash \neg\psi_1, ..., \neg\psi_k, \varphi_1, ..., \varphi_i, \Rightarrow$, then by (Cut) on $\vdash \Rightarrow \psi_1, \neg\psi_1$ we get $\vdash \neg\psi_2, ..., \neg\psi_k, \varphi_1, ..., \varphi_i, \Rightarrow \psi_1$. We repeat such deductions $k - 1$ times and finally by permutation we obtain 1. ☐

Exercise 2.3. *Prove 1.* $\Longleftrightarrow$ *4.*

If we limit our attention to sequents with one formula in the succedent only we can additionally show the relation between $\Rightarrow$ and $\rightarrow$ in the following lemma:

Lemma 2.2. *For any sequent* $\varphi_1, ..., \varphi_i \Rightarrow \psi$ $(i > 0)$, *the following forms are equivalent:*

  1. $\vdash \varphi_1, ..., \varphi_i \Rightarrow \psi$

  2. $\vdash \Rightarrow \varphi_1 \rightarrow (\varphi_2 \rightarrow ..., (\varphi_i \rightarrow \psi)...)$

  3. $\vdash \Rightarrow \varphi_1 \wedge .... \wedge \varphi_i \rightarrow \psi$

PROOF:

  1. $\implies$ 2. From 1. by means of $(P \Rightarrow)$ we get $\vdash \varphi_i, ..., \varphi_1 \Rightarrow \psi$, and by i applications of $(\Rightarrow\rightarrow)$ we obtain 2.

  2. $\implies$ 1. Note that for any φ, ψ it holds $\vdash \varphi \rightarrow \psi, \varphi \Rightarrow \psi$, in particular $\vdash \varphi_1 \rightarrow (\varphi_2 \rightarrow ..., (\varphi_i \rightarrow \psi)...), \varphi_1 \Rightarrow \varphi_2 \rightarrow (\varphi_3 \rightarrow ..., (\varphi_i \rightarrow \psi)...)$. Hence by 2. and (Cut) we get $\vdash \varphi_1 \Rightarrow \varphi_2 \rightarrow (\varphi_3 \rightarrow ..., (\varphi_i \rightarrow \psi)...)$. We repeat such a deduction $i - 1$ times (with successive instances of $\vdash \varphi_k \rightarrow \chi, \varphi_k \Rightarrow \chi, k \leq i$) until we get 1.

  1. $\implies$ 3. By lemma 2.1 and $(\Rightarrow\rightarrow)$

  3. $\implies$ 1. Since $\vdash \varphi_1 \wedge .... \wedge \varphi_i \rightarrow \psi, \varphi_1 \wedge .... \wedge \varphi_i \Rightarrow \psi$, so by (Cut) on 3. we have $\vdash \varphi_1 \wedge .... \wedge \varphi_i \Rightarrow \psi$ which, by lemma 2.1 implies 1. ☐

Clearly both lemmata imply that if some sequent which is an instance of some of the considered schemata is not provable, then its equivalents are not provable either.

2.2.2 Equivalent Rules

Both in K and LK logical rules have specific forms which we will analyse in detail at the end of chapter 3. On the other hand in section 1.11 we have opened the door for making possible extensions of SC with additional axiomatic sequents. This solution makes necessary the application of (restricted) cut in proofs and one can ask if it is possible to use some specific rules instead of axiomatic sequents and in this way to avoid the applications of cut. In the second volume this question will be treated in detail but now in a preliminary way we will present two results concerning the possible shape of rules replacing sequents in SC. The first lemma is rather concrete and is concerned with the generation of rules equivalent to sequents of the form $\varphi \Rightarrow \psi$, $\varphi, \psi \Rightarrow \chi$ or $\varphi \Rightarrow \psi, \chi$. Taking into account that quite often we deal with sequents or axioms (with $\rightarrow$ instead of $\Rightarrow$ and $\wedge$ or $\vee$ added) of this form this lemma is practically sufficient for most cases. It allows to establish the equivalence of different SC formalizations of some logics/theories quickly, and for the generation of new variants with desirable properties. Despite the practical sufficiency of this result we will formulate and prove also a generalised version of this lemma providing equivalent rules for any finite sequent. For unification of results and simplification of proofs we do not mention applications of structural rules and formulate all rules with more than one premiss in the multiplicative versions, but they may be proved also for additive versions.

Lemma 2.3. *The following schemata of sequents and rules collected in three groups are interderivable in LK:*

A: for (1) $\varphi \Rightarrow \psi$:

(2) $\dfrac{\psi, \Gamma \Rightarrow \Delta}{\varphi, \Gamma \Rightarrow \Delta}$
(3) $\dfrac{\Gamma \Rightarrow \Delta, \varphi}{\Gamma \Rightarrow \Delta, \psi}$
(4) $\dfrac{\Gamma \Rightarrow \Delta, \varphi \quad \psi, \Pi \Rightarrow \Sigma}{\Gamma, \Pi \Rightarrow \Delta, \Sigma}$

B: for (1) $\varphi, \psi \Rightarrow \chi$:

(2) $\dfrac{\chi, \Gamma \Rightarrow \Delta}{\varphi, \psi, \Gamma \Rightarrow \Delta}$
(3) $\dfrac{\Gamma \Rightarrow \Delta, \varphi}{\psi, \Gamma \Rightarrow \Delta, \chi}$
(4) $\dfrac{\Gamma \Rightarrow \Delta, \psi}{\varphi, \Gamma \Rightarrow \Delta, \chi}$

(5) $\dfrac{\Gamma \Rightarrow \Delta, \varphi \quad \Pi \Rightarrow \Sigma, \psi}{\Gamma, \Pi \Rightarrow \Delta, \Sigma, \chi}$
(6) $\dfrac{\Gamma \Rightarrow \Delta, \varphi \quad \chi, \Pi \Rightarrow \Sigma}{\psi, \Gamma, \Pi \Rightarrow \Delta, \Sigma}$

(7) $\dfrac{\Gamma \Rightarrow \Delta, \psi \quad \chi, \Pi \Rightarrow \Sigma}{\varphi, \Gamma, \Pi \Rightarrow \Delta, \Sigma}$
(8) $\dfrac{\Gamma \Rightarrow \Delta, \varphi \quad \Pi \Rightarrow \Sigma, \psi \quad \chi, \Lambda \Rightarrow \Theta}{\Gamma, \Pi, \Lambda \Rightarrow \Delta, \Sigma, \Theta}$

C: for (1) $\varphi \Rightarrow \psi, \chi$:

(2) $\dfrac{\Gamma \Rightarrow \Delta, \varphi}{\Gamma \Rightarrow \Delta, \psi, \chi}$
(3) $\dfrac{\psi, \Gamma \Rightarrow \Delta}{\varphi, \Gamma \Rightarrow \Delta, \chi}$
(4) $\dfrac{\chi, \Gamma \Rightarrow \Delta}{\varphi, \Gamma \Rightarrow \Delta, \psi}$

(5) $\dfrac{\psi, \Gamma \Rightarrow \Delta \quad \chi, \Pi \Rightarrow \Sigma}{\varphi, \Gamma, \Pi \Rightarrow \Delta, \Sigma}$
(6) $\dfrac{\Gamma \Rightarrow \Delta, \varphi \quad \chi, \Pi \Rightarrow \Sigma}{\Gamma, \Pi \Rightarrow \Delta, \Sigma, \psi}$

(7) $\dfrac{\Gamma \Rightarrow \Delta, \varphi \quad \psi, \Pi \Rightarrow \Sigma}{\Gamma, \Pi \Rightarrow \Delta, \Sigma, \chi}$
(8) $\dfrac{\Gamma \Rightarrow \Delta, \varphi \quad \psi, \Pi \Rightarrow \Sigma \quad \chi, \Lambda \Rightarrow \Theta}{\Gamma, \Pi, \Lambda \Rightarrow \Delta, \Sigma, \Theta}$

PROOF: For A.
1. $\Longrightarrow$ 2.: It is sufficient to apply (Cut) to 1 and to the premiss of 2.

2. $\implies$ 3.: From axiom $\psi \Rightarrow \psi$ by 2 we obtain $\varphi \Rightarrow \psi$ which by (Cut) with the premiss of 3 yields the conclusion.

3. $\implies$ 4.: the following schema shows its derivability:

$$\cfrac{\Gamma \Rightarrow \Delta, \varphi \qquad \cfrac{(3.)\cfrac{\varphi \Rightarrow \varphi}{\varphi \Rightarrow \psi} \qquad \psi, \Pi \Rightarrow \Sigma}{\varphi, \Pi \Rightarrow \Sigma}(Cut)}{\Gamma, \Pi \Rightarrow \Delta, \Sigma}(Cut)$$

4. $\implies$ 1.: From $\varphi \Rightarrow \varphi$ and $\psi \Rightarrow \psi$ by 4 we get 1.

For B.

1. $\implies$ 2.: It is enough to apply (Cut) to 1 and the premiss of 2 to obtain the conclusion.

2. $\implies$ 3.: From $\chi \Rightarrow \chi$ by 2 we get $\varphi, \psi \Rightarrow \chi$ which by (Cut) with the premiss of 3 yields the conclusion.

3. $\implies$ 4.: analogous, with 3 applied to $\varphi \Rightarrow \varphi$.

4. $\implies$ 5.: analogous, but 4 is applied not to an axiom but to the premiss of 5 of the shape $\Pi \Rightarrow \Sigma, \psi$ in order to get the conclusion by (Cut) on the second premiss.

5. $\implies$ 6.: the following schema shows derivability:

$$(5.)\cfrac{\cfrac{\Gamma \Rightarrow \Delta, \varphi \qquad \psi \Rightarrow \psi}{\psi, \Gamma \Rightarrow \Delta, \chi}(Cut) \qquad \chi, \Pi \Rightarrow \Sigma}{\psi, \Gamma, \Pi \Rightarrow \Delta, \Sigma}$$

6. $\implies$ 7.: the following schema shows derivability:

$$\cfrac{\Gamma \Rightarrow \Delta, \psi \qquad \cfrac{\varphi \Rightarrow \varphi \qquad \chi, \Pi \Rightarrow \Sigma}{\psi, \varphi, \Pi \Rightarrow \Sigma}(6.)}{\varphi, \Gamma, \Pi \Rightarrow \Delta, \Sigma}(Cut)$$

7. $\implies$ 8.: the following schema shows derivability:

$$\cfrac{\Gamma \Rightarrow \Delta, \varphi \qquad \cfrac{\Pi \Rightarrow \Sigma, \psi \qquad \chi, \Lambda \Rightarrow \Theta}{\varphi, \Pi, \Lambda \Rightarrow \Sigma, \Theta}(7.)}{\Gamma, \Pi, \Lambda \Rightarrow \Delta, \Sigma, \Theta}(Cut)$$

8. $\implies$ 1.: From $\varphi \Rightarrow \varphi$, $\psi \Rightarrow \psi$ and $\chi \Rightarrow \chi$ by 8 we deduce 1. $\qquad\square$

Exercise 2.4. *Prove part C.*

Prove an analogue of this lemma but for context-sharing rules.

Let us illustrate with the case of formalising $\vee$ the possible applications of this lemma. One rule $(\Rightarrow \vee)$ for (additive) disjunction is captured by schema $(A3)$, hence we can use the following equivalents:

$(A1)$ $\varphi \Rightarrow \varphi \vee \psi$

$(A2) \ \dfrac{\varphi \vee \psi, \Gamma \Rightarrow \Delta}{\varphi, \Gamma \Rightarrow \Delta}$
$\qquad\qquad\qquad\qquad (A4) \ \dfrac{\Gamma \Rightarrow \Delta, \varphi \quad \varphi \vee \psi, \Pi \Rightarrow \Sigma}{\Gamma, \Pi \Rightarrow \Delta, \Sigma}$

On the other hand $(\vee \Rightarrow)$ is captured by schema $(C5)$ (with $\varphi := \psi \vee \chi$) which generates the following equivalents:

$(C1) \ \psi \vee \chi \Rightarrow \psi, \chi$

$(C2) \ \dfrac{\Gamma \Rightarrow \Delta, \psi \vee \chi}{\Gamma \Rightarrow \Delta, \psi, \chi} \qquad (C3) \ \dfrac{\psi, \Gamma \Rightarrow \Delta}{\psi \vee \chi, \Gamma \Rightarrow \Delta, \chi} \qquad (C4) \ \dfrac{\chi, \Gamma \Rightarrow \Delta}{\psi \vee \chi, \Gamma \Rightarrow \Delta, \psi}$

$(C6) \ \dfrac{\Gamma \Rightarrow \Delta, \psi \vee \chi \quad \chi, \Pi \Rightarrow \Sigma}{\Gamma, \Pi \Rightarrow \Delta, \Sigma, \psi} \qquad (C7) \ \dfrac{\Gamma \Rightarrow \Delta, \psi \vee \chi \quad \psi, \Pi \Rightarrow \Sigma}{\Gamma, \Pi \Rightarrow \Delta, \Sigma, \chi}$

$(C8) \ \dfrac{\Gamma \Rightarrow \Delta, \psi \vee \chi \quad \psi, \Pi \Rightarrow \Sigma \quad \chi, \Lambda \Rightarrow \Theta}{\Gamma, \Pi, \Lambda \Rightarrow \Delta, \Sigma, \Theta}$

Some of the rules in both lists may seem 'unnatural' when compared with LK rules for $\vee$ (the same remark applies to other connectives and possible rules for them). In particular, $(A2), (A4), (C2), (C6), (C7)$ and $(C8)$ do not satisfy the subformula property, and $(A4), (C8)$ look like some special versions of cut. Remember that the possibility of the generation of all these rules is not so important for defining logical rules in standard SC for propositional logics but rather for obtaining systems of rules working for nonlogical theories. This topic will be dealt with in the second volume and the importance of this result, as well as of strong completeness from subsection 1.11.4, will be appreciated better in this place. Yet even rules for connectives which are of different characters than Gentzen's rules may find interesting applications. For example, one can easily recognise in $(C3)$ and $(C4)$, rules which were introduced in subsection 1.11.3 as possible SC counterparts of suitable KE rules. Recall that such a system has all logical rules with one premiss only but the price is that cut must be taken as primitive (fortunately in its analytic version). As a result a branching factor of KE is smaller and one may obtain for many sequents even exponentially smaller proofs than in cut-free K.

Exercise 2.5. *Provide the equivalents for $(\wedge \Rightarrow)$ using part A, and for $(\Rightarrow \wedge)$ using part B.*

Lemma 2.3 deals only with three possible cases. We can generalise this result in the following way:

Lemma 2.4. *For any sequent $\Gamma \Rightarrow \Delta$ with $\Gamma = \{\varphi_1, ..., \varphi_k\}$ and $\Delta = \{\psi_1, ..., \psi_n\}$, $k \geq 0, n \geq 0, k+n \geq 1$ there are $2^{k+n} - 1$ equivalent rules captured by the general schema:*

$$\dfrac{\Pi_1, \Rightarrow \Sigma_1, \varphi_1, \ ... \ \Pi_i \Rightarrow \Sigma_i, \varphi_i \qquad \psi_1, \Pi_{i+1} \Rightarrow \Sigma_{i+1}, \ ... \ \psi_j, \Pi_{i+j} \Rightarrow \Sigma_{i+j}}{\Gamma^{-i}, \Pi_1, ..., \Pi_i, \Pi_{i+1}, ..., \Pi_{i+j} \Rightarrow \Sigma_1, ..., \Sigma_i, \Sigma_{i+1}, ..., \Sigma_{i+j}\Delta^{-j}}$$

where $\Gamma^{-i} = \Gamma - \{\varphi_1, ..., \varphi_i\}$ and $\Delta^{-j} = \Delta - \{\psi_1, ..., \psi_j\}$ for $0 \leq i \leq k, \ 0 \leq j \leq n$.

A comment on the schema may be helpful before we provide a proof. In general we define rules by taking an arbitrary number of formulae from the antecedent or succedent, and for every such formula we create a premiss sequent where an item from the antecedent (φ_i) is put in the succedent, and an element of the succedent (ψ_j) is put in the antecedent of premiss sequent. The remaining formulae from the input sequent are collected into sets Γ^{-i} and Δ^{-j} of the conclusion sequent. Extreme cases are empty sets Γ^{-k} and Δ^{-n}. A rule has $k + n$ premisses, one for every formula from the input sequent, and the conclusion contains only the unions of the parameters from the premisses. On the other hand, if we consider a situation with Γ^{-0} and Δ^{-0}, then our schema captures also a case with no premisses at all, i.e. our input sequent.

PROOF: The proof goes by induction on $k + n$. For basic cases with $k = 1, n = 0$ or $k = 0, n = 1$ (sequents of the form $\varphi \Rightarrow$ and $\Rightarrow \psi$) there is only one equivalent rule: $\Gamma \Rightarrow \Delta, \varphi / \Gamma \Rightarrow \Delta$ and $\psi, \Gamma \Rightarrow \Delta / \Gamma \Rightarrow \Delta$, i.e. we have $2^1 - 1$ rules. Also the case of $n = k = 1$ (as well as two cases for $k + n = 3$) was shown above to satisfy the claim.

For the inductive step we assume that for some $S = \varphi_1, ..., \varphi_k \Rightarrow \psi_1, ..., \psi_n$ the claim holds and we will show that it holds if we add some χ to the antecedent or succedent of S. First consider the addition of χ to the antecedent which we schematize as χ, S. Let $i = 2^{k+n}$, so we have S and $i - 1$ equivalent rules by assumption. From each rule we generate 2 new rules in the following way: (a) either add χ to the antecedent of the conclusion or (b) add the additional premiss $\Pi \Rightarrow \Sigma, \chi$ with arbitrary Π, Σ. So if the rule equivalent to S has the shape: $S_1, ..., S_j / S_{j+1}$ for $i \geq j \geq 1$ we get either: (a) $S_1, ..., S_j / \chi, S_{j+1}$ or (b) $\Pi \Rightarrow \Sigma, \chi, S_1, ..., S_j / \Pi, S_{j+1}, \Sigma$. In case of S we do the same so we obtain (a) a sequent χ, S and (b) a rule $\Pi \Rightarrow \Sigma, \chi / S$. In total we obtain $2i - 1$ rules and a sequent χ, S, i.e. $2^{k+n+1} - 1$ rules.

It is easy to show that the new rules are interderivable with χ, S in the way illustrated in the proof of the preceding lemma. First, if we take χ, S, then by j applications of (Cut) to all premisses of (a) we will get χ, S_{j+1}. This is because j elements of S are distributed as active formulae in the premisses in the following way: all active formulae from antecedents of premisses are in the succedent of S and all from the succedents are in the antecedent. After j applications of (Cut) the remaining elements of S (if any, i.e. if $k + n > j$) with the union of all parameters from j premisses are in χ, S_{j+1}. This way we demonstrate the provability of (a) on the basis of χ, S.

(b) is derivable from (a) by one application of (Cut) in the following way:

$$\dfrac{\Pi \Rightarrow \Sigma, \chi \qquad \dfrac{S_1, ..., S_j}{\chi, S_{j+1}} \, (a)}{\Pi, S_{j+1}, \Sigma} \, (Cut)$$

Eventually, we derive χ, S by means of (b) from axiom $\chi \Rightarrow \chi$ and j axioms of the form $\varphi_l \Rightarrow \varphi_l, \psi_m \Rightarrow \psi_m$, for $l \leq k, m \leq n$.

If we add χ to the succedent, i.e. we consider a sequent S, χ, the procedure is symmetric: (a) either addition of χ to the succedent of the conclusion or (b) addition of the extra premiss $\chi, \Pi \Rightarrow \Sigma$. □

Exercise 2.6. *Write down all 15 schemata of rules which may be generated from a sequent of the form $\varphi_1, \varphi_2 \Rightarrow \psi_1, \psi_2$.*

2.3   Syntactical Invertibility

In chapter 1 we have shown that invertibility of rules is a very important feature, at least in K. In case of LK the situation is worse. One may easily check that in contrast to K not all rules of LK are invertible. We can prove:

Lemma 2.5. *In LK the following rules are not invertible:* $(W \Rightarrow), (\Rightarrow W),$ $(\rightarrow\Rightarrow), (\wedge \Rightarrow), (\Rightarrow \vee), (Cut)$.

PROOF: We demonstrate two cases:

$(\wedge \Rightarrow)$: To show that $\models \varphi \wedge \psi, \Gamma \Rightarrow \Delta$ but $\not\models \varphi, \Gamma \Rightarrow \Delta$, it is enough to consider a $\mathfrak{M}$ falsifying $\varphi, \Gamma \Rightarrow \Delta$ and assume that $\mathfrak{M} \not\models \psi$. But then $\mathfrak{M} \not\models \varphi \wedge \psi$ and $\mathfrak{M} \models \varphi \wedge \varphi, \Gamma \Rightarrow \Delta$. Analogously for $\psi, \Gamma \Rightarrow \Delta$.

$(\rightarrow\Rightarrow)$: Similarly let $\mathfrak{M} \not\models \Gamma \Rightarrow \Delta, \varphi$, hence $\mathfrak{M} \models \Gamma$ and $\mathfrak{M} \not\models \Delta, \varphi$. It is sufficient to assume that some element of Π is false in $\mathfrak{M}$ or some element of Σ is true for having $\mathfrak{M} \models \varphi \rightarrow \psi, \Gamma, \Pi \Rightarrow \Delta, \Sigma$. Analogously for the second premiss $\psi, \Pi \Rightarrow \Sigma$. □

Exercise 2.7. *Show the failure of invertibility for the remaining 4 rules.*

So far we were talking about invertibility of rules (or its failure) in the semantical sense. However, we can consider it also in the syntactical sense and, as we will see in the next chapter, this may be very helpful. Similarly as in the case of semantical invertibility we may consider it in a stronger or a weaker form. Thus syntactical invertibility may mean not only that the inversion of a primitive rule is an admissible rule (i.e. leading from provable sequents to provable sequents) but also its derivability (i.e. that the premisses are deducible from the conclusion). Clearly, if SC is proved adequate with respect to some semantical characterisation of the respective logic, then we may easily prove that every semantically invertible rule is also invertible in syntactical sense.

Claim 2.1. *If SC is an adequate formalisation of some semantically characterised logic, then a rule (r) is semantically invertible iff it is syntactically invertible.*

PROOF: We may show it even in the weaker sense of normality and derivability. If a rule (r) is not syntactically invertible, then for some premiss S_1 and the conclusion S_2 we have $S_2 \not\vdash S_1$. By strong completeness $S_2 \not\models S_1$, so there is a model satisfying S_2 but falsifying S_1 which means that (r) is not normal, hence not semantically invertible. If (r) is syntactically invertible, then $S_2 \vdash S_i$, for any premiss S_i. Hence, by soundness $S_2 \models S_i$ which means that (r) is semantically invertible. □

What about a direct syntactical proof of these results? Having cut as a primitive rule one may easily prove derivability of invertible rules in LK, or any other version of SC with cut, in a direct way. In fact, the first such proof was provided by Ketonen for SC where all logical rules are like in K. Here we will demonstrate this result for LK:

Theorem 2.1. *The inversion of every semantically invertible rule of LK is derivable in LK.*

PROOF: Inverses of (P) are trivial and inverses of (C) are derivable by (W) so we are left with only logical rules. We will demonstrate one example: $\Gamma \Rightarrow \Delta, \varphi \wedge \psi \vdash \Gamma \Rightarrow \Delta, \varphi$. First of all if $\varphi \wedge \psi \in \Delta$, then we obtain the result by $(\Rightarrow C)$ and $(\Rightarrow W)$ (note that this applies to all considered cases). Otherwise we proceed as follows:

$$\cfrac{\Gamma \Rightarrow \Delta, \varphi \wedge \psi \qquad \cfrac{\varphi \Rightarrow \varphi}{\varphi \wedge \psi \Rightarrow \varphi} \; (\wedge \Rightarrow)}{\Gamma \Rightarrow \Delta, \varphi} \; (Cut)$$

□

Exercise 2.8. *Prove derivability of the inverses of* $(\neg \Rightarrow), (\Rightarrow \neg), (\vee \Rightarrow)$ *and* $(\Rightarrow \rightarrow)$.

Moreover, one may show for LK that also in the case of primitive noninvertible rules we may prove the derivability of inverses of suitable rules from K. For example, although $(\wedge \Rightarrow)$ in LK is not invertible (it is a one-premiss A-rule) one may prove for LK that $\cfrac{\varphi \wedge \psi, \Gamma \Rightarrow \Delta}{\varphi, \psi, \Gamma \Rightarrow \Delta}$ is a derivable rule.

Theorem 2.2. *The following derivability results hold in LK:*

- $\varphi \wedge \psi, \Gamma \Rightarrow \Delta \vdash \varphi, \psi, \Gamma \Rightarrow \Delta$

- $\Gamma \Rightarrow \Delta, \varphi \vee \psi \vdash \Gamma \Rightarrow \Delta, \varphi, \psi$

- $\varphi \rightarrow \psi, \Gamma \Rightarrow \Delta \vdash \psi, \Gamma \Rightarrow \Delta$

- $\varphi \rightarrow \psi, \Gamma \Rightarrow \Delta \vdash \Gamma \Rightarrow \Delta, \varphi$

PROOF: We demonstrate the first case. Again if $\varphi \wedge \psi \in \Gamma$ we obtain the result by contraction and weakening, otherwise:

$$\cfrac{(\Rightarrow \wedge) \cfrac{(W \Rightarrow) \cfrac{\varphi \Rightarrow \varphi}{\varphi, \psi \Rightarrow \varphi} \qquad \cfrac{\psi \Rightarrow \psi}{\varphi, \psi \Rightarrow \psi} \; (W \Rightarrow)}{(Cut) \quad \varphi, \psi \Rightarrow \varphi \wedge \psi} \qquad \varphi \wedge \psi, \Gamma \Rightarrow \Delta}{\varphi, \psi, \Gamma \Rightarrow \Delta}$$

□

Exercise 2.9. *Prove the remaining cases. Prove lemma 2.1 and 2.2 by the above invertibility lemma.*

Notice that invertibility of A-cut is derivable in LK by applications of W.

Clearly, derivability implies admissibility (see subsection 1.4.2) but we can consider also the problem of direct proof of admissibility of inverted rules. As we will see in subsections 2.4.4, 2.5.2 and in chapter 3 such a proof may be useful as a preliminary result for proving admissibility of cut. Below we provide two proofs of essentially the same result. The first is based on the global transformation of a proof and was originally provided by Curry [56] for his version of LK with an additive rule for implication; a similar proof may be also found in Negri and von Plato [185] for SC with all rules multiplicative. Here we apply it to LK and prove in this way the admissibility of all rules which were shown derivable in subsection 1.4.3 (including inverses of K rules).

Theorem 2.3. *The following admissibility results hold in LK:*

1. *If* $\vdash \Gamma \Rightarrow \Delta, \neg\varphi$, *then* $\vdash \varphi, \Gamma \Rightarrow \Delta$

2. *If* $\vdash \neg\varphi, \Gamma \Rightarrow \Delta$, *then* $\vdash \Gamma \Rightarrow \Delta, \varphi$

3. *If* $\vdash \varphi \wedge \psi, \Gamma \Rightarrow \Delta$, *then* $\vdash \varphi, \psi, \Gamma \Rightarrow \Delta$

4. *If* $\vdash \Gamma \Rightarrow \Delta, \varphi \wedge \psi$, *then* $\vdash \Gamma \Rightarrow \Delta, \varphi$

5. *If* $\vdash \Gamma \Rightarrow \Delta, \varphi \wedge \psi$, *then* $\vdash \Gamma \Rightarrow \Delta, \psi$

6. *If* $\vdash \varphi \vee \psi, \Gamma \Rightarrow \Delta$, *then* $\vdash \varphi, \Gamma \Rightarrow \Delta$

7. *If* $\vdash \varphi \vee \psi, \Gamma \Rightarrow \Delta$, *then* $\vdash \psi, \Gamma \Rightarrow \Delta$

8. *If* $\vdash \Gamma \Rightarrow \Delta, \varphi \vee \psi$, *then* $\vdash \Gamma \Rightarrow \Delta, \varphi, \psi$

9. *If* $\vdash \varphi \rightarrow \psi, \Gamma \Rightarrow \Delta$, *then* $\vdash \Gamma \Rightarrow \Delta, \varphi$

10. *If* $\vdash \varphi \rightarrow \psi, \Gamma \Rightarrow \Delta$, *then* $\vdash \psi, \Gamma \Rightarrow \Delta$

11. *If* $\vdash \Gamma \Rightarrow \Delta, \varphi \rightarrow \psi$, *then* $\vdash \varphi, \Gamma \Rightarrow \Delta, \psi$

PROOF: We consider two cases:

Case 3: If $\vdash \varphi \wedge \psi, \Gamma \Rightarrow \Delta$, then $\vdash \varphi, \psi, \Gamma \Rightarrow \Delta$.

Consider a proof $\mathcal{D}$ of $\varphi \wedge \psi, \Gamma \Rightarrow \Delta$. We consider one by one in a root-first manner all sequents above $\varphi\wedge\psi, \Gamma \Rightarrow \Delta$ in $\mathcal{D}$ which contain the occurrence of $\varphi\wedge\psi$ in the antecedent and successively replace it with φ, ψ. Of course we track only these occurrences of $\varphi\wedge\psi$ which are ancestors of our formula under consideration; this means in particular that we leave intact such occurrences of $\varphi \wedge \psi$ which are cut-formulae or ancestors of cut-formula, and ancestors of other occurrences of $\varphi\wedge\psi$ (if any) in Γ. Due to the context independency of all rules all such replacements keep the application of rules correct. If $(C \Rightarrow)$ was applied to $\varphi \wedge \psi$ on some branch we do the same with both occurrences and obtain $\varphi, \psi, \varphi, \psi, \Gamma' \Rightarrow \Delta'$, then apply $(P \Rightarrow)$ and $(C \Rightarrow)$ twice. We repeat this procedure until we find all first occurrences of $\varphi \wedge \psi$ in $\mathcal{D}$ and we make the following changes:

1. If $\varphi \wedge \psi$ was introduced by $(W \Rightarrow)$ we apply $(W \Rightarrow)$ twice introducing φ, ψ instead.

2. If $\varphi \wedge \psi$ is active in the axiom, then it has the form $\varphi \wedge \psi \Rightarrow \varphi \wedge \psi$. We change it into $\varphi, \psi \Rightarrow \varphi \wedge \psi$ deduced from $\varphi \Rightarrow \varphi$ and $\psi \Rightarrow \psi$ by $(W \Rightarrow)$ and $(\Rightarrow \wedge)$.

3. If $\varphi \wedge \psi$ was deduced by $(\wedge \Rightarrow)$ from, say φ, we instead apply $(W \Rightarrow)$ and add ψ.

This way we obtain a proof $\mathcal{D}'$ of $\varphi, \psi, \Gamma \Rightarrow \Delta$.

We proceed in such a way in all other cases. Let us consider

Case 10: If $\vdash \varphi \rightarrow \psi, \Gamma \Rightarrow \Delta$, then $\vdash \psi, \Gamma \Rightarrow \Delta$.

This time we replace all occurrences of $\varphi \rightarrow \psi$ which are ancestors of the formula under consideration with ψ. The way of proceeding with contraction on $\varphi \rightarrow \psi$ and introduction of it by $(W \Rightarrow)$ are the same. In the case of an axiom $\varphi \rightarrow \psi \Rightarrow \varphi \rightarrow \psi$ we add a proof of $\psi \Rightarrow \varphi \rightarrow \psi$ from $\psi \Rightarrow \psi$ by $(W \Rightarrow)$ and $(\Rightarrow \rightarrow)$. Finally if $\varphi \rightarrow \psi$ was introduced by $(\rightarrow \Rightarrow)$ then we have $\varphi \rightarrow \psi, \Pi, \Lambda \Rightarrow \Sigma, \Theta$ deduced from $\Pi \Rightarrow \Sigma, \varphi$ and $\psi, \Lambda \Rightarrow \Theta$. In this case we delete the left premiss (and the whole proof of it) and obtain $\psi, \Pi, \Lambda \Rightarrow \Sigma, \Theta$ from the right premiss by successive applications of W and P. $\qquad \square$

Exercise 2.10. *Prove the remaining cases, including inverses of K rules.*

The second proof is based on local transformations and was originally provided by Schütte [232] for some related form of the calculus with one-sided sequents, which are finite sets of formulae. It goes by induction on the height and in essence it is similar to a proof of Dragalin [64] of a somewhat stronger result presented in the next chapter. The main difference is that in case of LK we are not able to prove height-preserving admissibility whereas Dragalin's proof preserves height. In order to deal with contraction we formulate the result in a somewhat more general manner. Let Γ_φ denote Γ with all occurrences of φ (if any) deleted.

Theorem 2.4. *The following admissibility results hold in LK:*

    1. If $\vdash \Gamma \Rightarrow \Delta$, then $\vdash \varphi, \Gamma \Rightarrow \Delta_{\neg\varphi}$

    2. If $\vdash \Gamma \Rightarrow \Delta$, then $\vdash \Gamma_{\neg\varphi} \Rightarrow \Delta, \varphi$

    3. If $\vdash \Gamma \Rightarrow \Delta$, then $\vdash \varphi, \psi, \Gamma_{\varphi\wedge\psi} \Rightarrow \Delta$

    4. If $\vdash \Gamma \Rightarrow \Delta$, then $\vdash \Gamma \Rightarrow \Delta_{\varphi\wedge\psi}, \varphi$

    5. If $\vdash \Gamma \Rightarrow \Delta$, then $\vdash \Gamma \Rightarrow \Delta_{\varphi\wedge\psi}, \psi$

6. *If* $\vdash \Gamma \Rightarrow \Delta$, *then* $\vdash \varphi, \Gamma_{\varphi \lor \psi} \Rightarrow \Delta$

7. *If* $\vdash \Gamma \Rightarrow \Delta$, *then* $\vdash \psi, \Gamma_{\varphi \lor \psi} \Rightarrow \Delta$

8. *If* $\vdash \Gamma \Rightarrow \Delta$, *then* $\vdash \Gamma \Rightarrow \Delta_{\varphi \lor \psi}, \varphi, \psi$

9. *If* $\vdash \Gamma \Rightarrow \Delta$, *then* $\vdash \Gamma_{\varphi \to \psi} \Rightarrow \Delta, \varphi$

10. *If* $\vdash \Gamma \Rightarrow \Delta$, *then* $\vdash \psi, \Gamma_{\varphi \to \psi} \Rightarrow \Delta$

11. *If* $\vdash \Gamma \Rightarrow \Delta$, *then* $\vdash \varphi, \Gamma \Rightarrow \Delta_{\varphi \to \psi}, \psi$

PROOF: We begin with two remarks: 1) In all cases if there is no occurrence of a formula which is supposed to be deleted, then the results follow trivially by weakening. Hence for the rest of a proof we assume that there is at least one occurrence of the deleted formula in the antecedent or succedent of sequent $\Gamma \Rightarrow \Delta$ and accordingly we will write it explicitly with superscript denoting the number of occurrences. 2) Each case corresponds strictly to respective case in theorem 2.3 only if there is exactly one occurrence of a compound formula in Γ (Δ) and this one occurrence is deleted in the resulting sequent. If there are more occurrences, we must restore them by applications of W to get a case which really expresses admissibility.

This time we make an induction on the height of a proof. Consider again a proof $\mathcal{D}$ of $\varphi \land \psi, \Gamma \Rightarrow \Delta$. In the basis we have two subcases: 1. If an axiom is of the form $\varphi \land \psi \Rightarrow \varphi \land \psi$, then (similarly as in the previous proof) we change it into a proof of $\varphi, \psi \Rightarrow \varphi \land \psi$ from $\varphi \Rightarrow \varphi$ and $\psi \Rightarrow \psi$. 2. If we have $\chi \Rightarrow \chi$ with χ different from $\varphi \land \psi$, then we just apply $(W \Rightarrow)$ twice.

For the induction step we assume that the claim holds for any $\varphi \land \psi^k, \Gamma' \Rightarrow \Delta'$ with a proof of height lower than n and consider a case with $\mathcal{D}$ having the height n. Again the context independence of all rules does the work if all occurrences of $\varphi \land \psi$ are parametric. For example, suppose we have our sequent deduced by some two-premiss rule—let it be $(\Rightarrow \land)$:

$$\frac{\varphi \land \psi^k, \Gamma \Rightarrow \Delta', \delta \qquad \varphi \land \psi^k, \Gamma \Rightarrow \Delta', \gamma}{\varphi \land \psi^k, \Gamma \Rightarrow \Delta}$$

where Δ is $\Delta', \delta \land \gamma$.

By the induction hypothesis the claim holds for both premisses and we obtain:

$$\frac{\varphi, \psi, \Gamma_{\varphi \land \psi} \Rightarrow \Delta', \delta \qquad \varphi, \psi, \Gamma_{\varphi \land \psi} \Rightarrow \Delta', \gamma}{\varphi, \psi, \Gamma_{\varphi \land \psi} \Rightarrow \Delta}$$

If one occurrence of $\varphi \land \psi$ is a side formula of a rule application leading to the conclusion we must restore it by W to obtain the result. Consider the following application of $(\lor \Rightarrow)$:

$$\frac{\varphi \land \psi^k, \Gamma \Rightarrow \Delta \qquad \chi, \varphi \land \psi^{k-1}, \Gamma \Rightarrow \Delta}{(\varphi \land \psi) \lor \chi, \varphi \land \psi^{k-1}, \Gamma \Rightarrow \Delta}$$

By the induction hypothesis the claim holds for both premises and we obtain:

$$(W \Rightarrow) \frac{\dfrac{\varphi, \psi, \Gamma_{\varphi \wedge \psi} \Rightarrow \Delta}{\varphi \wedge \psi, \varphi, \psi, \Gamma_{\varphi \wedge \psi} \Rightarrow \Delta} \qquad \dfrac{\varphi, \psi, \chi, \Gamma_{\varphi \wedge \psi} \Rightarrow \Delta}{\chi, \varphi, \psi, \Gamma_{\varphi \wedge \psi} \Rightarrow \Delta} (P \Rightarrow)}{(\varphi \wedge \psi) \vee \chi, \varphi, \psi, \Gamma_{\varphi \wedge \psi} \Rightarrow \Delta}$$

In case of $\varphi \wedge \psi$ introduced by $(W \Rightarrow)$ or $(\wedge \Rightarrow)$ we apply $(W \Rightarrow)$ twice or once. Finally, in case the last rule was $(C \Rightarrow)$ on our $\varphi \wedge \psi$ the result is already present in the premiss by the induction hypothesis. $\qquad \square$

Exercise 2.11. *Prove the remaining cases.*

Remark 2.1. Note that a difference between the two proofs is not only connected with the fact that the former is based on global transformations whereas the latter is based on local ones. Notice also that in the first case we are dealing only with those occurrences of a formula in transformed proof which are ancestors of its occurrences in the last sequent of the tree. In consequence only necessary replacements are made in the analysed proof. On the other hand, the second proof is based on brute force; we delete all occurrences of a formula in question which causes more global transformations, in particular a lot of applications of W and P. However the kind of transformations we have applied is not connected with global/local manner of proof. Notice also that both lemmata may also be proved for LK without cut.

Exercise 2.12. *Prove theorem 2.3 by induction on the height of proofs and theorem 2.4 by global transformation.*

2.4 Local Proofs of Cut Elimination

In the previous chapter we have shown that cut is an admissible rule of K. Gentzen, on the contrary, introduced cut as a primitive rule of LK and demonstrated how to eliminate it from every proof. His proof, although essentially similar to the proof exhibited in subsection 1.8.2 is considerably more complicated. It is not surprising that many logicians were trying to simplify his proof. We present a number of proofs in this chapter (in particular Gentzen's original proof) which apply to LK or some similar SC with primitive structural rules including cut. The greater complexity of these proofs follows from the fact that they must deal somehow with the applications of structural rules. In particular, two things must be taken into consideration:

1. How to take care of the number of cut applications in analysed proofs?

2. How to deal with applications of contraction?

Concerning the former question there are two strategies in proofs of eliminability:

1. Gentzen [95]: eliminability of the topmost cut (induction on the number of cut applications);

2. Tait [257]: eliminability of the maximal cut (induction on the cut-degree).

The last notion, i.e. the cut-degree of the proof, is the maximal complexity of the cut-formulae present in the proof. The strategy of Gentzen is simpler but the strategy of Tait allows for proving additional complexity results concerning cut-free proofs. Note that Tait's strategy was originally introduced for an infinitary version of SC and in this case the strategy of Gentzen simply does not work.

As for the latter question we start with an explanation of why contraction is troublesome in proving cut elimination, and how Gentzen solved the problem by introduction of a generalised form of cut called mix (or multicut) which allows for the deletion of all occurrences of the same formula in one step.

Example 2.4. *Here is an example of an application of mix.*

$$(Mix) \quad \frac{r \wedge t, q \Rightarrow p, \neg s, q \vee r, \neg s, t \qquad \neg s, q \rightarrow r, \neg s, \neg s \Rightarrow \neg q, t}{r \wedge t, q, q \rightarrow r \Rightarrow p, q \vee r, t, \neg q, t}$$

In the example two occurrences of $\neg s$ in the succedent of the left premiss and three occurrences in the antecedent of the right premiss were deleted in one step. Note that the exact place of the deleted occurrences of a formula does not matter. We will show the application of mix either by the following schema:

$$(Mix) \quad \frac{\Gamma \Rightarrow \Delta[\varphi] \qquad \Pi[\varphi] \Rightarrow \Sigma}{\Gamma, \Pi_\varphi \Rightarrow \Delta_\varphi, \Sigma}$$

where $\Pi[\varphi]$ denotes that φ has at least one occurrence in Π, and Π_φ denotes the result of deleting all occurrences (if any) of φ from Π. Sometimes we will be using a notation like this:

$$(Mix) \quad \frac{\Gamma \Rightarrow \Delta, \varphi^i \qquad \varphi^k, \Pi \Rightarrow \Sigma}{\Gamma, \Pi \Rightarrow \Delta, \Sigma}$$

where φ^k means that φ occurs k times and we assume that Π and Δ do not contain any occurrences of φ. Note that this is just a schema and it does not mean that we have all occurrences of the cut-formula in the left or the rightmost position—they may be placed everywhere in the list.

In order to understand why mix is proposed instead of cut consider the following schema:

$$\frac{\Gamma \Rightarrow \Delta, \varphi \qquad \dfrac{\varphi, \varphi, \Pi \Rightarrow \Sigma}{\varphi, \Pi \Rightarrow \Sigma} (C \Rightarrow)}{\Gamma, \Pi \Rightarrow \Delta, \Sigma} (Cut)$$

after reduction of the height we get:

$$\frac{\Gamma \Rightarrow \Delta, \varphi \qquad \dfrac{\Gamma \Rightarrow \Delta, \varphi \qquad \varphi, \varphi, \Pi \Rightarrow \Sigma}{\varphi, \Gamma, \Pi \Rightarrow \Delta, \Sigma} (Cut)}{\dfrac{\Gamma, \Gamma, \Pi \Rightarrow \Delta, \Delta, \Sigma}{\Gamma, \Pi, \Rightarrow \Delta, \Sigma} (C)(P)} (Cut)$$

Notice that only the first cut has lower height so the induction hypothesis does not apply to the second cut.

In what follows we will use LK' to denote LK with mix instead of cut. It is easy to show that LK with mix (LK') is equivalent to LK, i.e. $\vdash_{LK} S$ iff $\vdash_{LK'} S$. It is sufficient to show that cut is derivable in LK' and mix in LK. Any application of cut:

$$\frac{\Gamma \Rightarrow \Delta, \varphi \qquad \varphi, \Pi \Rightarrow \Sigma}{\Gamma, \Pi \Rightarrow \Delta, \Sigma} \ (Cut)$$

is replaced with:

$$\cfrac{\cfrac{\Gamma \Rightarrow \Delta, \varphi \qquad \varphi, \Pi \Rightarrow \Sigma}{\Gamma, \Pi_\varphi \Rightarrow \Delta_\varphi, \Sigma} \ (Mix)}{\Gamma, \Pi \Rightarrow \Delta, \Sigma} \ (W)(P)$$

where W is possibly used to restore some other occurrences of φ in Π, Δ. On the contrary, any application of mix:

$$\frac{\Gamma \Rightarrow \Delta, \varphi^i \qquad \varphi^k, \Pi \Rightarrow \Sigma}{\Gamma, \Pi \Rightarrow \Delta, \Sigma} \ (Mix)$$

is replaced with:

$$(C), (P) \cfrac{\cfrac{\Gamma \Rightarrow \Delta, \varphi^i}{\Gamma \Rightarrow \Delta, \varphi} \qquad \cfrac{\varphi^k, \Pi \Rightarrow \Sigma}{\varphi, \Pi \Rightarrow \Sigma} \ (C), (P)}{\Gamma, \Pi \Rightarrow \Delta, \Sigma} \ (Cut)$$

However, introduction of mix forced Gentzen to apply a more complicated measure for induction; cut-rank instead of cut-height. Consider the following:

$$\cfrac{\Gamma \Rightarrow \varphi \wedge \psi \qquad \cfrac{\cfrac{\varphi, \psi \Rightarrow \Sigma}{\varphi \wedge \psi, \psi \Rightarrow \Sigma} \ (\wedge \Rightarrow)}{\varphi \wedge \psi, \varphi \wedge \psi \Rightarrow \Sigma} \ (P \Rightarrow)(\wedge \Rightarrow)}{\Gamma \Rightarrow \Sigma} \ (Mix)$$

After the standard transformation we get:

$$\cfrac{\Gamma \Rightarrow \varphi \wedge \psi \qquad \cfrac{\cfrac{\Gamma \Rightarrow \varphi \wedge \psi \qquad \cfrac{\varphi, \psi \Rightarrow \Sigma}{\varphi \wedge \psi, \psi \Rightarrow \Sigma} \ (\wedge \Rightarrow)}{\psi, \Gamma \Rightarrow \Sigma} \ (Mix)}{\varphi \wedge \psi, \Gamma \Rightarrow \Sigma} \ (\wedge \Rightarrow)}{\cfrac{\Gamma, \Gamma_{\varphi \wedge \psi} \Rightarrow \Sigma}{\Gamma \Rightarrow \Sigma} \ (C)(P)} \ (Mix)$$

where the height of the proof of the right premiss of the second mix is even bigger than before. But note that in the second mix the cut-formula is introduced for the first time. So why not count occurrences of cut-formula above (called rank by Gentzen) instead of the height? It is natural but has a drawback since a definition of the rank of a mix is more complicated. We can put it as follows:

Definition 2.1. Let $\mathcal{D}_l$ denote a proof of the left premiss of an application of mix and $\mathcal{B}$ denote any branch terminating in this left premiss. Let $l\mathcal{B}$ be the number of sequents on this branch containing a cut-formula in the succedent counting upwards from the left premiss of mix.

The rank of $\mathcal{D}_l$ (or l-rank) is the maximum of $l\mathcal{B}$, where $\mathcal{B}$ is any branch terminating in the left premiss of mix.

The rank of $\mathcal{D}_r$ (or r-rank) is defined analogously but for the right premiss of an application of mix and counting the occurrences of cut-formulae in the antecedent.

The rank of an application of mix is the sum of l-rank and r-rank of that mix.

Note that the minimal rank of a mix is 2, in the case when both cut-formulae are principal and there are no other occurrences of cut-formula in the succedent (resp. antecedent) of the left (right) premiss.

2.4.1   Gentzen's Proof

The original version of the proof goes by triple induction:

1. the number of applications of mix in a proof;

2. the complexity of cut-formula;

3. the rank of mix.

Since it is a triple induction requiring a number of combinatorial moves it is worth to analyse its structure in detail. This way we can avoid a situation where the number of cases obscures the purpose of its consideration. What is even more important, we can assure ourselves that all cases were dealt with and that there is no flaw in the proof.

Here is a general schema of Gentzen's proof:

I. Induction on the number of applications of mix
1.1. Basis: proofs with one application of mix; eliminability proved by:
    II. Induction on the complexity of cut-formula
    2.1. Basis: mix on atomic formulae is eliminable; proved by:
        III. Induction on the rank of mix
        3.1. Basis: mix with rank $= 2$ on atomic cut-formula is eliminable
        3.2. Ind. step: if mix on an atomic cut-formula has rank $n > 2$,
        then it may be substituted with mix of lesser rank.
        Conclusion of 3.1., 3.2: mix on atomic cut-formulae is eliminable.
    2.2. Ind. step: mix on formulae of complexity n is eliminable,
    if it is eliminable on smaller formulae; proved (again) by:
        III. Induction on the rank of mix
        4.1. Basis: mix with rank $= 2$ on a complex cut-formula may be
        substituted with mix on a cut-formula of lesser complexity.

4.2. Ind. step: if mix on a complex cut-formula has rank $n > 2$, then it may be substituted with mix of lesser rank.

Conclusion of 4.1., 4.2: mix on complex cut-formulae may be substituted with mix on cut-formulae of lesser complexity.

Conclusion of 2.1., 2.2: mix on complex cut-formulae is eliminable from proofs with one applications of mix.

1.2. Ind. step: If mix is eliminable in proofs with n applications, then it is eliminable in proofs with $n + 1$ applications.

There are two points concerning the proof structure worth underlining. Note that induction III is applied twice as a subsidiary induction for both, the proof of the basis and of the induction step of induction II. This strategy is generally followed in other proofs of cut elimination, although sometimes with induction III applied on the height instead of the rank (compare the proof in section 1.8). Since the differences between these two embedded applications of induction III are not significant, it is usually extracted as just an induction on the rank (or height) without paying attention to the assumed complexity of the cut-formula. Below we will follow this custom. Note however, that in contrast to the proof from section 1.8, we cannot change (and simplify) the structure of the proof by changing the order of inductions, in the way described in remark 1.6. This is not possible because a reduction of rank cannot change the complexity of the cut-formula but a reduction of the complexity may increase the rank. Hence if the induction on the complexity is carried as subsidiary, we cannot correctly refer to the inductive hypothesis of the induction on the rank. This is not a problem in case of the height since the reduction of complexity cannot increase the height.

The second thing is that the proper proof of cut elimination (consisting of induction II and III) is included as a justification only for the basis of induction I. Consequently it may be treated in isolation as the proof of admissibility of cut since it is assumed that proofs of both premises are cut-free. Hence we can extract it and first take for granted:

Theorem 2.5. (1-Mix elimination). *Any $\mathcal{D}$ with exactly one application of mix as the last rule may be transformed into a cut-free $\mathcal{D}'$.*

This is sufficient for proving the lemma corresponding to the induction step of induction I:

Lemma 2.6. (Mix number reduction). *If $\mathcal{D}$ has any application of mix, then it may be transformed into a cut-free $\mathcal{D}'$.*

PROOF: By induction on the number of mix applications in $\mathcal{D}$:

Basis: $\mathcal{D}$ with one application of mix may be transformed into a cut-free proof—it holds by 1-Mix elimination theorem.

Induction step: If any $\mathcal{D}$ with n applications of mix may be transformed into a cut-free proof, then $\mathcal{D}$ with $n + 1$ applications of mix may be transformed into a cut-free proof. We can prove it as follows. Take the highest leftmost application of mix. The part of the proof ending with it may be replaced by a cut-free proof by 1-Mix elimination theorem. The resulting proof has n applications of mix, hence—by the induction hypothesis—it may be transformed into cut-free proof.   □

This is how Gentzen's strategy for dealing with the multiplicity of cuts works. It remains to prove 1-Mix elimination theorem:

PROOF: We consider first induction III (in both applications, i.e. as subsidiary to the basis or to the induction step of induction II).
The basis: rank = 2.

Point 3.1. The cut-formula is atomic, hence each premiss is either an axiom or a sequent where the cut-formula was introduced by weakening. In each of the 4 possible combinations mix (which is cut indeed) is trivially eliminable. For example, a proof

$$\cfrac{\varphi \Rightarrow \varphi \qquad \cfrac{\Gamma \Rightarrow \Delta}{\varphi, \Gamma \Rightarrow \Delta}\,(W \Rightarrow)}{\varphi, \Gamma \Rightarrow \Delta}\,(Mix)$$

is replaced with a proof of $\varphi, \Gamma \Rightarrow \Delta$ in the right premiss.

In case of applications of W in both premisses we have:

$$\begin{array}{c}(\Rightarrow W)\\ (Cut)\end{array}\cfrac{\cfrac{\Gamma \Rightarrow \Delta}{\Gamma \Rightarrow \Delta, \varphi} \qquad \cfrac{\Pi \Rightarrow \Sigma}{\varphi, \Pi \Rightarrow \Sigma}\,(W \Rightarrow)}{\Gamma, \Pi \Rightarrow \Delta, \Sigma}$$

which is replaced with:

$$(W)(P)\cfrac{\Gamma \Rightarrow \Delta}{\Gamma, \Pi \Rightarrow \Delta, \Sigma}$$

Point 4.1. Any premiss that is an axiom or that has the cut-formula introduced by W is eliminated as in point 3.1. So only 4 cases remain where the cut-formula is introduced as principal in both premisses by a suitable logical rule. In each case we show that a proof with the mix under consideration may be replaced by a proof with mix(es) on shorter cut-formulae.

Case of cut-formula $:= \varphi \wedge \psi$. A part of a proof:

$$\begin{array}{c}(\Rightarrow \wedge)\\ (Mix)\end{array}\cfrac{\cfrac{\Gamma \Rightarrow \Delta, \varphi \qquad \Gamma \Rightarrow \Delta, \psi}{\Gamma \Rightarrow \Delta, \varphi \wedge \psi} \qquad \cfrac{\varphi, \Pi \Rightarrow \Sigma}{\varphi \wedge \psi, \Pi \Rightarrow \Sigma}\,(\wedge \Rightarrow)}{\Gamma, \Pi \Rightarrow \Delta, \Sigma}$$

is replaced with:

$$\begin{array}{c}(Mix)\\ (W)(P)\end{array}\cfrac{\cfrac{\Gamma \Rightarrow \Delta, \varphi \qquad \varphi, \Pi \Rightarrow \Sigma}{\Gamma, \Pi_\varphi \Rightarrow \Delta_\varphi, \Sigma}}{\Gamma, \Pi \Rightarrow \Delta, \Sigma}$$

Note that since the rank $= 2$, $\varphi \wedge \psi$ occurs neither in Δ nor in Π and analogously for $(\wedge \Rightarrow)$ with ψ as side formula.

Exercise 2.13. *Prove reduction steps for the cases where the cut-formula is a negation and a disjunction.*

Case of cut-formula $:= \varphi \rightarrow \psi$. A proof:

$$(\Rightarrow\rightarrow)\ \dfrac{\dfrac{\varphi, \Gamma \Rightarrow \Delta, \psi}{\Gamma \Rightarrow \Delta, \varphi \rightarrow \psi} \quad \dfrac{\Pi \Rightarrow \Sigma, \varphi \quad \psi, \Lambda \Rightarrow \Xi}{\varphi \rightarrow \psi, \Pi, \Lambda \Rightarrow \Sigma, \Xi}\ (\rightarrow\Rightarrow)}{\Gamma, \Pi, \Lambda \Rightarrow \Delta, \Sigma, \Xi}\ (Mix)$$

is replaced with:

$$\dfrac{\Pi \Rightarrow \Sigma, \varphi \quad \dfrac{\dfrac{\varphi, \Gamma \Rightarrow \Delta, \psi \quad \psi, \Lambda \Rightarrow \Xi}{\varphi, \Gamma, \Lambda_\psi \Rightarrow \Delta_\psi, \Xi}\ (Mix)}{}}{\dfrac{\Pi, \Gamma_\varphi, \Lambda_{\psi,\varphi} \Rightarrow \Sigma_\varphi, \Delta_\psi, \Xi}{\Gamma, \Pi, \Lambda \Rightarrow \Delta, \Sigma, \Xi}\ (W)(P)}\ (Mix)$$

where $\Lambda_{\psi,\varphi}$ denotes Λ with all occurrences (if any) of ψ and φ deleted.

Note that we obtain a part of a proof with two applications of Mix. This may seem problematic since we are proving a lemma where only one mix, namely the last applied rule is eliminable. But the highest one is eliminable by the induction hypothesis of induction II. Hence we have a proof of $\varphi, \Gamma, \Lambda_\psi \Rightarrow \Delta_\psi, \Xi$ with no application of mix and then the lower one is the sole application of mix and also satisfies the induction hypothesis of induction II and as such is eliminable.

Induction step of induction III (points 3.2 and 4.2).

This part of a proof requires much more work. As the induction hypothesis we assume that mix of rank $< n$ is eliminable. We must prove that every mix of rank $= n$ may be replaced with applications of mix on the same cut-formula but of lower rank. In principle we have two parts:

A. We assume that r-rank > 1 and consider all possible cases leading to the proof of the right premiss of mix.

B. We assume that l-rank > 1 and consider all possible cases leading to the proof of the left premiss of mix.

In each part we must distinguish between situations where the cut-formula is parametric, side or principal. In fact, for proving point 3.2. it may be (as it is atomic) only parametric or a side formula, but for proving point 4.2 all cases must be considered. Moreover, in each situation we must consider all rules as they lead to the proof of the respective premiss. We prove A.

A1. Let the cut-formula be parametric.

A11. The right premiss is deduced by a rule operating only on the succedent.

A111. One-premiss rules:

A1111. $(\Rightarrow \vee)$. A part of a proof:

$$\cfrac{\Gamma \Rightarrow \Delta[\varphi] \qquad \cfrac{\Pi[\varphi] \Rightarrow \Sigma, \psi}{\Pi[\varphi] \Rightarrow \Sigma, \psi \vee \chi}\ (\Rightarrow \vee)}{\Gamma, \Pi_\varphi \Rightarrow \Delta_\varphi, \Sigma, \psi \vee \chi}\ (Mix)$$

is replaced with:

$$\cfrac{\cfrac{\Gamma \Rightarrow \Delta[\varphi] \qquad \Pi[\varphi] \Rightarrow \Sigma, \psi}{\Gamma, \Pi_\varphi \Rightarrow \Delta_\varphi, \Sigma, \psi}\ (Mix)}{\Gamma, \Pi_\varphi \Rightarrow \Delta_\varphi, \Sigma, \psi \vee \chi}\ (\Rightarrow \vee)$$

Exercise 2.14. *Prove the cases of* $(\Rightarrow W)$, $(\Rightarrow C)$, $(\Rightarrow P)$.

A112. Two-premiss rules—only $(\Rightarrow \wedge)$. A part of a proof:

$$\cfrac{\Gamma \Rightarrow \Delta[\varphi] \qquad \cfrac{\Pi[\varphi] \Rightarrow \Sigma, \psi \qquad \Pi[\varphi] \Rightarrow \Sigma, \chi}{\Pi[\varphi] \Rightarrow \Sigma, \psi \wedge \chi}\ (\Rightarrow \wedge)}{\Gamma, \Pi_\varphi \Rightarrow \Delta_\varphi, \Sigma, \psi \wedge \chi}\ (Mix)$$

is replaced with:

$$(Mix)\ \cfrac{\cfrac{\Gamma \Rightarrow \Delta[\varphi] \qquad \Pi[\varphi] \Rightarrow \Sigma, \psi}{\Gamma, \Pi_\varphi \Rightarrow \Delta_\varphi, \Sigma, \psi} \qquad \cfrac{\Gamma \Rightarrow \Delta[\varphi] \qquad \Pi[\varphi] \Rightarrow \Sigma, \chi}{\Gamma, \Pi_\varphi \Rightarrow \Delta_\varphi, \Sigma, \chi}\ (Mix)}{\Gamma, \Pi_\varphi \Rightarrow \Delta_\varphi, \Sigma, \psi \wedge \chi}\ (\Rightarrow \wedge)$$

where both applications of (Mix) are of lower rank hence eliminable by the induction hypothesis.

A12. Cases where the right premiss is deduced by a rule operating only on the antecedent.

A121. One-premiss rules.

A1211. Case of $(\wedge \Rightarrow)$. We have:

$$\cfrac{\Gamma \Rightarrow \Delta[\varphi] \qquad \cfrac{\psi, \Pi[\varphi] \Rightarrow \Sigma}{\psi \wedge \chi, \Pi[\varphi] \Rightarrow \Sigma}\ (\wedge \Rightarrow)}{\Gamma, \psi \wedge \chi, \Pi_\varphi \Rightarrow \Delta_\varphi, \Sigma}\ (Mix)$$

and change it into:

$$\cfrac{\cfrac{\Gamma \Rightarrow \Delta[\varphi] \qquad \psi, \Pi[\varphi] \Rightarrow \Sigma}{\Gamma, \psi, \Pi_\varphi \Rightarrow \Delta_\varphi, \Sigma}\ (Mix)}{\Gamma, \psi \wedge \chi, \Pi_\varphi \Rightarrow \Delta_\varphi, \Sigma}\ (\wedge \Rightarrow)(P \Rightarrow)$$

Exercise 2.15. *Prove it for* $(W \Rightarrow)$, $(C \Rightarrow)$, $(P \Rightarrow)$ *and also for the only case of a two-premiss rule*—$(\vee \Rightarrow)$.

A13. Cases where the right premiss is deduced by a rule operating on both sides of a sequent.

A131. One-premiss rules.

A1311. Case of $(\Rightarrow\rightarrow)$. We have:

$$\dfrac{\Gamma \Rightarrow \Delta[\varphi] \qquad \dfrac{\psi, \Pi[\varphi] \Rightarrow \Sigma, \chi}{\Pi[\varphi] \Rightarrow \Sigma, \psi \rightarrow \chi}\ (\Rightarrow\rightarrow)}{\Gamma, \Pi_\varphi \Rightarrow \Delta_\varphi, \Sigma, \psi \rightarrow \chi}\ (Mix)$$

and change it into:

$$\dfrac{\dfrac{\Gamma \Rightarrow \Delta[\varphi] \qquad \psi, \Pi[\varphi] \Rightarrow \Sigma, \chi}{\Gamma, \psi, \Pi_\varphi \Rightarrow \Delta_\varphi, \Sigma, \chi}\ (Mix)}{\Gamma, \Pi_\varphi \Rightarrow \Delta_\varphi, \Sigma, \psi \rightarrow \chi}\ (\Rightarrow\rightarrow)(P \Rightarrow)$$

Exercise 2.16. *Prove the cases of* $(\neg\Rightarrow)$, $(\Rightarrow\neg)$.

A132. Two-premiss rules—only $(\rightarrow\Rightarrow)$:

$$\dfrac{\Gamma \Rightarrow \Delta[\varphi] \qquad \dfrac{\Pi[\varphi] \Rightarrow \Sigma, \psi \qquad \chi, \Lambda[\varphi] \Rightarrow \Xi}{\psi \rightarrow \chi, \Pi[\varphi], \Lambda[\varphi] \Rightarrow \Sigma, \Xi}\ (\rightarrow\Rightarrow)}{\Gamma, \psi \rightarrow \chi, \Pi_\varphi, \Lambda_\varphi \Rightarrow \Delta_\varphi, \Sigma, \Xi}\ (Mix)$$

is replaced with:

$$\dfrac{(Mix)\ \dfrac{\Gamma \Rightarrow \Delta[\varphi] \qquad \Pi[\varphi] \Rightarrow \Sigma, \psi}{\Gamma, \Pi_\varphi \Rightarrow \Delta_\varphi, \Sigma, \psi} \qquad \dfrac{\dfrac{\dfrac{\Gamma \Rightarrow \Delta[\varphi] \qquad \chi, \Lambda[\varphi] \Rightarrow \Xi}{\Gamma, \chi, \Lambda_\varphi \Rightarrow \Delta_\varphi, \Xi}\ (Mix)}{\chi, \Gamma, \Lambda_\varphi \Rightarrow \Delta_\varphi, \Xi}\ (P \Rightarrow)}{\psi \rightarrow \chi, \Gamma, \Pi_\varphi, \Gamma, \Lambda_\varphi \Rightarrow \Delta_\varphi, \Sigma, \Delta_\varphi, \Xi}\ (\rightarrow\Rightarrow)}{\Gamma, \psi \rightarrow \chi, \Pi_\varphi, \Lambda_\varphi \Rightarrow \Delta_\varphi, \Sigma, \Xi}\ (P), (C)$$

where both applications of (Mix) are of lower rank.

A2. Cases where the cut-formula is a side formula of a rule applied to the right premiss.

A21. One-premiss rules.

A211. Case of $(\Rightarrow\rightarrow)$:

$$\dfrac{\Gamma \Rightarrow \Delta[\varphi] \qquad \dfrac{\varphi, \Pi[\varphi] \Rightarrow \Sigma, \psi}{\Pi[\varphi] \Rightarrow \Sigma, \varphi \rightarrow \psi}\ (\Rightarrow\rightarrow)}{\Gamma, \Pi_\varphi \Rightarrow \Delta_\varphi, \Sigma, \varphi \rightarrow \psi}\ (Mix)$$

is replaced with:

$$\dfrac{\dfrac{\dfrac{\Gamma \Rightarrow \Delta[\varphi] \qquad \varphi, \Pi[\varphi] \Rightarrow \Sigma, \psi}{\Gamma, \Pi_\varphi \Rightarrow \Delta_\varphi, \Sigma, \psi}\ (Mix)}{\varphi, \Gamma, \Pi_\varphi \Rightarrow \Delta_\varphi, \Sigma, \psi}\ (W \Rightarrow)}{\Gamma, \Pi_\varphi \Rightarrow \Delta_\varphi, \Sigma, \varphi \rightarrow \psi}\ (\Rightarrow\rightarrow)$$

Exercise 2.17. *Prove the cases of* $(\neg \Rightarrow)$ *and* $(\wedge \Rightarrow)$.

A22. Two-premiss rules.

A221. Case of $(\rightarrow\Rightarrow)$:

$$\cfrac{\Gamma \Rightarrow \Delta[\varphi] \qquad \cfrac{\Pi \Rightarrow \Sigma, \psi \qquad \varphi, \Lambda \Rightarrow \Xi}{\psi \rightarrow \varphi, \Pi, \Lambda \Rightarrow \Sigma, \Xi}(\rightarrow\Rightarrow)}{\Gamma, \psi \rightarrow \varphi, \Pi_\varphi, \Lambda_\varphi \Rightarrow \Delta_\varphi, \Sigma, \Xi}(Mix)$$

This time we have two possibilities:

A221a. if φ is not in Π ($\Pi_\varphi = \Pi$), then we introduce:

$$\cfrac{\Pi \Rightarrow \Sigma, \psi \qquad \cfrac{\cfrac{\Gamma \Rightarrow \Delta[\varphi] \qquad \varphi, \Lambda \Rightarrow \Xi}{\Gamma, \Lambda_\varphi, \Rightarrow \Delta_\varphi, \Xi}(Mix)}{\varphi, \Gamma, \Lambda_\varphi, \Rightarrow \Delta_\varphi, \Xi}(W \Rightarrow)}{\cfrac{\psi \rightarrow \varphi, \Pi, \Gamma, \Lambda_\varphi \Rightarrow \Sigma, \Delta_\varphi, \Xi}{\Gamma, \psi \rightarrow \varphi, \Pi_\varphi, \Lambda_\varphi \Rightarrow \Delta_\varphi, \Sigma, \Xi}(P)}(\rightarrow\Rightarrow)$$

A221b. if φ is in Π, then we introduce:

$$\cfrac{\cfrac{(Mix)\cfrac{\Gamma \Rightarrow \Delta[\varphi] \qquad \Pi[\varphi] \Rightarrow \Sigma, \psi}{\Gamma, \Pi_\varphi \Rightarrow \Delta_\varphi, \Sigma, \psi} \qquad \cfrac{\cfrac{\Gamma \Rightarrow \Delta[\varphi] \qquad \varphi, \Lambda \Rightarrow \Xi}{\Gamma, \Lambda_\varphi, \Rightarrow \Delta_\varphi, \Xi}(Mix)}{\varphi, \Gamma, \Lambda_\varphi, \Rightarrow \Delta, \Xi}(W \Rightarrow)}{\psi \rightarrow \varphi, \Gamma, \Pi_\varphi, \Gamma, \Lambda_\varphi \Rightarrow \Delta_\varphi, \Sigma, \Delta_\varphi, \Xi}(\rightarrow\Rightarrow)}{\Gamma, \psi \rightarrow \varphi, \Pi_\varphi, \Lambda_\varphi \Rightarrow \Delta_\varphi, \Sigma, \Xi}(P)(C)$$

Exercise 2.18. *Prove the case of* $(\vee \Rightarrow)$.

A3 Cases where the cut-formula is principal (note that in this case the cut-formula also must have parametric occurrences, by our assumption concerning r-rank).

A31. One-premiss rule.

A311. Case of $(\wedge \Rightarrow)$:

$$\cfrac{\Gamma \Rightarrow \Delta[\varphi \wedge \psi] \qquad \cfrac{\varphi, \Pi[\varphi \wedge \psi] \Rightarrow \Sigma}{\varphi \wedge \psi, \Pi[\varphi \wedge \psi] \Rightarrow \Sigma}(\wedge \Rightarrow)}{\Gamma, \Pi_{\varphi \wedge \psi} \Rightarrow \Delta_{\varphi \wedge \psi}, \Sigma}(Mix)$$

is replaced with:

$$\cfrac{\Gamma \Rightarrow \Delta[\varphi \wedge \psi] \qquad \cfrac{\cfrac{\cfrac{\Gamma \Rightarrow \Delta[\varphi \wedge \psi] \qquad \varphi, \Pi[\varphi \wedge \psi] \Rightarrow \Sigma}{\Gamma, \varphi, \Pi_{\varphi \wedge \psi} \Rightarrow \Delta_{\varphi \wedge \psi}, \Sigma}(Mix)}{\varphi, \Gamma, \Pi_{\varphi \wedge \psi} \Rightarrow \Delta_{\varphi \wedge \psi}, \Sigma}(P \Rightarrow)}{\varphi \wedge \psi, \Gamma, \Pi_{\varphi \wedge \psi} \Rightarrow \Delta_{\varphi \wedge \psi}, \Sigma}(\wedge \Rightarrow)}{\cfrac{\Gamma, \Gamma, \Pi_{\varphi \wedge \psi} \Rightarrow \Delta_{\varphi \wedge \psi}, \Delta_{\varphi \wedge \psi}, \Sigma}{\Gamma, \Pi_{\varphi \wedge \psi} \Rightarrow \Delta_{\varphi \wedge \psi}, \Sigma}(P), (C)}(Mix)$$

Notice that the second application of (Mix) has r-rank=1 since the cut-formula is for the first time in the right premiss; occurrences of the cut-formula above are not taken into consideration because the first application of (Mix) is first eliminated by the induction hypothesis. The same remark applies to every lower application of mix in other similar transformations.

Exercise 2.19. *Prove the case of* $(\neg \Rightarrow)$.

A32. Two-premiss rules.
A321. Case of $(\rightarrow \Rightarrow)$:

$$
\dfrac{\Gamma \Rightarrow \Delta[\varphi \rightarrow \psi] \qquad \dfrac{\Pi \Rightarrow \Sigma, \varphi \qquad \psi, \Lambda \Rightarrow \Xi}{\varphi \rightarrow \psi, \Pi, \Lambda \Rightarrow \Sigma, \Xi} \, (\rightarrow \Rightarrow)}{\Gamma, \Pi_{\varphi \rightarrow \psi}, \Lambda_{\varphi \rightarrow \psi} \Rightarrow \Delta_{\varphi \rightarrow \psi}, \Sigma, \Xi} \, (Mix)
$$

Let us notice that we do not know where exactly we may have occurrences of $\varphi \rightarrow \psi$ in Π and Λ. Accordingly we must consider three subcases:

A321a. $\varphi \rightarrow \psi$ is in Π and Λ; we introduce:

$$
\dfrac{\Gamma \Rightarrow \Delta[\varphi \rightarrow \psi] \qquad \dfrac{(Mix)\dfrac{\Gamma \Rightarrow \Delta[\varphi \rightarrow \psi] \qquad \Pi \Rightarrow \Sigma, \varphi}{\Gamma, \Pi_{\varphi \rightarrow \psi} \Rightarrow \Delta_{\varphi \rightarrow \psi}, \Sigma, \varphi} \qquad \dfrac{\dfrac{\Gamma \Rightarrow \Delta[\varphi \rightarrow \psi] \qquad \psi, \Lambda \Rightarrow \Xi}{\Gamma, \psi, \Lambda_{\varphi \rightarrow \psi} \Rightarrow \Delta_{\varphi \rightarrow \psi}, \Xi}(Mix)}{\psi, \Gamma, \Lambda_{\varphi \rightarrow \psi} \Rightarrow \Delta_{\varphi \rightarrow \psi}, \Xi}(P \Rightarrow)}{\varphi \rightarrow \psi, \Pi_{\varphi \rightarrow \psi}, \Gamma, \Lambda_{\varphi \rightarrow \psi} \Rightarrow \Delta_{\varphi \rightarrow \psi}, \Sigma, \Delta_{\varphi \rightarrow \psi}, \Xi}(\rightarrow \Rightarrow)}{\dfrac{\Gamma, \Gamma_{\varphi \rightarrow \psi}, \Pi_{\varphi \rightarrow \psi}, \Gamma_{\varphi \rightarrow \psi}, \Lambda_{\varphi \rightarrow \psi} \Rightarrow \Delta_{\varphi \rightarrow \psi}, \Delta_{\varphi \rightarrow \psi}, \Sigma, \Delta_{\varphi \rightarrow \psi}, \Xi}{\Gamma, \Pi_{\varphi \rightarrow \psi}, \Lambda_{\varphi \rightarrow \psi} \Rightarrow \Delta_{\varphi \rightarrow \psi}, \Sigma, \Xi}(P)(C)}(Mix)
$$

A321b. $\varphi \rightarrow \psi$ is in Π but not in Λ. We obtain:

$$
\dfrac{\Gamma \Rightarrow \Delta[\varphi \rightarrow \psi] \qquad \dfrac{(Mix)\dfrac{\Gamma \Rightarrow \Delta[\varphi \rightarrow \psi] \qquad \Pi[\varphi \rightarrow \psi] \Rightarrow \Sigma, \varphi}{\Gamma, \Pi_{\varphi \rightarrow \psi} \Rightarrow \Delta_{\varphi \rightarrow \psi}, \Sigma, \varphi} \qquad \psi, \Lambda \Rightarrow \Xi}{\varphi \rightarrow \psi, \Gamma, \Pi_{\varphi \rightarrow \psi}, \Lambda \Rightarrow \Delta_{\varphi \rightarrow \psi}, \Sigma, \Xi}(\rightarrow \Rightarrow)}{\dfrac{\Gamma, \Gamma_{\varphi \rightarrow \psi}, \Pi_{\varphi \rightarrow \psi}, \Lambda \Rightarrow \Delta_{\varphi \rightarrow \psi}, \Delta_{\varphi \rightarrow \psi}, \Sigma, \Xi}{\Gamma, \Pi_{\varphi \rightarrow \psi}, \Lambda \Rightarrow \Delta_{\varphi \rightarrow \psi}, \Sigma, \Xi}(P)(C)}(Mix)
$$

A321c. $\varphi \rightarrow \psi$ is in Λ but not in Π. We introduce:

$$
\dfrac{\Gamma \Rightarrow \Delta[\varphi \rightarrow \psi] \qquad \dfrac{\Pi \Rightarrow \Sigma, \varphi \qquad \dfrac{\dfrac{\Gamma \Rightarrow \Delta[\varphi \rightarrow \psi] \qquad \psi, \Lambda[\varphi \rightarrow \psi] \Rightarrow \Xi}{\Gamma, \psi, \Lambda_{\varphi \rightarrow \psi} \Rightarrow \Delta_{\varphi \rightarrow \psi}, \Xi}(Mix)}{\psi, \Gamma, \Lambda_{\varphi \rightarrow \psi} \Rightarrow \Delta_{\varphi \rightarrow \psi}, \Xi}(P \Rightarrow)}{\varphi \rightarrow \psi, \Pi, \Gamma, \Lambda_{\varphi \rightarrow \psi} \Rightarrow \Sigma, \Delta_{\varphi \rightarrow \psi}, \Xi}(\rightarrow \Rightarrow)}{\dfrac{\Gamma, \Pi, \Gamma_{\varphi \rightarrow \psi}, \Lambda_{\varphi \rightarrow \psi} \Rightarrow \Delta_{\varphi \rightarrow \psi}, \Sigma, \Delta_{\varphi \rightarrow \psi}, \Xi}{\Gamma, \Pi, \Lambda_{\varphi \rightarrow \psi} \Rightarrow \Delta_{\varphi \rightarrow \psi}, \Sigma, \Xi}(P)(C)}(Mix)
$$

$\square$

Exercise 2.20. *Prove the case of* $(\vee \Rightarrow)$; *no subcases here.*

    Prove part B.

Few remarks are in order. Gentzen's proof is brilliant but has a rather complicated structure. There are plenty of cases to consider; in particular when performing induction on the rank of compound formulae we must distinguish cases where the cut-formula is a principal-, side- or parameter-formula of the last inference step. These complexities suggest numerous later improvements. In particular we can consider two questions:

    1) can we use the height instead of rank in the presence of mix?

    2) can we provide a direct proof of cut (instead of mix) elimination?

The positive answers for these questions were provided by some logicians and we review some proofs of this kind below.

2.4.2   Cross-Cuts Technique

One of the proposals is due to Girard [99]. It is based on solutions applied by Tait [257] in his proof of cut elimination for some nonstandard (infinitary) SC for arithmetic. In Girard's proof [99] we can find three significant differences with Gentzen's proof:

1. a different strategy for dealing with the multiplicity of cuts due to Tait (elimination of the maximal cuts);

2. a subsidiary induction on the height not on the rank;

3. applications of the technique of cross-cuts.

    To deal with contraction Girard in fact also uses mix but its application is restricted to one lemma which just shows mix admissibility in LK (with cut), hence the proof generally goes for LK, not for LK'. However we can formulate it for LK' as well (see the remark 2.2). His original proof consists of the following steps:

1. Mix reduction lemma;

2. Cut-degree reduction lemma;

3. Cut-elimination theorem.

    Let us define the notions of cut-degree and proof degree:

1. Cut-degree is the complexity of the cut-formula φ (the number of occurrences of constants—see subsection 1.1.1—$d\varphi$

2. Proof-degree (dD) is the maximal cut-degree in D, i.e. the complexity of the most complex cut-formula.

Lemma 2.7. (Mix reduction). *If* $D_1 \vdash \Gamma \Rightarrow \Delta, \varphi^i$ *and* $D_2 \vdash \varphi^k, \Pi \Rightarrow \Sigma$ *and* $dD_1, dD_2 < d\varphi$, *then we can construct a proof* D *such that* $D \vdash \Gamma, \Pi \Rightarrow \Delta, \Sigma$ *and* $dD < d\varphi$

PROOF: It goes by induction on the sum of the heights of $\mathcal{D}_1$ and $\mathcal{D}_2$. Similarly as in Gentzen's proof we consider different cases. There are two differences:

1. One must consider the case of cut (of lower degree) as leading to the left (right) premiss of the mix that is being analysed. This does not cause any difficulties; we just permute cuts.

2. The cases where both cut-formulae are principal but not unique (i.e. some other occurrences are parametric) is treated in a different way, by the application of the cross-cuts technique which was introduced first by Martin-Löf [172] (see also Pfenning [196]). This corresponds to Gentzen's reduction of the rank where the cut-formula is parametric; the difference is that we take care of the second premiss as well. We consider some cases:

a) Let $\Gamma \Rightarrow \Delta, \varphi^i$ be an axiom $\psi \Rightarrow \psi$. If $\psi = \varphi$, then Γ is just φ and we derive $\Gamma, \Pi \Rightarrow \Delta, \Sigma$ from the second premiss by means of structural rules. Otherwise we derive $\Gamma, \Pi \Rightarrow \Delta, \Sigma$ from $\psi \Rightarrow \psi$.

b) $\Gamma \Rightarrow \Delta, \varphi^i$ was deduced by cut. Since $d\mathcal{D}_1 < d\varphi$, the cut-formula must be different from and shorter than φ. By the induction hypothesis we delete all occurrences of φ in both premisses of cut and the new conclusion of this cut application is $\Gamma \Rightarrow \Delta$. This yields a proof of $\Gamma, \Pi \Rightarrow \Delta, \Sigma$ by W and P with the same degree.

All other applications of structural and logical rules with φ parametric are transformed in the same way due to the context independence of all rules. Similarly, if φ is an active formula of C or W, we just delete all occurrences of φ in the premiss by the induction hypothesis and obtain the result by P and W.

Exercise 2.21. *Provide a few examples of the transformations sketched above.*

The original part of the proof is concerned with cases where both $\Gamma \Rightarrow \Delta, \varphi^i$ and $\varphi^k, \Pi \Rightarrow \Sigma$ were obtained by the application of logical rules with one of the occurrence of φ principal. The most complicated case is that of implication.

c) Let $\varphi := \psi \to \chi$. We have the following deductions as the last steps of $\mathcal{D}_1$ and $\mathcal{D}_2$:

$$\frac{\psi, \Gamma \Rightarrow \Delta, \varphi^{i-1}, \chi}{\Gamma \Rightarrow \Delta, \varphi^i} \ (\Rightarrow\to)$$

and

$$\frac{\varphi^m, \Pi' \Rightarrow \Sigma', \psi \qquad \chi, \varphi^n, \Pi'' \Rightarrow \Sigma''}{\varphi^k, \Pi \Rightarrow \Sigma} \ (\to\Rightarrow)$$

where $m + n = k - 1, \Pi$ is a concatenation of Π' and Π'' and similarly for Σ. By the induction hypothesis we obtain proofs of the following three sequents:

(a) $\psi, \Gamma, \Pi \Rightarrow \Delta, \Sigma, \chi$ – from $\psi, \Gamma \Rightarrow \Delta, \varphi^{i-1}, \chi$ and $\varphi^k, \Pi \Rightarrow \Sigma$;

(b) $\Gamma, \Pi' \Rightarrow \Delta, \Sigma', \psi$ – from $\Gamma \Rightarrow \Delta, \varphi^i$ and $\varphi^m, \Pi' \Rightarrow \Sigma', \psi$;

(c) $\chi, \Gamma, \Pi'' \Rightarrow \Delta, \Sigma''$ – from $\Gamma \Rightarrow \Delta, \varphi^i$ and $\chi, \varphi^n, \Pi'' \Rightarrow \Sigma''$.

Now we can combine them to derive the desired effect:

$$
(Cut) \cfrac{(Cut) \cfrac{\Gamma, \Pi' \Rightarrow \Delta, \Sigma', \psi \qquad \psi, \Gamma, \Pi \Rightarrow \Delta, \Sigma, \chi}{\Gamma, \Pi', \Gamma, \Pi \Rightarrow \Delta, \Sigma', \Delta, \Sigma, \chi} \qquad \chi, \Gamma, \Pi'' \Rightarrow \Delta, \Sigma''}{(C), (P) \cfrac{\Gamma, \Pi', \Gamma, \Pi, \Gamma, \Pi'', \Rightarrow \Delta, \Sigma', \Delta, \Sigma, \Delta, \Sigma''}{\Gamma, \Pi \Rightarrow \Delta, \Sigma}}
$$

These cuts are not eliminable but since all are applied on subformulae of φ the degree of a new proof is smaller than $d\varphi$ and we are done. □

Note that to obtain (a), (b), (c) we have used the technique of cross-cuts. Note also that the special induction on the degree (complexity) of the cut-formula is not necessary as it is involved into conditions specified in the lemma.

Exercise 2.22. *Prove the cases where the principal formula φ is a negation, conjunction and disjunction.*

Lemma 2.8. (Cut-degree reduction). *If $\mathcal{D} \vdash \Gamma \Rightarrow \Delta$ and $d\mathcal{D} > 0$, then we can construct a proof $\mathcal{D}'$ such that $\mathcal{D}' \vdash \Gamma \Rightarrow \Delta$ and $d\mathcal{D}' < d\mathcal{D}$*

PROOF: Induction on the height of $\mathcal{D}$. There are two cases:
    a) the last inference is not a maximal cut;
    b) the last inference is a maximal cut.
    In case a) we consider the proof(s) of the premiss(es) of $\Gamma \Rightarrow \Delta$ which is/are of lower height hence by the induction hypothesis may be replaced with a proof (or proofs) of the premiss(es) of lower degree. We apply suitable rules to obtain a proof of $\Gamma \Rightarrow \Delta$ of lower degree.
    In case b) we have the following last step:

$$
(Cut) \cfrac{\begin{array}{cc} \mathcal{D}_1 & \mathcal{D}_2 \\ \Gamma \Rightarrow \Delta, \varphi & \varphi, \Pi \Rightarrow \Sigma \end{array}}{\Gamma, \Pi \Rightarrow \Delta, \Sigma}
$$

Both proofs of the premisses may contain other maximal cuts but by the induction hypothesis they may be replaced with the proofs of lower degree. Hence the mix reduction lemma applies to them and we get a proof of $\Gamma, \Pi_\varphi \Rightarrow \Delta_\varphi, \Sigma$ of lower degree. By applications of W, if necessary, we get $\Gamma, \Pi \Rightarrow \Delta, \Sigma$. □

Note that the above proof may be used also as an alternative technique for showing how to prove cut elimination on the basis of an admissibility result. Gentzen's method is top-down whereas this one is a bottom-up procedure.
    Finally, by successive applications of the above lemma we get:

Theorem 2.6. (Cut elimination). *If $\mathcal{D} \vdash \Gamma \Rightarrow \Delta$ and $d\mathcal{D} > 0$, then we can construct a proof $\mathcal{D}'$ such that $\mathcal{D}' \vdash \Gamma \Rightarrow \Delta$ and $d\mathcal{D}' = 0$*

Remark 2.2. Girard's proof may be applied also to LK' to show eliminability of Mix (see e.g. Dowek [63]). In this case we can construct suitable proof figures showing transformations of parts of a proof in the mix reduction lemma. For example in case of negation we have:

$$(Mix) \frac{(\Rightarrow \neg) \dfrac{\varphi, \Gamma \Rightarrow \Delta, \neg\varphi^k}{\Gamma \Rightarrow \Delta, \neg\varphi^{k+1}} \quad \dfrac{\neg\varphi^n, \Pi \Rightarrow \Sigma, \varphi}{\neg\varphi^{n+1}, \Pi \Rightarrow \Sigma} (\neg \Rightarrow)}{\Gamma, \Pi \Rightarrow \Delta, \Sigma}$$

and this is replaced with:

$$\frac{(Mix) \dfrac{(\Rightarrow \neg) \dfrac{\varphi, \Gamma \Rightarrow \Delta, \neg\varphi^k}{\Gamma \Rightarrow \Delta, \neg\varphi^{k+1}} \quad \neg\varphi^n, \Pi \Rightarrow \Sigma, \varphi}{\Gamma, \Pi \Rightarrow \Delta, \Sigma, \varphi} \quad (Mix) \dfrac{\varphi, \Gamma \Rightarrow \Delta, \neg\varphi^k \quad \dfrac{\neg\varphi^n, \Pi \Rightarrow \Sigma, \varphi}{\neg\varphi^{n+1}, \Pi \Rightarrow \Sigma} (\neg \Rightarrow)}{\varphi, \Gamma, \Pi \Rightarrow \Delta, \Sigma}}{\dfrac{\Gamma_\varphi, \Gamma, \Pi_\varphi, \Pi \Rightarrow \Delta_\varphi, \Delta, \Sigma_\varphi, \Sigma}{\Gamma, \Pi \Rightarrow \Delta, \Sigma} (C)(P)}$$

where two uppermost mixes are eliminable by the induction hypothesis. The last one is not eliminable but it has lower degree. As we can see even in the simplest case the transformations obtained are significantly complicated.

Exercise 2.23. *Provide transformations for the case of $\wedge, \vee$ and $\rightarrow$ as principal.*

2.4.3 Reductive Proof

We can extract one more proof of cut elimination of a similar character from the more general construction applied by Metcalfe, Olivetti and Gabbay [176] in the framework of hypersequent calculi, a generalised version of SC, particularly useful for non-classical logics (see chapter 4 for more information). In fact, this construction and similar ones for other generalised forms of SC for non-classical logics (e.g. Belnap [25] for display calculus, or Restall [214]) all are somewhat influenced by Curry's proof which will be presented in section 2.5.1.

In the context of the approach developed in [176], a general categorization of rules is proposed into substitutive and reductive. Roughly speaking, the former allow for height reduction, the latter for complexity reduction. Such an approach in [176] is particularly useful for developing a general schema of proof which applies to any extension of the basic system provided the new rules are substitutive and reductive. Moreover, it is easier to develop a global proof of cut elimination on the basis of such preliminary categorization of proof (proofs from [176] are of this sort)[2]. However, we leave the problem of defining general schemata for rules yielding at least sufficient criteria for proving cut elimination aside and provide a local proof. Such a proof is working even for systems using rules of different characters (e.g. rules for modal logics which in general are not substitutive). One

---

[2]Several proposals of this sort were offered for different classes of SC. For example by Avron [14] or Ciabattoni [50], where the notion of substitutive and reductive rules was introduced. Ciabattoni and Terui [51] provide a survey of such attempts.

can find similar proofs in Ciabattoni, Metcalfe, Montagna [52], Indrzejczak [135], Kurokawa [154], Lellmann [160], all in the hypersequent framework. Indrzejczak [137, 141, 142] provide examples of its application for standard SC. Here we adapt this strategy to LK.

The proof is based on two lemmata; the first of which is:

Lemma 2.9. (Right reduction). *Let $\mathcal{D}_1 \vdash \Gamma \Rightarrow \Delta, \varphi$ and $\mathcal{D}_2 \vdash \varphi^k, \Pi \Rightarrow \Sigma$ with $d\mathcal{D}_1, d\mathcal{D}_2 < d\varphi$, and φ principal in $\Gamma \Rightarrow \Delta, \varphi$, then we can construct a proof $\mathcal{D}$ such that $\mathcal{D} \vdash \Gamma^k, \Pi \Rightarrow \Delta^k, \Sigma$ and $d\mathcal{D} < d\varphi$.*

PROOF: By induction on the height of $\mathcal{D}_2$. We must consider all cases. In most of them we simply apply the induction hypotheses to the premises of $\varphi^k, \Pi \Rightarrow \Sigma$ and then apply the respective rule. We consider some cases:

a) The basis: $\varphi^k, \Pi \Rightarrow \Sigma$ is an axiom $\varphi \Rightarrow \varphi$, hence $k = 1, \Pi$ is empty and Σ consists of one occurrence of φ. Then $\Gamma \Rightarrow \Delta, \varphi$ is our $\Gamma^k, \Pi \Rightarrow \Delta^k, \Sigma$ and it is already proved.

b) $\varphi^k, \Pi \Rightarrow \Sigma$ is derived by $(\vee \Rightarrow)$ with one occurrence of φ as side formula and the remaining ones as parametric. So we have:

$$(\vee \Rightarrow), (P \Rightarrow) \frac{\varphi^{k+1}, \Pi' \Rightarrow \Sigma \qquad \psi, \varphi^k, \Pi' \Rightarrow \Sigma}{\varphi^k, \Pi \Rightarrow \Sigma}$$

where Π consists of Π' and $\varphi \vee \psi$. By the induction hypothesis we get $\vdash \Gamma^{k+1}, \Pi' \Rightarrow \Delta^{k+1}, \Sigma$ and $\vdash \psi, \Gamma^k, \Pi' \Rightarrow \Delta^k, \Sigma$ both with proofs having proof degree $d\mathcal{D} < d\varphi$. Then we continue with them:

$$(C), (P), (W \Rightarrow) \frac{\dfrac{\Gamma^{k+1}, \Pi' \Rightarrow \Delta^{k+1}, \Sigma}{\varphi, \Gamma^k, \Pi' \Rightarrow \Delta^k, \Sigma} \qquad \psi, \Gamma^k, \Pi' \Rightarrow \Delta^k, \Sigma}{\Gamma^k, \Pi \Rightarrow \Delta^k, \Sigma}$$
$$(\vee \Rightarrow), (P \Rightarrow)$$

Exercise 2.24. *Provide transformations for the other logical rules with φ being parametric or with some occurrences being side formulae.*

c) $\varphi^k, \Pi \Rightarrow \Sigma$ is derived by (Cut) on some ψ such that $d\psi < d\varphi$. So we have:

$$(Cut) \frac{\varphi^i, \Pi' \Rightarrow \Sigma', \psi \qquad \psi, \varphi^j, \Pi'' \Rightarrow \Sigma''}{\varphi^k, \Pi \Rightarrow \Sigma}$$

where $k = i + j, \Pi$ consists of Π' and Π'' and similarly for Σ. By the induction hypothesis we get $\vdash \Gamma^i, \Pi' \Rightarrow \Delta^i, \Sigma', \psi$ and $\vdash \psi, \Gamma^j, \Pi'' \Rightarrow \Delta^j, \Sigma''$ both with proofs having $d\mathcal{D} < d\varphi$. Then we continue:

$$(Cut) \frac{\Gamma^i, \Pi' \Rightarrow \Delta^i, \Sigma', \psi \qquad \psi, \Gamma^j, \Pi'' \Rightarrow \Delta^j, \Sigma''}{\Gamma^k, \Pi \Rightarrow \Delta^k, \Sigma}$$

Note that φ cannot be a cut-formula due to our assumption that $d\mathcal{D}_1, d\mathcal{D}_2 < d\varphi$.

Exercise 2.25. *Provide transformations for other structural rules with φ being either parametric or with some occurrences being side- and principal formulae.*

d) In cases where one occurrence of φ in $\varphi^k, \Pi \Rightarrow \Sigma$ is principal we make use of the fact that φ in the left premiss is principal too. Let us consider an example with $\varphi = \psi \wedge \chi$. By the induction hypothesis we get $\mathcal{D}'_2 \vdash \psi, \Gamma^{k-1}, \Pi \Rightarrow \Delta^{k-1}, \Sigma$ with $d\mathcal{D}'_2 < d\varphi$. Then we continue with one of the premisses of $\Gamma \Rightarrow \Delta, \varphi$:

$$(Cut) \ \frac{\Gamma \Rightarrow \Delta, \psi \qquad \psi, \Gamma^{k-1}, \Pi \Rightarrow \Delta^{k-1}, \Sigma}{\Gamma^k, \Pi \Rightarrow \Delta^k, \Sigma}$$

The new proof has obviously a degree lower than $d\varphi$, since the new cut is on a subformula of φ. $\qquad\qquad\square$

Exercise 2.26. *Provide proofs for the other logical rules with one occurrence of φ being principal.*

Lemma 2.10. (Left reduction). *Let $\mathcal{D}_1 \vdash \Gamma \Rightarrow \Delta, \varphi^k$ and $\mathcal{D}_2 \vdash \varphi, \Pi \Rightarrow \Sigma$ with $d\mathcal{D}_1, d\mathcal{D}_2 < d\varphi$, then we can construct a proof $\mathcal{D}$ such that $\mathcal{D} \vdash \Gamma, \Pi^k \Rightarrow \Delta, \Sigma^k$ and $d\mathcal{D} < d\varphi$.*

PROOF: Similarly, now by induction on the height of $\mathcal{D}_1$, but with some important differences. First note that we do not require φ to be principal in $\varphi, \Pi \Rightarrow \Sigma$, so it includes the case where φ is atomic. In cases where no occurrence of φ is principal in the left premiss we just apply the induction hypothesis. Note that the only transformations which introduce new cuts (but of lower degree) are for φ compound. Thus if φ is atomic, then—by the assumption of the lemma—there are no other cuts above, and the proof obtained is cut-free.

Now, in cases where one occurrence of φ in $\Gamma \Rightarrow \Delta, \varphi^k$ is principal we make use of the Right reduction lemma. For example: let $\varphi = \psi \vee \chi$, then the premiss is $\Gamma \Rightarrow \Delta, \varphi^{k-1}, \psi$ (or (χ) and by the induction hypothesis we get $\mathcal{D}'_1 \vdash \Gamma, \Pi^{k-1} \Rightarrow \Delta, \Sigma^{k-1}, \psi$ with $d\mathcal{D}'_1 < d\varphi$, and by $(\Rightarrow \vee)$ we get $\Gamma, \Pi^{k-1} \Rightarrow \Delta, \Sigma^{k-1}, \varphi$. Since φ is principal in this sequent, the Right reduction lemma applies to it and to $\mathcal{D}_2$ and, if there is no φ in Π, we obtain $\mathcal{D} \vdash \Gamma, \Pi^k \Rightarrow \Delta, \Sigma^k$ with $d\mathcal{D} < d\varphi$. Otherwise, we obtain this result after some applications of structural rules. $\qquad\square$

Exercise 2.27. *Provide at least some other cases in the proof of this lemma.*

Now we are ready to prove the cut-elimination theorem:

PROOF: by double induction: primary on $d\mathcal{D}$ and subsidiary on the number of maximal cuts (in the basis and in the inductive step of the primary induction). We always take the topmost maximal cut, hence we have the following situation:

$$(Cut) \ \frac{\Gamma \Rightarrow \Delta, \varphi \qquad \varphi, \Pi \Rightarrow \Sigma}{\Gamma, \Pi \Rightarrow \Delta, \Sigma}$$

If φ is not atomic, then by the Left reduction lemma we obtain a proof of $\Gamma, \Pi^k \Rightarrow \Delta_\varphi, \Sigma^k$ of lower degree; if it is atomic, then the new proof is cut-free, as we observed above. This, by applications of $(\Rightarrow W)$ and (C) (if necessary) yields $\Gamma, \Pi \Rightarrow \Delta, \Sigma$. By successive repetition of this procedure we diminish either the degree of the proof or the number of maximal cuts in it until we obtain a proof with $d = 0$.   □

Remark 2.3. Note that we can interchange the present proof of the cut-elimination theorem with the proof given in the preceding subsection, i.e. we can prove it as above (by double induction) on the basis of Girard's reduction lemmata (taking the cut reduction lemma instead of the left reduction lemma) or we can prove it only by induction on the height (but using the left reduction lemma instead of the cut reduction lemma).

Note also that the problem of dealing with the multiplicity of cuts in eliminability proofs does not exist for proofs of the admissibility of cut. Moreover, all the proofs of eliminability presented in this subsection may be reworked as proofs of admissibility. However in this case we cannot compute the bounds on the complexity of output proofs, like in Girard's proof.

The interesting thing is that the above proof applies directly to LK with cut although it may be argued that some generalised forms of cut are implicitly involved in it. Such rules are extracted by Restall [214] as succedent multiple cut and antecedent multiple cut and may be formally defined as follows:

$$(AMCut) \ \frac{\Gamma \Rightarrow \Delta, \varphi \quad \varphi^k, \Pi \Rightarrow \Sigma}{\Gamma^k, \Pi \Rightarrow \Delta^k, \Sigma} \qquad (SMCut) \ \frac{\Gamma \Rightarrow \Delta, \varphi^k \quad \varphi, \Pi \Rightarrow \Sigma}{\Gamma, \Pi^k \Rightarrow \Delta, \Sigma^k}$$

Exercise 2.28. *Prove that LK is equivalent to LK with any of these rules instead of cut.*

2.4.4   Contraction-Sensitive Proof

The proofs of cut elimination described in two preceding subsections have many advantages. In particular, Girard additionally provides, on the basis of his proof, a complexity result concerning the hyperexponential growth of the transformed proof trees. Still in both cases we seem to be halfway with respect to the problem of eliminating cut instead of mix. In both cases we finally prove elimination of cut but it is based on the previous work with mix (Girard's proof) or some kind of one-sided mix. So the question may be posed once more: Can we provide a proof similar to Gentzen's original one but directly for cut?

The crucial point is how to deal with contraction in the course of the proof. Different strategies were proposed for that aim which we are not going to consider here. One of them is based on the introduction of an additional induction parameter – the number of applications of contraction above cut (Bimbo [31]).

Another one is connected with a different interpretation of multiset unions in sequents (Indrzejczak [134]). However, the simplest solution may be obtained if we take advantage of our invertibility results from section 2.3. We can provide a direct proof of the admissibility of cut for LK on the basis of a careful analysis of the part of a proof where contraction was applied to the cut-formula. Such a proof was provided by Negri and von Plato [185] for a kind of SC similar to LK. In fact, due to invertibility, we can restrict our interest to the case of atomic cut-formulae. The proof is by induction on the complexity of the cut-formula. Only in the basis do we need a subsidiary induction on the height of a proof of one premiss. Let LK$^-$ denote LK without cut.

Theorem 2.7. (Admissibility of cut in LK$^-$). *If* $\vdash \Gamma \Rightarrow \Delta, \varphi$ *and* $\varphi, \Pi \Rightarrow \Sigma$, *then* $\vdash \Gamma, \Pi \Rightarrow \Delta, \Sigma$.

PROOF: In the basis we consider an atomic cut-formula and perform an additional induction on the height of the left (or right, if we wish) premiss. Cases of axioms and W are trivial, and cases of logical rules follow from context independence of rules (the cut-formula may only be parametric). Now consider a case where the cut-formula was derived by $(\Rightarrow C)$ on φ. Even if there was a series of such applications of C there must be a first one so we have above a sequent $\Gamma \Rightarrow \Delta, \varphi^k$. It cannot be an axiom, so it was deduced by some structural or logical rule. If it was $(\Rightarrow W)$ applied to φ we just delete this application of $(\Rightarrow W)$, the height is reduced and we are done. Otherwise φ is parametric and we may permute the application of this rule with the series of applications of $(\Rightarrow C)$ to φ and suitable cut and again obtain the effect of height reduction. Let the last rule be $(\rightarrow\Rightarrow)$ and we have the following proof figure:

$$
(\rightarrow\Rightarrow) \cfrac{
  \cfrac{\Gamma', \Rightarrow \Delta', \varphi^i, \psi \qquad \chi, \Gamma'', \Rightarrow \Delta'', \varphi^j}{
    (\Rightarrow C) \cfrac{\psi \rightarrow \chi, \Gamma \Rightarrow \Delta, \varphi^k}{
      (Cut) \cfrac{\psi \rightarrow \chi, \Gamma \Rightarrow \Delta, \varphi \qquad\qquad \varphi, \Pi \Rightarrow \Sigma}{\psi \rightarrow \chi, \Gamma, \Pi \Rightarrow \Delta, \Sigma}
    }
  }
}{}
$$

where $k = i + j$ and Γ is a concatenation of Γ', Γ'', similarly for Δ. It is transformed as follows:

$$
(\rightarrow\Rightarrow) \cfrac{
  (\Rightarrow P) \cfrac{
    (Cut) \cfrac{
      (P), (\Rightarrow C) \cfrac{\Gamma', \Rightarrow \Delta', \varphi^i \psi}{\Gamma' \Rightarrow \Delta', \psi, \varphi} \qquad \varphi, \Pi \Rightarrow \Sigma
    }{\Gamma', \Pi \Rightarrow \Delta', \psi, \Sigma}
  }{\Gamma', \Pi \Rightarrow \Delta', \Sigma, \psi}
  \qquad
  (Cut) \cfrac{
    (\Rightarrow C) \cfrac{\chi, \Gamma'', \Rightarrow \Delta'', \varphi^j}{\chi, \Gamma'' \Rightarrow \Delta'', \varphi} \qquad \varphi, \Pi \Rightarrow \Sigma
  }{\chi, \Gamma'', \Pi \Rightarrow \Delta'', \Sigma}
}{
  (P)(C) \cfrac{\psi \rightarrow \chi, \Gamma', \Pi, \Gamma'', \Pi \Rightarrow \Delta', \Sigma, \Delta'', \Sigma}{\psi \rightarrow \chi, \Gamma, \Pi \Rightarrow \Delta, \Sigma}
}
$$

Clearly the height of both cuts is reduced.

In the inductive step we make a reduction of the complexity of the cut-formula. Let $\varphi = \psi \to \chi$. By invertibility (Theorem 2.3) we obtain from both premisses of cut:

(a) $\psi, \Gamma \Rightarrow \Delta, \chi$

(b) $\Pi \Rightarrow \Sigma, \psi$

(c) $\chi, \Pi \Rightarrow \Sigma$

and combine them to derive the desired effect:

$$(Cut) \dfrac{\dfrac{\Pi \Rightarrow \Sigma, \psi \qquad \psi, \Gamma \Rightarrow \Delta, \chi}{(Cut) \dfrac{\Pi, \Gamma \Rightarrow \Sigma, \Delta, \chi \qquad\qquad \chi, \Pi \Rightarrow \Sigma}{(P)(C) \dfrac{\Pi, \Gamma, \Pi, \Rightarrow \Sigma, \Delta, \Sigma}{\Gamma, \Pi \Rightarrow \Delta, \Sigma}}}}{}$$

All cuts are of lower complexity.   □

One should notice that this proof contains transformations which are not wholly local. Hence it may be treated as being halfway in the direction of proofs based on global transformations.

2.5   Global Proofs of Cut Elimination

In contrast to local proofs this kind of proof is based on global transformations, defined on the whole proofs of the premisses of cut. Such an approach allows for a convenient treatment of contraction. On the other hand, proofs of this kind are strongly connected with the fact that rules are context independent. Proofs based on global transformations are seldom. We provide two such proofs of different characters, the first due to Curry [56], and the second due to Buss [43]. Curry's proof is in fact the first proof of admissibility (not eliminability) of cut. Buss' proof is again for the cut-elimination theorem and it is based on Tait strategy, and similar to Girard's proof.

2.5.1   Curry's Substitutive Proof

This kind of a proof of admissibility of cut (called by Curry the Elimination Theorem) was introduced first in [55] and later [56] extended to a number of SC similar to (and including) LK. Curry's proof was quite influential due to a detailed analysis of the properties of rules and its impact on the construction of many other proofs. It paved the way for the search for general (sufficient and necessary) conditions for rules that enable such a proof. Some of them (e.g. Belnap's analysis for display calculus [25] or Restall [214]) follow Curry's analysis and strategy quite closely. Other proposals, like the one in Ciabattoni and Terui [51], Metcalfe, Olivetti and Gabbay [176], or Avron and Lev [14] provide different solutions.

The differences with Gentzen's proof are significant. First of all, Curry shows admissibility not eliminability of cut and no mix is involved in his proof. Essentially it is by a double induction on the complexity of the cut-formula and on the heights of the proofs of the premises. However, the most original features of the proof are the following two:

1. the proof is divided into three (partly) independent stages;

2. proofs of two (dual) stages 1 and 2 are based on global transformations.

We will formulate the stages in the form of lemmata for easier comparison with other solutions. But first we introduce some abbreviations. Let φ^p mean that the cut-formula φ is principal in a sequent and the following capital letters represent:

$$A : \vdash \Gamma \Rightarrow \Delta, \varphi$$
$$B : \vdash \varphi, \Pi \Rightarrow \Sigma$$
$$C : \vdash \Gamma, \Pi \Rightarrow \Delta, \Sigma$$
$$A' : \vdash \Gamma \Rightarrow \Delta, \varphi^p$$
$$B' : \vdash \varphi^p, \Pi \Rightarrow \Sigma$$

The lemmata corresponding to Curry's stages may be formulated as follows (H1, H2, H3 refer to the antecedents of displayed implications):

Lemma 2.11. (Stage 1 reduction). *If (H1) A' and B imply C, then A and B imply C.*

Lemma 2.12. (Stage 2 reduction). *If (H2) A and B' imply C, then A and B imply C.*

Lemma 2.13. (Stage 3 reduction). *If (H3) cut is admissible for each subformula of φ, then A' and B' imply C.*

Stage 3 is proven in the same way as complexity reduction steps in the proof of cut admissibility from section 1.8 (case 2) or in Gentzen's proof from section 2.4 (the case of rank=2 from part 4.1). Stages 1 and 2 are proven by induction on the height of the proof of A and B, respectively. Finally we prove $\vdash \Gamma, \Pi \Rightarrow \Delta, \Sigma$ by induction on the complexity of the cut-formula on the basis of these three lemmata. Belnap [25] has already noticed that any informal explanation (like those provided by Curry) rather obscure the structure of the whole proof and demonstrated it (for display calculus) by means of natural deduction proof. We follow his approach below in this respect although our reconstruction is a bit different (but closer to Curry's original presentation). The following proof in Jaśkowski-style natural deduction shows precisely the structure of Curry's proof:

1. Cut on subformulae of cut-formula φ is admissible ($H3$)
2. $\vdash \Gamma \Rightarrow \Delta, \varphi$
3. $\vdash \varphi, \Pi \Rightarrow \Sigma$

3.1. $\vdash \Gamma \Rightarrow \Delta, \varphi^p$

    3.1.1. $\vdash \varphi^p, \Pi \Rightarrow \Sigma$

    3.1.2. $\vdash \Gamma, \Pi \Rightarrow \Delta, \Sigma$ 1., 3.1., 3.1.1., stage 3 reduction lemma

3.2. if $\vdash \varphi^p, \Pi \Rightarrow \Sigma$, then $\vdash \Gamma, \Pi \Rightarrow \Delta, \Sigma$ 3.1.1 – 3.1.2.

3.3. $H2$ 3.2.

3.4. A and B imply C 3.3, stage 2 reduction lemma

3.5. $\vdash \Gamma, \Pi \Rightarrow \Delta, \Sigma$ 2, 3, 3.4.

4. if $\vdash \Gamma \Rightarrow \Delta, \varphi^p$, then $\vdash \Gamma, \Pi \Rightarrow \Delta, \Sigma$ 3.1 – 3.5.

5. $H1$ 4.

6. A and B imply C 5, stage 1 reduction lemma

7. $\vdash \Gamma, \Pi \Rightarrow \Delta, \Sigma$ 2, 3, 6.

Note that it is possible to simplify slightly both lemmata corresponding to stages 1 and 2 and to obtain a slightly simpler proof.

Lemma 2.14. (Stage 1 reduction). *If (H1) A' implies C, then A and B imply C.*

Lemma 2.15. (Stage 2 reduction). *If (H2) B' implies C, then A and B imply C.*

Now $\vdash \Gamma, \Pi \Rightarrow \Delta, \Sigma$ is derivable in the following way:

1. Cut on subformulae of cut-formula φ is admissible $(H3)$
2. $\vdash \Gamma \Rightarrow \Delta, \varphi$
3. $\vdash \varphi, \Pi \Rightarrow \Sigma$

    3.1. $\vdash \Gamma \Rightarrow \Delta, \varphi^p$

        3.1.1. $\vdash \varphi^p, \Pi \Rightarrow \Sigma$

        3.1.2. $\vdash \Gamma, \Pi \Rightarrow \Delta, \Sigma$ 1., 3.1., 3.1.1., stage 3 reduction lemma

    3.2. $H2$ 3.1.1–3.1.2.

    3.3. $\vdash \Gamma, \Pi \Rightarrow \Delta, \Sigma$ 2, 3, 3.2, stage 2

4. $H1$ 3.1–3.3.

5. $\vdash \Gamma, \Pi \Rightarrow \Delta, \Sigma$ 2, 3, 4, stage 1

It remains to prove stages 1 and 2. They are proven independently by global transformation of the whole proof of the respective premiss. It is possible due to some properties of rules (context independence)—hence not possible in SC for many non-classical logics. We will provide a proof of the stage 2 reduction lemma.

PROOF: Let $\mathcal{D}$ be a proof of the right premiss, i.e. $\varphi, \Pi \Rightarrow \Sigma$ and $S_1, ..., S_n$ an enumeration of sequents in this proof preserving the order of inferences. Clearly $S_n := \varphi, \Pi \Rightarrow \Sigma$. We define recursively $S_k := \Phi_k, \Pi_k \Rightarrow \Sigma_k$ for $k \leq n$, starting with the root S_n:

1. Φ_n contains one occurrence of $\varphi, \Pi_n = \Pi, \Sigma_n = \Sigma$

2. If S_k is a premiss for S_m, then:

    (a) all parameters from Φ_m which are in S_k are in Φ_k

(b) if the principal formula of the application of $(C \Rightarrow)$ or $(W \Rightarrow)$ is in Φ_m, then either both side formulae are in Φ_k (the first case) or this formula is absent in Φ_k (the second case).

Clearly Φ_k is defined for each k and it is a multiset (possibly empty) of occurrences of φ. Let $\mathcal{D}_1$ denote the list of these sequents from $\mathcal{D}$, where Φ_k is empty, and $\mathcal{D}_2$ the remaining part of the list. Notice that all premises for elements of $\mathcal{D}_1$ also belong to it by definition of Φ_k (condition 2).

For every S_k we define $S'_k := \Gamma^n, \Pi_k \Rightarrow \Delta^n, \Sigma_k$, where $n \geq 0$ corresponds to the number of occurrences of φ. Note that if S_k is in $\mathcal{D}_1$, then $S_k = S'_k$ and that S'_n is the conclusion of application of cut under consideration, i.e. $S'_n := \Gamma, \Pi \Rightarrow \Delta, \Sigma$.

We show by induction on k, that if the hypothesis $H2$ holds, then every S'_k is provable, in particular S'_n. We have 5 cases:

1. S_k belongs to $\mathcal{D}_1$, hence $S'_k := S_k$ and it is provable.

2. S_k is an axiom and belongs to $\mathcal{D}_2$, so, because Φ_k is not empty, it has a form $\varphi \Rightarrow \varphi$. Then $S'_k := \Gamma \Rightarrow \Delta, \varphi$ but this is just the left premiss of cut which is provable by assumption.

3. S_k belongs to $\mathcal{D}_2$ and is derivable from S_i (and S_j in case of two-premiss rules) by means of an instance of a rule where all elements of Φ_k are parameters. By the induction hypothesis S'_i (and S'_j) is provable. Since all rules are context independent S'_k is derivable from S'_i (and S'_j) by means of the same rule.

4. S_k belongs to $\mathcal{D}_2$, the principal formula is in Φ_k and S_k is derivable from S_i by means of $(C \Rightarrow)$ or $(W \Rightarrow)$. By the induction hypothesis S'_i is provable. In case of $(C \Rightarrow)$ the difference of S'_k and S'_i is only the matter of one occurrence of Γ (and possibly Δ if nonempty) less, so it is derivable from S'_i by applications of C. In case of $(W \Rightarrow)$, Φ_k in S_k has one more occurrence of φ than Φ_i, so S'_k is derivable from S'_i by applications of W for getting Γ in the antecedent and possibly Δ (if nonempty) in the succedent.

5. S_k is in $\mathcal{D}_2$, the principal formula is in Φ_k and S_k is derivable from S_i (and possibly S_j) by means of some logical rule. Let S''_k be S_k in which all parameters from Φ_k (i.e. other occurrences of φ, if there are any) are replaced with occurrences of Γ and Δ added in the succedent. Then S''_k is derivable from S'_i in the same way as under point 3. Moreover, in S''_k φ is principal, so S''_k satisfies the assumption of (1a) and S'_k is derivable from it. $\square$

Curry's technique may be applied also to K (with additive cut as well) yielding very simple proofs. But admissibility of W (derivability of axioms from the second premiss) and C (in stage 3 or, if we prefer additive cut, in reducing S_k from $\Gamma, \Gamma \Rightarrow \Delta, \Delta$ to $\Gamma \Rightarrow \Delta$) is still necessary.

Exercise 2.29. *Prove the Stage 1 reduction lemma.*

2.5.2   Buss' Invertive Proof

Quite a different global proof was provided by Buss [43] for a version of LK but with invertible rules, additive cut and atomic axioms. Here, at the price of some extra complications, we adapt it for the original LK with multiplicative cut but with atomic axioms. As for the latter, we know that this is an inessential restriction thanks to Lemma 1.8 (section 1.5) which holds also for LK.

Exercise 2.30. *Prove Lemma 1.8 for LK.*

The proof of cut elimination may be developed in a similar way as Girard's proof, however, for the sake of demonstrating a different technique of proof as well as for the possible further use, we separate first the special case. Let us call all applications of (Cut) where the cut-formula is propositional symbol atomic cuts. We will show that:

Lemma 2.16. (Atomic Cut Admissibility). *In cut-free LK atomic cuts are admissible.*

PROOF: We take a proof $\mathcal{D}_l$ of the left premiss $\Gamma \Rightarrow \Delta, \varphi$ with φ atomic and transform it into $\mathcal{D}'_l$ in the following way.

Each sequent $\Lambda \Rightarrow \Theta$ in $\mathcal{D}_l$ is replaced with $\Pi, \Lambda \Rightarrow \Sigma, \Theta_\varphi$, where Θ_φ means that all occurrences of φ in Θ were deleted. As a result every leaf $\psi \Rightarrow \psi$ of $\mathcal{D}_l$ becomes in $\mathcal{D}'_l$ a sequent of the form $\Pi, \psi \Rightarrow \psi, \Sigma$ if ψ is not φ, or $\Pi, \psi \Rightarrow \Sigma$, if $\psi = \varphi$.

Generally, thanks to the context independence of the rules, $\mathcal{D}'_l$ is a derivation or almost a derivation (see points (d)–(e) below) of $\Gamma, \Pi, \Rightarrow \Delta_\varphi, \Sigma$ from some nonaxiomatic sequents. Finally from $\mathcal{D}'_l$ we get a proof $\mathcal{D}'$ of $\Gamma, \Pi \Rightarrow \Delta, \Sigma$ as follows:

(a) A final sequent is either complete or obtained from the root of $\mathcal{D}'_l$ by applications of W (in case there were some occurrences of φ in Δ).

(b) Every leaf of the shape $\Pi, \psi \Rightarrow \psi, \Sigma$, we get from $\psi \Rightarrow \psi$ by applications of W.

(c) Every leaf of the shape $\Pi, \varphi \Rightarrow \Sigma$ is just the right premiss so we add its proof above it.

(d) In case of the application of $(\rightarrow\Rightarrow)$ in $\mathcal{D}_l$ we obtain a figure:

$$\frac{\Pi, \Lambda' \Rightarrow \Sigma, \Theta'_\varphi, \psi \qquad \chi, \Pi, \Lambda'' \Rightarrow \Sigma, \Theta''_\varphi}{\psi \rightarrow \chi, \Pi, \Lambda', \Lambda'' \Rightarrow \Sigma, \Theta'_\varphi, \Theta''_\varphi}$$

which is not a proper application of $(\rightarrow\Rightarrow)$. We must insert it between the premisses and the conclusion of this figure a sequent $\psi \rightarrow \chi, \Pi, \Pi, \Lambda', \Lambda'' \Rightarrow \Sigma, \Sigma.\Theta'_\varphi, \Theta''_\varphi$ and successive applications of C to Π and Σ to get a correct proof.

(e) In case some occurrence of φ is a side formula of $(\neg \Rightarrow)$ or $(\Rightarrow \vee)$ we justify an inference by W introducing a principal formula. In case it is a side formula of $(\Rightarrow\rightarrow)$ we must insert a sequent recovering φ by means of W in the succedent to allow for an application of $(\Rightarrow\rightarrow)$.

(f) In case of $(\Rightarrow C)$ or $(\Rightarrow W)$ applied to φ in $\mathcal{D}_l$, we delete one sequent in $\mathcal{D}'_l$. $\qquad\square$

Exercise 2.31. *Provide a proof of Atomic Cut Admissibility for Buss' original calculus— observe in what respects and why it is simpler.*

In Buss' approach, instead of the mix reduction lemma we first prove:

Lemma 2.17. (Cut reduction). *If $\mathcal{D}_l \vdash \Gamma \Rightarrow \Delta, \varphi$ and $\mathcal{D}_r \vdash \varphi, \Pi \Rightarrow \Sigma$ and $d\mathcal{D}_l, d\mathcal{D}_r < d\varphi$, then we can construct a proof $\mathcal{D}$ such that $\mathcal{D} \vdash \Gamma, \Pi \Rightarrow \Delta, \Sigma$ and $d\mathcal{D} < d\varphi$*

PROOF: The proof goes by induction on d. In the basis φ is atomic and, by assumption concerning $d\mathcal{D}$, proofs of both premises must be cut-free, hence the atomic cut admissibility lemma applies. We must demonstrate the induction step. The proof provided by Buss is essentially based on a global argument for invertibility. Since we abstracted this result in section 2.3 we can just use it here. Let us consider one case as an example.

The case where the cut-formula $\varphi := \psi \to \chi$. On the basis of the proof of the left premiss $\mathcal{D}_l$ we obtain $\mathcal{D}'_l$ for $\psi, \Gamma \Rightarrow \Delta, \chi$ and instead of $\mathcal{D}_r$ we construct two proofs: $\mathcal{D}_\psi$ for $\Pi \Rightarrow \Sigma, \psi$ and $\mathcal{D}_\chi$ for $\chi, \Pi \Rightarrow \Sigma$. All are obtained by invertibility. Finally we build from these three proofs a new $\mathcal{D}'$ in the following way:

$$
\cfrac{
\mathcal{D}_\psi \qquad
\cfrac{
\cfrac{\mathcal{D}'_l}{\psi, \Gamma \Rightarrow \Delta, \chi} \qquad \cfrac{\mathcal{D}_\chi}{\chi, \Pi \Rightarrow \Sigma}
}{\psi, \Gamma, \Pi \Rightarrow \Delta, \Sigma} (Cut)
}{
\cfrac{\Pi \Rightarrow \Sigma, \psi \qquad\qquad\qquad\qquad}{\Pi, \Gamma, \Pi \Rightarrow \Sigma, \Delta, \Sigma}(Cut)
}
$$
$$
\cfrac{\Pi, \Gamma, \Pi \Rightarrow \Sigma, \Delta, \Sigma}{\Gamma, \Pi \Rightarrow \Delta, \Sigma} (P)(C)
$$

Both applications of cut are of lower degree. $\qquad\square$

From this lemma the cut-elimination theorem follows via the cut-degree reduction lemma, like in Girard's proof.

Exercise 2.32. *Complete the proof where φ is a negation and a disjunction. Prove it for Buss' original system with A-cut and rules like in K.*

Both Curry's and Buss' proofs are global but based on significantly different transformations. Let us note some differences with Curry's proof:

1. Buss' proof is a proof of eliminability but also avoids induction on the number of cut applications (Tait's technique).

2. In Curry's proof a transformation goes by the addition of parameters from the second premiss whereas in Buss' strategy there is a replacement of the cut-formula with its subformulae. Thus the former was named substitutive since it generalises a local strategy of height reduction, whereas in Buss' proof a similar transformation is applied only to atomic cuts.

3. In Curry's proof only sequents with the cut-formula are changed in input proofs, whereas in Buss' proof all sequents are replaced with new ones. However, this may be changed; we may define the transformation only for sequents where the cut-formula in question occurs and leave untouched all sequents above.

If we analyse the proof of the cut-degree reduction lemma it is obvious why we call Buss' proof invertive. A transformation applied in this proof is a generalisation of Schütte strategy of using invertibility instead of height reduction (with the exception of the basic step based on the atomic cut admissibility lemma). It shows that if we separate the atomic cut admissibility lemma for LK and the invertbility lemma as preliminary results, we obtain an immediate proof of eliminability of cut based only on induction on the complexity of the cut-formula (degree of cut) and using Tait's strategy.

2.6 Decidability

As we already noticed one of the most important applications of cut elimination is for proving decidability. At first sight it may seem that cut elimination for LK yields decidability of **CPL** in a similar way as in K. It is true that if S is provable in LK, then, by cut elimination, we are assured that there is a proof satisfying the subformula property and there is only a finite number of such proofs to consider. However the specific features of LK and its rules make proof search much harder. We will discuss first the sources of difficulties before we provide a solution.

1. Not all rules of LK are invertible. By Theorem 1.3 (section 1.6) this implies that LK is not confluent which makes proving decidability much harder. In particular, if we are trying systematic proof search in LK we are forced to backtracking. Let us consider the following:

$$\textbf{Example 2.5.} \quad \cfrac{\cfrac{\cfrac{p \Rightarrow p \qquad q \Rightarrow r}{p \to q, p \Rightarrow r}\,(\to\Rightarrow)}{p \to q \Rightarrow p \to r}\,(\Rightarrow\to)}{(p \to q) \land (p \lor q \to r) \Rightarrow p \to r}\,(\land\Rightarrow)$$

Although the sequent is obviously provable we failed to find a proof because of a wrong choice of a side formula in the first (at the bottom) application of $(\land \Rightarrow)$. What we should do is to backtrack to the beginning and try with $p \lor q \to r$ as side formula. Similar problems are connected with $(\Rightarrow \lor)$ whereas applications of $(\to\Rightarrow)$ can be even more troublesome due to the number of possible choices concerning parameters; we must decide how parameters from the conclusion will be divided between two premises. Here is an illustration:

$$\textbf{Example 2.6.} \quad \cfrac{\Rightarrow q, p \qquad q, p \Rightarrow}{p \to q, p \Rightarrow q}\,(\to\Rightarrow)$$

The example is so trivial that it is hard to believe that someone would make such a choice from 4 possibilities but in nontrivial cases we cannot be sure in advance if we correctly divided parameters.

2. We have mentioned in section 2.1. that contraction is indispensable in LK (see example 2.2.) This fact is not only troublesome for proofs of cut elimination but may lead to problems in performing backward proof search. We never know in advance if we should apply contraction to some formula before we apply the logical rule with it as a principal formula. In fact we could use contraction to avoid backtracking in case of the application of three non-invertible rules. One may postulate that before any (backward) application of $(\Rightarrow \vee)$ and $(\wedge \Rightarrow)$ the second occurrence of the principal formula should be added by contraction, and before application of $(\rightarrow \Rightarrow)$ all parametric formulae should be duplicated and then one occurrence of each is put into both premises. This way we indeed introduce K-like counterparts of these rules into action. But generally contraction may lead to uncontrolled production of new sequents on the branch. In the context of K the subformula property is enough, but in LK it is not, since it is not formulae but their occurrences that count.

3. Proofs in LK may contain redundancies in the sense that duplications of the same sequent occur on the branch. In K, due to the subformula property and the lack of structural rules, this was not possible, but in LK we can easily obtain something like this:

Example 2.7.

$$\frac{\dfrac{\vdots}{\dfrac{\varphi, \varphi, \Gamma \Rightarrow \Delta}{\varphi, \Gamma \Rightarrow \Delta} (C \Rightarrow)}}{\varphi, \varphi, \Gamma \Rightarrow \Delta} (W \Rightarrow)$$

According to König's lemma a tree may be infinite either by having infinitely many branches starting from a node or by having at least one infinite branch. The first possibility is excluded in LK since for any sequent there is always only a finite number of possible premises for it, but the second option is allowed in backward proof search. Again the example is trivial, but this does not change the fact that there is a real risk of building infinite proof trees.

The problems briefly described show that in LK the issues of systematic proof search and proving decidability are not so closely connected as in K and we can prove the latter in a way not based on the former. Below we will present Gentzen's original proof of decidability of **CPL** on the basis of LK, which is an example of such an approach.

In fact, the last problem (from the three discussed above) is rather easy to eliminate.

Claim 2.2. *Every redundant proof may be transformed into a proof with exactly one* *occurrence of any sequent on the same branch.*

PROOF: Suppose we have the following segment on the branch:

$$\frac{B_1}{\Gamma \Rightarrow \Delta}$$

$$\vdots$$

$$\frac{\Gamma \Rightarrow \Delta}{B_2}$$

We just replace it with:

$$\frac{B_1}{\frac{\Gamma \Rightarrow \Delta}{B_2}}$$

This way we eliminate one by one every redundant part of the original proof.
□

Thanks to this claim we can restrict further consideration to nonredundant proofs. In order to deal with the second problem (uncontrolled contractions) we introduce additional notions:

Definition 2.2. *A sequent $\Gamma \Rightarrow \Delta$ is 3-reduced iff any formula in it occurs at most three times in the antecedent or in the succedent. It is 1-reduced iff any formula in it occurs exactly once in the antecedent or in the succedent. A proof is reduced iff all sequents in it are 3-reduced.*

Now we prove:

Lemma 2.18. *If $\Gamma \Rightarrow \Delta$ is provable and $\Gamma^1 \Rightarrow \Delta^1$ is any 1-reduction of it, then there exists a cut-free, nonredundant, reduced proof of $\Gamma^1 \Rightarrow \Delta^1$.*

PROOF: We proceed by induction on the height of a proof of $\Gamma \Rightarrow \Delta$ and restrict our considerations to cut-free and nonredundant proofs. The basis is trivial since every axiom is 1-reduced. So suppose that the claim holds for any $k < n$ and that $\Gamma \Rightarrow \Delta$ has a proof of height n. We consider the last applied rule and proofs of the premiss(es). Take as an example an application of $(\rightarrow\Rightarrow)$ to $\Gamma \Rightarrow \Delta$, then it has the form $\varphi \rightarrow \psi, \Pi, \Sigma \Rightarrow \Lambda, \Theta$ and the premisses are $\Pi \Rightarrow \Lambda, \varphi$ and $\psi, \Sigma \Rightarrow \Theta$. Let $\Pi^1 \Rightarrow \Lambda^1, \varphi$ and $\psi, \Sigma^1 \Rightarrow \Theta^1$ be any 1-reductions of the premisses, then by the induction hypothesis both have reduced cut-free proofs. Applying $(\rightarrow\Rightarrow)$ to $\Pi^1 \Rightarrow \Lambda^1, \varphi$ and $\psi, \Sigma^1 \Rightarrow \Theta^1$ we obtain a proof of $\varphi \rightarrow \psi, \Pi^1, \Sigma^1 \Rightarrow \Lambda^1, \Theta^1$. If necessary we apply to this sequent P and C to obtain a 1-reduction of it. The resulting proof is reduced since the proofs of both premisses were reduced, the new premisses are both 1-reduced, and $\varphi \rightarrow \psi, \Pi^1, \Sigma^1 \Rightarrow \Lambda^1, \Theta^1$ is 3-reduced. The last claim, and in general the restriction to 3-reduced sequents in the definition of reduced proofs is just dictated by the form of $(\rightarrow\Rightarrow)$. A formula $\varphi \rightarrow \psi$ may already appear once in Π^1 and Σ^1 (since both premisses are 1-reduced) and in this case it occurs thrice in $\varphi \rightarrow \psi, \Pi^1, \Sigma^1 \Rightarrow \Lambda^1, \Theta^1$ but not more!
□

Exercise 2.33. *Check the remaining cases of rule applications in the proof (what if* $\Gamma \Rightarrow \Delta$ *is deduced by contraction?).*

This way we have obtained an upper bound on the number of sequents required in proof search. If we want to establish whether $\Gamma \Rightarrow \Delta$ is provable it is sufficient to consider if any of it is provable. And to do this it is enough to check if any tree consisting of only 3-reduced sequents built from subformulae of Γ^1, Δ^1 yields a proof of it. If no such tree exists, then $\Gamma \Rightarrow \Delta$ is unprovable.

The above considerations provide a theoretical proof of the decidability result, but the separate problem is how to organise a specific algorithm for checking. The original solution sketched by Gentzen is not based on root-first proof search; on the contrary. Since $SF(\Gamma \cup \Delta)$ is finite, the set $S^3(\Gamma \cup \Delta)$ of all 3-reduced sequents built from these formulae is also finite. Take as $S^3(\Gamma \cup \Delta)_0$ the set of all axiomatic sequents from $S^3(\Gamma \cup \Delta)$, then as $S^3(\Gamma \cup \Delta)_1$ the set of these sequents from $S^3(\Gamma \cup \Delta) - S^3(\Gamma \cup \Delta)_0$ which are deducible from elements of $S^3(\Gamma \cup \Delta)_0$. In general for every stage $n + 1$ take as $S^3(\Gamma \cup \Delta)_{n+1}$ the set of these sequents from $S^3(\Gamma \cup \Delta) - S^3(\Gamma \cup \Delta)_n$ which are deducible from $S^3(\Gamma \cup \Delta)_n$. Every stage is running in finite time and the whole procedure is terminating. Either at some stage k we put $\Gamma \Rightarrow \Delta$ to $S^3(\Gamma \cup \Delta)_k$ which means that it is provable, or we finish with $S^3(\Gamma \cup \Delta) - S^3(\Gamma \cup \Delta)_k$ containing only sequents which are not provable and including $\Gamma \Rightarrow \Delta$.

Such a procedure is not really practical and efficient so we still may think about some better behaving algorithm based on systematic backward proof search. This may be done in many different ways and in this case the problem is not only with the choice between depth-first versus breath-first strategies. We do not aim to go deeply into the details, since K is much better for such an enterprise in case of **CPL**. However, in chapter 4 and 5 we will find logics which do not admit sequent calculi as convenient as K with respect to projecting proof search procedures. So we sketch here some solutions which may be profitable there.

One strategy is to develop an algorithm of proof search based on backtracking. In case of essential choice (like applications of $(\Rightarrow \vee)$, $(\wedge \Rightarrow)$ or $(\rightarrow\Rightarrow)$) we store in the memory the conclusion sequent and if we fail with the construction of a proof, we backtrack to this sequent and repeat application of this rule but with different side formula or parameters divided in a different way. As a result we are building not one but many trees for $\Gamma \Rightarrow \Delta$. If at least one is a proof we can stop, otherwise we obtain a finite collection of proof trees for $\Gamma \Rightarrow \Delta$. Of course in order to keep the procedure finite we must use only 3-reduced sequents build from $SF(\Gamma \cup \Delta)$.

Exercise 2.34. *Define an algorithm according to remarks given. Restrict W only to cases where the application of some rule introduces a side formula which is the 4th occurrence of this formula on the same side of a sequent and to the deletion of all parameters from axiomatic sequents in the sense of K.*

We can apply a slightly different solution and introduce more general trees instead of proof trees. Let us call them proof-search trees. Intuitively these are trees of all possible choices of rule applications in the proof search. Thus such a tree may contain proofs as subtrees (in case a root sequent is provable). Let us consider the following (meta-)rule of disjunctive branching:

$$(DB) \quad \frac{S_1, ..., S_n}{S}$$

where for each $i \leq n$ either $\dfrac{S_i}{S}$ is an instance of some one-premiss rule, or for some $j \leq n$ $\dfrac{S_i \ S_j}{S}$ is an instance of some two-premiss rule.

Two things should be noted. First, in this schema we do not require that all rule applications possible with respect to S should be reflected above the line. We may restrict the list of premisses to some essential choices in order to keep the branching factor of our tree reasonable. In particular, we should remember that applications of C to any formula in S should not exceed 2. More focused version of such a rule will be introduced in chapter 4.

The second thing is connected with the character of branching represented by this rule. All two-premiss rules of LK represent a branching which may be called conjunctive, since both branches must provide proofs of premisses for the conclusion to be provable. In the case of (DB) a provability of at least one of S_i (or S_i and S_j) is sufficient for having a proof of S. Such a solution, taking into account two kinds of branching, was first applied by Beth [28] in the context of a tableau system for intuitionistic logic. Now, the introduction of two kinds of branching makes the control over the result harder, since the occurrence of some nonaxiomatic leaves is not sufficient for falsification of a root sequent.

In [125], to keep track of these differences, we introduced the following machinery. Let us call a *grade* of a proof-search tree the number of applications of (DB) in it. We may now define recursively the notion of a closed proof-search tree as representing proofs of its root sequent.

1. A proof-search tree of grade 0 is closed iff it is a proof of its root sequent.

2. A proof-search tree of grade n is closed iff at least one of its subtrees of smaller degree is closed.

In practice (see e.g. Bilkova [30]) one may mark all axioms as positive and other leaves as negative and continue with marking sequents in the following way:

(a) mark a conclusion of (DB) as positive if any S_i (or S_i, S_j in case of two-premiss rule) is marked as positive;

(b) in case of other rule applications mark a conclusion as positive if all premisses are positive.

In this way we see that a root sequent is proved if it is marked as positive; we can delete all negative sequents and a proof is distilled from a proof-search tree.

Of course some other solutions are also possible. For example, instead of introducing two kinds of branching we can generalise the notion of a sequent and keep only conjunctive branching. The price is that we are using finite sets of sequents called hypersequents. This strategy will be considered in chapter 4.

Exercise 2.35. *Try to develop an algorithm for proof search on the basis of proof-search trees and (DB). Think of reasonable restrictions on the application of DB.*

Remark 2.4. In general in case of LK there is a problem with applications of P. In fact, we already treated P badly in cut-elimination proofs and this suggests an even more drastic move which will be considered in the next chapter; use finite multisets instead of lists in sequents. However, there is a good opportunity to point out in this place that it is possible to dispense with P and keep sequents built from lists if we suitably reformulate rules. A solution proposed by Gallier [93] consists in placing side and principal formulae in suitable places. In case of rules which do not switch their position it is just the same place in the list; in case of rules where there is some switch we put side formulae at the beginning of suitable list. Thus rules for $\wedge$ look like this:

$$(\wedge \Rightarrow) \ \frac{\Gamma, \varphi, \psi, \Delta \Rightarrow \Pi}{\Gamma, \varphi \wedge \psi, \Delta \Rightarrow \Pi} \qquad\qquad (\Rightarrow \wedge) \ \frac{\Gamma \Rightarrow \Delta, \varphi, \Pi \quad \Gamma \Rightarrow \Delta, \psi, \Pi}{\Gamma \Rightarrow \Delta, \varphi \wedge \psi, \Pi}$$

whereas rules for $\neg$ and $\rightarrow$ look like that:

$$(\neg \Rightarrow) \ \frac{\Gamma, \Delta \Rightarrow \varphi, \Pi}{\Gamma, \neg \varphi, \Delta \Rightarrow \Pi} \qquad\qquad (\Rightarrow \neg) \ \frac{\varphi, \Gamma \Rightarrow \Delta, \Pi}{\Gamma \Rightarrow \Delta, \neg \varphi, \Pi}$$

$$(\rightarrow \Rightarrow) \ \frac{\Gamma, \Delta \Rightarrow \varphi, \Pi \quad \psi, \Gamma, \Delta \Rightarrow \Pi}{\Gamma, \varphi \rightarrow \psi, \Delta \Rightarrow \Pi} \qquad (\Rightarrow \rightarrow) \ \frac{\varphi, \Gamma \Rightarrow \psi, \Delta, \Pi}{\Gamma \Rightarrow \Delta, \varphi \rightarrow \psi, \Pi}$$

Notice that Gallier introduces also K-like rules (multiplicative $(\wedge \Rightarrow)$, additive $(\rightarrow \Rightarrow)$) for better proof search. This solution makes P dispensable even in case of sequents made of lists, and notice that using such sequents may be convenient for designing fully automated proof search procedures, since one can always just take the leftmost (or the rightmost) compound formula to deal with in the root-first procedure.

2.7 Permutability of Rules

As we have observed, in local proofs of cut elimination the possibility of permuting cut upwards plays an essential role. It is worth noting that all steps made for height reduction are special cases of a more general result due to Kleene [151] concerning permutability of rules in SC. Suppose we have a proof of a sequent S with two occurrences of compound formulae φ and ψ. Let φ be a principal formula of an application of r and ψ be a principal formula of an application of r'. If r' was the last logical rule applied in the proof of S and r was a rule performed immediately above, can we show that the order of the application of both rules may be reversed, i.e. r' applied immediately above r in a proof of the same sequent? There are 64 cases to consider in four groups: both rules one-premiss $(25)^3$, the first two-premiss, the second one-premiss (15), reverse order (15), both two-premiss rules (9). Kleene has shown for a variant of LK with additive $(\to\Rightarrow)$ that all rules may be permuted in such a way. We state this result here for LK:

Lemma 2.19. (Permutability). *The successive applications of any two logical rules r and r' of LK may be permuted if the principal formula of r is not a side formula of r'.*

PROOF: Let us consider one example in each group.

(A) If both rules are one-premiss, for example r is $(\wedge \Rightarrow)$ (additive) and r' is $(\Rightarrow\to)$ (multiplicative):

$$
\begin{array}{l}
(\wedge\Rightarrow) \\
(P\Rightarrow) \\
(\Rightarrow\to)
\end{array}
\cfrac{\cfrac{\cfrac{\varphi,\chi,\Gamma\Rightarrow\Delta,\gamma}{\varphi\wedge\psi,\chi,\Gamma\Rightarrow\Delta,\gamma}}{\chi,\varphi\wedge\psi,\Gamma\Rightarrow\Delta,\gamma}}{\varphi\wedge\psi,\Gamma\Rightarrow\Delta,\chi\to\gamma}
$$

is changed into:

$$
\begin{array}{l}
(P\Rightarrow) \\
(\Rightarrow\to) \\
(\wedge\Rightarrow)
\end{array}
\cfrac{\cfrac{\cfrac{\varphi,\chi,\Gamma\Rightarrow\Delta,\gamma}{\chi,\varphi,\Gamma\Rightarrow\Delta,\gamma}}{\varphi,\Gamma\Rightarrow\Delta,\chi\to\gamma}}{\varphi\wedge\psi,\Gamma\Rightarrow\Delta,\chi\to\gamma}
$$

(B) r is two-premiss and r' one-premiss rule. For example, $(\Rightarrow\wedge)$ (additive) versus $(\Rightarrow\to)$ (multiplicative):

$$
\begin{array}{l}
(\Rightarrow\wedge) \\
\\
(\Rightarrow P) \\
(\Rightarrow\to)
\end{array}
\cfrac{\cfrac{\cfrac{\chi,\Gamma\Rightarrow\Delta,\gamma,\varphi \qquad \chi,\Gamma\Rightarrow\Delta,\gamma,\psi}{\chi,\Gamma\Rightarrow\Delta,\gamma,\varphi\wedge\psi}}{\chi,\Gamma\Rightarrow\Delta,\varphi\wedge\psi,\gamma}}{\Gamma\Rightarrow\Delta,\varphi\wedge\psi,\chi\to\gamma}
$$

is changed into:

$$
\begin{array}{l}
(\Rightarrow P) \\
(\Rightarrow\to) \\
(\Rightarrow P) \\
(\Rightarrow\wedge)
\end{array}
\cfrac{\cfrac{\cfrac{\cfrac{\chi,\Gamma\Rightarrow\Delta,\gamma,\varphi}{\chi,\Gamma\Rightarrow\Delta,\varphi,\gamma}}{\Gamma\Rightarrow\Delta,\varphi,\chi\to\gamma}}{\Gamma\Rightarrow\Delta,\chi\to\gamma,\varphi} \qquad \cfrac{\cfrac{\cfrac{\chi,\Gamma\Rightarrow\Delta,\gamma,\psi}{\chi,\Gamma\Rightarrow\Delta,\psi,\gamma}}{\Gamma\Rightarrow\Delta,\psi,\chi\to\gamma}}{\Gamma\Rightarrow\Delta,\chi\to\gamma,\psi} \begin{array}{l}(\Rightarrow P)\\(\Rightarrow\to)\\(\Rightarrow P)\end{array}}{\begin{array}{c}\Gamma\Rightarrow\Delta,\chi\to\gamma,\varphi\wedge\psi\\\hline\Gamma\Rightarrow\Delta,\varphi\wedge\psi,\chi\to\gamma\end{array}\ (\Rightarrow P)}
$$

(C) r is one-premiss and r' two-premiss rule. For example, $(\Rightarrow\to)$ (multiplicative) versus $(\Rightarrow\wedge)$ (additive):

---

^{3}We treat both variants of $(\wedge\Rightarrow)$ and $(\Rightarrow\vee)$ as one rule.

$$
\begin{array}{c}
(\Rightarrow \to) \\
(\Rightarrow P) \\
(\Rightarrow \wedge)
\end{array}
\dfrac{\dfrac{\dfrac{\chi, \Gamma \Rightarrow \Delta, \varphi, \gamma}{\Gamma \Rightarrow \Delta, \varphi, \chi \to \gamma}}{\Gamma \Rightarrow \Delta, \chi \to \gamma, \varphi} \qquad \Gamma \Rightarrow \Delta, \chi \to \gamma, \psi}{\Gamma \Rightarrow \Delta, \chi \to \gamma, \varphi \wedge \psi}
$$

Now because the rule permuted upwards is additive and we must unify all parameters a transformation is considerably more complicated:

$$
\begin{array}{c}
(P)(W) \\
(\Rightarrow \wedge) \\
\\
(\Rightarrow P) \\
(\Rightarrow \to) \\
(\Rightarrow P) \\
(\Rightarrow C) \\
(\Rightarrow P)
\end{array}
\dfrac{\dfrac{\dfrac{\dfrac{\dfrac{\dfrac{\dfrac{\chi, \Gamma \Rightarrow \Delta, \varphi, \gamma}{\chi, \Gamma \Rightarrow \Delta, \chi \to \gamma, \gamma, \varphi} \qquad \dfrac{\Gamma \Rightarrow \Delta, \chi \to \gamma, \psi}{\chi, \Gamma \Rightarrow \Delta, \chi \to \gamma, \gamma, \psi}}{\chi, \Gamma \Rightarrow \Delta, \chi \to \gamma, \gamma, \varphi \wedge \psi}}{\chi, \Gamma \Rightarrow \Delta, \chi \to \gamma, \varphi \wedge \psi, \gamma}}{\Gamma \Rightarrow \Delta, \chi \to \gamma, \varphi \wedge \psi, \chi \to \gamma}}{\Gamma \Rightarrow \Delta, \varphi \wedge \psi, \chi \to \gamma, \chi \to \gamma}}{\Gamma \Rightarrow \Delta, \varphi \wedge \psi, \chi \to \gamma}}{\Gamma \Rightarrow \Delta, \chi \to \gamma, \varphi \wedge \psi}
\begin{array}{c}
(P), (W) \\
\\
\end{array}
$$

We have considered the case with one-premiss rule in the left branch but the case where their location is in the right branch is similar.

(D) Finally consider the case where both rules are two-premiss. For example, multiplicative $(\to \Rightarrow)$ versus additive $(\vee \Rightarrow)$:

$$
\begin{array}{c}
(\to \Rightarrow) \\
(P) \\
(\vee \Rightarrow)
\end{array}
\dfrac{\dfrac{\dfrac{\Gamma \Rightarrow \Delta, \varphi \qquad \psi, \chi, \Pi \Rightarrow \Sigma}{\varphi \to \psi, \chi, \Gamma, \Pi \Rightarrow \Delta, \Sigma}}{\chi, \varphi \to \psi, \Gamma, \Pi \Rightarrow \Delta, \Sigma} \qquad \gamma, \varphi \to \psi, \Gamma, \Pi \Rightarrow \Delta, \Sigma}{\chi \vee \gamma, \varphi \to \psi, \Gamma, \Pi \Rightarrow \Delta, \Sigma}
$$

is changed into:

$$
\dfrac{\Gamma \Rightarrow \Delta, \varphi \qquad \dfrac{\dfrac{\dfrac{\dfrac{\psi, \chi, \Pi \Rightarrow \Sigma}{\chi, \psi, \varphi \to \psi, \Gamma, \Pi \Rightarrow \Delta, \Sigma} \qquad \dfrac{\gamma, \varphi \to \psi, \Gamma, \Pi \Rightarrow \Delta, \Sigma}{\gamma, \psi, \varphi \to \psi, \Gamma, \Pi \Rightarrow \Delta, \Sigma}}{\chi \vee \gamma, \psi, \varphi \to \psi, \Gamma, \Pi \Rightarrow \Delta, \Sigma}}{\psi, \chi \vee \gamma, \varphi \to \psi, \Gamma, \Pi \Rightarrow \Delta, \Sigma}}{\varphi \to \psi, \Gamma, \chi \vee \gamma, \varphi \to \psi, \Gamma, \Pi \Rightarrow \Delta, \Sigma, \Delta}}{\chi \vee \gamma, \varphi \to \psi, \Gamma, \Pi \Rightarrow \Delta, \Sigma}
$$

with labels $(P), (W)$; $(P), (W)$; $(\vee \Rightarrow)$; (P); $(\to \Rightarrow)$; $(P), (C)$.

In the last group we should notice that if we assume additionally that: (a) the formula permuted down (i.e. the principal formula of the conclusion of r) is principal in both premisses of the application of r' and (b) that permuted rules are additive (i.e. invertible), then we can get rid of the additional applications of structural rules (except possibly P). An example:

$$
\begin{array}{c}
(\Rightarrow \wedge)
\end{array}
\dfrac{\dfrac{\chi, \Gamma \Rightarrow \Delta, \varphi \qquad \chi, \Gamma \Rightarrow \Delta, \psi}{\chi, \Gamma \Rightarrow \Delta, \varphi \wedge \psi} \qquad \dfrac{\gamma, \Gamma \Rightarrow \Delta, \varphi \qquad \gamma, \Gamma \Rightarrow \Delta, \psi}{\gamma, \Gamma \Rightarrow \Delta, \varphi \wedge \psi}}{\chi \vee \gamma, \Gamma \Rightarrow \Delta, \varphi \wedge \psi}
\begin{array}{c}
(\Rightarrow \wedge) \\
(\vee \Rightarrow)
\end{array}
$$

is changed into:

$$(\vee \Rightarrow) \dfrac{\chi, \Gamma \Rightarrow \Delta, \varphi \qquad \gamma, \Gamma \Rightarrow \Delta, \varphi}{\chi \vee \gamma, \Gamma \Rightarrow \Delta, \varphi} \qquad \dfrac{\dfrac{\chi, \Gamma \Rightarrow \Delta, \psi \qquad \gamma, \Gamma \Rightarrow \Delta, \psi}{\chi \vee \gamma, \Gamma \Rightarrow \Delta, \psi} (\vee \Rightarrow)}{\chi \vee \gamma, \Gamma, \Rightarrow \Delta, \varphi \wedge \psi} (\Rightarrow \wedge)$$

This fact is of importance for Zeman's [276] proof of admissibility of contraction which we will consider in the next chapter.

Exercise 2.36. *Check at least two examples in each group. Try to select different combinations of rules (e.g. both additive, both multiplicative) to see what structural rules are needed in the transformation.*

Remark 2.5. It should be noted that problems of permutability and of inversion of rules are closely related, as was already noted by Curry in connection with his proof of invertibility based on global transformation (see the proof of theorem 2.3). Let us consider a proof ending with the application of r' but with r introducing φ as a principal formula immediately above. If φ is parametric in this application of r' and we apply Curry's transformation to φ (i.e. we replace it in the proof with its side(s) formula(e)), then the last sequent of the transformed proof contains its side(s) formula(e) and we can apply r below r'.

In this sense we may roughly say that invertibility implies permutability. It also provides a better explanation for the fact noted in section 1.6 that one can explain confluency of rules either in terms of invertibility or in terms of permutability.

Chapter 3

Purely Logical Sequent Calculus

In this chapter we consider a system very similar to K but with one significant difference: sequents are ordered pairs of finite multisets. We briefly discuss why such a choice seems to be a better solution than using lists or sets as building blocks. A specific calculus, commonly called G3, is apparently identical with K but we need to provide for it some auxiliary results like e.g. admissibility of contraction rules. This will be done in section 3.2. In section 3.3, we focus on four different strategies of proving admissibility of cut either in multiplicative or in additive version. The most popular approach is due to Dragalin and it is essentially the proof we provided for K in chapter 1. However, the fact that sequents are built from multisets introduces some additional complications which require applications of contraction. Smullyan-style proof is similar but devised for multiplicative cut. It was originally presented for tableau system but we provide here its modification for G3. Both proofs work in general for SC with invertible rules but invertibility itself is not used in proofs. Schütte provided a different kind of proof which explicitly applies invertibility of rules. We present two variants of his proof for multiplicative and additive version of cut. Section 3.4 is devoted to additional topics connected with different ways of interpretation of sequents. As a by-product of these considerations we will provide a Post-style completeness proof via conjunctive normal forms in the setting of G3. In section 3.5 a general survey of different types of SC, including nonstandard and generalised systems, will be sketched. Some of them will be applied in the remaining chapters. Summary of kinds of rules and their properties is provided for better characterisation of standard SCs. In the next section we provide three different proofs of the interpolation theorem, due to Maehara, Smullyan, Wintein and Muskens. All are constructive and serve as another application of the cut-elimination theorem. In the last section we consider some other rule, called tautology elimination rule, which is equivalent to cut. We will prove admissibility of this rule for G3 as an alternative way of proving admissibility of cut for SC with invertible rules. Finally, a proof of a refined

© Springer Nature Switzerland AG 2021

A. Indrzejczak, *Sequents and Trees*, Studies in Universal Logic,
https://doi.org/10.1007/978-3-030-57145-0_3

version of the strong completeness theorem for nonstandard SC with elimination rules is provided. This result, due to Přenosil, shows that all structural rules may be restricted to atomic instances and in particular, cuts are only on atoms occurring in the initial nonaxiomatic sequents. The resulting proofs resemble normal natural deduction proofs and additionally give us interpolants.

3.1   The System G3

The reader certainly noticed that although sequents of LK were defined as pairs of finite lists we quite often disregard the order of formulae in sequents. In this chapter we officially use sequents built from multisets. However, the special calculus dealt with here is not just like LK with permutation rules deleted. In this chapter we will be concerned with one of the most popular versions of SC commonly called G3[1]. It consists of axioms of the form $\varphi, \Gamma \Rightarrow \Delta, \varphi$ with $\varphi \in PROP$ and the following rules:

$$(\neg\Rightarrow) \quad \frac{\Gamma \Rightarrow \Delta, \varphi}{\neg\varphi, \Gamma \Rightarrow \Delta} \qquad\qquad (\Rightarrow\neg) \quad \frac{\varphi, \Gamma \Rightarrow \Delta}{\Gamma \Rightarrow \Delta, \neg\varphi}$$

$$(\wedge\Rightarrow) \quad \frac{\varphi, \psi, \Gamma \Rightarrow \Delta}{\varphi\wedge\psi, \Gamma \Rightarrow \Delta} \qquad\qquad (\Rightarrow\wedge) \quad \frac{\Gamma \Rightarrow \Delta, \varphi \quad \Gamma \Rightarrow \Delta, \psi}{\Gamma \Rightarrow \Delta, \varphi\wedge\psi}$$

$$(\Rightarrow\vee) \quad \frac{\Gamma \Rightarrow \Delta, \varphi, \psi}{\Gamma \Rightarrow \Delta, \varphi\vee\psi} \qquad\qquad (\vee\Rightarrow) \quad \frac{\varphi, \Gamma \Rightarrow \Delta \quad \psi, \Gamma \Rightarrow \Delta}{\varphi\vee\psi, \Gamma \Rightarrow \Delta}$$

$$(\Rightarrow\rightarrow) \quad \frac{\varphi, \Gamma \Rightarrow \Delta, \psi}{\Gamma \Rightarrow \Delta, \varphi\rightarrow\psi} \qquad\qquad (\rightarrow\Rightarrow) \quad \frac{\Gamma \Rightarrow \Delta, \varphi \quad \psi, \Gamma \Rightarrow \Delta}{\varphi\rightarrow\psi, \Gamma \Rightarrow \Delta}$$

The notions of a proof, derivation, principal, side formula, etc. remain the same as in the preceding chapters. It is obvious that we defined G3 in exactly the same way as K (or rather K' from section 1.5). The only difference is that sequents consist of finite multisets and we restrict axioms to atomic, i.e. an active formula on both sides of an axiomatic sequent is a propositional symbol. The last restriction is inessential—recall lemma 1.8 for K; its proof is the same for G3. It follows also that G3 is an adequate formalisation of **CPL**; the reader may apply some strategies from chapter 1 and reprove it for G3 to see that application of multisets does not call for any substantial changes.

However, the lack of any structural rules may seem surprising. Permutation is dispensable and weakening as well, due to generalised form of axioms and context independence of rules. But we can expect that, due to the replacement of sets with multisets, at least one structural rule will be necessary—contraction. However, it is not so—as we will see in the next section one may prove that contraction is admissible in G3. In fact, one may wonder if we really need contraction in proofs of sequents in G3. In chapter 2 we have shown that contraction is required in

---

[1]The calculus was essentially introduced by Ketonen and in the exact form and under this name was introduced by Troelstra and Schwichtenberg [264].

(cut-free) LK proofs because of using context-free two-premiss rule for → and one-premiss rules for ∧ and ∨ with one-side formula. But K and G3 avoid the problem by using only context-sharing two-premiss rules and one-premiss rules with both side formulae. In fact, we do not need contraction for constructing proofs in G3 but it does not mean that is not needed at all. As we will see contraction is necessary for proving metatheorems about G3. The problem arises in particular with proving admissibility of cut for G3 and we will see that admissibility of this rule is essential, at least for multiplicative cut. Due to the lack of primitive structural rules G3 is called purely logical calculus (only logical rules are primitive).

One may ask why we prefer to use multisets as the basic data structures instead of sets. As far as we are concerned with classical logic the problem may seem apparent. Many authors (for example, Avron, Smullyan, Fitting) prefer sets, and some even argue (e.g. Bimbo [31]) that the use of multisets is not necessary. On the other hand, authors like Negri and von Plato [185, 186] provide strong arguments for using multisets instead of sets, mainly on formal grounds. For example, in rules like (∧ ⇒) or (⇒ ∨), we do not have a formal way for representing a suitable inference if both side formulae are identical. Such a possibility is usually treated (also in chapter 1) implicitly as an obvious case but if considered formally it leads to further complications in the definition of a calculus (see [185]). In fact, it was noticed very early by Curry [55] that formally adequate application of sets must complicate rules. However, the most important rationale for using multisets is connected with SC formalizations of many non-classical logics which will be discussed in chapter 5. It appeared that in the framework of SC, for dealing with several logics weaker than **CPL**, like relevant or many-valued logics, the application of more refined data structures (at least multisets) in sequents is sometimes necessary. As we will see in chapter 5, for some of these logics we may obtain a satisfactory formalisation by deleting or modifying structural rules. This phenomenon led in consequence to the introduction of the term "substructural logics" to cover some of the families of non-classical logics [2].

3.2 Preliminary Results

We describe here some results of auxiliary character which are needed for proofs of cut admissibility presented in the next section. We start with an almost trivial result:

Lemma 3.1 (Height-preserving admissibility of W for G3). *If* $\vdash_n \Gamma \Rightarrow \Delta$, *then* $\vdash_n \Gamma' \Rightarrow \Delta'$ *for* $\Gamma \subseteq \Gamma', \Delta \subseteq \Delta'$

PROOF: By induction on the height exactly as for K (lemma 1.4 in subsection 1.4.2). □

---

[2]Interestingly enough Negri and von Plato who strongly insist on using multisets are rather not convinced with this argument since some of these logics may be formalised with structural rules as well but on the ground of generalised forms of SC.

It is obvious that this result holds also for G3 with axioms not restricted to atomic active formula. What is more interesting is that we can prove admissibility of weakening for even more strongly restricted SC with axioms strictly atomic, i.e. having all formulae atomic: active and parametric. However, a proof is more complicated since we must make an additional induction on the complexity of added formulae (see Ershow and Palyutin [74]). Moreover, it is not height-preserving admissibility.

Exercise 3.1. *Prove admissibility of W in G3 with strictly atomic axioms.*

There is also no problem with the following result:

Lemma 3.2. $\vdash \varphi, \Gamma \Rightarrow \Delta, \varphi$, *for any* φ.

PROOF: The same as a proof of lemma 1.8 in section 1.5.   □

3.2.1 Invertibility Again

We know (see subsection 1.4.3) that all rules of K are (semantically) invertible and this result applies to G3 as well. Also, in section 2.3 we have shown that it is possible to prove invertibility of rules syntactically in several ways:

1. by cut applications—the simplest method due to Ketonen [148];

2. by tracking the proof of the conclusion (Curry [56], Negri and von Plato [185]);

3. by induction on the height (Schütte [232], Pogorzelski [199]).

The last method in case of G3 may be strengthened in the sense that for inversions of all rules we may prove its height-preserving invertibility. We will recall here this result after Dragalin [64]:

Lemma 3.3 (Height-preserving invertibility). *For any instance of the application of any logical rule, if the conclusion has a proof of height n, then premises have proofs of height $\leq n$.*

PROOF: by induction on the height. We will illustrate the proof with the case of $(\wedge \Rightarrow)$. The basis is provable because axioms are generalised and atomic—let $\varphi \wedge \psi, \Gamma, p \Rightarrow p, \Delta$ be an axiom, then $\varphi, \psi, \Gamma, p \Rightarrow p, \Delta$ is also an axiom.

In induction step let $\mathcal{D} \vdash_n \varphi \wedge \psi, \Gamma \Rightarrow \Delta$. There are two cases:

a. $\varphi \wedge \psi$ is principal

b. $\varphi \wedge \psi$ is not principal.

The case (a) is trivial; just delete the last sequent to get $\mathcal{D} \vdash_{n-1} \varphi, \psi, \Gamma \Rightarrow \Delta$. In case (b) we must consider the last rule application. Due to context independence of all rules we may demonstrate it schematically. E.g. the last rule is $(\vee \Rightarrow)$ and we have:

$$\frac{\gamma, \varphi \wedge \psi, \Gamma \Rightarrow \Delta \qquad \delta, \varphi \wedge \psi, \Gamma \Rightarrow \Delta}{\gamma \vee \delta, \varphi \wedge \psi, \Gamma \Rightarrow \Delta}$$

By the induction hypothesis both $\gamma, \varphi, \psi, \Gamma \Rightarrow \Delta$ and $\delta, \varphi, \psi, \Gamma \Rightarrow \Delta$ are provable with lower height. By the same rule we infer $\gamma \vee \delta, \varphi, \psi, \Gamma \Rightarrow \Delta$. $\qquad\square$

Exercise 3.2. *Prove the remaining subcases of the case b.*

It is instructive to compare this proof with the proof of theorem 2.4 from section 2.3. Despite the similarities in case of G3 we can both simplify a proof and obtain a stronger result. The reader should consider why this is possible for G3 but not for LK.

3.2.2 Admissibility of Contraction

All the results stated so far were possible to obtain for K and invertibility also for LK although in a slightly restricted form. Now we focus on the result which does not hold for LK (and for K is implicit)—admissibility of contraction. Again, the most popular form of the proof of this result is due to Dragalin [64] (see also Schwichtenberg and Troelstra [264], Negri and von Plato [185]) although sometimes the name Curry's lemma is used. Below we explain why we do not follow this custom here.

Lemma 3.4 (Height-preserving admissibility of C for G3). *If $\vdash_n \Gamma, \varphi, \varphi \Rightarrow \Delta$, then $\vdash_n \Gamma, \varphi \Rightarrow \Delta$ and if $\vdash_n \Gamma \Rightarrow \Delta, \varphi, \varphi$, then $\vdash_n \Gamma \Rightarrow \Delta, \varphi$*

PROOF: By induction on the height of proof of $\varphi, \varphi, \Gamma \Rightarrow \Delta$ (for $\Gamma \Rightarrow \Delta, \varphi, \varphi$ analogous). In the basis: $\vdash_0 \varphi, \varphi, \Gamma \Rightarrow \Delta$, hence it is an axiom and $\varphi, \Gamma \Rightarrow \Delta$ is also an axiom.

The induction hypothesis claims that the lemma holds for any sequent with the proof having the height $< n$. We must show that it holds if the proof has the height n. There are two cases:

   a. φ is not principal

   b. φ is principal

If φ is not principal, then $\varphi, \varphi, \Gamma \Rightarrow \Delta$ is derived from $\varphi, \varphi, \Gamma' \Rightarrow \Delta'$ by one-premiss rule or additionally from $\varphi, \varphi, \Gamma'' \Rightarrow \Delta''$ by two-premiss rule. In each case proofs of premisses are covered by the induction hypothesis so $\varphi, \Gamma' \Rightarrow \Delta'$ (and $\varphi, \Gamma'' \Rightarrow \Delta''$) has a proof of the height $< n$, hence by the application of the same rule we get a proof of $\varphi, \Gamma \Rightarrow \Delta$ with the height at most n. Note that this step holds by context independence of rules.

If φ is principal, then we must consider all cases of φ.

Let $\varphi := \psi \wedge \chi$, then our sequent is of the form $\psi \wedge \chi, \psi \wedge \chi, \Gamma \Rightarrow \Delta$ and the premiss of the form $\psi, \chi, \psi \wedge \chi, \Gamma \Rightarrow \Delta$ has a proof of the height $n - 1$. By height-preserving invertibility lemma $\psi, \chi, \psi, \chi, \Gamma \Rightarrow \Delta$ has a proof of the height not exceeding $n - 1$. By the induction hypothesis applied twice we get a proof of $\psi, \chi, \Gamma \Rightarrow \Delta$ of the height not greater than $n - 1$, so by ($\wedge \Rightarrow$) we get a proof of $\psi \wedge \chi, \Gamma \Rightarrow \Delta$ of the height at most n.   □

Exercise 3.3. *Complete the proof for other cases of principal φ and prove the lemma for succedent contraction. Note that in all cases of φ principal we need to apply the height-preserving invertibility lemma.*

A careful analysis of this proof shows that we can extract from it some weaker result which does not require appealing to the invertibility lemma. It is:

Lemma 3.5 (Height-preserving admissibility of atomic C for G3). *If $\vdash_n \Gamma, \varphi, \varphi \Rightarrow \Delta$, then $\vdash_n \Gamma, \varphi \Rightarrow \Delta$ and if $\vdash_n \Gamma \Rightarrow \Delta, \varphi, \varphi$, then $\vdash_n \Gamma \Rightarrow \Delta, \varphi$, for φ atomic.*

Appealing to height-preserving invertibility during the proof of contraction admissibility is also not required if we use a variant of G3 with rules modified in a way described in subsection 1.7.2. (a system KK). We mean rules involving so called Kleene's trick, i.e. having a copy of the principal formula in premisses (see e.g. Pfenning [196]). Nowadays such rules are often called contraction-absorbing for this reason. In fact, such a result was indeed first proved by Curry [55] but it should be noted that he did not appeal to invertibility and did not explicitly mention height-preservation. Curry noted that such a result holds when standard rules are replaced with contraction-absorbing rules. Later, in [56] he has taken into account also invertibility of ordinary rules. On the other hand, Dragalin first proved it for G3 on the basis of height-preserving invertibility.

In fact, height-preserving admissibility of contraction is indeed too much for most of our purposes; simple admissibility is enough and one may prove it by using Zeman's method. Originally Zeman [276] proved the eliminability of C as a prerequisite for decidability proof for a version of LK but with all rules invertible. Invertibility of rules as such is not required in the proof but anyway his proof works for SC with invertible rules only and atomic axioms, since it is not possible to eliminate contraction in the presence of noninvertible rules (see examples of proofs for Peirce law and excluded middle in section 2.1). A detailed exposition of such proof but for labelled SC may be also found in Vigano [274]. Below we apply Zeman's proof for showing admissibility of C in G3.

The proof goes by induction on the complexity of contracted formula and subsidiary induction on the so called rank of C. The latter notion is defined similarly as the rank of cut so we informally describe it as the highest (different branches may be taken into account) number of sequents above the root which contain at least one occurrence of a formula in question.

Lemma 3.6 (Admissibility of C). *If $\vdash \Gamma, \varphi, \varphi \Rightarrow \Delta$, then $\vdash \Gamma, \varphi \Rightarrow \Delta$ and if $\vdash \Gamma \Rightarrow \Delta, \varphi, \varphi$, then $\vdash \Gamma \Rightarrow \Delta, \varphi$*

PROOF: In the basis we consider atomic φ. Even if it is an active formula in axiomatic sequents only one occurrence is necessary so we can delete the second occurrence in all sequents without destroying a proof. Note that this is possible because all two-premiss rules are additive. If we have at least one multiplicative rule we could obtain two occurrences of φ in the conclusion of such rule application coming from different premisses and deletion of one of them could destroy a proof.

In the induction step we consider all compound formulae and as the induction hypothesis we assume that the claim holds for any shorter formulae. First of all notice that if at least one occurrence of φ was not introduced by logical rule it must be present in all axiomatic sequents as a parameter (since only atomic formulae may be active in axioms) and it may be just deleted from the whole proof. So we assume that both occurrences were introduced by (the same) logical rule.

First we assume that one occurrence of φ was introduced by the last applied rule and perform an induction on the rank. Note that under this additional assumption the rank is just a distance between a sequent with two occurrences of φ and the introduction of the first one. In the basis the rank is 2. We illustrate a proof with two cases: of $\varphi \vee \psi$ in the succedent and $\varphi \to \psi$ in the antecedent.

In the first case we have:

$$(\Rightarrow \vee) \cfrac{(\Rightarrow \vee) \cfrac{\Gamma \Rightarrow \Delta, \varphi, \psi, \varphi, \psi}{\Gamma \Rightarrow \Delta, \varphi, \psi, \varphi \vee \psi}}{\Gamma \Rightarrow \Delta, \varphi \vee \psi, \varphi \vee \psi}$$

We apply the induction hypothesis twice and obtain:

$$(\Rightarrow \vee) \frac{\Gamma \Rightarrow \Delta, \varphi, \psi}{\Gamma \Rightarrow \Delta, \varphi \vee \psi}$$

In the second case we have:

$$(\to \Rightarrow) \cfrac{(\to \Rightarrow) \cfrac{\Gamma \Rightarrow \Delta, \varphi, \varphi \qquad \psi, \Gamma \Rightarrow \Delta, \varphi}{\varphi \to \psi, \Gamma \Rightarrow \Delta, \varphi} \qquad \cfrac{\psi, \Gamma \Rightarrow \Delta, \varphi \qquad \psi, \psi, \Gamma \Rightarrow \Delta}{\varphi \to \psi, \psi, \Gamma \Rightarrow \Delta} (\to \Rightarrow)}{\varphi \to \psi, \varphi \to \psi, \Gamma \Rightarrow \Delta}$$

again we apply the induction hypothesis twice to the leftmost and to the rightmost premiss and obtain:

$$(\to \Rightarrow) \frac{\Gamma \Rightarrow \Delta, \varphi \qquad \psi, \Gamma \Rightarrow \Delta}{\varphi \to \psi, \Gamma \Rightarrow \Delta}$$

In the induction step we consider rank $= k + 1$ and assume that the claim holds for rank $= k$. Since rank is higher than 2 at least one logical rule r' is applied below the application of a rule r introducing φ. Moreover φ is a parameter in the application of r' since it is parametric in all sequents below. Hence we can apply the permutability lemma and the claim holds by the induction hypothesis. Note that in case of r' being two-premiss rule due to the fact that all such rules are additive our φ is present in both premisses. Moreover, it must be introduced by r immediately above r' in both premisses of r' due to our assumption that r' is the next rule. In such a situation we have a simplified proof of permutability which was illustrated in the comment to the proof of the permutability lemma in section 2.7.

Thus we have proved that the claim holds if the second occurrence of φ was introduced as the last step in the proof of $\Gamma \Rightarrow \Delta, \varphi, \varphi$ (or $\varphi, \varphi, \Gamma \Rightarrow \Delta$). Now assume that it is not, so above we have some (at least one) sequent(s) on different branches of the shape $\Gamma' \Rightarrow \Delta', \varphi, \varphi$ (or $\varphi, \varphi, \Gamma' \Rightarrow \Delta'$) where one occurrence of φ was just introduced. So our previous consideration applies and we obtain $\Gamma' \Rightarrow \Delta', \varphi$ (or $\varphi, \Gamma' \Rightarrow \Delta'$) in all cases. Now proofs in which some parametric formula occurs twice may be replaced with proofs where there is only one occurrence of it due to context independence. So we may safely delete the second occurrence of φ in all sequents below $\Gamma' \Rightarrow \Delta', \varphi$ (or $\varphi, \Gamma' \Rightarrow \Delta'$) and obtain the proof of $\Gamma \Rightarrow \Delta, \varphi$ (or $\varphi, \Gamma \Rightarrow \Delta$). $\qquad\square$

Exercise 3.4. *Complete the proof for other cases.*
Prove the permutability lemma for G3.

Such a proof is sufficient—as we will see—as the basis for some cut admissibility proofs. Note again its essential limitations, namely that this proof does not work for SC with noninvertible rules. We have noticed that the basis cannot be proved in the presence of multiplicative two-premiss rules. But additive one-premiss rule also does not admit such a proof in the induction step. Consider again the case of $\varphi \vee \psi$ in the succedent but this time obtained by additive $(\Rightarrow \vee)$. It may be that each occurrence of $\varphi \vee \psi$ was deduced from a different side formula and we have:

$$(\Rightarrow \vee) \cfrac{(\Rightarrow \vee) \cfrac{\Gamma \Rightarrow \Delta, \varphi, \psi,}{\Gamma \Rightarrow \Delta, \varphi, \varphi \vee \psi}}{\Gamma \Rightarrow \Delta, \varphi \vee \psi, \varphi \vee \psi}$$

In such a situation we cannot apply the induction hypothesis and the proof breaks down.

On the other hand, it should be noted that the restriction to atomic axioms is not necessary at the cost of little complication in the proof. Notice that the only point where the assumption concerning atomic axioms was needed is at the beginning of the induction step, on the basis of the induction on rank. In brief, if both occurrences of (compound) contracted formula are not introduced by respective logical rule, then one must be a parameter and may be safely deleted. If we admit nonatomic axioms it is possible that one occurrence is introduced by logical rule but the other is active in axiom. But it does not make any problems. Consider again the first example but now in this shape, so we have:

$$(\Rightarrow \vee) \cfrac{\varphi \vee \psi, \Gamma \Rightarrow \Delta, \varphi, \psi, \varphi \vee \psi}{\varphi \vee \psi, \Gamma \Rightarrow \Delta, \varphi \vee \psi, \varphi \vee \psi}$$

and instead we can simply start with an axiom $\varphi \vee \psi, \Gamma \Rightarrow \Delta, \varphi \vee \psi$. Other cases are eliminated similarly.

Zeman's original proof is a proof of elimination which leads to some small differences with the proof displayed above. He requires that:

1. C is the last applied rule;

2. a proof contains no other application of C;

3. one of the occurrences of the contracted formula is introduced immediately above C.

The first two restrictions play the same role as in Gentzen's proof of cut elimination; we can always apply his top-down method of elimination of more applications of C in any proof or just treat our proof as the proof of the basis of induction on the number of applications of C in arbitrary proof. The third is added also for simplifying a proof and is easy to explain as well. Proofs in which some parametric formula occurs twice for several steps and then it is contracted may be replaced with proofs where contraction is applied immediately after the second occurrence was introduced. It is justified because correctness of all inferences is preserved if only one occurrence of this parameter is present instead of two.

3.3 Admissibility of Cut

Now we focus on proofs of admissibility of cut for G3. All of them are local. The most popular one is due to Dragalin [64] (see excellent presentations in Troelstra and Schwichtenberg [264] or Negri and von Plato [185]) and it is formulated exactly for G3 with M-cut. Some other proofs which will be presented in this section are due to Schütte [232] and Smullyan [245]. They may be classified as of essentially similar character although in both cases they were formulated for systems of quite a different character. Schütte proved the result for a system with one-sided sequents which are finite sets of formulae in the negation normal form, moreover he applied explicitly the notion of antecedent and succedent occurrence of a formula. Smullyan's formulation is for Hintikka's style tableau system also operating on finite sets of formulae. This way in both cases the problems with C are avoided from the start. We will show in what way their strategies may be applied to G3 at the price of departure from many details of a peculiar character. There are two significant differences between presented proofs: a) Dragalin and Schütte prove the admissibility of M-cut, whereas Smullyan's proof is for A-cut. b) Dragalin and Smullyan do not apply explicitly the invertibility of rules but refer to admissibility of C instead, whereas Schütte's proof explicitly applies invertibility of rules. These seemingly small differences lead to further changes in the list of prerequisites but also in the proof technique. We additionally examine the proof which combines some features of Schütte's and Smullyan's approach by explicit application of invertibility but for A-cut.

As for similarities we should note that all these proofs are local, all apply directly to cut and proceed by double induction on the complexity of cut-formula (main) and the cut-height (subsidiary).

3.3.1   Dragalin's Proof

This is a very elegant proof and in some sense the most similar to Gentzen's original proof but much simpler in details due to the fact that structural rules, mix, and rank are avoided. In particular, the induction on the cut-height does not require distinct transformations according to the status of cut-formula (principal, side, etc.). This proof in fact was provided for K in section 1.8 so we do not need to record it here again in full but only consider what must be changed due to the fact that G3 uses sequents built from multisets.

Let us recall that it is handy to divide a proof into three parts:

1. at least one premiss is axiomatic;

2. the cut-formula is principal in both premisses;

3. the cut-formula is not principal in at least one premiss.

One may go even further and make an abstraction of these parts as partial results which together yield the cut admissibility theorem (similarly as in Curry's proof). This will be useful in the next subsections since we organise all proofs in this way for easier comparison. So let us call them, respectively:

1. The trivial cuts elimination lemma—a result to the effect that if one premiss of M-cut is axiomatic, then it is eliminable.

2. The principal cut reduction lemma—a result (corresponding to Curry's stage 3 reduction lemma) to the effect that M-cut on compound formula may be always replaced with M-cut(s) on their subformulae.

3. The height reduction lemma—a result stating that if cut-formula is not principal in one premiss of M-cut, then we may lift up an application of M-cut on this premiss.

As for 1 and 3, there are no differences between a proof for K and for G3 so one may repeat for G3 the proof presented in section 1.8. However in case of 2, there is a difference which requires the application of contraction. Let us illustrate the problem:

$$(\Rightarrow \vee) \dfrac{\dfrac{\Gamma \Rightarrow \Delta, \varphi, \psi}{\Gamma \Rightarrow \Delta, \varphi \vee \psi} \qquad \dfrac{\varphi, \Pi \Rightarrow \Sigma \qquad \psi, \Pi \Rightarrow \Sigma}{\varphi \vee \psi, \Pi \Rightarrow \Sigma} (\vee \Rightarrow)}{\Gamma, \Pi \Rightarrow \Delta, \Sigma} (Cut)$$

is replaced with:

$$(Cut) \dfrac{\dfrac{\dfrac{\Gamma \Rightarrow \Delta, \varphi, \psi \qquad \varphi, \Pi \Rightarrow \Sigma}{\Gamma, \Pi \Rightarrow \Delta, \Sigma, \psi} \qquad \psi, \Pi \Rightarrow \Sigma}{\Gamma, \Pi, \Pi \Rightarrow \Delta, \Sigma, \Sigma}}{\Gamma, \Pi \Rightarrow \Delta, \Sigma} \begin{matrix} \\ (Cut) \\ \\ (C) \end{matrix}$$

Note that not only two cuts are introduced (both eliminable by the induction hypothesis) but we must additionally apply contraction to the last sequent to get a correct result.

So in order to apply the strategy of Dragalin in proving cut admissibility for G3 we must have contraction at our disposal and this fact requires some other preliminary results. We may list necessary prerequisites:

1. The calculus requires axioms restricted to atomic active formulae. Clearly we must also show that this restriction does not preclude a proof of such sequents for any compound φ (see lemma 1.8).

2. We must show height-preserving admissibility of W (necessary for trivial cuts elimination)

3. We must show also height-preserving invertibility of logical rules (presupposes 1).

4. Finally we must show height-preserving admissibility of C (presupposes 3, hence also 1).

This is how it is usually presented. The striking simplicity of proof is strongly dependent on the rich collection of auxiliary results. On the other hand, for K we need only point 2, i.e. admissibility of W. Is it possible to reduce the list of preliminary results in case of G3?

Point 2 is necessary for proving part 1 (i.e. at least one premiss of cut axiomatic) for all proofs of M-cut admissibility in K or G3. It is not necessary to have W height-preserving admissible but it does not matter. As for the rest, we have seen that point 4 is essential for proving part 2, however one may drop point 3 if point 4 is obtained by Zeman's proof of the weaker result (just admissibility of C). It is possible because: (a) invertibility of rules is not directly applied in Dragalin's proof; (b) The proof uses admissibility of C but not necessarily in height-preserving version. (c) Zeman's proof, although it also holds only for the system with atomic axioms and invertible rules, does not apply invertibility as well and provides only a sheer admissibility of C. So we can resign from proving invertibility of rules for a system if our purpose is only to show M-cut admissibility; points 1, 2 and 4 in weaker version (just admissibility of C) are enough for that. In fact, for Zeman's proof even 1 is not needed, as we noticed in the previous section. So only 2 and 4 (in the weakened version, not requiring h-p admissibility) are necessary as preliminaries for this proof.

One should also notice that it is possible to avoid application of contraction if instead of using the ordinary version of M-cut we take the version due to Herbelin [110] and mentioned in subsection 2.1.1. For G3 with such 'intelligent' form of cut we do not need to prove admissibility of C and, in consequence, also invertibility of rules. Restriction to atomic axioms is also not necessary and the proof of the admissibility of cut goes simply like for K.

3.3.2   Smullyan's Proof

In case of Dragalin's proof we put stress on the fact that all three parts (or corresponding lemmata) hold for M-cut. Is it necessary? Are there any differences if we prove admissibility of A-cut instead? Here we focus on this problem while presenting a proof of A-cut admissibility due to Smullyan [245]. We have already mentioned that Smullyan's original proof has apparently many differences with other proofs. In particular, it was provided for tableau system with A-cut and with axioms not restricted to atomic. C is dispensable because the system is defined for sets, but height-preserving admissibility of W is essential. The original proof of Smullyan is in fact organised in the same way so, after reformulation for G3, we may find it essentially similar to Dragalin's proof. The only serious difference results from using A-cut. This time both reduction steps are strongly dependent on having W as height-preserving admissible. Moreover, introducing multisets calls for C and atomic axioms which were dispensable in Smullyan's system operating on sets. Summing up the prerequisites for the proof include:

1. atomic generalised axioms;

2. height-preserving admissibility of W (required for the reduction of height and of complexity, but not for axiom case);

3. admissibility (not necessarily height-preserving) of C (required for the case of cut on axiom and of height reduction).

So again the invertibility of rules is not required in the proof, if point 3 is obtained by Zeman's proof. As we noted in this case axioms also do not need to be atomic since (improved) Zeman's proof does not require this condition. Summing up, necessary prerequisites are exactly the same as for Dragalin's proof. But notice that W must be h-p admissible, not just admissible (which is sufficient for Dragalin's proof).

In part 1, we again have two subcases. Let the left premiss of cut be axiomatic, if φ is not active in it, then $\Gamma \Rightarrow \Delta$ is an axiom and we are done. Otherwise $\Gamma = \varphi, \Gamma'$ and we obtain $\Gamma \Rightarrow \Delta$ from the right premiss by contraction on φ. Note that in contrast to elimination of M-cut, we do not need weakening here but contraction.

Part 2 of the proof is straightforward although the reduction of complexity in the presence of A-cut requires admissibility of W for unification of premisses. Here is an example:

$$(\Rightarrow \vee) \ \frac{\dfrac{\Gamma \Rightarrow \Delta, \varphi, \psi}{\Gamma \Rightarrow \Delta, \varphi \vee \psi} \qquad \dfrac{\varphi, \Gamma \Rightarrow \Delta \qquad \psi, \Gamma \Rightarrow \Delta}{\varphi \vee \psi, \Gamma \Rightarrow \Delta} (\vee \Rightarrow)}{\Gamma \Rightarrow \Delta} (Cut)$$

is replaced with:

$$(Cut) \cfrac{\Gamma \Rightarrow \Delta, \varphi, \psi \qquad \cfrac{\varphi, \Gamma \Rightarrow \Delta}{\varphi, \Gamma \Rightarrow \Delta, \psi}(W)}{(Cut) \cfrac{\Gamma \Rightarrow \Delta, \psi \qquad\qquad \psi, \Gamma \Rightarrow \Delta}{\Gamma \Rightarrow \Delta}}$$

The most complicated part of Smullyan's proof is part 3. Let us consider an example:

$$\cfrac{\varphi \vee \psi, \Gamma \Rightarrow \Delta, \chi \qquad \cfrac{\chi, \varphi, \Gamma \Rightarrow \Delta \qquad \chi, \psi, \Gamma \Rightarrow \Delta}{\chi, \varphi \vee \psi, \Gamma \Rightarrow \Delta}(\vee \Rightarrow)}{\varphi \vee \psi, \Gamma \Rightarrow \Delta}(Cut)$$

it is replaced with:

$$(C) \cfrac{(\vee \Rightarrow) \cfrac{(W) \cfrac{\varphi \vee \psi, \Gamma \Rightarrow \Delta, \chi}{(Cut) \cfrac{\varphi, \varphi \vee \psi, \Gamma \Rightarrow \Delta, \chi \qquad \chi, \varphi, \Gamma \Rightarrow \Delta}{\varphi, \varphi \vee \psi, \Gamma \Rightarrow \Delta}}}{\cfrac{\cfrac{\varphi \vee \psi, \Gamma \Rightarrow \Delta, \chi \qquad \chi, \psi, \Gamma \Rightarrow \Delta}{\psi, \varphi \vee \psi, \Gamma \Rightarrow \Delta, \chi \qquad \chi, \psi, \varphi \vee \psi, \Gamma \Rightarrow \Delta}}{\psi, \varphi \vee \psi, \Gamma \Rightarrow \Delta}}}{\cfrac{\varphi \vee \psi, \varphi \vee \psi, \Gamma \Rightarrow \Delta}{\varphi \vee \psi, \Gamma \Rightarrow \Delta}}$$

Both cuts are of lower height hence eliminable. Note however that in this part we need W not just admissible, but height-preserving admissible for induction to work.

Exercise 3.5. *Provide proofs of the remaining cases in all three parts of the proof.*

Note that C is derivable by A-cut but in general it is not sufficient for avoiding a proof of C admissibility; without it we have only a system with noneliminable trivial cuts of the form:

$$(Cut) \cfrac{\varphi, \Gamma \Rightarrow \Delta, \varphi \qquad \varphi, \varphi, \Gamma \Rightarrow \Delta}{\varphi, \Gamma \Rightarrow \Delta}$$

Our exposition of Smullyan's proof differs significantly from his original presentation since we tried to focus on the essential points of his strategy. One may find in Bimbo [31] a thorough analysis so we finish with only brief comments concerning one aspect of his original proof. In particular, Smullyan applies SC similar to K, i.e. he uses sequents built from sets so he did not bother about admissibility of contraction. Accordingly he did not provide any proof of invertibility of rules but instead he proves the following lemma:

Lemma 3.7. *For any* $i, j < k,$

1. *if* $\vdash \Gamma \Rightarrow_k^{\neg\varphi} \Delta, \neg\varphi,$ *then* $\vdash \Gamma, \varphi \Rightarrow_i \Delta, \neg\varphi$

2. *if* $\vdash \Gamma, \neg\varphi \Rightarrow_k^{\neg\varphi} \Delta,$ *then* $\vdash \Gamma, \neg\varphi \Rightarrow_i \Delta, \varphi$

3. *if* $\vdash \Gamma \Rightarrow_k^{\varphi\wedge\psi} \Delta, \varphi \wedge \psi,$ *then* $\vdash \Gamma \Rightarrow_i \Delta, \varphi, \varphi \wedge \psi$ *and* $\vdash \Gamma \Rightarrow_j \Delta, \psi, \varphi \wedge \psi$

4. *if* $\vdash \Gamma, \varphi \wedge \psi \Rightarrow_k^{\varphi\wedge\psi} \Delta,$ *then* $\vdash \Gamma, \varphi, \psi, \varphi \wedge \psi \Rightarrow_i \Delta$

5. *if* $\vdash \Gamma \Rightarrow_k^{\varphi \vee \psi} \Delta, \varphi \vee \psi$, *the* $\vdash \Gamma \Rightarrow_i \Delta, \varphi, \psi, \varphi \vee \psi$

6. *if* $\vdash \Gamma, \varphi \vee \psi \Rightarrow_k^{\varphi \vee \psi} \Delta$, *then* $\vdash \Gamma, \varphi, \varphi \vee \psi \Rightarrow_i \Delta$ *and* $\vdash \Gamma, \psi, \varphi \vee \psi \Rightarrow_j \Delta$

7. *if* $\vdash \Gamma \Rightarrow_k^{\varphi \to \psi} \Delta, \varphi \to \psi$, *then* $\vdash \Gamma, \varphi \Rightarrow_i \Delta, \psi, \varphi \to \psi$

8. *if* $\vdash \Gamma, \varphi \to \psi \Rightarrow_k^{\varphi \to \psi} \Delta$, *then* $\vdash \Gamma, \psi, \varphi \to \psi \Rightarrow_i \Delta$ *and* $\vdash \Gamma, \varphi \to \psi \Rightarrow_j \Delta, \varphi$

The notation $\vdash \Gamma \Rightarrow_k^{\varphi} \Delta$ means that a proof of $\Gamma \Rightarrow \Delta$ has height k and $\varphi \in \Gamma \cup \Delta$ is a principal formula of this sequent. It is obvious that this lemma provides a weaker version of the inversion lemma, sufficient for proving the main theorem.

Exercise 3.6. *Prove this lemma.*

3.3.3   Schütte's Proof

One of the most interesting proofs of admissibility of cut is due to Schütte [232]. Although developed for a version of SC rather different from G3 (see subsection 3.4.2 and 3.5.1) it may be applied for it without many changes. It has two special features:

a) It is heavily based on the application of invertibility of rules.
b) Induction on the height is performed for only one premiss, and only on the basis of induction on the complexity.

Exactly because of the explicit application of the invertibility lemma we do not need to perform induction on the height twice. The general schema will show the structure of proof better:

   I. Induction on the complexity of cut-formula
   1.1. Basis: cut on atomic cut-formulae is admissible
     II. Induction on the height of the left (or right) premiss of cut
     2.1. Basis: cut with the left premiss of height 0 is admissible
     2.2. Inductive step: if cut with the left height $< n$ is admissible, then with the height n is admissible too.
     Conclusion of 2.1., 2.2: cut on atomic cut-formulae is admissible
   1.2. Inductive step: if cut on the cut-formula of the complexity $< n$ is admissible, then on n is admissible.
   Conclusion of 1.1., 1.2: cut on every formula is admissible.

The proof of part 1 and 3 (axiom case and reduction of height) is exactly as in Dragalin's proof. The fact that we are reducing the height only in case of cut-formula atomic does not make any changes in the proof. On the other hand, a proof of part 2 (reduction of complexity) is totally different since we do not require that the cut-formula is principal in both (or even in one) premiss. We are just applying the invertibility lemma[3]. In fact, using the name principal cut reduction lemma for this part of Schütte's proof is inadequate. It is rather the complexity reduction lemma assuming invertibility of rules instead of the principality of cut-formulae. Here is an example:

$$\frac{\Gamma \Rightarrow \Delta, \varphi \vee \psi \qquad \varphi \vee \psi, \Pi \Rightarrow \Sigma}{\Gamma, \Pi \Rightarrow \Delta, \Sigma} \; (Cut)$$

is replaced with:

$$\frac{\dfrac{Inv \dfrac{\Gamma \Rightarrow \Delta, \varphi \vee \psi}{\Gamma \Rightarrow \Delta, \varphi, \psi} \quad \dfrac{\varphi \vee \psi, \Pi \Rightarrow \Sigma}{\varphi, \Pi \Rightarrow \Sigma} Inv}{(Cut) \; \Gamma, \Pi \Rightarrow \Delta, \Sigma, \psi} \qquad \dfrac{\dfrac{\varphi \vee \psi, \Pi \Rightarrow \Sigma}{\psi, \Pi \Rightarrow \Sigma} Inv}{} (Cut)}{\dfrac{\Gamma, \Pi, \Pi \Rightarrow \Delta, \Sigma, \Sigma}{\Gamma, \Pi \Rightarrow \Delta, \Sigma} \; (C)}$$

where Inv denotes an application of the invertibility lemma.

Now, we do not require that the cut-formula is principal, only compound. Note that also C is needed.

Exercise 3.7. *Provide transformations for other compound cut-formulae.*

For easier comparison we organised a presentation of Schütte's proof in the same three parts as for Dragalin's and Smulyann's proof but it may be misleading. In fact, Schütte's proof is naturally composed of two parts which may be called:

1. The atomic cuts admissibility lemma.

2. The complexity reduction lemma.

Note that although we have proved the basis of induction on the complexity (i.e. the atomic cuts admissibility lemma) by induction on the height of one premiss, it may be avoided. In subsection 2.5.2 such a lemma (lemma 2.16) was proved for LK by means of a global proof. It may be done also for G3.

Exercise 3.8. *Prove the atomic cuts admissibility lemma for G3 using Buss' strategy.*

This shows possible simplification of the organisation of proof: one may do it by only one induction, on the complexity of cut-formula, if the basis is proven by Buss' method.

The analysis of the essential prerequisites for Schütte proof is also instructive— we need:

1. Admissibility of W (required for axiom case as in Dragalin's proof).

2. Invertibility of rules (not necessarily height-preserving—applied only for complexity reduction).

---

[3]Situation is a bit different in case of logics where full invertibility fails like in modal logic (see subsection 4.3.2) or first-order logic which will be discussed in the second volume.

3. C admissibility (not necessarily height-preserving—only for complexity reduction and axiom case).

Notice that the inductive step is performed by using invertibility of rules; we do not need to assume that the cut-formula is principal. Also height-preserving invertibility is not necessary because we reduce only the complexity of cut formula in this step. Summing up, this method may be applied also to LK (C primitive and invertibility proved by Schütte's method (see also Pogorzelski [199] or Negri and von Plato [185] for illustrations of how this method works in the presence of noninvertible rules). This shows that Schütte's strategy of proof has a rather wide scope of application.

3.3.4   Schütte-style Proof for Admissibility of A-cut

One may consider also a variant of Schütte's proof for G3 with A-cut. Part 1 of the proof goes exactly as in Smullyan's proof.

On the other hand, in part 2 (reduction of complexity) we do not need C. For example:

$$\frac{\Gamma \Rightarrow \Delta, \varphi \vee \psi \qquad \varphi \vee \psi, \Gamma \Rightarrow \Delta}{\Gamma \Rightarrow \Delta} \ (Cut)$$

is replaced with:

$$Inv \frac{\Gamma \Rightarrow \Delta, \varphi \vee \psi}{\Gamma \Rightarrow \Delta, \varphi, \psi} \quad (W)\frac{Inv\dfrac{\varphi \vee \psi, \Gamma \Rightarrow \Delta}{\varphi, \Gamma \Rightarrow \Delta}}{\varphi, \Gamma \Rightarrow \Delta, \psi}$$
$$(Cut)\frac{\qquad\qquad}{\Gamma \Rightarrow \Delta, \psi} \qquad Inv\frac{\varphi \vee \psi, \Gamma \Rightarrow \Delta}{\psi, \Gamma \Rightarrow \Delta}$$
$$(Cut)\frac{}{\Gamma \Rightarrow \Delta}$$

A comparison of this example with the analogous one in the previous subsection shows that admissibility of W is required.

Exercise 3.9. *Prove the remaining cases.*

This time part 3 of the proof (the reduction of height) is the most involved. Let us analyse an example:

$$\frac{\varphi \vee \psi, \Gamma \Rightarrow \Delta, \chi \qquad (\vee \Rightarrow)\dfrac{\chi, \varphi, \Gamma \Rightarrow \Delta \qquad \chi, \psi, \Gamma \Rightarrow \Delta}{\chi, \varphi \vee \psi, \Gamma \Rightarrow \Delta}}{\varphi \vee \psi, \Gamma \Rightarrow \Delta} \ (Cut)$$

is replaced with:

$$(\vee \Rightarrow)\frac{Inv\,(Cut)\dfrac{\dfrac{\varphi \vee \psi, \Gamma \Rightarrow \Delta, \chi}{\varphi, \Gamma \Rightarrow \Delta, \chi} \quad \chi, \varphi, \Gamma \Rightarrow \Delta}{\varphi, \Gamma \Rightarrow \Delta} \qquad \dfrac{Inv\,(Cut)\dfrac{\varphi \vee \psi, \Gamma \Rightarrow \Delta, \chi}{\psi, \Gamma \Rightarrow \Delta, \chi} \quad \chi, \psi, \Gamma \Rightarrow \Delta}{\psi, \Gamma \Rightarrow \Delta}}{\varphi \vee \psi, \Gamma \Rightarrow \Delta}$$

Note that now we must rely on height-preserving invertibility of rules for induction to work!

Exercise 3.10. *Prove the remaining cases.*

This proof in general shows a different strategy for dealing with the problem of unifying premisses of A-cut when doing reductions of height or complexity. In Smullyan's proof this unification was based on the application of height-preserving admissibility of W and then on the application of C to delete superfluous occurrences of some formulae. In Schütte-like proof we apply invertibility instead and no new formulae are added hence C is not required in parts 2 and 3. At first the prerequisites for this kind of proof seem to be stronger:

1. Height-preserving invertibility of rules (essential for height reduction, requires atomic axioms).

2. W admissible (not necessarily height-preserving—only for complexity reduction).

3. C admissible (in part 1)

But only the requirement for invertibility is stronger. On the other hand, note that C is not needed at all unless we do not apply Buss' proof on the basis of induction on complexity. It is enough to observe that in the atomic axiom case we can always delete the second occurrence of the cut-formula—it was extracted in subsection 3.2.2 as the atomic contraction admissibility (lemma 3.5).

Summing up: this is the only proof where we do not need C at all, however, when compared with Schütte's proof, it has more rigid scope of application. Similarly as in Dragalin's and Smullyan's proof it may be applied only to G3, because it is based on the application of the height-preserving invertibility lemma. Also, in contrast to Schütte's proof, we cannot get rid of the induction on the height, because Buss' proof requires C.

One may notice that we can also avoid admissibility of C as well as proving the invertibility of rules, at the price of small complications in figures showing transformations, if we prove for G3 the admissibility of Herbelin's cut (see subsection 2.1.1) combining the features of M- and A-cut. There are also some other ways of reducing the list of preliminaries, in particular the admissibility of C, which are not discussed here (but see Indrzejczak [134]).

3.4 Interpretations of Sequents

In subsection 2.2.1 we have proved lemma 2.1 and 2.2. Provided proofs required the application of cut in many places but in G3 we can reprove them much simpler with the help of the invertibility lemma. Moreover, we do not need to apply contraction since we use the additive versions of $(\Rightarrow \lor)$ and $(\land \Rightarrow)$.

Exercise 3.11. *Prove lemma 2.1 and 2.2 for G3.*

In fact, it is useful to strengthen lemma 2.1 by the addition of the following equivalences:

Lemma 3.8. *The following are equivalent:*

1. $\vdash \Rightarrow \varphi_1 \wedge ... \wedge \varphi_i \to \psi_1 \vee ... \vee \psi_k$
2. $\vdash \varphi_1 \wedge ... \wedge \varphi_i \wedge \neg \psi_1 \wedge ... \wedge \neg \psi_k \Rightarrow$
3. $\vdash \Rightarrow \neg \varphi_1 \vee ..., \vee \neg \varphi_i \vee \psi_1 \vee ... \vee \psi_k$

PROOF: $2 \Longrightarrow 3$. By invertibility of $(\wedge \Rightarrow)$ we get $\vdash \varphi_1, ...\varphi_i, \neg \psi_1, ...\neg \psi_k \Rightarrow$ which, by lemma 2.1, is equivalent to $\vdash \Rightarrow \neg \varphi_1, ..., \neg \varphi_i, \psi_1, ...\psi_k$, and this implies $\vdash \Rightarrow \neg \varphi_1 \vee ..., \vee \neg \varphi_i \vee \psi_1 \vee ... \vee \psi_k$ by $(\Rightarrow \vee)$.   □

Exercise 3.12. *Prove the remaining parts of the lemma.*
Prove this lemma for LK.

Note that in the proofs of the equivalences from lemma 2.1, 2.2 and lemma 3.8 (either in LK or in G3) the assumption that sequents of the specified form are provable was not required. Hence in the more general setting of proof trees we can prove the same result in a weaker form for any sequents, showing (in the same way) that they are all interderivable, i.e. that $\varphi_1, ...\varphi_i \Rightarrow \psi_1, ...\psi_k \vdash \varphi_1 \wedge ... \wedge \varphi_i \Rightarrow \psi_1 \vee ... \vee \psi_k$, $\varphi_1 \wedge ... \wedge \varphi_i \Rightarrow \psi_1 \vee ... \vee \psi_k \vdash \varphi_1, ...\varphi_i \Rightarrow \psi_1, ...\psi_k$ etc.

Exercise 3.13. *Prove the content of lemma 2.1, 2.2 and 3.8 in the form of derivability results.*

On the basis of lemma 3.8 we may introduce three interpretations of sequents in terms of formulae:

1. $I_G(\varphi_1, ..., \varphi_i \Rightarrow \psi_1, ..., \psi_k) = \varphi_1 \wedge ... \wedge \varphi_i \to \psi_1 \vee ... \vee \psi_k$
2. $I_C(\varphi_1, ..., \varphi_i \Rightarrow \psi_1, ..., \psi_k) = \neg(\varphi_1 \wedge ... \wedge \varphi_i \wedge \neg \psi_1 \wedge ... \wedge \neg \psi_k)$
3. $I_D(\varphi_1, ..., \varphi_i \Rightarrow \psi_1, ..., \psi_k) = \neg \varphi_1 \vee ..., \vee \neg \varphi_i \vee \psi_1 \vee ... \vee \psi_k$

One can prove for each interpretation $I \in \{I_G, I_C, I_D\}$:

Lemma 3.9. *The following claims are equivalent:*

1. $S_1, ..., S_n \vdash S$
2. $\Rightarrow I(S_1), ..., \Rightarrow I(S_n) \vdash \Rightarrow I(S)$
3. $\vdash I(S_1), ..., I(S_n) \Rightarrow I(S)$

We will demonstrate this lemma specifically for I_D. But first we need some supporting result:

Lemma 3.10. *For every primitive rule* $\dfrac{S_1, ..., S_n}{S}$ *of G3, a corresponding rule* $\dfrac{\Rightarrow I_D(S_1), ..., \Rightarrow I_D(S_n)}{\Rightarrow I_D(S)}$ *(called I_D-transform of suitable rule) is admissible and invertible in G3.*

PROOF: We must show that it holds for all primitive rules of G3 except $(\Rightarrow \vee)$ where in its I_D-transform the premiss is identical with the conclusion. We examine the cases of $(\wedge \Rightarrow)$ and $(\Rightarrow \wedge)$.

In the first case I_D-transform has the form: $\dfrac{\Rightarrow \neg\Gamma \vee \neg\varphi \vee \neg\psi \vee \Delta}{\Rightarrow \neg\Gamma \vee \neg(\varphi \wedge \psi) \vee \Delta}$ From the premiss we obtain $\varphi, \psi \Rightarrow \neg\Gamma \vee \Delta$ by invertibility of $(\Rightarrow \vee)$ and $(\Rightarrow \neg)$. Then, successively applying $(\wedge \Rightarrow), (\Rightarrow \neg)$ and $(\Rightarrow \vee)$ we obtain the conclusion. Note that all steps are carried out by invertible rules, hence we can obtain the proof of the premiss from the conclusion as well.

In the second case I_D-transform has the form: $\dfrac{\Rightarrow \neg\Gamma \vee \Delta \vee \varphi \quad \Rightarrow \neg\Gamma \vee \Delta \vee \psi}{\Rightarrow \neg\Gamma \vee \Delta \vee \varphi \wedge \psi}$ From the premisses we obtain $\Rightarrow \neg\Gamma \vee \Delta, \varphi$ and $\Rightarrow \neg\Gamma \vee \Delta, \psi$ by invertibility of $(\Rightarrow \vee)$. Then, by $(\Rightarrow \wedge)$ and $(\Rightarrow \vee)$, we obtain the conclusion. Since all steps are obtained by invertible rules we can obtain the proof of each premiss from the conclusion as well. $\qquad\square$

Note that although this proof may be displayed as a proof in G3 it does not prove derivability, since we are using invertible rules which were shown to be admissible, not derivable.

Exercise 3.14. *Show admissibility of the remaining I_D-transforms.*
Prove this lemma for I_G- and I_C-transforms of primitive rules of G3.

Now we can prove lemma 3.9 for the case of I_D.

PROOF: $1 \Longrightarrow 2$: By lemma 3.10, to each primitive rule of G3 there corresponds its I_D-transform. Hence we rewrite step by step a proof-tree $\mathcal{D}$ for $S_1, ..., S_n \vdash S$ as a proof tree $\mathcal{D}'$ for $\Rightarrow I_D(S_1), ..., \Rightarrow I_D(S_n) \vdash \Rightarrow I_D(S)$ treating I_D-transforms of the applications of G3 rules in $\mathcal{D}$ as macros comprising sequences of the applications of primitive rules and their inversions.

$2 \Longrightarrow 3$: First note that G3 enriched with I_D-transforms is still closed under weakening, so we can add any parameters to any side of the premisses and conclusion without losing their admissibility for G3. Now take a derivation of $\Rightarrow I_D(S_1), ..., \Rightarrow I_D(S_n) \vdash \Rightarrow I_D(S)$ and add $I_D(S_1), ..., I_D(S_n)$ to the antecedent of each sequent in it. This way all leaves are axiomatic and the root sequent $I_D(S_1), ..., I_D(S_n) \Rightarrow I_D(S)$ is a provable sequent.

$3 \Longrightarrow 1$: Start with each S_i and derive from it $\Rightarrow I_D(S_i)$ by successive applications of $(\Rightarrow \neg)$ and $(\Rightarrow \vee)$. Now by n applications of cut to $I_D(S_1), ..., I_D(S_n) \Rightarrow I_D(S)$ and these n sequents of the form $\Rightarrow I_D(S_i)$ we obtain $\Rightarrow I_D(S)$ and then, by invertibility of $(\Rightarrow \neg)$ and $(\Rightarrow \vee)$, we get S. $\qquad\square$

Exercise 3.15. *Prove this lemma for interpretations I_G and I_C.*

Note that each I_D-transform corresponds to some provable equivalence. In case of two-premiss rule we take the conjunction of two premisses. We state it as:

Lemma 3.11. *For each one-premiss I_D-transform of a primitive rule we have $\vdash \Rightarrow$
$I_D(S_1) \leftrightarrow I_D(S_2)$.*

*For each two-premiss I_D-transform of a primitive rule we have $\vdash \Rightarrow I_D(S_1) \leftrightarrow$
$I_D(S_2) \wedge I_D(S_3)$.*

PROOF: By lemma 3.9 and 3.10. Take the case of some two-premiss rule. Since
$\Rightarrow I(S_1), \Rightarrow I(S_2) \vdash \Rightarrow I(S)$, then by $2 \Longrightarrow 3$ and $(\wedge \Rightarrow)$ we get $\vdash I(S_1) \wedge I(S_2) \Rightarrow$
$I(S)$. By invertibility of this rule and $(\Rightarrow \wedge)$ we obtain $\vdash I(S) \Rightarrow I(S_1) \wedge I(S_2)$. $\square$

Note that the proof applies also to the remaining transforms.

One may prove the equivalences corresponding to rules directly; In fact it was
done for I_G in chapter 1. But now we obtain this as a simple corollary of theorem
3.10 in one step for any kind of transform. Otherwise we must prove this for each
rule directly; in case of I_D it means that we must prove 6 equivalences of the form:

1. $\vdash \Rightarrow \gamma \vee \varphi \leftrightarrow \gamma \vee \neg\neg\varphi$

2. $\vdash \Rightarrow (\gamma \vee \neg\varphi) \wedge (\gamma \vee \neg\psi) \leftrightarrow \gamma \vee \neg(\varphi \vee \psi)$

3. $\vdash \Rightarrow \gamma \vee \neg\varphi \vee \neg\psi \leftrightarrow \gamma \vee \neg(\varphi \wedge \psi)$

4. $\vdash \Rightarrow (\gamma \vee \varphi) \wedge (\gamma \vee \psi) \leftrightarrow \gamma \vee \varphi \wedge \psi$

5. $\vdash \Rightarrow \gamma \vee \neg\varphi \vee \psi \leftrightarrow \gamma \vee (\varphi \rightarrow \psi)$

6. $\vdash \Rightarrow (\gamma \vee \varphi) \wedge (\gamma \vee \neg\psi) \leftrightarrow \gamma \vee \neg(\varphi \rightarrow \psi)$

(to simplify things we let γ replace $\neg\Gamma \vee \Delta$ occurring in I_D-transforms)

Exercise 3.16. *Provide proofs for these equivalences.*

Display and prove equivalences corresponding to I_C-transforms.

3.4.1   Post Theorem

We have focused on I_D since it allows us for showing that G3 (or any other
SC) may produce a conjunctive normal form (CNF) for any formula. There are
different ways of proving this fact and it may be interesting to compare them. We
will show some more general result:

Theorem 3.1. *If $S_1, ..., S_n \vdash \Rightarrow \varphi$, then $\vdash \Rightarrow \varphi \leftrightarrow I_D(S_1) \wedge ... \wedge I_D(S_n)$*

PROOF 1: We show two implications:

(a) $\vdash \Rightarrow \varphi \rightarrow I_D(S_1) \wedge ... \wedge I_D(S_n)$

and

(b) $\vdash \Rightarrow I_D(S_1) \wedge ... \wedge I_D(S_n) \rightarrow \varphi$.

ad (a), by invertibility of all rules we get: $\Rightarrow \varphi \vdash S_1, ..., \Rightarrow \varphi \vdash S_n$ from our assumption that $S_1, ..., S_n \vdash \Rightarrow \varphi$. By the fact that $\varphi = I_D(\varphi)$ and lemma 3.9. (1 $\implies$ 3) we get $\vdash \varphi \Rightarrow I_D(S_1), ..., \vdash \varphi \Rightarrow I_D(S_n)$. Hence by successive applications of ($\Rightarrow \wedge$) and ($\Rightarrow \rightarrow$) we obtain the result.

ad (b), directly from lemma 3.9. (1 $\implies$ 3), ($\wedge \Rightarrow$) and ($\Rightarrow \rightarrow$). $\qquad\square$

The above proof was based on lemma 3.9 and has an internal character in the sense of being proved in G3. It is interesting to show how the same result may be obtained in the external way by using the resources of axiomatic system, namely the extensionality principle (point 15 of lemma 1.1) and the equivalence of G3 to axiomatic formalisation of **CPL** (theorem 1.8). It goes by induction on the stages of the systematic root-first construction of a proof tree for φ. We prove that for each stage φ is equivalent to the conjunction of I_D-transforms of all leaves.

PROOF 2 The basis is trivial since φ is the only leaf and $\varphi = I_D(\varphi)$ so we have directly $\vdash \Rightarrow \varphi \leftrightarrow I_D(\varphi)$.

Assume that the claim holds for the nth stage of the construction of a proof tree, i.e. that we have $\vdash \Rightarrow \varphi \leftrightarrow I_D(S_1) \wedge ... \wedge I_D(S_n)$ for $S_1, ..., S_n$ which are all leaves of the actual proof-search tree. Let the stage $n+1$ be obtained by the application of some two-premiss rule to, say S_n, hence we obtain a proof tree with two new leaves S_{n1} and S_{n2} attached over S_n. We know by lemma 3.11 that $\vdash \Rightarrow I_D(S_n) \leftrightarrow I_D(S_{n1}) \wedge I_D(S_{n2})$ By theorem 1.8 in Hilbert system we have $\vdash_H \varphi \leftrightarrow I_D(S_1) \wedge ... \wedge I_D(S_n)$ and $\vdash_H I_D(S_n) \leftrightarrow I_D(S_{n1}) \wedge I_D(S_{n2})$, hence by the extensionality principle also $\varphi \leftrightarrow I_D(S_1) \wedge ... \wedge I_D(S_{n1}) \wedge I_D(S_{n2})$ is provable. Thus by theorem 1.8 $\vdash \Rightarrow \varphi \leftrightarrow I_D(S_1) \wedge ... \wedge I_D(S_{n1}) \wedge I_D(S_{n2})$. $\qquad\square$

Exercise 3.17. *Provide a direct proof of admissibility of extensionality rule for G3*

From theorem 3.1 we obtain:

Theorem 3.2 (Post). *For any φ there is an equivalent formula in CNF*

PROOF: It is enough to apply the procedure from chapter 1 for constructing completed proof trees to $\Rightarrow \varphi$ and take the conjunction of I_D-transforms of all atomic leaves (including axiomatic ones). By theorem 3.1 it is equivalent to φ. $\qquad\square$

This fact allows for another completeness proof based on the following fact concerning CNF-formulae

Lemma 3.12 *For any φ in CNF, $\models \varphi$ iff any clause in φ covers an instance of the law of excluded middle (LEM).*

PROOF: $\implies$ Suppose that in at least one clause we do not have any pair of literals that form an instance of LEM. We can provide a valuation which gives 0 to all atoms and 1 to all negated atoms. This makes the clause false and so φ is a false conjunction under this valuation.

$\impliedby$ Obvious—the conjunction of valid formulae is valid. $\qquad\square$

Now we are in the position to prove:

Theorem 3.3 (Completeness). *If $\models \varphi$, then $\vdash \Rightarrow \varphi$*

1. $\models \varphi$
2. $\vdash \Rightarrow \varphi \leftrightarrow I_D(S_1) \wedge ... \wedge I_D(S_n)$ Post theorem
3. $\models \varphi \leftrightarrow I_D(S_1) \wedge ... \wedge I_D(S_n)$ 2, Soundness
4. $\models I_D(S_1) \wedge ... \wedge I_D(S_n)$ 1, 3
5. $\vdash \Rightarrow I_D(S_1) \wedge ... \wedge I_D(S_n)$ 4, lemma 3.12, by $(\Rightarrow \neg), (\Rightarrow \vee), (\Rightarrow \wedge)$
6. $\vdash \Rightarrow I_D(S_1) \wedge ... \wedge I_D(S_n) \rightarrow \varphi$ 2
7. $\vdash I_D(S_1) \wedge ... \wedge I_D(S_n) \Rightarrow \varphi$ 6, invertibility of $(\Rightarrow \rightarrow)$
8. $\vdash \Rightarrow \varphi$ 5, 7 Cut

Note that $I_D(S_1) \wedge ... \wedge I_D(S_n)$ denotes here a conjunction of clauses not just arbitrary disjunctions.

3.4.2   SC and Other Kinds of Systems

We have already mentioned in subsection 1.3.2 (remark 1.5) that there is a close relationship between SC and tableau calculi. On the basis of the results presented in this section we can state these observations in more general and precise terms.

Let us recall that Hintikka [116] provided one of the first versions of tableau calculus for classical logic. His version is an indirect one in the sense that if we want to prove that $\Gamma \vdash \varphi$ we start with $\Gamma, \neg\varphi$ and decompose this set by means of suitable rules until we construct a tree (with sets of formulae as nodes) where each leaf is contradictory, i.e. contains some $\psi, \neg\psi$. The main difference with K is that instead of sequents simply sets are used and that a tree is built upside-down. One can easily check that in Hintikka's tableau system there is a step-wise simulation of every rule of K for $\wedge, \vee, \rightarrow$:

$$\frac{\Gamma \Rightarrow \Delta}{\Gamma' \Rightarrow \Delta'} \qquad \Longleftrightarrow \qquad \frac{\Gamma', \neg\Delta'}{\Gamma, \neg\Delta}$$

$$\frac{\Gamma \Rightarrow \Delta \quad \Gamma' \Rightarrow \Delta'}{\Gamma'' \Rightarrow \Delta''} \qquad \Longleftrightarrow \qquad \frac{\Gamma'', \neg\Delta''}{\Gamma, \neg\Delta \mid \Gamma', \neg\Delta'}$$

This simulation is based on lemma 3.10 but applies interpretation I_C. In the case of negation a simulation of $(\neg \Rightarrow)$ is trivial and the counterpart of the second rule is a double negation elimination. On the basis of this translation we obtain:

Theorem 3.4 (Equivalence of Hintikka tableaux and K).

- $\Gamma \vdash_{Tab} \varphi \ (:= \Gamma, \neg\varphi \vdash_{Tab} \bot) \ \Longrightarrow \ \vdash_K \Gamma \Rightarrow \varphi$

- $\vdash_K \Gamma \Rightarrow \Delta \ \Longrightarrow \ \Gamma, \neg\Delta \vdash_{Tab} \bot$

There is a dual version of tableaux devised for classical logic by Rasiowa and Sikorski [210] and developed significantly for several logics and theories by Orłowska and her collaborators (see in particular Orłowska and Golińska-Pilarek

[192]). It is based on disjunctive reading of sets, i.e. on I_D. In fact, as far as we are concerned with tableau systems for classical logic, this dual version is just an upside-down version of Schütte's SC [233] with one-sided sequents, referred to in subsection 3.3.3.

Exercise 3.18. *Show schemata of step-wise simulation for rules of dual tableaux and state equivalence theorem with K.*

SC is also in close relationship to resolution calculi of Robinson [219], in particular on the propositional level[4]. Note that propositional resolution rule is just a cut performed on atomic sequents. The main difference is that in ordinary resolution systems we operate on clauses but we know that they correspond to atomic sequents by I_D so this is only a minor notational variant. What is more important is how we obtain a set of clauses (atomic sequents). Standard resolution is again an indirect method; to prove that $\vdash \varphi$ we transform $\neg \varphi$ into its CNF and then make resolutions on its member clauses until we derive an empty clause, i.e. $\bot$. In the framework of K (with cut) we can divide the job into two stages. We start with $\varphi \Rightarrow$ and build a completed proof tree for it (see section 1.3) which by lemma 1.9 must terminate. All nonaxiomatic leaves of this tree provide a set of sequents from which we derive $\Rightarrow$ by cuts only. In general case of proving $\Gamma \vdash \varphi$ we can establish the equivalence of both methods in the following:

Theorem 3.5. *For any sequent the following are equivalent:*

1. $\vdash \varphi_1, ..., \varphi_i \Rightarrow \psi_1, ..., \psi_k$

2. $\Rightarrow \varphi_1, ..., \Rightarrow \varphi_i, \psi_1 \Rightarrow, ..., \psi_k \Rightarrow \vdash \Rightarrow$

Exercise 3.19. *Prove the above theorem using cut. Hint: look at the proof of claim 1.6 in subsection 1.9.2.*

3.5   Variants of SC

Before we focus on the applications of SC to non-classical logics it is instructive to make a more detailed characterisation of several variants of SC. It helps to understand why logics other than **CPL** often require several changes in the basic apparatus. In fact, even in the case of **CPL** there is a plenty of SC which are very different from the three calculi we presented. Indeed all SCs introduced so far, although slightly different in some respects, are in fact quite similar to each other. The most impressive feature of all these systems is the fact that all are progressive in the sense that only logical rules of introduction are primitive. In what follows such calculi will be called *standard SC*. Two main aims of this section are to provide a closer characterisation of standard systems and to sketch a general typology of sequent calculi.

---

[4]More on the relations between SC and resolution, for example, in Avron [10], Gallier [93], or Fitting [85].

Let us recall that our preliminary characterisation of SC (in section 1.2) was very general—just a collection of (schemata) of rules composed from sequents, including rules with empty set of premisses i.e. axiomatic sequents. Such a characterisation admits significantly different realisations depending on the notion of a sequent and special requirements concerning the shape of rules.

3.5.1   Types of Sequents

In case of the notion of a sequent we only (implicitly) decided on two things: that sequents are ordered pairs and that both arguments consists of finite collections of formulae. The first difference between three SCs (and some small variations of them) presented so far concerned the way of specifying what exactly is this collection. We admitted sets, multisets, sequences, and after Casari [45] we can call them the main alternatives of SC. But this nomination does not preclude a possibility of other solutions. In particular, one is free to consider also different data structures as arguments of a sequent—we will mention some proposals of this kind in the remaining chapters. It is possible to introduce even more radical changes in the definition of a sequent. We consider briefly three such variations: the number of formulae (or other objects) included in arguments of a sequent, the very structure of a sequent (for example, the arity—the number of its arguments), the character of objects included in arguments of a sequent.

The number of (occurrences of) formulae in the succedent or antecedent may be regulated. In particular, introduction of single-succedent sequents, i.e. with (at most) one formula in the succedent seems to be a very natural solution, corresponding to Tarski-style consequence relations. Gentzen made it a characteristic feature of his SC for intuitionistic logic although it is not necessary for providing SCs adequate for this logic (see section 5.1). In general a variety of different options is possible, for example, sequents having exactly one formula in the antecedent and the succedent (like in the system of Rieger [217] for **CPL**, or SC for **FDE** in subsection 5.3.3), or at most one formula in the antecedent (see Wansing and Kamide [145]). We have mentioned also sequents with empty antecedent (or succedent)—so called one-sided sequents in Schütte [232] or in tableau transformations of SC. These one-sided sequents are formally justified by interpretations I_D and I_C. On the other hand, even sequents with an infinite number of formulae were admitted in proof theoretical considerations (e.g. in Tait [257]).

In what follows we will say that sequents being ordered pairs of finite (multi)sets or sequences of formulae, possibly with numerical restrictions (including one-sided, i.e. with empty antecedent or succedent), are *ordinary sequents* and all systems with rules operating on such kind of sequents will be called *ordinary SC*. Any system defined by means of some other notion of a sequent will be called a *generalised SC*[5].

---

[5]In Troelstra and Schwichtenberg [264] such calculi are called varying G-systems.

3.5.2 Generalised SC

During the past five decades researchers who tried to apply Gentzen's formalism to several non-classical logics faced many serious problems. In order to overcome the difficulties they provided a lot of ingenious solutions, mainly based on the changes in the notion of basic items on which rules are defined. It is possible to describe at least four main types of generalised SC:

- structured sequent calculi;

- labelled sequent calculi;

- many-sequent calculi;

- multisequent calculi.

The above list of different generalised SC is by no means exhaustive. The categories are also certainly not disjoint, hence it is only a proposal of some typology and surely a different conceptual organisation of this domain may be proposed. A detailed presentation of several generalised SCs may be found in Wansing [270] and Poggiolesi [198] with slightly different conceptual order of solutions.

In the first group the very notion of a sequent is generalised somehow. We can find here at least three solutions where ordinary distinction between the antecedent and the succedent is saved but with the additional division of both arguments into smaller parts, or with more refined structure occurring in arguments of a sequent, and even such where we have sequents with n-arguments, for $n > 2$.

The first approach may be applied either for expression of some essential features of formalised logics which is rather not possible in ordinary sequents, or as a kind of technical device with no deeper motivation. The former approach is evident in systems for modal (Sato [224], Blamey and Humberstone [34]) and temporal logics (Nishimura [188]) where a division of arguments is introduced for making distinctions between modal status, or temporal localizations of formulae. We briefly characterise Sato's system in section 4.8 and provide SC of this kind for some many-valued logics in section 5.5. In the latter case an introduction of additional collections of formulae into antecedent or succedent has rather a technical character needed for improvement of a proof search. As an example we can mention systems with displayed 'head formulae' in sequents considered in Troelstra and Schwichtenberg [264]. We will illustrate this approach in section 4.4 by means of some calculi developed for modal logics and mention also some other ones in section 5.1 for intuitionistic logic.

The next approach in this group was introduced independently by Mints [178] and Dunn [67], and is commonly called consecution calculus for relevant logics. Several developments of this approach are analysed, for example, in Restall [214] and Bimbo [31] so we only briefly describe its main features in section 5.3. Certainly the most general realisation of this idea is provided by *display calculus* constructed by Belnap [25]. In this approach arguments of a sequent are built from special structures where formulae are combined by means of additional structural constants. Different variants of display calculi were extensively studied and applied to a variety of non-classical logics (see e.g. Wansing [269] or a survey of Ciabattoni, Ramanayake, Wansing [53]).

The last solution in the group of structured SC is popular in several systems for many-valued logics (Rousseau [221], Carnielli [44]), where $n > 2$ correspond to n logical values, or – in a more refined version—to collections of values (Hähnle [106]). The former approach is strongly dependent on the kind of interpretation we put on n-sided sequents. Reading in terms of verification or falsification of such a sequent leads to the construction of significantly different calculi. The latter provides a more unified perspective, closer to standard SC and computationally better behaving. We provide in sections 5.4 and 5.5 an exposition of these different realisations of the same idea for some many-valued logics. Moreover, systems presented there will be formalised in the way which makes them essentially similar to systems from the first group. i.e. as defined on sequents with additionally divided arguments. In fact, Wansing [270] and Poggiolesi [198] treat these systems as essentially of the same form under the heading higher-arity SC.

As for the character of elements in arguments of sequents the most popular approach is based on the application of several kinds of labels added to formulae. A general theory of labelled systems was presented by Gabbay [91]; a survey of solutions applied to modal logics may be found in Indrzejczak [130]. Several SCs with term-labels added to formulae to express the so called Curry–Howard isomorphism, as well as systems developed in the spirit of Martin-Löf [172] type theory, may be also treated as belonging to this group. It should be underlined that expressive power of labels is so strong that usually labelled SC are built with sequents which are close to ordinary ones but with formulae enriched with labels. However, there are also systems with labels attached to whole sequents like in Mints [182]. Labelled approaches of some kinds will be dealt with in the second volume; here we introduce only extremely simple system of this sort for modal logic **S5** in section 4.6.

The name many-sequent calculi is proposed here for the unification of all systems using collections of (usually ordinary) sequents as the basic items. It covers two main families of calculi operating on hyper- or nested sequents. Hypersequents are structures of the form $\Gamma_1 \Rightarrow \Delta_1 \mid \ldots \mid \Gamma_i \Rightarrow \Delta_i$. They have found applications to variety of non-classical logics (e.g. Avron [10]). We will discuss them in some detail in section 4.7 with the application to modal logics. Nested sequents are a slightly more complicated structures where, in addition to formulae, the elements of a sequent may be other sequents, containing other sequents. This approach in general form was initiated by Došen [60] and it was extensively applied, under different names (deep inference calculi, tree-hypersequent calculi), in the field of modal and temporal logics (e.g. Bull [42], Kashima [147], Stouppa [246], Poggiolesi [198]). Note that if we define hypersequents as (finite) sequences of sequents we obtain in fact a special variant of nested sequents, with only one sequent admitted as an element of a given sequent (Lellmann [161] calls them linear nested sequents whereas in Indrzejczak [136] they are called non-commutative hypersequents).

The last type, so called *multisequent calculi*, contains systems with different kinds of sequents, for which possibly different collections of rules are defined. This approach started with Curry [56] and Zeman [276] and was developed by Indrzejczak [126]. A system of this kind based on a different interpretation of two kinds of sequents was also presented by Avron [10]. In section 4.8 we introduce a simple double sequent calculus for **S5** as an illustration of this approach.

We limited the list of generalised SCs only to these solutions which will be presented in the remaining chapters. In particular, several systems belonging to the first type will be treated in more detail and one system belonging to the third type, namely hypersequent calculus. Other ones will be only incidentally invoked for the sake of illustration.

3.5.3 Ordinary SC

We stipulated that all kinds of SCs which are defined on sequents being ordered pairs of collections of formulae, even with several numerical restrictions, will be called *ordinary SC*. This group includes standard SCs but also many other calculi with rules of different characters. A survey of ordinary SCs presented in Indrzejczak [133] proposes a division into three different types according to the priority given either to primitive rules or to initial (axiomatic) sequents:

1. Gentzen's type based on rules with a small number (usually one) of axiomatic sequents;

2. Hertz's type based on axiomatic sequents with a small number of rules (usually structural);

3. Mixed type with balanced participation of axioms and rules.

In particular, the first type contains, as a subtype, a standard SC containing LK, G3, K and other systems being their variations. Gentzen's type and Hertz's type represent extreme solutions. In the former type logical constants are usually characterised (mainly) by rules with axiomatic sequents usually having a structural character. On the contrary, in the latter type rules have structural character and logical constants are characterised by axiomatic sequents. Such a solution was proposed by Hertz [114] who in fact introduced the very notion of a (single-succedent) sequent, sequent rule, tree representation of proofs and started the program developed later by Gentzen.[6] But it should be noted that Hertz did not present any specific system for concrete logic. His approach was abstract; he defined rather a schema of the system in which rules have purely structural character. In addition to rules of contraction, permutation and weakening in the antecedent, Hertz used also the rule of syllogism which was a special kind of cut. Gentzen, in his earlier work [94], developed to some extent a theory of Hertz. In particular, he presented purely structural criteria which must be satisfied by any SC of this type with an

---

[6]One can find a detailed overview of Hertz's system in Schroeder-Heister [227].

independent set of axiomatic sequents. Also he replaced Hertz's rule of syllogism with cut.

In fact, the approach of Hertz to characterisation of concrete logics did not find many applications, except two significant examples: one due to Suszko [253] (see also [252] and [254]) and the other due to Smullyan [244]. Suszko proposed in 1940s the original SC of Hertz's type operating on single-succedent sequents with sequences in the antecedents and with complete set of sequents for **CPL** and some non-classical logics. We do not describe it here; one may find a detailed analysis in Indrzejczak [129]. Smullyan devised a system for classical logic where the only rule is cut, moreover in analytic form. Since sequents in this system are built from sets no other structural rules are needed. In the propositional part the following sequents are primitive:

- $\varphi \Rightarrow \varphi \quad \Rightarrow \varphi, \neg\varphi \quad \neg\varphi, \varphi \Rightarrow$

- $\varphi \wedge \psi \Rightarrow \varphi \quad \varphi \wedge \psi \Rightarrow \psi \quad \varphi, \psi \Rightarrow \varphi \wedge \psi$

- $\varphi \Rightarrow \varphi \vee \psi \quad \psi \Rightarrow \varphi \vee \psi \quad \varphi \vee \psi \Rightarrow \varphi, \psi$

- $\varphi \rightarrow \psi, \varphi \Rightarrow \psi \quad \psi \Rightarrow \varphi \rightarrow \psi \quad \Rightarrow \varphi, \varphi \rightarrow \psi$

That analytic cut is sufficient for adequacy follows from lemma 2.4, or even 2.3. One can easily check that all applications of cut required for proving admissibility of Gentzen's rules use only subformulae of the proved conclusion sequent. In the other direction it is routine to provide cut-free proofs of all the above sequents in K.

Exercise 3.20. *Demonstrate the equivalence of Smullyan's system with K.*

In fact, the first solution of this kind is already in Gentzen [95], where he noticed that many rules of LK may be replaced with primitive sequents. Gentzen used exactly the same sequents except for the last two with implication in the succedent. Thus he eliminated all rules for $\neg, \wedge, \vee$ but has left a logical rule $(\Rightarrow\rightarrow)$ [7]. Also structural rules W, C, P were present due to the fact that sequents were made from lists, like in LK. Gentzen noticed that cut is not eliminable from such a system but he did not notice that it may be restricted to analytic form.

Due to the fact that some rules in Gentzen's system are logical it belongs to the mixed type. This group includes a plethora of SC where logical constants are characterised either by means of sequents or rules to gain more flexibility in proof search. There are a great many possible combinations with some constants characterised only by means of rules and other only by means of sequents, or the same constants partly characterised by rules and partly by sequents. One can find considerations on these matters already in Bernays [26], Popper [200] and many other logicians. Some other systems of this kind for **CPL** were provided, among others, by Kleene [149], Hasenjaeger [109], Surma [250] and Rieger [217]. All are

---

[7]He also left two rules for quantifiers whereas Smullyan found a way to replace all of them with sequents.

using single-succedent sequents except Rieger who applies sequents with exactly one formula on both sides of a sequent.

3.5.4 Gentzen's Type of Ordinary SC

We mentioned that Gentzen's original LK and other standard SC discussed so far belong to the first type of ordinary SC but satisfy additional important requirements concerning logical rules. Roughly, logical constants are characterised only by means of rules of introduction to the conclusion; usually a pair introducing a constant to the antecedent or to the succedent. However, even in Gentzen's type we can find a lot of systems which are not standard because different kind of logical rules is used in them.

We provide as an example of nonstandard SC of Gentzen's type his system of sequent natural deduction (ND). It should not be confused with the well-known system of natural deduction NK (or NJ for intuitionistic logic) which was presented by Gentzen in [95]; the latter is not SC since its rules are defined on formulae not on sequents. However, the system introduced in [96] which was applied in the consistency proof of Peano's arithmetic is a combination of SC and ND. It is devised for classical logic but uses sequents with sequences of formulae in antecedents but exactly one formula in the succedent. There are only rules of introduction of logical constants to the succedent; instead of rules for introduction to the antecedent it has the rules of elimination of constants in the succedent. Hence it seems like a kind of a compromise between his system NK of natural deduction and his standard sequent system LK in the sense that all inference rules are basically as in his NK, but items which are operated on in proofs are not formulae but sequents. Antecedents of sequents do not involve any logical operations; they provide only a record of active assumptions.

In fact, such a system was implicitly present in [95] in the proof of correctness of NK. Gentzen shows there an equivalence of his NK with Hilbert's system via LK. One part of the proof shows how to transform every NK proof tree into LK proof tree. For this aim every inference rule of NK is rewritten with addition of all active assumptions, in this way we obtain as a by-product the system which in [96] is defined explicitly.

The only primitive sequents are of the form $\varphi \Rightarrow \varphi$, exactly as in LK. He also applies structural rules of weakening, contraction and permutation in antecedents. Logical rules for propositional part are the following:

$$(\Rightarrow \neg) \quad \frac{\Gamma, \varphi \Rightarrow \psi \quad \Delta, \varphi \Rightarrow \neg\psi}{\Gamma, \Delta \Rightarrow \neg\varphi} \qquad\qquad (\Rightarrow \neg E) \quad \frac{\Gamma \Rightarrow \neg\neg\varphi}{\Gamma \Rightarrow \varphi}$$

$$(\Rightarrow \wedge E1) \quad \frac{\Gamma \Rightarrow \varphi \wedge \psi}{\Gamma \Rightarrow \varphi} \qquad\qquad (\Rightarrow \wedge E2) \quad \frac{\Gamma \Rightarrow \varphi \wedge \psi}{\Gamma \Rightarrow \psi}$$

$$(\Rightarrow \wedge) \quad \frac{\Gamma \Rightarrow \varphi \quad \Delta \Rightarrow \psi}{\Gamma, \Delta \Rightarrow \varphi \wedge \psi} \qquad\qquad (\Rightarrow \vee E) \quad \frac{\Gamma, \varphi \Rightarrow \chi \quad \Delta, \psi \Rightarrow \chi \quad \Pi \Rightarrow \varphi \vee \psi}{\Gamma, \Delta, \Pi \Rightarrow \chi}$$

$(\Rightarrow\vee)\quad \dfrac{\Gamma\Rightarrow\varphi}{\Gamma\Rightarrow\varphi\vee\psi}$ $\qquad\qquad$ $(\Rightarrow\vee)\quad \dfrac{\Gamma\Rightarrow\psi}{\Gamma\Rightarrow\varphi\vee\psi}$

$(\Rightarrow\rightarrow E)\quad \dfrac{\Gamma\Rightarrow\varphi\quad \Delta\Rightarrow\varphi\rightarrow\psi}{\Gamma,\Delta\Rightarrow\psi}$ $\qquad$ $(\Rightarrow\rightarrow)\quad \dfrac{\Gamma,\varphi\Rightarrow\psi}{\Gamma\Rightarrow\varphi\rightarrow\psi}$

It is easily seen that this calculus when compared with LK has standard rules for introduction in the succedent taken from his system LJ for intuitionistic logic (see section 5.1); two-premiss form of $(\Rightarrow\neg)$ follows from the fact that $\bot$ (or empty succedent) is not allowed in the language. Instead of rules of introduction in the antecedent we have elimination rules (in the succedent) taken from Gentzen's system NK (or NJ for intuitionistic logic). In particular, the rules $(\Rightarrow\neg)$, $(\Rightarrow\rightarrow)$ and $(\Rightarrow\vee E)$ correspond to proof construction rules in NK, i.e. to these rules which introduce subderivations based on the additional assumptions to be discharged later (after deduction of a suitable formula). It is represented here as a subtraction of some formula (additional assumption) from the antecedent of a sequent. All other rules correspond to inference rules of NK (NJ) so the only operation on antecedents of premisses is their concatenation.

It is easy to note that cut is derivable in this system by means of $(\Rightarrow\rightarrow)$ and $(\Rightarrow\rightarrow E)$ but of course it can be proved directly as an admissible rule of the system, independently of the special rules for implication.

Exercise 3.21. *Prove derivability of rules of this SC in LK.*

As an ND system this calculus is not very practical in its original form. The drawbacks follow from the fact that proofs are defined as trees of sequents so all assumptions must be rewritten in each inference step. But this very fact allows us to resign from defining the proof as a tree of sequents and use linear proofs (i.e. sequences of sequents) which are much easier to deal with from the standpoint of actual proof search. Moreover, in contrast to ND in Jaśkowski-style[8], it does not have to use any bookkeeping devices, like lines or boxes, for showing the dependency of formulae in the proof on their assumptions. Also the process of rewriting of all assumptions in each inference step may be significantly simplified since the only operations on them have structural character. Instead of rewriting formulae we may rewrite the numbers of lines where the respective assumptions were introduced. This solution was first introduced in Feys' and Ladriere [77] translation of [95] into French. It was made popular due to Suppes' [249] and its later simplifications in Lemmon [162], Forbes [89] and many other textbooks. Usually in all these systems authors simplify matters by introducing sets instead of sequences of formulae/numbers in antecedents. It makes structural rules of permutation and contraction dispensable[9].

Note also that Gentzen's sequent ND is based on the rigid conception of natural deduction as characterising logical constants only by means of rules operating

---

[8]In Anglo-American tradition usually called Fitch-style ND.

[9]Although one should observe that in some cases, e.g. in modal logics, it may be better to keep sequences—see Garson [92].

only on succedents of sequents. The only operations admissible on antecedents have—as we noted above—structural character. But this restriction is by no means necessary and one can find also a variety of systems representing greater flexibility with respect to the shape of rules. In particular, also rules for introduction and elimination of constants in the antecedent are applied. Of course this excludes the solution mentioned above, of replacing collections of formulae in the antecedent with the numerals of respective lines, but in many cases allows for construction of shorter proofs. One can mention here systems of Hermes [113], Ebbinghaus, Flum and Thomas [73], Ershow and Palyutin [74], Lawrow and Maksimowa [167]), Andrews [5], Leblanc [157, 158] or symmetric system with rules for negated formulae due to Smullyan [245] and some other of this sort due to Wiśniewski [272] (also Leszczyńska-Jasion, Urbański and Wiśniewski [163]). One of the particularly important SC of this type is Došen's SC [61] with invertible rules for constants based on the earlier ideas of Popper (e.g. [200, 201]). According to Došen, in order to claim that an expression is a logical constant it is necessary to provide such a double validity-preserving rule which after addition to structural rules allows for obtaining a full characterisation of this constant. Rules of this system satisfy this requirement in contrast to earlier Poppers' proposal (a detailed analysis of Popper's approach is provided by Schroeder-Heister [226] (see also [228]).

3.5.5 Standard SC

We close this section with a more detailed analysis of standard SC focusing not so much on the differences between K, LK, G3 and their minor variants, but rather on the similarities between them.

There are two main differences between SCs investigated so far concerning the notion of a sequent and the role of structural rules. The first one was already discussed and in the case of classical logic it is not very important due to the fact that structural rules ensure identical behaviour of the collections of formulae in all these calculi. Things will change radically for some non-classical logics described in chapter 5. The second difference between SCs investigated so far was connected with the role of structural rules. If they are primitive we may say about structural variant of SC, otherwise it is a logical variant[10]. Note that even SC using sequents built from sequences may be presented as purely logical, as the example of Gallier variant of LK shows (see remark 2.4 in section 2.6) so this distinction does not depend on the chosen alternative. There were also differences concerning the choice of rules (additive versus multiplicative, invertible versus noninvertible) or axioms (simple versus contextual, atomic versus general) and we can easily provide more variants of these SCs by further combinatorial changes. Moreover, in the class of structural SC we can distinguish between analytic (no cut and contraction primitive) and nonanalytic (logical SCs are all analytic).

---

[10]Poggiolesi [198] is using the same distinction in a slightly different way—she reserved the term structural for SC of quite a different character due to Došen [61] and remarked above.

What is characteristic and common for all these variants of SC is the fact that all primitive logical rules are only introduction rules and on this basis we call them standard[11]. Concerning the shape of the rules we did not formulate any requirements although we made a lot of comments on some features of presented rules. The requirement of progressivity of all logical rules which we proposed in this section for preliminary characterisation of standard SC is quite general and allow for further specifications. When we are analysing concrete rules of LK or G3 we can find some more interesting properties. Some of them, like context independence, the subformula property or invertibility were already discussed (see in particular subsections 1.2.2, 1.4.2 and 1.4.3) Following Wansing [269] (see also Poggiolesi [198]) we may list some other desiderata for well-defined logical rules of standard SC:

- Separation: a rule for a constant should not exhibit any other constants in its schema.

- Weak symmetry: each rule should either introduce a constant to the antecedent or to the succedent; if we have both such rules the calculus is (simply) symmetric with respect to this constant.

- Weak explicitness: a constant should be present only in the conclusion; if only one occurrence of it is present a rule is (simply) explicit.

Rules satisfying these properties and the subformula property are called *canonical* by Avron [14]. One may easily check that all logical rules of K, LK, G3 are canonical in this sense. But not all standard systems are canonical—as we will see soon.

Some of these properties are strongly connected with different approaches to anti-realist and proof-theoretic approaches to the semantics of logical constants which we mentioned in subsection 1.4.1. In these contexts two additional conditions are sometimes required:

- Uniqueness: rules for a constant c_1 provide a unique characterisation of it if when we add to the language of this calculus a constant c_2 which is characterised in this extended calculus by the same rules as those for c_1, then S is provable iff S' is provable, when S' is just as S but with all occurrences of c_1 replaced with c_2.

- Avron's property: rules should be independent of any particular semantics.

Some other desiderata are of interest when we consider SCs for families of logics in the same language, as will be done in the remaining chapters.

- Modularity: in the family of logics all extensions should be characterised by separate rules. In particular, if we consider H-systems as a point of reference, then each new axiom should be represented by a new rule.

---

[11]Avron [14] uses a naming standard SC for all systems having standard structural rules without any requirements concerning logical rules.

- Došen property: the rules for constants should be stable; all extensions for stronger logics in the same language should be obtained by structural rules only.

Once the former is accepted rather with no doubts, the latter is criticised by Poggiolesi [198], where some reformulation of it is proposed.

Many researchers were interested in the characterisation of such properties of rules which are at least sufficient for ensuring cut admissibility for some strictly specified classes of SC. As we already remarked Curry [56] was the first author who undertaken such research. On the basis of his considerations Belnap [25] provided such a list for a rather specific form of generalised SC, namely display calculus. More recently Restall [214] presented a similar analysis but in the context of a more general version of SC admitting also (after slight reformulation) standard SCs. These analyses are rather complicated and when we focus on the most important features of different proofs of cut admissibility we can separate just two such desiderata. These were called substitutivity and reductivity (by Ciabattoni [52], in Avron [14] these are called purity and coherence conditions[12]). The former is connected with possibility of performing height reduction steps and the name refers to the fact that a (multi)set of formulae is substituted for the occurrence(s) of cut-formula in the last applied step where cut-formula is parametric. Roughly speaking if all rules are context independent (pure) they are substitutive, but the latter may hold even in the presence of some rules which are impure.

Reductivity (coherence) is abstracted from these features of logical rules which enable complexity reduction steps. Informally it allows for deriving the same conclusion from a series of cuts made on subformulae of some complex cut-formula. Note that all rules of standard SC which were considered so far are reductive but it is not the same as being canonical. Again we may use as an example Prior's rules for 'tonk' introduced in section 1.4. Since one of these rules was elimination rule it is not standard but with the help of lemma 2.3.A we can easily replace it with suitable introduction rule, thus we obtain:

$$\Gamma \Rightarrow \Delta, \varphi \ / \ \Gamma \Rightarrow \Delta, \varphi \ tonk \ \psi \quad \text{and}$$
$$\psi, \Gamma \Rightarrow \Delta \ / \ \varphi \ tonk \ \psi, \Gamma \Rightarrow \Delta$$

One can easily check that both rules are canonical but not reductive.

Although for some specific classes of SC it was possible to show that discussed properties characterise cut elimination, in the sense that in such a class they are also necessary conditions for this result, it does not preclude the possibility that

---

[12]Note that different authors introduce their terminology in the context of investigation on specific classes of SC and by means of specific machinery, so their results are different and not simply reducible to each other. When we refer to their terminology we rather want to point out the common conceptual core of their considerations, not to suggest that they are using different terms for the same things. In fact, one may consider not rules but formulae which admit characterisation terms of suitable rules; for example, Restall [214] is talking on regular formulae to the same effect.

(standard) SC with nonreductive, or even noncanonical rules, cannot be cut-free and analytic. So for many non-classical logics ordinary SCs fail to satisfy some of the properties we discussed. On the other hand, in generalised SCs very often we can observe that their rules satisfy some generalised counterparts of features satisfied by canonical and coherent SC. We will see examples of both kinds in the remaining two chapters.

3.6   Constructive Proofs of Interpolation Theorem

In section 1.11, we have proved Craig's interpolation theorem. However, that proof was not constructive, partly semantical, and based on the application of analytic cut. Now we are in a position to provide a proof which is constructive and based on the cut-elimination theorem. One may find several constructive proofs of this result based on the application of cut-free SC or tableaux. The common feature of all of them is some way of separation of everything which is connected with the antecedent and with the succedent. We have seen this already in the proof from section 1.11 where two symmetric forms of cut were needed as well as relative distinctions in building a model were necessary to keep track of two parts. All constructive proofs of the interpolation theorem must also use some devices to save this separation. In particular, rules in which a formula is transferred from the antecedent to the consequent or vice versa are troublesome in this respect. Just consider an instance of the application of $(\Rightarrow \neg)$ of the form:

$$\frac{p, \Gamma \Rightarrow \Delta}{\Gamma \Rightarrow \Delta, \neg p}$$

and assume that we have an interpolant for the premiss which contains p but no atoms from Γ and that $p \notin PROP(\Gamma)$. It is easily seen that it cannot be an interpolant for the conclusion. So the crucial thing is to take care of the separation of what is connected with the antecedent and what with the succedent part. Different ways of attacking this problem were utilised in the framework of SC or tableau systems. One of the possible strategies is to introduce some bookkeeping device (e.g. labels) for marking in the proof which formula is essentially antecedent and which is essentially succedent despite its actual position in the current sequent. Such an approach was applied for example by Kleene [150] or Fitting [85].

We consider below three different proofs in which rather strictly syntactic solution is applied. The first is due to Maehara (see Takeuti [261], Ono [191] or Troelstra and Schwichtenberg [264]) and justifies a slightly generalised version of the interpolation theorem. This proof is rather involved but it is very popular and based on the interesting construction so it is important to know it. Moreover, it may be developed for any standard version of SC with only minor modifications. The second proof due to Smullyan [245] (see also Fitting [85]) is relatively simple but based on noncanonical version of SC. The last proof due to Wintein and Muskens [271], does not demand any changes in the system but splits proofs instead.

3.6.1 Maehara's Proof

We mentioned that the proof provided first by Maehara [169] may be carried for any kind of standard cut-free SC. We will use slightly modified LK. First we extend the language with $\top$ and $\bot$ and add to LK axioms $\Rightarrow \top$ and $\bot \Rightarrow$. Moreover, to simplify matters we will change lists into multisets. In fact, we were tacitly doing this many times in chapter 2 when disregarding the order of formulae in sequents. However, for this proof we must consider unions of collections of data which is simpler for multisets than for lists. Let $\sqcup$ denote a union of multisets which is like ordinary set union $\cup$ but with additional counting of the number of occurrences of the same formula. Thus, e.g. $[\varphi, \psi, \psi, \chi] \sqcup [\psi, \chi] = [\varphi, \psi, \psi, \psi, \chi, \chi]$, where $[$ and $]$ are used for enumeration of multisets.

For any $\Gamma \Rightarrow \Delta$ we define the notion of its partition. Let $\Gamma_1 \sqcup \Gamma_2 = \Gamma$ and $\Delta_1 \sqcup \Delta_2 = \Delta$, then $((\Gamma_1, \Delta_1), (\Gamma_2, \Delta_2))$ is a partition of $\Gamma \Rightarrow \Delta$.

Note: in partitions we admit empty multisets, e.g. $(\Gamma, \varnothing), (\varnothing, \Delta)$.

Maehara's generalised interpolation theorem claims:

Theorem 3.6. *If* $\vdash \Gamma \Rightarrow \Delta$, *then for any partition* $((\Gamma_1, \Delta_1), (\Gamma_2, \Delta_2))$ *we can find* φ, *such that:*

1. $\vdash \Gamma_1 \Rightarrow \Delta_1, \varphi$

2. $\vdash \varphi, \Gamma_2 \Rightarrow \Delta_2$

3. $PROP(\varphi) \subseteq PROP(\Gamma_1 \sqcup \Delta_1) \cap PROP(\Gamma_2 \sqcup \Delta_2)$

PROOF: By induction on the height of a proof of $\Gamma \Rightarrow \Delta$ in cut-free LK.

Case $k = 0$, so $\Gamma \Rightarrow \Delta$ is an axiom. If it is $\varphi \Rightarrow \varphi$, then we have 4 partitions:

1. $(([\varphi], [\varphi]), (\varnothing, \varnothing))$

2. $((\varnothing, \varnothing), ([\varphi], [\varphi]))$

3. $(([\varphi], \varnothing), (\varnothing, [\varphi]))$

4. $((\varnothing, [\varphi]), ([\varphi], \varnothing))$.

In 1, the interpolant is $\bot$, since $\vdash \varphi \Rightarrow \varphi, \bot$ and $\vdash \bot \Rightarrow$. In 2, the interpolant is $\top$, since $\vdash \Rightarrow \top$ and $\vdash \top, \varphi \Rightarrow \varphi$. Finally in 3 it is φ, and in 4 it is $\neg\varphi$, since $\vdash \Rightarrow \varphi, \neg\varphi$ and $\vdash \neg\varphi, \varphi \Rightarrow$.

For added axioms: $\Rightarrow \top$ has two partitions $((\varnothing, [\top]), (\varnothing, \varnothing))$ and $((\varnothing, \varnothing), (\varnothing, [\top]))$. In the first case the interpolant is $\bot$, since $\vdash \Rightarrow \top, \bot$ and $\vdash \bot \Rightarrow$; in the second it is $\top$. For axiom $\bot \Rightarrow$ a proof is dual.

In case $k > 0$, we consider all possible cases of the last rule applied in the proof. Let us take as an example the case of $(\wedge \Rightarrow)$:

$\Gamma \Rightarrow \Delta := \psi \wedge \chi, \Gamma' \Rightarrow \Delta$ and the premiss is $\psi, \Gamma' \Rightarrow \Delta$ or $\chi, \Gamma' \Rightarrow \Delta$. Take the first one, there are two possible situations: (a) a partition with $\psi \wedge \chi \in \Gamma_1$ and (b) with $\psi \wedge \chi \in \Gamma_2$. For the first case, i.e. $\Gamma_1 := \psi \wedge \chi, \Gamma'_1$ in the premiss, we consider the respective partition with $\psi \in \Gamma_1$ i.e. $\Gamma_1 := \psi, \Gamma'_1$. Since $\psi, \Gamma' \Rightarrow \Delta$ has a proof of height $k - 1$, therefore by the induction hypothesis there is φ such that:

1. $\vdash \psi, \Gamma'_1 \Rightarrow \Delta_1, \varphi$

2. $\vdash \varphi, \Gamma_2 \Rightarrow \Delta_2$

3. $PROP(\varphi) \subseteq PROP(\Gamma'_1 \sqcup [\psi] \sqcup \Delta_1) \cap PROP(\Gamma_2 \sqcup \Delta_2)$

By 1, we have $\psi \wedge \chi, \Gamma'_1 \Rightarrow \Delta_1, \varphi$. By 2, and $PROP(\varphi) \subseteq PROP(\Gamma'_1 \sqcup [\psi \wedge \chi] \sqcup \Delta_1) \cap PROP(\Gamma_2 \sqcup \Delta_2)$ we conclude that φ is also an interpolant for $\psi \wedge \chi, \Gamma' \Rightarrow \Delta$.

Now take any partition such that $\psi \wedge \chi \in \Gamma_2$ (i.e. $\Gamma_2 := \psi \wedge \chi, \Gamma'_2$) and for the premiss consider respective partition with $\psi \in \Gamma_2$ (i.e. $\Gamma_2 := \psi, \Gamma'_2$). Again by the induction hypothesis there is φ such that:

1. $\vdash \Gamma_1 \Rightarrow \Delta_1, \varphi$

2. $\vdash \varphi, \Gamma'_2, \psi \Rightarrow \Delta_2$

3. $PROP(\varphi) \subseteq PROP(\Gamma_1 \sqcup \Delta_1) \cap PROP(\Gamma'_2 \sqcup [\psi] \sqcup \Delta_2)$

By 2, we get $\psi \wedge \chi, \varphi, \Gamma'_2 \Rightarrow \Delta_2$, and by 1, and $PROP(\varphi) \subseteq PROP(\Gamma_1 \sqcup \Delta_1) \cap PROP(\Gamma'_2 \sqcup [\psi \wedge \chi] \sqcup \Delta_2)$ we conclude that φ is also an interpolant for $\psi \wedge \chi, \Gamma' \Rightarrow \Delta$.

Consider the case of $(\Rightarrow \wedge)$: then $\Gamma \Rightarrow \Delta := \Gamma \Rightarrow \Delta', \psi \wedge \chi$, and both premisses have the form $\Gamma \Rightarrow \Delta', \psi$ and $\Gamma \Rightarrow \Delta', \chi$. We take any partition such that $\psi \wedge \chi \in \Delta_1$ (i.e. $\Delta_1 := \Delta'_1, \psi \wedge \chi$) and for premisses we consider respective partitions with $\psi \in \Delta_1$ and $\chi \in \Delta_1$ (i.e. $\Delta_1 := \Delta'_1, \psi$ and $\Delta_1 := \Delta'_1, \chi$). By the induction hypothesis for both premisses we have interpolants φ_1, φ_2 such that:

1. $\vdash \Gamma_1 \Rightarrow \Delta'_1, \psi, \varphi_1$

2. $\vdash \varphi_1, \Gamma_2 \Rightarrow \Delta_2$

3. $PROP(\varphi_1) \subseteq PROP(\Gamma_1 \sqcup \Delta'_1 \sqcup [\psi]) \cap PROP(\Gamma_2 \sqcup \Delta_2)$

4. $\vdash \Gamma_1 \Rightarrow \Delta'_1, \chi, \varphi_2$

5. $\vdash \varphi_2, \Gamma_2 \Rightarrow \Delta_2$

6. $PROP(\varphi_2) \subseteq PROP(\Gamma_1 \sqcup \Delta'_1 \sqcup [\chi]) \cap PROP(\Gamma_2 \sqcup \Delta_2)$

From 1 and 4, we derive by $(\Rightarrow \wedge)$ and $(\Rightarrow \vee)$ $\Gamma_1 \Rightarrow \Delta'_1, \psi \wedge \chi, \varphi_1 \vee \varphi_2$, and from 2 and 5, we derive $\varphi_1 \vee \varphi_2, \Gamma_2 \Rightarrow \Delta_2$ by $(\vee \Rightarrow)$. Hence $\varphi_1 \vee \varphi_2$ is a desired interpolant for $\Gamma \Rightarrow \Delta', \psi \wedge \chi$ since $PROP(\varphi_1 \vee \varphi_2) \subseteq PROP(\Gamma_1 \sqcup \Delta'_1 \sqcup [\psi \wedge \chi]) \cap PROP(\Gamma_2 \sqcup \Delta_2)$.

Now, we take any partition such that $\psi \wedge \chi \in \Delta_2$ (i.e. $\Delta_2 := \Delta'_2, \psi \wedge \chi$) and we consider for premisses respective partitions with $\psi \in \Delta_2$ and $\chi \in \Delta_2$ (i. e. $\Delta_2 := \Delta'_2, \psi$ i $\Delta_2 := \Delta'_2, \chi$). By the induction hypothesis for both premisses we have interpolants φ_1, φ_2 such that:

1. $\vdash \Gamma_1 \Rightarrow \Delta_1, \varphi_1$

2. $\vdash \varphi_1, \Gamma_2 \Rightarrow \Delta'_2, \psi$

3. $PROP(\varphi_1) \subseteq PROP(\Gamma_1 \sqcup \Delta_1) \cap PROP(\Gamma_2 \sqcup \Delta'_2 \sqcup [\psi])$

4. $\vdash \Gamma_1 \Rightarrow \Delta_1, \varphi_2$

5. $\vdash \varphi_2, \Gamma_2 \Rightarrow \Delta'_2, \chi$

6. $PROP(\varphi_2) \subseteq PROP(\Gamma_1 \sqcup \Delta_1) \cap PROP(\Gamma_2 \sqcup \Delta'_2 \sqcup [\chi])$

Then from 1 and 4 we derive by means of $(\Rightarrow \wedge)$ $\Gamma_1 \Rightarrow \Delta_1, \varphi_1 \wedge \varphi_2$, and from 2 and 5 by $(\Rightarrow \wedge)$ and $(\wedge \Rightarrow)$ we obtain $\varphi_1 \wedge \varphi_2, \Gamma_2 \Rightarrow \Delta'_2, \psi \wedge \chi$. Therefore $\varphi_1 \wedge \varphi_2$ is an interpolant for $\Gamma \Rightarrow \Delta', \psi \wedge \chi$, since $PROP(\varphi_1 \wedge \varphi_2) \subseteq PROP(\Gamma_1 \sqcup \Delta_1) \cap PROP(\Gamma_2 \sqcup \Delta'_2 \sqcup [\psi \wedge \chi])$.

□

Exercise 3.22. *Prove the cases of rules for disjunction, implication and negation and the cases of contraction and weakening.*

One can easily notice that Maehara's theorem implies Craig's theorem. Assume that: $\models \varphi \rightarrow \psi$ and $PROP(\varphi) \cap PROP(\psi) \neq \varnothing$. By adequacy of LK and the inversion lemma we get $\vdash \varphi \Rightarrow \psi$. Hence taking a partition $\Gamma_1 := [\varphi]$, $\Delta_2 := [\psi]$, $\Gamma_2 = \Delta_1 = \varnothing$ we derive from Maehara's theorem an interpolant χ, such that $\vdash \varphi \Rightarrow \chi$ and $\vdash \chi \Rightarrow \psi$.

3.6.2 Smullyan's Proof

Smullyan's proof may be seen as a refinement of Maehara proof since it does not need any partition of sequents. We also demonstrate for each rule how to obtain an interpolant for the conclusion on the basis of interpolant(s) for premiss(es). But the same strategy is presented in a significantly simpler way at the price of changing the basic calculus[13]. It is based on the idea that all rules must operate on the same side of a sequent. So it may be treated as similar in spirit to the proof from section 1.11, where ordinary cut was replaced with symmetric cuts. Here we do not need cut at all but symmetric versions of all rules and moreover, additional rules for negated compound formulae. Let us call SG3 a symmetrical variant of

---

[13]Note however that it is less general than Maehara's strategy—the example will be discussed in section 5.1.

G3 where all rules satisfy the condition that no formula is shifted from one side of the premiss to the other side of the conclusion. The list of axioms consists of:

$$p, \Gamma \Rightarrow \Gamma, p \qquad \neg p, \Gamma \Rightarrow \Gamma, \neg p \qquad p, \neg p, \Gamma \Rightarrow \Delta \qquad \Gamma \Rightarrow \Delta, p, \neg p$$
$$\bot, \Gamma \Rightarrow \Delta \qquad \Gamma \Rightarrow \Delta, \top \qquad \neg\top, \Gamma \Rightarrow \Delta \qquad \Gamma \Rightarrow \Delta, \neg\bot$$

In addition to standard rules for $\wedge$ and $\vee$ we have:

$$(\neg\neg\Rightarrow) \ \frac{\varphi, \Gamma \Rightarrow \Delta}{\neg\neg\varphi, \Gamma \Rightarrow \Delta} \qquad\qquad (\Rightarrow\neg\neg) \ \frac{\Gamma \Rightarrow \Delta, \varphi}{\Gamma \Rightarrow \Delta, \neg\neg\varphi}$$

$$(\Rightarrow\rightarrow) \ \frac{\Gamma \Rightarrow \Delta, \neg\varphi, \psi}{\Gamma \Rightarrow \Delta, \varphi \rightarrow \psi} \qquad (\rightarrow\Rightarrow) \ \frac{\neg\varphi, \Gamma \Rightarrow \Delta \quad \psi, \Gamma \Rightarrow \Delta}{\varphi \rightarrow \psi, \Gamma \Rightarrow \Delta}$$

$$(\neg\vee\Rightarrow) \ \frac{\neg\varphi, \neg\psi, \Gamma \Rightarrow \Delta}{\neg(\varphi \vee \psi), \Gamma \Rightarrow \Delta} \qquad (\Rightarrow\neg\vee) \ \frac{\Gamma \Rightarrow \Delta, \neg\varphi \quad \Gamma \Rightarrow \Delta, \neg\psi}{\Gamma \Rightarrow \Delta, \neg(\varphi \vee \psi)}$$

$$(\Rightarrow\neg\wedge) \ \frac{\Gamma \Rightarrow \Delta, \neg\varphi, \neg\psi}{\Gamma \Rightarrow \Delta, \neg(\varphi \wedge \psi)} \qquad (\neg\wedge\Rightarrow) \ \frac{\neg\varphi, \Gamma \Rightarrow \Delta \quad \neg\psi, \Gamma \Rightarrow \Delta}{\neg(\varphi \wedge \psi), \Gamma \Rightarrow \Delta}$$

$$(\neg\rightarrow\Rightarrow) \ \frac{\varphi, \neg\psi, \Gamma \Rightarrow \Delta}{\neg(\varphi \rightarrow \psi), \Gamma \Rightarrow \Delta} \qquad (\Rightarrow\neg\rightarrow) \ \frac{\Gamma \Rightarrow \Delta, \varphi \quad \Gamma \Rightarrow \Delta, \neg\psi}{\Gamma \Rightarrow \Delta, \neg(\varphi \rightarrow \psi)}$$

Exercise 3.23. *Check that the new rules are normal.*

Prove that general versions (with any φ active) of the first four axioms are derivable.

Prove that the new rules are derivable in G3 with cut.

Clearly this SG3 (symmetric G3) is standard but not canonical. In particular, the subformula property fails but it holds in a slightly more general way: rules are closed under subformulae and their negations. Soundness is obvious. To show that it is complete without any form of cut it is sufficient to demonstrate:

Theorem 3.7. *If $\vdash_{G3} \Gamma \Rightarrow \Delta$, then $\vdash_{SG3} \Gamma \Rightarrow \Delta$.*

PROOF: By induction on the height of the proof. Every axiom of G3 is axiomatic in SG3 and all rules for $\wedge$ and $\vee$ are also primitive in SG3. So it is enough to show that both rules for $\neg$ and $\rightarrow$ are admissible in SG3. This will follow from two lemmata stated below.   □

Lemma 3.13. $(\neg\neg \Rightarrow)$ *and* $(\Rightarrow \neg\neg)$ *are invertible in SG3.*

PROOF: By induction on the height of a proof of $\neg\neg\varphi, \Gamma \Rightarrow \Delta$ ($\Gamma \Rightarrow \Delta, \neg\neg\varphi$ respectively). In the basis $\neg\neg\varphi$ is a parameter so we can change it into φ. Since all rules are context independent we can make identical substitutions in all places where $\neg\neg\varphi$ is parametric. Eventually, if it is principal, the premiss is what we want.   □

Now instead of proving admissibility of both G3 rules for $\neg$ we provide a more general result. Let $\Gamma^\neg \Rightarrow \Delta^\neg$ denote any $\neg$-variant of $\Gamma \Rightarrow \Delta$ obtained by transferring any formula from one side of the sequent to the other with the addition of negation. It holds:

Lemma 3.14. *If $\vdash_{SG3} \Gamma \Rightarrow \Delta$, then $\vdash_{SG3} \Gamma^{\neg} \Rightarrow \Delta^{\neg}$ is provable.*

PROOF: By induction on the height of the proof. In case a transferred formula is parametric in the axiom or in the application of any rule it follows from context independency of rules, so we need to check only the cases where it is a principal formula. In the basis either we directly obtain (different forms of) axioms or sequents derivable from axioms by $(\neg\neg \Rightarrow)$ or $(\Rightarrow \neg\neg)$. For the induction step consider the case of implication. There are four subcases:

1. We have $\vdash_n \varphi \to \psi, \Gamma \Rightarrow \Delta$ where n is the height of a proof and we want to show that $\vdash \Gamma \Rightarrow \Delta, \neg(\varphi \to \psi)$. Since $\varphi \to \psi$ is principal we have $\vdash \neg\varphi, \Gamma \Rightarrow \Delta$ and $\vdash \psi, \Gamma \Rightarrow \Delta$, both with lower height. By the induction hypothesis $\vdash \Gamma \Rightarrow \Delta, \neg\neg\varphi$ and $\vdash \Gamma \Rightarrow \Delta, \neg\psi$, hence, by the preceding lemma, $\vdash \Gamma \Rightarrow \Delta, \varphi$. Therefor we infer $\vdash \Gamma \Rightarrow \Delta, \neg(\varphi \to \psi)$.

2. In case of $\vdash_n \Gamma \Rightarrow \Delta, \varphi \to \psi$, the preceding sequent is $\vdash_{n_1} \Gamma \Rightarrow \Delta, \neg\varphi, \psi$. By the induction hypothesis we get $\vdash \neg\neg\varphi, \neg\psi, \Gamma \Rightarrow \Delta$ which, by lemma 3.13, implies $\vdash \varphi, \neg\psi, \Gamma \Rightarrow \Delta$ and then $\vdash \neg(\varphi \to \psi), \Gamma \Rightarrow \Delta$.

3. If we have $\vdash_n \neg(\varphi \to \psi), \Gamma \Rightarrow \Delta$, we want to show that $\vdash \Gamma \Rightarrow \Delta, \neg\neg(\varphi \to \psi)$. Now, the premiss is $\vdash \varphi, \neg\psi, \Gamma \Rightarrow \Delta$ and by the induction hypothesis $\vdash \Gamma \Rightarrow \Delta, \neg\varphi, \neg\neg\psi$, hence, by the preceding lemma, $\vdash \Gamma \Rightarrow \Delta, \neg\varphi, \psi$. From this we obtain $\vdash \Gamma \Rightarrow \Delta, \neg\neg(\varphi \to \psi)$ by $(\Rightarrow\to)$ and $(\Rightarrow \neg\neg)$.

4. In case of $\vdash_n \Gamma \Rightarrow \Delta, \neg(\varphi \to \psi)$ the premisses are: $\vdash \Gamma \Rightarrow \Delta, \varphi$ and $\vdash \Gamma \Rightarrow \Delta, \neg\psi$. By the induction hypothesis we get $\vdash \neg\varphi, \Gamma \Rightarrow \Delta$ and $\neg\neg\psi, \Gamma \Rightarrow \Delta$ which yields $\vdash \psi, \Gamma \Rightarrow \Delta$. Eventually we derive $\vdash \neg\neg(\varphi \to \psi), \Gamma \Rightarrow \Delta$. □

Exercise 3.24. *Check other cases.*
Show that lemma 3.14 implies admissibility of G3 rules for $\neg$ and $\to$ in SG3.

Remark 3.1. One may also prove directly for SG3 the admissibility of cut by any of the methods we used for G3. We postpone this task to chapter 5 where some weaker versions of this calculus will be examined. But the reader may want to try this now.

Having established the adequacy of SG3 we can prove the interpolation theorem in a direct way for sequents:

Theorem 3.8. *If $\vdash \Gamma \Rightarrow \Delta$, then for some φ such that $PROP(\varphi) \subseteq PROP(\Gamma \sqcap \Delta)$, $\vdash \Gamma \Rightarrow \varphi$ and $\varphi \Rightarrow \Delta$.*

PROOF: Again by the induction on the height of proof. It is straightforward to check that for axioms the respective interpolants are:

$$\varphi, \bot, \top, \bot, \top, \bot, \top.$$

For all one-premiss rules if we have established an interpolant for the premiss, it still holds for the conclusion.

Exercise 3.25. *Check three cases of interpolant-preservation for one-premiss rules.*

Interpolants for two-premiss rules are built from interpolants established for premisses. Let χ_1 and χ_2 be interpolants of the left and the right premiss, respectively. Then, for $(\Rightarrow \wedge), (\Rightarrow \neg\vee), (\Rightarrow \neg \rightarrow)$ the interpolant is $\chi_1 \wedge \chi_2$ whereas for $(\vee \Rightarrow).(\rightarrow\Rightarrow),(\neg\wedge \Rightarrow)$ it is $\chi_1 \vee \chi_2$.

Let us check the cases of $(\Rightarrow \neg \rightarrow)$ and $(\neg\wedge \Rightarrow)$. In the first case we assume that χ_1 is an interpolant of $\Gamma \Rightarrow \Delta, \varphi$ and χ_2 is an interpolant for $\Gamma \Rightarrow \Delta, \neg\psi$. Hence we have: $\vdash \Gamma \Rightarrow \chi_1, \vdash \chi_1 \Rightarrow \Delta, \varphi, \vdash \Gamma \Rightarrow \chi_2$ and $\vdash \chi_2 \Rightarrow \Delta, \neg\psi$. Hence by $(\Rightarrow \wedge)$ we have $\vdash \Gamma \Rightarrow \chi_1 \wedge \chi_2$ and by $(W \Rightarrow), (\Rightarrow \neg \rightarrow)$ and $(\wedge \Rightarrow)$ we get $\vdash \chi_1 \wedge \chi_2 \Rightarrow \Delta, \neg(\varphi \rightarrow \psi)$. It is easy to verify that atom containment condition for $\chi_1 \wedge \chi_2$ holds if they hold for χ_1 and χ_1, therefore it is an interpolant for $\Gamma \Rightarrow \Delta, \neg(\varphi \rightarrow \psi)$.

For $(\neg\wedge \Rightarrow)$ we prove that $\chi_1 \vee \chi_2$ is an interpolant. Now we have: $\vdash \neg\varphi, \Gamma \Rightarrow \chi_1, \vdash \chi_1 \Rightarrow \Delta, \vdash \neg\psi, \Gamma \Rightarrow \chi_2$ and $\vdash \chi_2 \Rightarrow \Delta$. We get $\chi_1 \vee \chi_2 \Rightarrow \Delta$ and $\neg(\varphi \wedge \psi), \Gamma \Rightarrow \chi_1 \vee \chi_2$ by $(\Rightarrow W), (\neg\wedge \Rightarrow)$ and $(\Rightarrow \vee)$. Verification that it is an interpolant for $\neg(\varphi \wedge \psi), \Gamma \Rightarrow \chi_1 \vee \chi_2$ is straightforward. $\qquad \square$

Exercise 3.26. *Prove the remaining cases.*

Remark 3.2. One may wonder why not to use a simpler version of SC in which only the four rules for negation and implication are replaced with their symmetric forms. Unfortunately such a solution does not work; without rules for negated compound formulae such a system is not complete. Try to prove, for example, $\Rightarrow (p \rightarrow q) \rightarrow (\neg q \rightarrow \neg p)$. After the first application of $(\Rightarrow\rightarrow)'$ and $(\Rightarrow \neg\neg)$ we just stop with $\Rightarrow \neg(p \rightarrow q), q, \neg p$. Hence negated rules are really indispensable.

3.6.3   Interpolation by Splitting a Proof

The above proofs of the interpolation theorem may not necessarily provide the most economical interpolant. Moreover, Maehara's proof is rather complex and Smullyan's proof requires noncanonical SC. Also both are based on induction on the height of a proof. Can we do better? We present here an additional constructive proof quite recently proposed by Muskens and Wintein [271]. In fact, they provided a novel proof of interpolation for some non-classical logics based on direct construction of tableaux for both formulae from examined implication $\varphi \rightarrow \psi$.

We adapt their method here to **CPL** and use G3 instead of tableaux. In contrast to Maehara's or Smullyan's proof we do not need a proof by induction on the height of a proof. But instead of one proof tree we split the work and build separate trees for $\varphi \Rightarrow$ and $\Rightarrow \psi$. Finally we construct an interpolant on the basis of the chosen leaves in both trees. So interpolant is not built successively on the basis of all sequents in the proof but rather directly constructed from atoms which are derived from φ and ψ.

We express the interpolation theorem in the following way.

Theorem 3.9. *For any contingent formulae φ, ψ, if $\models \varphi \to \psi$, then we can construct an interpolant on the basis of proof trees for $\varphi \Rightarrow$ and $\Rightarrow \psi$.*

PROOF: Assume that $\models \varphi \to \psi$; hence by the completeness proof we have a cut-free proof of it. Now apply a proof-search procedure to $\varphi \Rightarrow$ and to $\Rightarrow \psi$ and produce completed trees for them. Let $\Gamma_1 \Rightarrow \Delta_1, ..., \Gamma_k \Rightarrow \Delta_k$ be the list of nonaxiomatic leaves of the tree for $\varphi \Rightarrow$ and $\Pi_1, ..., \Sigma_1, ..., \Pi_n \Rightarrow \Sigma_n$ such a list taken from the tree for $\Rightarrow \psi$. It holds:

Claim 3.1. *For any $i \leq k$ and $j \leq n$, $\Gamma_i, \Pi_j \Rightarrow \Delta_i, \Sigma_j$ is an axiomatic clause.*

To see this take a tree for $\varphi \Rightarrow$ and add ψ to succedents of all sequents in the tree. Due to context independence of all rule it is a correct derivation tree. Now for each leaf $\Gamma_i \Rightarrow \Delta, \psi$ append a tree of $\Rightarrow \psi$ but with Γ_i added to each antecedent and Δ_i added to each succedent. In the resulting tree we have leaves of the form $\Gamma_i, \Pi_j \Rightarrow \Delta_i, \Sigma_j$ for all $i \leq k$ and $j \leq n$. If at least one of them is not axiomatic, then $\nvDash \varphi \Rightarrow \psi$. □

For every $\Gamma_i \Rightarrow \Delta_i, i \leq k$, define $\Gamma'_i = \{p \in \Gamma_i : p \in \Sigma_j \text{ for some } \Pi_j \Rightarrow \Sigma_j, j \leq n\}$ and $\Delta'_i = \{p \in \Delta_i : p \in \Pi_j \text{ for some } \Pi_j \Rightarrow \Sigma_j, j \leq n\}$. Since every $\Gamma_i, \Pi_j \Rightarrow \Delta_i, \Sigma_j$ is axiomatic we are guaranteed that $\Gamma'_i \cup \Delta'_i \neq \varnothing$. Note also that $PROP(\Gamma'_i \cup \Delta'_i) \subseteq PROP(\{\varphi\} \cap \{\psi\})$.

Now we can show that:

Claim 3.2. $Int(\varphi, \psi) := \Gamma'_1 \wedge \neg \Delta'_1 \vee ... \vee \Gamma'_k \wedge \neg \Delta'_k$ *is an interpolant for $\varphi \to \psi$.*

Since for every $\Gamma'_i \wedge \neg \Delta'_i$ all (negated) atoms are by definition taken from $PROP(\{\varphi\} \cap \{\psi\})$ we must only prove that $\vdash \varphi \Rightarrow Int(\varphi, \psi)$ and $\vdash Int(\varphi, \psi) \Rightarrow \psi$.

Again take a tree for $\varphi \Rightarrow$ and add $Int(\varphi, \psi)$ to every succedent. For every $\Gamma_i \Rightarrow \Delta_i, Int(\varphi, \psi)$ apply $(\Rightarrow \vee)$ to get $\Gamma_i \Rightarrow \Delta_i, \Gamma'_i \wedge \neg \Delta'_i, Int(\varphi, \psi)^{-i}$, where $Int(\varphi, \psi)^{-i}$ is the rest of the disjunction (if any). By definition $\Gamma'_i \wedge \neg \Delta'_i$ is a conjunction of some $p \in \Gamma_i$ and $\neg q$ for some $q \in \Delta_i$. Hence by the application of $(\Rightarrow \wedge)$ and (possibly) $(\Rightarrow \neg)$ we obtain a set of axiomatic leaves for each conjunct. Hence we have a proof of $\varphi \Rightarrow Int(\varphi, \psi)$.

Do the same with a tree for $\Rightarrow \psi$ but now adding $Int(\varphi, \psi)$ to every antecedent. For every leaf $Int(\varphi, \psi), \Pi_j \Rightarrow \Sigma_j$ apply $(\vee \Rightarrow)$ for each disjunct until you get leaves: $\Gamma'_1 \wedge \neg \Delta'_1, \Pi_j \Rightarrow \Sigma_j ... \Gamma'_k \wedge \neg \Delta_k, \Pi_j \Rightarrow \Sigma_j$. In each such leaf apply $(\wedge \Rightarrow)$ and (possibly) $(\neg \Rightarrow)$. Since any $\Gamma_i, \Pi_j \Rightarrow \Delta_i, \Sigma_j$ is axiomatic, in every leaf either there is some $p \in \Gamma'_i \cap \Sigma_j$ or some $q \in \Delta'_i \cap \Pi_j$. Hence $Int(\varphi, \psi) \Rightarrow \psi$. □

3.7 Some Related Results

Eventually we take a look at two additional results which are strongly connected with admissibility of cut. The first is concerned with some rule which is equivalent to cut. The second is a refinement of theorem 1.13 in the spirit of normalisation for natural deduction systems.

3.7.1   Tautology Elimination Rule

Let us consider the following rules of tautology elimination (TE):

$$\frac{\top, \Gamma \Rightarrow \Delta}{\Gamma \Rightarrow \Delta} \qquad \text{or} \qquad \frac{\varphi \to \varphi, \Gamma \Rightarrow \Delta}{\Gamma \Rightarrow \Delta}$$

Although it is a valid schema of rule for elimination of any thesis from the antecedent of any sequent —hence the name tautology elimination—we prefer for our purposes the more specific instance on the right. One may easily prove the following:

Lemma 3.15 (Equivalence of TE and Cut). *TE and cut are interderivable in G3.*

$$(Cut) \frac{(\Rightarrow\to) \dfrac{\varphi \Rightarrow \varphi}{\Rightarrow \varphi \to \varphi} \qquad \varphi \to \varphi, \Gamma \Rightarrow \Delta}{\Gamma \Rightarrow \Delta}$$

$$(TE) \frac{(\to\Rightarrow) \dfrac{\Gamma \Rightarrow \Delta, \varphi \qquad \varphi, \Gamma \Rightarrow \Delta}{\varphi \to \varphi, \Gamma \Rightarrow \Delta}}{\Gamma \Rightarrow \Delta}$$

Exercise 3.27. *Prove the equivalence also for A-cut and for multiplicative form of* $(\to\Rightarrow)$.

Such a rule was first introduced in the late 1950s by Davis and Putnam [57] in their automated theorem prover for **CPL** in the form:

$$\frac{C_1, ..., C_{k-1}, C[\top]_k, C_{k+1}, ..., C_n}{C_1, ..., C_{k-1}, C_{k+1}, ..., C_n}$$

where C denotes a clause.

It is well known that Davis and Putnam procedure is one of the most efficient for **CPL** and in particular TE was also applied in many variants of resolution as an additional technique for improvement of performance. In the framework of sequent calculi TE was introduced by Lyaletsky [166] under the name *tautology rule* and it was used rather for building proof-search procedures. As a device for proving cut admissibility it was applied recently by Brighton [40] and then by Tourlakis and Gao [90]. In both cases admissibility of TE was proved for some modal logics of provability and provided proofs were slightly complicated because of taking into account not only proof trees but also proof-search trees. In case of **CPL** all such complications may be avoided. Below we provide a simple proof of admissibility of TE in G3 by induction on the complexity (of eliminated formula) and the height of the proof.

We shall prove:

Theorem 3.10 (Admissibility of TE). *For any* φ, Γ, Δ, *if* $\vdash \varphi \to \varphi, \Gamma \Rightarrow \Delta$, *then* $\vdash \Gamma \Rightarrow \Delta$.

PROOF: Assume that:

$$\text{(a)} \vdash \varphi \to \varphi, \Gamma \Rightarrow \Delta.$$

Immediately, by invertibility we obtain:

$$\text{(b)} \vdash \Gamma \Rightarrow \Delta, \varphi \quad \text{and}$$
$$\text{(c)} \vdash \varphi, \Gamma \Rightarrow \Delta.$$

We prove that $\vdash \Gamma \Rightarrow \Delta$ by induction on the complexity of φ.

Basis: φ is atomic.

Consider (c); we prove the claim that $\vdash \Gamma \Rightarrow \Delta$ by subsidiary induction on the height of $\vdash \varphi, \Gamma \Rightarrow \Delta$.

If it is an axiom, then either $\Gamma \Rightarrow \Delta$ is an axiom, or $\varphi, \Gamma \Rightarrow \Delta$ is of the form $\varphi, \Gamma \Rightarrow \Delta', \varphi$. Then, by (b) we have $\vdash \Gamma \Rightarrow \Delta', \varphi, \varphi$ which, by contraction, reduces to $\Gamma \Rightarrow \Delta', \varphi$ and we are done.

For the induction step of subsidiary induction assume that the claim holds for any proof of $\vdash \varphi, \Gamma \Rightarrow \Delta$ of the height $k < n$ and prove it for the height $= n$. The proof is trivial since φ may be only a parameter and its deletion (by the induction hypothesis) in premisses does not affect the application of any rule.

The induction step (of the main induction): Assume as the induction hypothesis that the lemma holds for all formulae of lower complexity than φ. The proof goes by cases. For all types of formulae it is similar and based on invertibility of respective rules. We consider only one case as an example:

$\varphi := \psi \vee \chi$:

By invertibility we obtain from (b) and (c): $\vdash \Gamma \Rightarrow \Delta, \psi, \chi, \vdash \psi, \Gamma \Rightarrow \Delta$ and $\vdash \chi, \Gamma \Rightarrow \Delta$ and we build the following proof, where IH is marking the application of induction hypothesis:

$$
(\to\Rightarrow) \cfrac{\Gamma \Rightarrow \Delta, \psi, \chi \qquad IH \cfrac{\chi, \Gamma \Rightarrow \Delta \qquad (\Rightarrow W) \cfrac{\chi, \Gamma \Rightarrow \Delta}{\chi, \Gamma \Rightarrow \Delta, \psi}}{\chi \to \chi, \Gamma \Rightarrow \Delta, \psi}}{(\to\Rightarrow) \cfrac{\Gamma \Rightarrow \Delta, \psi \qquad \psi, \Gamma \Rightarrow \Delta}{IH \cfrac{\psi \to \psi, \Gamma \Rightarrow \Delta}{\Gamma \Rightarrow \Delta}}}
$$

Exercise 3.28. *Provide proofs for other compound formulae.*

From lemma 3.15 follows that admissibility of TE implies admissibility of cut. Moreover, the proof is very simple and similar to Schütte's proof of the admissibility of cut.

3.7.2   Strong Completeness in Normal Form

In section 1.11 we have proved strong completeness of K and of course this result holds also for G3. However, on the basis of results concerning admissibility of structural rules and invertibility of logical rules we can provide even more sharpened version where the only cuts are not only analytic but atomic in the strong sense that both premises of all cuts are clauses. Moreover, this novel proof of strong completeness shows that every proof of S from some $S_1, ..., S_k$ may be put into a normal form in the sense specified for natural deduction systems. Such a result was provided by Přenosil [204] for some non-classical logics which will be discussed in chapter 5. It holds for classical logic as well, moreover, we obtain here much simpler proof for G3, simply as a corollary of results stated so far.

Let us recall that in the context of natural deduction, the normalisation theorem plays the same role as the cut-elimination theorem in the framework of SC. In fact, Gentzen proved cut elimination for LK in order to show that natural deduction proofs in his system NK may be transformed into normal form[14]. Very roughly, in normal natural deduction proofs elimination rules are applied before introduction rules in order to avoid detours in proofs[15]. Due to the complicated structure of many proof construction rules connected with discharge of assumptions and several irregularities in the shape of rules this result is both hard to obtain and complicated to express even for systems for which it does hold. For ND systems with uniform elimination rules it is easier to prove normal form theorem and the structure is more transparent (see Negri and von Plato [185]). Here we show that for every $S_1, ..., S_k \vdash S$ we can obtain such a normal proof which consists of three separated parts:

1. Elimination part.

2. Structural part.

3. Introduction part.

in concrete proofs at least one of these parts must be nonempty.

Let us consider G3 enriched with structural rules and inverses of all logical rules. Moreover, instead of generalised atomic axioms of G3 we will use simple atomic axioms of the form $\varphi \Rightarrow \varphi$ for any atomic formula φ. We will call this system G3E. Note that despite the fact that we have proved for G3 admissibility of all structural rules and inverses of logical rules, these results fail for a calculus with nonaxiomatic sequents allowed as leaves. As for cut we have commented on this in section 1.11. but the same applies to W and C. In other words, we cannot prove that, e.g. if $S_1, ..., S_n \vdash \Gamma \Rightarrow \Delta$, then $S_1, ..., S_n \vdash \varphi, \Gamma \Rightarrow \Delta$. The only thing we can prove is that $\varphi, S_1, ..., \varphi, S_n \vdash \varphi, \Gamma \Rightarrow \Delta$, where φ, S_i denotes a sequent S_i with φ added to the antecedent, similarly for the other weakening and contractions.

---

[14]Von Plato [197] found and published the original direct proof of Gentzen obtained for intuitionistic system only.

[15]Locus classicus is still Prawitz [203], see also Negri and von Plato [185].

Concerning the inverses of logical rules, one may prove their derivability by means of cut on compound formulae but this is not sufficient for our purposes, since we want to show that all cuts may be replaced by atomic cuts. So we must admit that all structural and elimination rules (i.e. inverses of logical introduction rules) are primitive.

It is obvious that every proof in G3 may be easily changed into a proof in G3E just by adding over any axiomatic leaf a segment starting with simple axiom and a series of weakenings. Thus in particular, every proof of S from $S_1, ..., S_n$ with analytic applications of cut (see theorem 1.13 in section 1.11) may be immediately transformed into a proof in G3E, also analytic. In what follows we consider only such proofs. Let us say that a segment of a branch satisfies ESI order iff all applications of elimination rules (if any) precede all applications of structural rules (if any), and the latter precedes all applications of introduction rules (if any). We will say that an application of a structural rule (including cut) is atomic based iff all parametric formulae of premises and conclusion are atomic, and that it is atomic iff also active formulae are atomic. A proof is in the normal form iff all its branches satisfy ESI order and all applications of structural rules are atomic.

We will show that on any analytic proof $\mathcal{D}$ of $S_1, ..., S_n \vdash S$ we can systematically perform local transformations which eventually yield a normal proof of $S_1, ..., S_n \vdash S$. First we prove:

Lemma 3.16. *Every application of a structural rule may be changed into atomic based and satisfying ESI order.*

PROOF: By induction on the number of connectives in all parametric formulae occurring in a sequent being the premiss of the application of some structural rule. Assume that we analyse an application of $(W \Rightarrow)$ and that some conjunction is parametric, there are two subcases depending on the position of this conjunction in the sequent.

1) We have:

$$(W \Rightarrow) \ \frac{\varphi \wedge \psi, \Gamma \Rightarrow \Delta}{\chi, \varphi \wedge \psi, \Gamma \Rightarrow \Delta}$$

and it is replaced by:

$$\begin{array}{l} (\wedge E \Rightarrow) \ \dfrac{\varphi \wedge \psi, \Gamma \Rightarrow \Delta}{\varphi, \psi, \Gamma \Rightarrow \Delta} \\ (W \Rightarrow) \ \dfrac{}{\chi, \varphi, \psi, \Gamma \Rightarrow \Delta} \\ (\wedge \Rightarrow) \ \dfrac{}{\chi, \varphi \wedge \psi, \Gamma \Rightarrow \Delta} \end{array}$$

Note that ESI order is preserved in the resulting segment and the same applies to the next case:

2)

$$(W \Rightarrow) \frac{\Gamma \Rightarrow \Delta, \varphi \wedge \psi}{\chi, \Gamma \Rightarrow \Delta, \varphi \wedge \psi}$$

and it is replaced with:

$$(\Rightarrow \wedge E) \frac{\Gamma \Rightarrow \Delta, \varphi \wedge \psi}{(W \Rightarrow) \frac{\Gamma \Rightarrow \Delta, \varphi}{\chi, \Gamma \Rightarrow \Delta, \varphi}} \qquad \frac{\Gamma \Rightarrow \Delta, \varphi \wedge \psi}{\frac{\Gamma \Rightarrow \Delta, \psi}{\chi, \Gamma \Rightarrow \Delta, \psi} (\Rightarrow \wedge E)}{\chi, \Gamma \Rightarrow \Delta, \varphi \wedge \psi}$$

Exercise 3.29. *Provide proofs for other compound formulae on both sides of a sequent. Check that for* $(\Rightarrow W)$ *and contraction the proof goes in the same way.*

For cut and parametric conjunction we have:

$$(Cut) \frac{\varphi \wedge \psi, \Gamma \Rightarrow \Delta, \chi \qquad \chi, \Pi \Rightarrow \Sigma}{\varphi \wedge \psi, \Gamma, \Pi \Rightarrow \Delta, \Sigma}$$

or

$$(Cut) \frac{\Gamma \Rightarrow \Delta, \varphi \wedge \psi, \chi \qquad \chi, \Pi \Rightarrow \Sigma}{\Gamma, \Pi \Rightarrow \Delta, \Sigma, \varphi \wedge \psi}$$

(for the right premiss the proof is exactly the same)

The former is replaced with:

$$(\wedge E \Rightarrow) \frac{\varphi \wedge \psi, \Gamma \Rightarrow \Delta, \chi}{(Cut) \frac{\varphi, \psi, \Gamma \Rightarrow \Delta, \chi \qquad \chi, \Pi \Rightarrow \Sigma}{(\wedge \Rightarrow) \frac{\varphi, \psi, \Gamma, \Pi \Rightarrow \Delta, \Sigma}{\varphi \wedge \psi, \Gamma, \Pi \Rightarrow \Delta, \Sigma}}}$$

and the latter with:

$$(\Rightarrow \wedge E) \frac{\Gamma \Rightarrow \Delta, \varphi \wedge \psi, \chi}{(Cut) \frac{\Gamma \Rightarrow \Delta, \varphi, \chi \qquad \chi, \Pi \Rightarrow \Sigma}{(\Rightarrow \wedge) \frac{\Gamma, \Pi \Rightarrow \Delta, \Sigma, \varphi}{\Gamma, \Pi \Rightarrow \Delta, \Sigma, \varphi \wedge \psi}}} \qquad \frac{\frac{\Gamma \Rightarrow \Delta, \varphi \wedge \psi, \chi}{\Gamma \Rightarrow \Delta, \psi, \chi} (\Rightarrow \wedge E) \qquad \chi, \Pi \Rightarrow \Sigma}{\Gamma, \Pi \Rightarrow \Delta, \Sigma, \psi} (Cut)$$

Exercise 3.30. *Complete a proof for cut with other compound formulae on both sides of a sequent.*

Repeating this procedure we must finish with a proof where all applications of structural rules are atomic-based.   □

Next we must prove:

Lemma 3.17. *Every atomic-based application of a structural rule may be changed into atomic application satisfying ESI order.*

PROOF: Again we prove this by induction on the complexity, now of active formulae in the applications of rules. For example, weakening with implication may be dealt with as follows:

$$(\Rightarrow W) \frac{\Gamma \Rightarrow \Delta}{\Gamma \Rightarrow \Delta, \varphi \to \psi}$$

is replaced by:

$$(\Rightarrow W) \frac{\Gamma \Rightarrow \Delta}{\Gamma \Rightarrow \Delta, \psi}$$
$$(W \Rightarrow) \frac{}{\varphi, \Gamma \Rightarrow \Delta, \psi}$$
$$(\Rightarrow \to) \frac{}{\Gamma \Rightarrow \Delta, \varphi \to \psi}$$

and

$$(W \Rightarrow) \frac{\Gamma \Rightarrow \Delta}{\varphi \to \psi, \Gamma \Rightarrow \Delta}$$

with:

$$(\Rightarrow W) \frac{\Gamma \Rightarrow \Delta}{\Gamma \Rightarrow \Delta, \varphi} \quad \frac{\Gamma \Rightarrow \Delta}{\psi, \Gamma \Rightarrow \Delta} (W \Rightarrow)}{\varphi \to \psi, \Gamma \Rightarrow \Delta} (\to \Rightarrow)$$

Exercise 3.31. *Show other cases.*

For contraction let us examine a disjunction:

$$(\Rightarrow C) \frac{\Gamma \Rightarrow \Delta, \varphi \vee \psi, \varphi \vee \psi}{\Gamma \Rightarrow \Delta, \varphi \vee \psi}$$

is replaced by:

$$(\Rightarrow \vee E) \frac{\Gamma \Rightarrow \Delta, \varphi \vee \psi, \varphi \vee \psi}{\Gamma \Rightarrow \Delta, \varphi \vee \psi, \varphi, \psi}$$
$$(\Rightarrow \vee E) \frac{}{\Gamma \Rightarrow \Delta, \varphi, \psi, \varphi, \psi}$$
$$(\Rightarrow C) \frac{}{\Gamma \Rightarrow \Delta, \varphi, \psi, \psi}$$
$$(\Rightarrow C) \frac{}{\Gamma \Rightarrow \Delta, \varphi, \psi}$$
$$(\Rightarrow \vee) \frac{}{\Gamma \Rightarrow \Delta, \varphi \vee \psi}$$

and

$$(C \Rightarrow) \frac{\varphi \vee \psi, \varphi \vee \psi, \Gamma \Rightarrow \Delta}{\varphi \vee \psi, \Gamma \Rightarrow \Delta}$$

with:

$$(\vee E \Rightarrow) \frac{\varphi \vee \psi, \varphi \vee \psi, \Gamma \Rightarrow \Delta}{\varphi, \varphi \vee \psi, \Gamma \Rightarrow \Delta} \quad \frac{\varphi \vee \psi, \varphi \vee \psi, \Gamma \Rightarrow \Delta}{\psi, \varphi \vee \psi, \Gamma \Rightarrow \Delta} (\vee E \Rightarrow)$$
$$(\vee E \Rightarrow) \frac{\varphi, \varphi, \Gamma \Rightarrow \Delta}{} \quad \frac{\psi, \psi, \Gamma \Rightarrow \Delta}{} (\vee E \Rightarrow)$$
$$(C \Rightarrow) \frac{\varphi, \Gamma \Rightarrow \Delta}{} \quad \frac{\psi, \Gamma \Rightarrow \Delta}{} (C \Rightarrow)$$
$$\frac{\varphi \vee \psi, \Gamma \Rightarrow \Delta}{} (\vee \Rightarrow)$$

Exercise 3.32. *Show other cases.*

Finally, for cut with conjunction as the cut-formula:

$$\frac{\Gamma \Rightarrow \Delta, \varphi \wedge \psi \qquad \varphi \wedge \psi, \Pi \Rightarrow \Sigma}{\Gamma, \Pi \Rightarrow \Delta, \Sigma} \ (Cut)$$

we have:

$$(\Rightarrow \wedge E) \ \frac{\dfrac{(\Rightarrow \wedge E) \ \dfrac{\Gamma \Rightarrow \Delta, \varphi \wedge \psi}{\Gamma \Rightarrow \Delta, \varphi} \qquad \dfrac{\varphi \wedge \psi, \Pi \Rightarrow \Sigma}{\varphi, \psi, \Pi \Rightarrow \Sigma} \ (\wedge \Rightarrow)}{\psi, \Gamma, \Pi \Rightarrow \Delta, \Sigma} \ (Cut)}{\dfrac{\dfrac{\Gamma \Rightarrow \Delta, \varphi \wedge \psi}{\Gamma \Rightarrow \Delta, \psi} \qquad \qquad \qquad \qquad \qquad \qquad \qquad}{\Gamma, \Pi, \Pi \Rightarrow \Delta, \Sigma, \Sigma}} \ (Cut)}{\Gamma, \Pi, \Rightarrow \Delta, \Sigma} \ (C)$$

Exercise 3.33. *Show other cases.*

Note that all transformations preserve ESI order.   □

Now we can see why it was better to separate transformation into atomic-based applications from transformations into atomic ones. In the latter case a suitable transformation of cut may lead to the introduction of new contractions. If they were not atomic, then ESI order on the applications of structural rules is not preserved.

Proofs obtained as a consequence of application of two preceding lemmata are structurally atomic in the sense that all structural rules are made on atomic sequents only. But they are not necessarily normal since our transformations were local and we have made them in a nondeterministic fashion. It is obvious that there is no application of introduction rule immediately above an application of a structural rule and no application of elimination rule immediately below. But we cannot be sure that in segments consisting of the application of logical rules only there are no introduction rules immediately above elimination rules. In fact, we can have a situation where on a branch there are many segments with succession of logical rules intertwined with segments with atomic structural rules only. Hence we need:

Lemma 3.18. *All introduction rules may be permuted down with respect to all elimination rules.*

PROOF: In case where the principal formula of an application of some introduction rule is parametric in the premiss of elimination rule we make a permutation exactly as in the proof of lemma 2.19. Now if there is no structural part below, then after repeating this process we are done. If there is a structural part then, since it is atomic, we must finish with a detour where the principal formula of introduction rule is a side formula of elimination rule, like, e.g.:

$$\frac{\dfrac{\Gamma \Rightarrow \Delta, \varphi \qquad \psi, \Gamma \Rightarrow \Delta}{\varphi \rightarrow \psi, \Gamma \Rightarrow \Delta} \ (\rightarrow \Rightarrow)}{\psi, \Gamma \Rightarrow \Delta} \ (\rightarrow E \Rightarrow)$$

but then we simply erase the immediate sequent (and in this case also a proof of the other premiss). □

Exercise 3.34. *Show two cases of permutation.*

As a result of the application of these three lemmata we have:

Theorem 3.11. *Every proof of $S_1, ..., S_n \vdash S$ in G3E may be transformed into normal proof.*

Note that after the last transformation if any premiss of a cut is a descendant of (simple) axiomatic leaf it must be still a generalised axiom and we know that such applications of cut are eliminable so we additionally have:

Corollary 3.1. *All necessary cuts (if any) in a normal proof $S_1, ..., S_n \vdash S$ in G3E are made only on descendants of $S_1, ..., S_n$*

As a consequence of this result we may provide yet another constructive proof of the interpolation theorem by consideration of a normal proof for $\Rightarrow \varphi \vdash \Rightarrow \psi$ This time a structural part of the proof takes the role of separating device. But first we must prove that:

Lemma 3.19. *In every normal proof all applications of weakening may be permuted down with cuts and contractions.*

PROOF: This is obvious. Of course in case a weakening above cut introduces cut-formula, this cut and weakening is eliminable, and similarly with weakenings introducing one of the active formulae of contraction. All other applications of W may be pushed down immediately above the part with logical introduction rules. □

Again this small refinement is necessary to make sure that there is no atom which was introduced by W immediately after elimination part, such that it is not present in $PROP(\varphi)$ but it will be present in the interpolant. Now suppose we have such refined normal proof for $\Rightarrow \varphi \vdash \Rightarrow \psi$ and call a critical sequent a clause in the structural part of every branch starting with $\Rightarrow \varphi$ which is immediately above any applications of W. It is obvious that all atoms of any critical sequent must be in $PROP(\varphi) \cap PROP(\psi)$ since only elimination rules, cut and C were applied to $\Rightarrow \varphi$ and only W and introduction rules were applied to critical sequent on every such branch. Let $S_1, ..., S_k, 1 \leq k$ be all critical sequents of a proof and consider any of the interpretations defined in section 3.4, then $I(S_1) \wedge ... \wedge I(S_k)$ is an interpolant for $\varphi \rightarrow \psi$.

In fact, one may think also about a systematic procedure for constructing such normal proofs; we just sketch the idea. We start with the collection of assumption-sequents $S_1, ..., S_k$ and apply successively elimination rules to all compound formulae. Since the number of compound formulae is finite and for each we have at most two elimination rules we obtain in a finite way a collection of clauses (atomic sequents) $S'_1, ..., S'_n$, $n \geq k$ where each S'_i follows from some S_j, the ancestor of S'_i. This is the elimination part of the procedure.

On the other hand, we can apply root-first procedure from section 1.6 and obtain completed proof tree $\mathcal{T}$ for S. All leaves are also clauses. This is the introduction part of the procedure.

Now we must compare two collections of clauses: $S'_1, ..., S'_n$ and a set of leaves of $\mathcal{T}$ in order to provide matching between them. Clearly, applications of W, C and cut may be necessary for that. This is a structural part of the procedure.

Chapter 4

Sequent Calculi for Modal Logics

Advantages of using SC as the basic syntactic tool in proof theory soon became obvious for many logicians and inspired them to search for extensions and generalisations. One branch, starting with Gentzen, showed that SC may be profitably used as a basis of formalisation of mathematical theories.[1] The other, which we describe in this and the next chapter, is connected with the need for finding decent syntactic formulation of non-classical logics[2]. However, this kind of research on SC formulations of non-classical logics showed also several limitations of Gentzen's approach. It appeared that many non-classical logics cannot be successfully formalised by means of canonical rules and even the basic format of Gentzen's sequents may be in need of some changes and generalisations.

Gentzen already provided a calculus LJ as a formalisation of intuitionistic logic. Since treatment of non-classical logics being alternative to classical one is postponed until the next chapter, we present it in section 5.1. Here we will focus on the applications of SC machinery to some logics which are extensions of **CPL** in richer languages. Although there is a great diversity of such systems, we restrict our attention to only one family but certainly the most important—modal logics. In fact, many families of logics that are extensions of classical logic like epistemic, deontic or temporal logics are, at least in a technical sense, just specific kinds of modal logics.

We do not aim to provide a detailed survey of modal logic, since we are only interested in showing what kind of problems may arise when we want to extend our machinery of SC to some extensions of **CPL**. Nevertheless, to keep the text self-contained, we recall the basic information necessary for further reading

---

[1]This kind of research will be dealt with in the second volume.

[2]In fact, both trends may be unified by treating the semantics of non-classical logics as formal theories, e.g. via labelled systems as in Negri and von Plato [186].

© Springer Nature Switzerland AG 2021
A. Indrzejczak, *Sequents and Trees*, Studies in Universal Logic,
https://doi.org/10.1007/978-3-030-57145-0_4

in section 4.1. We will focus on the three most representative and popular modal logics: **T**, **S4** and **S5** and only mention in passing some other logics. In section 4.2, we describe extensions of LK and G3 to these modal logics. Although rules for modal constants are significantly different from canonical rules for boolean connectives, we still obtain reasonable standard formalisations of the discussed logics. In particular, for **T** and **S4** we obtain cut-free systems. Section 4.3 presents proofs of cut elimination/admissibility, whereas section 4.4, deals with the problem of proof search and decidability for these logics as formalised on the basis of G3. As a by-product, we obtain analytic proofs of completeness for both systems. The remaining sections are devoted to problems with obtaining a cut-free formalisation of **S5**. On this occassion, we survey some generalised forms of SC. First, it will be shown in section 4.5, that for the standard system a direct constructive proof of cut elimination is not in general possible. We survey some proposals based either on possible translation into other cut-free logics, or restriction put on cut or introducing additional nonstandard rules. As a different way of overcoming this problem, we will present some generalised SC for **S5** in the remaining sections. First, we present practically the simplest one based on the application of labelled formulae. Then we focus on the family of solutions introduced in the framework of hypersequent calculus. Eventually, we describe a simplified solution based on the application of bisequents which are numerically restricted hypersequents. It will be shown that some other generalised approaches based on the application of structured sequents or different sorts of sequents may be translated into this framework.

4.1   Basic Modal Logics

In the foreword, we explained the rationale behind our policy of avoiding general treatments of wide domains and focusing rather on case studies. Thus, in this chapter, we will not be interested in the general approach covering as many logics as possible. In particular, we do not consider either weaker families of modal logics, like congruent, regular or monotonic[3], or more generalised versions like multimodal systems. We will focus on the well known class of normal (alethic) modal logics and, in particular, on three well known systems **T**, **S4** and **S5**. However, we have chosen them not only because they are important as modal logics. From the standpoint of the methodology of SC these three logics represent three different levels of difficulties which are encountered if we try to obtain cut-free systems and decidability results. **T** is not particularly problematic; although it introduces a different type of rules which are context-dependent (impure), there are no problems with proving cut elimination or decidability and standard techniques developed for classical logic work pretty well. In case of **S4**, we still may prove cut elimination in a standard way but proving decidability is harder. Finally, **S5** is very problem-

---

[3]SC and cut elimination proofs for the most important of these weaker modal logics is provided in [128, 131].

atic in ordinary SC framework; we cannot provide a constructive cut elimination proof. However, there are many possibilities to overcome this problem and we will present some of them. In particular, we will see that some generalisations of the standard SC provide very good solutions.

Let us recall briefly a standard axiomatic and semantic characterisation of normal modal logics; the three logics we selected are particularly popular examples from this class[4].

We will use standard monomodal language with a countable set $PROP$ of propositional variables, $\Box$—the unary modal necessity operator, and ordinary boolean constants. Syntactically $\Box$ behaves exactly as $\neg$. The modal depth of a formula is defined as the maximal number of nested occurrences of $\Box$, thus, e.g. $\Box(p \wedge \Box(q \to r))$ has modal depth 2. Usually, a dual modal operator of possibility $\Diamond$ is used but we treat it here only as a definitional shortcut for $\neg\Box\neg$, i.e: $\Diamond\varphi := \neg\Box\neg\varphi$. Anyway, incidentally we will make some remarks on possible rules for $\Diamond$.

The following abbreviations will be used throughout:
$\Box\Gamma = \{\Box\varphi : \varphi \in \Gamma\}$;
$\Gamma^{\Box} = \{\varphi : \Box\varphi \in \Gamma\}$;
$\Gamma_{\Box} = \{\Box\varphi : \Box\varphi \in \Gamma\}$.

4.1.1 Hilbert Systems

One can axiomatize respective modal logics by adding to the Hilbert system for **CPL** from section 1.2, the following schemata:

K $\Box(\varphi \to \psi) \to (\Box\varphi \to \Box\psi)$

T $\Box\varphi \to \varphi$

4 $\Box\varphi \to \Box\Box\varphi$

5 $\neg\Box\varphi \to \Box\neg\Box\varphi$ or $\Diamond\varphi \to \Box\Diamond\varphi$

Sometimes it will be convenient to refer to their $\Diamond$-based converses (partial in case of K)

K' $\Box(\varphi \to \psi) \to (\Diamond\varphi \to \Diamond\psi)$

T' $\varphi \to \Diamond\varphi$

4' $\Diamond\Diamond\varphi \to \Diamond\varphi$

5' $\Diamond\Box\varphi \to \Box\varphi$

Exercise 4.1 *Prove that all axioms are equivalent with their 'primed'-versions on the basis of the interdefinability of $\Box$ and $\Diamond$.*

---

[4]For an extensive study of the subject, we advise the reader to look at some of the following textbooks: [33, 48, 101]

All the systems are closed under MP (modus ponens) and the additional rule GR (Gödel's rule) called also the necessitation rule

$$\varphi \; / \; \Box\varphi$$

Now, if we add K as the only modal axiom, we get the weakest normal modal logic **K**. It is not particularly interesting in itself and may seem even defective if we interpret $\Box$ as 'necessity' and $\Diamond$ as 'possibility', yet it is a convenient basis. By addition of K and T we obtain an axiomatization of **T**. Addition of 4 to **T** makes **S4**, and addition of 5 yields **S5**. Instead of 5 one can use

B $\neg\varphi \to \Box\neg\Box\varphi$ or $\varphi \to \Box\Diamond\varphi$ or
B' $\Diamond\Box\varphi \to \varphi$.

and dispense with 4 since it is provable from B and T. Clearly one may obtain different normal modal logics by different combinations of the respective axioms, e.g. by addition of 4 or B directly to **K** we obtain logics **K4** and **KB**, whereas the addition of B to **T** is commonly known as **B**. In what follows we will be concerned mainly with **T**, **S4** and **S5**, since other logics either do not generate specific problems in SC framework or the problems generated by them require application of more general approaches (like in **KB** or **B**). In fact, **S5** may be seen as a kind of a limit case, and we will introduce some generalised systems for it by the end of this chapter.

The closure under GR is a definitional condition for normal modal logics in contrast to weaker logics like congruent, regular or monotonic modal logics. We omit a presentation of these logics but record here two useful rules which define the class of regular logics and are admissible also in normal logics

RR $\varphi \to \psi \; / \; \Box\varphi \to \Box\psi$ or
RR' $\varphi \to \psi \; / \; \Diamond\varphi \to \Diamond\psi$.

Let L denote one of the modal logics under consideration. In contrast to **CPL** the relation of deducibility (provability) may be defined in two nonequivalent ways

Definition 4.1 (Local ($\vdash$) and global ($\Vdash$) deducibility).

- $\Gamma \vdash_L \varphi$ iff $\vdash_L \psi_1 \wedge \; .... \wedge \psi_n \to \varphi$, where $\{\psi_1, ..., \psi_n\} \subseteq \Gamma$
- $\Gamma \Vdash_L \varphi$ iff there is a proof of φ in **L**, where formulae from Γ are also used as premises for the application of rules.

If Γ is empty then φ is a thesis of respective logic. Although from the point of view of thesishood, it does not matter which notion is considered (in both cases the sets of theses coincide), in general $\Vdash$ is a stronger relation than $\vdash$. This is because the closure under GR means that for $\Vdash$ the *deduction theorem* in simple form does not hold[5]. For example, already in **K** we have $p \Vdash \Box p$ (due to closure under

---

[5]It does not mean that the deduction theorem does not hold in general for these logics. Many theorems of this kind were established by Perzanowski [194, 195].

GR),but $p \nvdash \Box p$ (since $\nvdash_K p \to \Box p$). We have only one directional dependence

$$\text{if } \Gamma \vdash_L \varphi, \text{ then } \Gamma \Vdash_L \varphi$$

For $\vdash_L$, a deduction theorem is satisfied by definition. In fact, all properties listed for $\vdash$ in lemma 1.1, for **CPL**, including extensionality principle, hold without changes for $\vdash_L$ in all considered logics. Moreover, $\vdash_L$ satisfies the following property:

$$\text{if } \Gamma \vdash_L \varphi, \text{ then } \Box\Gamma \vdash_L \Box\varphi.$$

Note that this property covers both GR (in case Γ is empty) and K (which is easily provable). Hence, it may be used instead of GR and K to characterise the weakest normal logic **K**. Since the rules of SCs naturally generate relations of the type $\vdash$ as corresponding to $\Rightarrow$, in what follows we will be interested only in the first (weaker) notion of deducibility. Accordingly, we define Γ as **L**-inconsistent iff $\Gamma \vdash_L \bot$; otherwise, Γ is **L**-consistent.

4.1.2 Relational Semantics

Standard semantic approach to normal modal logics is based on the use of relational frames (models) often called *possible worlds semantics* or Kripke frames, although independently of Saul Kripke [152], semantics of this kind were introduced by many other logicians like Kanger [146] or Hintikka [117], to mention just a few[6]. The popularity of this approach follows from the fact that it offers a very natural and philosophically motivated interpretation of modal operators. It is also a natural tool for the interpretation of many other non-classical logics, like e.g. intuitionistic logic which will be considered in section 5.1. The basic notions of a frame and a model are defined in the following way:

Definition 4.2 (Frame). A modal frame is a tuple $\mathfrak{F} = \langle \mathcal{W}, \mathcal{R} \rangle$, where $\mathcal{W} \neq \varnothing$ is a set of states (possible worlds), and $\mathcal{R}$ is a binary relation on $\mathcal{W}$, called the accessibility relation.

In modal logics $\mathcal{R}ww'$ means that w' is accessible from w (possible relative to w); $\mathcal{R}(w) = \{w' : \mathcal{R}ww'\}$ is the set of all alternatives for w.

Definition 4.3 (Model). A model on the frame $\mathfrak{F}$ is a structure $\mathfrak{M} = \langle \mathfrak{F}, V \rangle$, where V is a valuation this time defined as a function on atoms ($V : PROP \longrightarrow \mathcal{P}(\mathcal{W})$). The set of all points of a model $\mathfrak{M}$ will be denoted as $\mathcal{W}_{\mathfrak{M}}$; the set of all models based on a frame will be denoted as $MOD(\mathfrak{F})$.

Intuitively, V assigns to each atom a set of states where it is true or satisfied (in a model). Note that due to the more complicated character of the semantics, the notion of an interpretation of a formula (and related semantical concepts) may be defined on different levels. The most basic is the notion of satisfaction of a formula in a state of a model, which is defined as follows:

---

[6]A detailed history of these early investigations may be found in Copeland [54].

$\mathfrak{M}, w \vDash \varphi$	iff	$w \in V(\varphi)$ for any $\varphi \in PROP$
$\mathfrak{M}, w \vDash \neg\varphi$	iff	$\mathfrak{M}, w \nvDash \varphi$
$\mathfrak{M}, w \vDash \varphi \wedge \psi$	iff	$\mathfrak{M}, w \vDash \varphi$ and $\mathfrak{M}, w \vDash \psi$
$\mathfrak{M}, w \vDash \varphi \vee \psi$	iff	$\mathfrak{M}, w \vDash \varphi$ or $\mathfrak{M}, w \vDash \psi$
$\mathfrak{M}, w \vDash \varphi \rightarrow \psi$	iff	$\mathfrak{M}, w \nvDash \varphi$ or $\mathfrak{M}, w \vDash \psi$
$\mathfrak{M}, w \vDash \Box\varphi$	iff	$\mathfrak{M}, w' \vDash \varphi$ for any w' such that $\mathcal{R}ww'$
$\mathfrak{M}, w \vDash \Diamond\varphi$	iff	$\mathfrak{M}, w' \vDash \varphi$ for some w' such that $\mathcal{R}ww'$

In the case where a model is established, we will simply write $w \vDash \varphi$. The set of all states where φ is satisfied in a model will be denoted as $\|\varphi\|_{\mathfrak{M}}$. Formally, $\|\varphi\|_{\mathfrak{M}} = \{w \in W_{\mathfrak{M}} : w \vDash \varphi\}$. Usually, we will simply use $\|\varphi\|$ when $\mathfrak{M}$ is established or unimportant. $\|\varphi\|$ is sometimes called a *proposition* expressed by φ. Although there are serious obstacles for considering this as a representation of a logical proposition, we also follow this convenient habit. In case of a set of formulae, $\|\Gamma\|_{\mathfrak{M}} = \bigcap \|\psi\|_{\mathfrak{M}}$ for all $\psi \in \Gamma$.

The preceding definition states conditions for local (at a state in a model) truth/satisfiability. It may be naturally generalised to various concepts of global satisfiability, or universal truth: in a model (all its worlds), frame (all its models), or some specified classes of frames (all models based on all frames in this class). They may be denoted, respectively, by $\mathfrak{M} \vDash \varphi$, $\mathfrak{F} \vDash \varphi$ and $\mathcal{F} \vDash \varphi$, where $\mathcal{F}$ is a class of frames. It appears that the most important is the last notion which is accordingly called validity (in a class of frames). Although the remaining two notions of truth may be disregarded, we briefly comment on them. The reason for skipping the notion of truth in a model follows from the fact that semantical characterisation of (normal) modal logics is connected not with particular models but with frames or sets of frames. Otherwise, we do not secure closure under substitution which is an important feature of all logics we consider.

The reason for neglecting the notion of truth in a frame is a bit different. It is well known that a set of formulae true in any frame is a normal logic. But a domain of any frame (the number, or a character of objects) is of no importance for defining logics, only properties of accessibility relations play an essential role in this respect. So dealing with uniform (modulo accessibility relations) classes of frames gives us the proper level of abstraction. Since the set of validities of any $\mathcal{F}$ is a normal modal logic (although not every normal modal logic is characterised by a class of frames) in what follows we will be talking about classes of frames, or models, with the same sort of relation. The latter will be denoted as $MOD(\mathcal{F})$ or $MOD(\mathbf{L})$ if it is known that $\mathcal{F}$ determines logic $\mathbf{L}$ in the sense specified below. In particular, all normal modal logics which we are dealing with are determined by classes of frames satisfying some properties on accessibility relations. It follows from the fact that formulae of modal language correspond to well known relational conditions; more precisely

Definition 4.4 (Correspondence). φ defines the class of structures $\mathcal{F}$ iff $\mathcal{F} \vDash \varphi$

The following table displays well known correspondencies for axioms listed in subsection 4.1.1.

name	condition	axiom
reflexivity	$\forall x \mathcal{R}xx$	T
transitivity	$\forall xyz(\mathcal{R}xy \wedge \mathcal{R}yz \rightarrow \mathcal{R}xz)$	4
symmetry	$\forall xy(\mathcal{R}xy \rightarrow \mathcal{R}yx)$	B
Euclideaness	$\forall xyz(\mathcal{R}xy \wedge \mathcal{R}xz \rightarrow \mathcal{R}yz)$	5

Note that axiom K does not correspond to any condition. In fact, **K** is characterised (or determined) by the class of all frames in the sense that φ is valid (in K) if it is true in every world of any model. On the basis of the table, we may establish determination results for other normal logics, in particular: **T** is determined by the class of reflexive frames, **S4** is determined by the class of frames with a quasi-ordering accessibility relation (i.e. reflexive and transitive), since T defines reflexivity and 4 — transitivity. If we take B instead of 5 in characterising **S5**, we can see that it is determined by the class of equivalence relations and this allows one to simplify its semantics. In fact, the simplest semantical characterisation of **S5** may be obtained by means of Kripke frames without an accessibility relation. A Model for **S5** is thus any pair $\mathfrak{M} = \langle \mathcal{W}, V \rangle$, $\mathcal{W} \neq \varnothing$ and $V : PROP \longrightarrow \mathcal{P}(\mathcal{W})$. Satisfaction in a world of a model is inductively defined as in models for relational frames but the clause for $\square$ is simpler

- $\mathfrak{M}, w \models \square\varphi$ iff $\mathfrak{M}, v \models \varphi$ for any $v \in \mathcal{W}_{\mathfrak{M}}$

A formula is **S5**-valid iff it is true in every world of every such model.

We will say that Γ is $\mathcal{F}$-*satisfiable* iff there is a state in a model $\mathfrak{M} \in MOD(\mathcal{F})$ which locally satisfies Γ, otherwise Γ is $\mathcal{F}$-*unsatisfiable*. In a dual manner, we may define falsifiability—formally

Definition 4.5 (Satisfiability, falsifiability). Γ is $\mathcal{F}$-satisfiable in a model $\mathfrak{M} \in MOD(\mathcal{F})$ iff $\|\Gamma\| \neq \varnothing$;
Γ is $\mathcal{F}$-satisfiable iff there is a model $\mathfrak{M} \in MOD(\mathcal{F})$, where it is satisfiable;
Γ is $\mathcal{F}$-falsifiable in a model $\mathfrak{M} \in MOD(\mathcal{F})$ iff $\|\Gamma\| \neq \mathcal{W}_{\mathfrak{M}}$ (or: $\mathfrak{M}$ falsifies Γ);
Γ is $\mathcal{F}$-falsified iff, there is a model $\mathfrak{M} \in MOD(\mathcal{F})$ which falsifies it

The concept of an entailment (consequence relation) may also be defined in at least two nonequivalent ways but we restrict our interest only to local notion:

φ *follows locally* in $\mathcal{F}$ from Γ ($\Gamma \models_{\mathcal{F}} \varphi$) iff $\|\Gamma\|_{\mathfrak{M}} \subseteq \|\varphi\|_{\mathfrak{M}}$ for any $\mathfrak{M} \in MOD(\mathcal{F})$.

(or: for all $w \in \mathcal{W}_{\mathfrak{M}}$ of any $\mathfrak{M} \in MOD(\mathcal{F})$ it holds that if $\mathfrak{M}, w \models \Gamma$, then $\mathfrak{M}, w \models \varphi$)

Obviously, $\Gamma \models_{\mathcal{F}} \varphi$ iff $\Gamma \cup \{\neg\varphi\}$ is $\mathcal{F}$-unsatisfiable.

The link between syntactic formalisations and classes of frames is obtained via soundness and completeness theorems of the form

- (Soundness) if $\Gamma \vdash_L \varphi$, then $\Gamma \models_{\mathcal{F}} \varphi$

- (Completeness) if $\Gamma \models_{\mathcal{F}} \varphi$, then $\Gamma \vdash_L \varphi$

  The last one is often formulated equivalently

- if Γ is **L**-consistent, then Γ is $\mathcal{F}$-satisfiable

If the first theorem holds, then **L** is sound with respect to $\mathcal{F}$, if the second holds, then **L** is (strongly) complete with respect to $\mathcal{F}$. If **L** is adequate (i.e. both sound and complete) with respect to $\mathcal{F}$, then $\mathcal{F}$ characterises **L** or **L** is determined by $\mathcal{F}$. Note that if Γ is empty (in the first formulation) or finite, we have weak completeness, otherwise we have strong form (i.e. admitting infinite Γ).

Standard proofs of completeness for modal logics apply the well known construction of a canonical model which is also based on the Henkin/Lindenbaum result concerning maximalization of consistent sets. As a result, we obtain a unique infinite model belonging to $MOD(\mathbf{L})$ that falsifies all formulae unprovable in **L**. For questions of decidability and automated theorem proving, it is more important that for many logics under consideration there are constructive methods of proving completeness. They show how to find for any unprovable formula some finite falsifying model. In particular, for many normal logics, we obtain special falsifying models based on rooted frames in the sense specified below.

Definition 4.6 (Rooted Frames). A frame $\mathfrak{F} = \langle \mathcal{W}, \mathcal{R} \rangle$ is rooted if there is $w_0 \in \mathcal{W}$, such that $\mathcal{W} = \{w : \mathcal{R}^+ w_0 w\}$, where $\mathcal{R}^+$ denotes the transitive closure of $\mathcal{R}$. A model based on such frame is called a rooted model.

So a rooted frame is generated by w_0 with the help of $\mathcal{R}^+$. We will be dealing further with falsifying models of this sort. Moreover, for many logics, the frames of such models are in fact trees or they turn into trees if we take as nodes not single points but their *clusters*.

Definition 4.7 (Cluster). Let $\mathfrak{F} = \langle \mathcal{W}, \mathcal{R} \rangle$ be any transitive frame, a cluster $\mathcal{C}$ is a maximal subset of $\mathcal{W}$, such that $\mathcal{R}$ is the universal relation in it. A cluster is simple if it contains only one point w, such that $\mathcal{R}ww$ (w is reflexive), otherwise it is proper (contains at least two different states).

The following table displays completeness results for the most important monomodal normal logics. The middle column contains results obtainable by the canonical model method. The right column gives determination results in terms of finite frames. Detailed data concerning sources may be found in Goré [102] or Indrzejczak [130].

L	L-frames	finite L-frames
T	reflexive	intransitive and reflexive trees
S4	reflexive and transitive	trees of finite clusters
S5	equivalential	single, finite clusters

All modal logics under consideration are decidable. It follows from the fact that they satisfy finite model property and are axiomatisable in finite way (a profound result due to Segerberg [235]).

4.2 SC for Modal Logics

First applications of SC to modal logics were provided in the early 1950s, by Feys [76], Curry [55], Ohnishi and Matsumoto [189, 190][7]. This approach has shown that modal constants cannot be treated in a similar way as Boolean constants on the basis of standard SC. Although it did not generally preclude the possibility of proving results such as cut elimination for many systems, the important negative examples were found, amongst them **S5**, one of the most important modal logics. Troubles with providing cut-free systems or just SC systems with satisfactory rules led to the creation of generalised forms of SC. The most important generalisations of SC include: display calculi (Belnap [25], Wansing [269]), hypersequent calculi (Mints [177], Pottinger [202], Avron [10]), multisequent calculi (Curry [56], Zeman [276], Indrzejczak [126]), several variants of nested SC (Došen [60], Bull [42], Kashima [147], Stouppa [246], Brünnler [41], Poggiolesi [198]) or labelled SC (Kanger [146], Mints [182], Negri [187]). It is not possible here to characterise them even briefly. One can consult surveys of Fitting [87], Wansing [270] or Poggiolesi [198].

All these approaches are interesting and usually provide more general treatments of phenomena hardly tractable in standard SC framework. As an illustration of the difficulties and ways of overcoming them, we describe the problem of searching for cut-free formalisations of **S5** in the last three sections. But first, we consider a standard approach.

4.2.1 Modal Extensions of LK

One may extend any variant of SC provided for **CPL** to modal logics. Clearly, one may treat SC for **CPL**, but in the modal language, as the weakest modal logic and add suitable axiomatic sequents corresponding to modal axioms. Thus, for example, to obtain such a formalisation of **T**, we need to add sequents of the form $\Box(\varphi \to \psi), \Box\varphi \Rightarrow \Box\psi$ and $\Box\varphi \Rightarrow \varphi$. In this way, we treat modal logics as a kind of theories over **CPL**. This is not enough to obtain complete results, and we also need a counterpart of GR which in a sense destroys such a project. Since it is not possible to avoid the addition of specific modal rules, it is better to mix K and GR and express them both by means of a rule; the axiomatic sequent corresponding to T may also be easily changed into a rule by the application of lemma 2.3, from subsection 2.2.2. Moreover, the rule-based approach makes it possible to treat modal constants like genuine logical constants. In the case of LK, the standard

---

[7]Zeman [276] provides a good summary of this early stage of research.

solution is to use the rules proposed by Ohnishi and Matsumoto. To obtain LK for **T**, which we call LKT, we must add two rules

$$(\Box\Rightarrow) \quad \frac{\varphi,\Gamma\Rightarrow\Delta}{\Box\varphi,\Gamma\Rightarrow\Delta} \qquad\qquad (\Rightarrow\Box) \quad \frac{\Gamma\Rightarrow\varphi}{\Box\Gamma\Rightarrow\Box\varphi}$$

$(\Box\Rightarrow)$ is unproblematic, but $(\Rightarrow\Box)$ differs significantly from canonical SC rules. First of all, it is context-dependent in two respects. The first restriction concerns the succedent in which no parametric formulae are admitted. The second restriction concerns the antecedent in which all parametric formulae are also prefixed with $\Box$ in the conclusion. Moreover, in contrast to other rules considered so far, we introduce not one but possibly many occurrences of $\Box$ in the conclusion. Hence, in terms of properties explained in subsection 3.5.5, this rule is only weakly explicit and not even weakly symmetric. Since all formulae in the succedent are changed into modal ones, we treat them all as active formulae of this rule application.

One may prove an instance of the axiom K in the following way:

$$
\begin{array}{l}
(\rightarrow\Rightarrow) \\
(\Rightarrow\Box) \\
(\Rightarrow\rightarrow) \\
(\Rightarrow\rightarrow)
\end{array}
\cfrac{
\cfrac{
\cfrac{
\cfrac{p\Rightarrow p \qquad q\Rightarrow q}{p,p\rightarrow q\Rightarrow q}
}{\Box p,\Box(p\rightarrow q)\Rightarrow\Box q}
}{\Box(p\rightarrow q)\Rightarrow\Box p\rightarrow\Box q}
}{\Rightarrow\Box(p\rightarrow q)\rightarrow(\Box p\rightarrow\Box q)}
$$

A proof of axiom T is obvious and GR is simulated by $(\Rightarrow\Box)$ with empty Γ. Hence, we obtain

Theorem 4.1 *If $\Gamma\vdash_T\varphi$, then $\vdash\Gamma\Rightarrow\varphi$ in LKT for any finite Γ.*

The converse may be proved by adjusting a proof of lemma 1.18, from subsection 1.9.1. or by demonstrating in H-**T** the admissibility of two rules obtained by the application of Gentzen's transform I_G to both SC rules.

Exercise 4.2 *Complete a proof of lemma 1.18, by considering the cases of applications of $(\Box\Rightarrow)$ and $(\Rightarrow\Box)$.*

Alternatively, one may refer to the completeness of the Hilbert system and prove instead the soundness of LKT by showing that

Lemma 4.1 *Both modal rules of LKT are validity-preserving in* **T**.

PROOF: We will demonstrate it for $(\Rightarrow\Box)$ and leave the task of checking $(\Box\Rightarrow)$ to the reader.

Assume that $\models_T\Gamma\Rightarrow\varphi$ and consider a model $\mathfrak{M}$ and a state w such that $w\vdash\Box\Gamma$ but $w\nVDash\Box\varphi$. Hence, in some accessible state v, $v\nVDash\varphi$ but $v\vDash\Gamma$ which contradicts our assumption. $\qquad\Box$

Exercise 4.3 *Show the validity-preservation of $(\Box\Rightarrow)$.*

This lemma implies the soundness of LKT. Since the axiomatic formulation of **T** is strongly adequate and compact, LKT is also strongly adequate.

Note that showing validity-preservation for $(\Rightarrow \Box)$ does not require the reference to any frame-properties whereas in case of $(\Box \Rightarrow)$ we need to refer to reflexivity. It follows that if we drop this rule, we obtain an adequate LK for **K**, the weakest normal modal logic. It also shows that from the standpoint of the SC framework, **K** is in a sense defective since we do not have a specific rule for the introduction of $\Box$ into antecedent.

What about other semantic features of the rules? It may be easily checked that only $(\Box \Rightarrow)$ is normal, in the sense of preserving satisfiability in any world of an arbitrary reflexive model. However, it is not (semantically) invertible. As for $(\Rightarrow \Box)$ it is neither normal nor invertible.

Exercise 4.4 *Prove that* $(\Box \Rightarrow)$ *is normal.*

Provide countermodels showing that both rules are not invertible and that $(\Rightarrow \Box)$ *is not normal.*

As for **S4** and **S5**, we can again think of the addition of suitable axiomatic sequents but this time on the basis of SC for **T**, LKT in particular. But again, it is better to introduce suitable rules. By the lemma 2.3, (subsection 2.2.2) we could add something like

$$\frac{\Box\Box\varphi,\Gamma\Rightarrow\Delta}{\Box\varphi,\Gamma\Rightarrow\Delta} \qquad \text{or} \qquad \frac{\Gamma\Rightarrow\Delta,\Box\varphi}{\Gamma\Rightarrow\Delta,\Box\Box\varphi}$$

to obtain SC for **S4** (and similarly for **S5**). But this is not a good solution. The left rule is an elimination rule; the right rule may be seen as a kind of (additional) introduction rule for $\Box$, but LK for **S4** with such a rule does not admit cut elimination. It is better to keep $(\Box \Rightarrow)$ but replace $(\Rightarrow \Box)$ in LKT with

$$(\Rightarrow\Box^4)\ \frac{\Box\Gamma\Rightarrow\varphi}{\Box\Gamma\Rightarrow\Box\varphi} \qquad \text{or} \qquad (\Rightarrow\Box^5)\ \frac{\Box\Gamma\Rightarrow\Box\Delta,\varphi}{\Box\Gamma\Rightarrow\Box\Delta,\Box\varphi}$$

this way obtaining LKS4 or LKS5, respectively. The rules are also context-dependent but in a different way. Here, we must have all parametric formulae modalised already in the premiss so they are not even weekly explicit, in contrast to $(\Rightarrow \Box)$. Note also that $(\Rightarrow \Box^5)$ seems to be less restrictive since (modalised) parametric formulae are allowed in the succedent too. Surprisingly enough this relaxation leads to serious troubles in case of LKS5. Since in both cases the context is fixed, we count all formulae in $\Box\Gamma$ ($\Box\Delta$) as parametric and only φ as a side formula and $\Box\varphi$ as the principal formula of rule application.

The proof of axiom 4 in LKS4 is immediate, for 5 in LKS5 is a bit longer but still simple and cut-free

$$(\Rightarrow \neg) \frac{\Box p \Rightarrow \Box p}{\Rightarrow \neg \Box p, \Box p}$$
$$(\Rightarrow \Box) \frac{}{\Rightarrow \Box \neg \Box p, \Box p}$$
$$(\neg \Rightarrow) \frac{}{\neg \Box p \Rightarrow \Box \neg \Box p}$$
$$(\Rightarrow \rightarrow) \frac{}{\Rightarrow \neg \Box p \rightarrow \Box \neg \Box p}$$

However, when we try to prove B, we cannot do this without cut

$$(\Rightarrow \neg) \frac{\Box p \Rightarrow \Box p}{\Rightarrow \neg \Box p, \Box p}$$
$$(\Rightarrow \Box) \frac{}{\Rightarrow \Box \neg \Box p, \Box p} \qquad \frac{p \Rightarrow p}{\Box p \Rightarrow p} (\Box \Rightarrow)$$
$$(Cut) \frac{}{\Rightarrow \Box \neg \Box p, p}$$
$$(\neg \Rightarrow) \frac{}{\neg p \Rightarrow \Box \neg \Box p}$$
$$(\Rightarrow \rightarrow) \frac{}{\Rightarrow \neg p \rightarrow \Box \neg \Box p}$$

As a result of the above proofs, we have

Theorem 4.2 *If $\Gamma \vdash_L \varphi$, then $\vdash \Gamma \Rightarrow \varphi$ in LKL.*

where L is **S4** or **S5**. The converse may also be easily proved, either by proving the admissibility of both rules in Hilbert systems, or by showing soundness on the basis of the validity-preservation of both rules.

Exercise 4.5 *Prove validity-preservation of $(\Rightarrow \Box^4)$ with respect to transitive frames. Prove validity-preservation of $(\Rightarrow \Box^5)$ with respect to euclidean or to simple frames, or to symmetric and transitive frames.*

As can be expected both rules are not normal; the reader should provide countermodels. On the other hand, both rules are invertible in the presence of $(\Box \Rightarrow)$; one may easily prove a premiss of any of these rules by cut on the conclusion and $\Box \varphi \Rightarrow \varphi$ which is directly provable.

Exercise 4.6 *Show the invertibility of $(\Rightarrow \Box^4)$ and $(\Rightarrow \Box^5)$ semantically; note that the assumption of reflexivity is essential.*

4.2.2   Modal Extensions of G3

When one prefers to use G3 as a basis, one should use variants of the above rules which allow the preservation of our results concerning the admissibility of structural rules, as well as of smooth proof-search, in the absence of weakening and contraction. For G3T we need

$$(\Box\Rightarrow) \frac{\Box\varphi, \varphi, \Gamma \Rightarrow \Delta}{\Box\varphi, \Gamma \Rightarrow \Delta} \qquad\qquad (\Rightarrow\Box) \frac{\Box\Gamma \Rightarrow \varphi}{\Pi, \Box\Gamma \Rightarrow \Sigma, \Box\varphi}$$

Here $(\Box \Rightarrow)$ is an example of contraction-absorbing rule in the sense that one could obtain it as a derived rule in LKT by the application of $(C \Rightarrow)$. $(\Rightarrow \Box)$ is weakening-absorbing in the sense that one could obtain it as a derived rule in LKT with the help of weakening. Although both rules lack the same syntactic properties as their LK counterparts, in case of ($\square \Rightarrow$), we can notice that it is not only normal but also invertible, although trivially, being just an instance of ($W \Rightarrow$). Suitable rules for G3S4 and G3S5 are also weakening-absorbing variants of LK rules of the form

$$(\Rightarrow\square^4) \quad \frac{\square\Gamma \Rightarrow \varphi}{\Pi, \square\Gamma \Rightarrow \Sigma, \square\varphi} \qquad\qquad (\Rightarrow\square^5) \quad \frac{\square\Gamma \Rightarrow \square\Delta, \varphi}{\Pi, \square\Gamma \Rightarrow \square\Delta, \Sigma, \square\varphi}$$

Note that in contrast to both the rules devised as extensions of LK, these two variants are not invertible since weakening (which is involved here) is not invertible. However, all four rules are obviously validity-preserving, which is proved exactly as for their LK counterparts from the preceding section. Hence, soundness is guaranteed. As for completeness, we can easily apply the argument from the preceding section as well, but for the sake of variety, we rather extend the completeness proof from section 1.10, to cover modal logics formalised as SCs. It is a well known construction based on the application of so-called *canonical models*. We follow quite closely the strategy applied for Hilbert systems with the difference that instead of maximal consistent sets, we will use maximal consistent pairs of sets of formulae, similarly as in the proof of Chagrov and Zakharyaschev [48]. All definitions and results from section 1.10, apply without changes. Of course, cut is necessary either as a primitive rule (for G3S5) or as an admissible one (which will be proved in the next section).

In contrast to **CPL**, to build a model, we have to use a collection of all maximal consistent pairs, formally

Definition 4.8 (Canonical Model). $\mathfrak{M}_C = \langle \mathfrak{F}_C, V_C \rangle$ is based on the frame $\mathfrak{F}_C$ which contains a set $\mathcal{W}_C$ of all maximal consistent pairs and $\mathcal{R}_C$ defined:

$\langle\Gamma, \Delta\rangle \mathcal{R}_C \langle\Pi, \Sigma\rangle$ iff $\Gamma^\square \subseteq \Pi$ (recall that $\Gamma^\square = \{\psi : \square\psi \in \Gamma\}$);
$V_C(p) = \{\langle\Gamma, \Delta\rangle \in \mathcal{W}_C : p \in \Gamma\}$ for all atomic formulae p.

Before proving a truth lemma, we need some preliminary results

Lemma 4.2 $\vdash_L \Gamma \Rightarrow \Delta$ iff for every maximal consistent pair $\langle\Pi, \Sigma\rangle$, if $\Gamma \subseteq \Pi$, then $\Delta \cap \Pi \neq \varnothing$.

PROOF: $\Longrightarrow$ Assume to the contrary that for some $\langle\Pi, \Sigma\rangle$, $\Gamma \subseteq \Pi$ but $\Delta \cap \Pi$ is empty. By maximality $\Delta \subseteq \Sigma$, but this implies, by consistency of $\langle\Pi, \Sigma\rangle$, that $\nvdash_L \Gamma \Rightarrow \Delta$.

$\Longleftarrow$ If $\nvdash_L \Gamma \Rightarrow \Delta$, then $\langle\Gamma, \Delta\rangle$ is consistent, and by lemma 1.23 there is a maximal extension of it $\langle\Pi, \Sigma\rangle$ with $\Gamma \subseteq \Pi$ and $\Delta \subseteq \Sigma$, but then $\Delta \cap \Pi$ is empty. $\square$

Exercise 4.7 Prove that $\vdash_L \Gamma \Rightarrow \Delta$ iff for every maximal consistent pair $\langle\Pi, \Sigma\rangle$, if $\Delta \subseteq \Sigma$, then $\Gamma \cap \Sigma \neq \varnothing$.

Lemma 4.3 If $\square\varphi \in \Delta$ for some maximal consistent pair $\langle\Gamma, \Delta\rangle$, then $\langle\Gamma^\square, \{\varphi\}\rangle$ is consistent.

PROOF: Assume that $\langle \Gamma^\square, \{\varphi\}\rangle$ is inconsistent, then $\vdash_L \Gamma^\square \Rightarrow \varphi$. But then (by $(\Rightarrow \square)$ in G3T and by $(\square \Rightarrow)$ and $(\Rightarrow \square^4)$ or $(\Rightarrow \square^5)$ in the remaining systems) $\vdash_L \square(\Gamma^\square) \Rightarrow \square\varphi$. Since $\square(\Gamma^\square) \subseteq \Gamma$, by lemma 4.2 $\square\varphi \in \Gamma$, and the lemma follows by contraposition. $\square$

Lemma 4.4 (Truth Lemma). *For every formula φ and every $\langle \Gamma, \Delta\rangle \in \mathcal{W}_C$: $\varphi \in \Gamma$ iff $\langle \Gamma, \Delta\rangle \vDash \varphi$.*

PROOF: in case of atoms and boolean formulae, it is the same as a proof of lemma 1.24, so we need to consider only the case of $\square\psi$

$\Longrightarrow$ Assume that $\square\psi \in \Gamma$ but $\langle \Gamma, \Delta\rangle \nvDash \square\psi$. So there is some $\langle \Pi, \Sigma\rangle$ such that $\langle \Gamma, \Delta\rangle \mathcal{R}_C \langle \Pi, \Sigma\rangle$ and $\langle \Pi, \Sigma\rangle \nvDash \psi$. But by definition of $\mathcal{R}_C$ $\psi \in \Pi$, hence by the induction hypothesis $\langle \Pi, \Sigma\rangle \vDash \psi$—contradiction.

$\Longleftarrow$ If $\langle \Gamma, \Delta\rangle \vDash \square\psi$ but $\square\psi \notin \Gamma$, then by maximality $\square\psi \in \Delta$. Hence, by the preceding lemma $\langle \Gamma^\square, \{\psi\}\rangle$ is consistent. By lemma 1.23, it may be extended to a maximal consistent $\langle \Pi, \Sigma\rangle$ with $\psi \in \Sigma$. By the induction hypothesis $\langle \Pi, \Sigma\rangle \nvDash \psi$. On the other hand, $\langle \Pi, \Sigma\rangle \vDash \psi$ since $\langle \Gamma, \Delta\rangle \mathcal{R}_C \langle \Pi, \Sigma\rangle$. $\square$

The last thing we must prove is a demonstration that the canonical model for a logic **L** is really a model for this logic, i.e. that $\mathcal{R}_C$ is reflexive for **T**, reflexive and transitive for **S4** and additionally euclidean or symmetric for **S5**. We prove one case as an example

Lemma 4.5 *The canonical model for G3S4 is transitive.*

PROOF: Assume that $\langle \Gamma, \Delta\rangle \mathcal{R}_C \langle \Pi, \Sigma\rangle$ and $\langle \Pi, \Sigma\rangle \mathcal{R}_C \langle \Lambda, \Theta\rangle$. In order to show $\langle \Gamma, \Delta\rangle \mathcal{R}_C \langle \Lambda, \Theta\rangle$ we must show that $\Gamma^\square \subseteq \Lambda$. Consider an arbitrary $\square\varphi \in \Gamma$. Since $\vdash_{S4} \square\varphi \Rightarrow \square\square\varphi$, by lemma 4.2 $\square\square\varphi$ is also in Γ. Hence $\square\varphi \in \Pi$ and $\varphi \in \Lambda$ as required. $\square$

Exercise 4.8 *Show that for G3T $\mathcal{R}_C$ is reflexive. Show that for G3S5 $\mathcal{R}_C$ is euclidean (use the fact that $\vdash_{S5} \neg\square\varphi \Rightarrow \square\neg\square\varphi$).*

Hence completeness follows for all three SCs.

4.2.3 Rules for Diamonds

Sometimes also rules for $\Diamond$ are provided. In Ohnishi and Matsumoto [189], we have the following pair for LKT:

$$(\Diamond \Rightarrow) \quad \frac{\varphi \Rightarrow \Delta}{\Diamond\varphi \Rightarrow \Diamond\Delta} \qquad\qquad (\Rightarrow \Diamond) \quad \frac{\Gamma \Rightarrow \Delta, \varphi}{\Gamma \Rightarrow \Delta, \Diamond\varphi}$$

Clearly, without $(\Rightarrow \Diamond)$ we obtain a formalisation of **K**. It is worth remarking that although the addition of these rules to LK provides an adequate formalisation of **T** in a language with $\Diamond$ as the only primitive modal constant, the addition of rules for both $\square$ and $\Diamond$ is not sufficient for the completeness of LKT in the full modal language, i.e. with both modal constants as primitive. It was already noted by Kripke [153], that such rules are insufficient for proving the interdefinability of

$\square$ and $\lozenge$. In order to remedy the problem, one should rather use the following pair of rules:

$$(\Rightarrow \square') \quad \frac{\Gamma \Rightarrow \Delta, \varphi}{\square\Gamma \Rightarrow \lozenge\Delta, \square\varphi} \qquad\qquad (\lozenge \Rightarrow') \quad \frac{\varphi, \Gamma \Rightarrow \Delta}{\lozenge\varphi, \square\Gamma \Rightarrow \lozenge\Delta}$$

Both the rules are even more defective from the standpoint of well-behaved canonical rules; they are not even separated since two different constants are involved. On the other hand, they allow for proving suitable equivalences concerning $\square$ and $\lozenge$. For example

$$
\begin{array}{l}
(\Rightarrow \neg) \dfrac{p \Rightarrow p}{\Rightarrow \neg p, p} \\
(\Rightarrow \square') \dfrac{}{\Rightarrow \square\neg p, \lozenge p} \\
(\neg \Rightarrow) \dfrac{}{\neg\square\neg p \Rightarrow \lozenge p} \\
(\Rightarrow \rightarrow) \dfrac{}{\Rightarrow \neg\square\neg p \rightarrow \lozenge p}
\end{array}
$$

Exercise 4.9 *Prove the converse of this implication and $\neg\lozenge\neg p \leftrightarrow \square p$.*

A formalisation of stronger logics requires at least a modification of rules ($\Rightarrow \square'$) and ($\lozenge \Rightarrow'$), taking into account a type of operations admissible on parametric formulae. In order to get LK for **S4**, we add to **CPL** ($\Rightarrow \lozenge$), ($\square \Rightarrow$) and the following variants of ($\lozenge \Rightarrow'$) and ($\Rightarrow \square'$):

$$(\Rightarrow \square^4) \quad \frac{\square\Gamma \Rightarrow \lozenge\Delta, \varphi}{\square\Gamma \Rightarrow \lozenge\Delta, \square\varphi} \qquad\qquad (\lozenge \Rightarrow^4) \quad \frac{\varphi, \square\Gamma \Rightarrow \lozenge\Delta}{\lozenge\varphi, \square\Gamma \Rightarrow \lozenge\Delta}$$

And for LKS5

$$(\Rightarrow \square^5) \quad \frac{\square\Gamma, \lozenge\Pi \Rightarrow \square\Sigma, \lozenge\Delta, \varphi}{\square\Gamma, \lozenge\Pi \Rightarrow \square\Sigma, \lozenge\Delta, \square\varphi} \qquad (\lozenge \Rightarrow^5) \quad \frac{\varphi, \square\Gamma, \lozenge\Pi \Rightarrow \square\Sigma, \lozenge\Delta}{\lozenge\varphi, \square\Gamma, \lozenge\Pi \Rightarrow \square\Sigma, \lozenge\Delta}$$

The modification of these rules for G3 is obvious, and for ($\lozenge \Rightarrow'$) and ($\Rightarrow \square'$), it may be nicely summarised in general schemata due to Fitting [83]

$$(\Rightarrow \square^F) \quad \frac{\Gamma^\star \Rightarrow \Delta^\natural, \varphi}{\Gamma \Rightarrow \Delta, \square\varphi} \qquad\qquad (\lozenge \Rightarrow^F) \quad \frac{\varphi, \Gamma^\star \Rightarrow \Delta^\natural}{\lozenge\varphi, \Gamma \Rightarrow \Delta}$$

where $\Gamma^\star$ and $\Delta^\natural$ are defined as

logic	$\Gamma^\star$	$\Delta^\natural$
K, T	$\{\psi : \square\psi \in \Gamma\}$	$\{\psi : \lozenge\psi \in \Delta\}$
S4	$\{\square\psi : \square\psi \in \Gamma\}$	$\{\lozenge\psi : \lozenge\psi \in \Delta\}$
S5	$\{\square\psi : \square\psi \in \Gamma\} \cup \{\lozenge\psi : \lozenge\psi \in \Gamma\}$	$\{\square\psi : \square\psi \in \Delta\} \cup \{\lozenge\psi : \lozenge\psi \in \Delta\}$

Of course, in case we are using only $\square$, the rule ($\lozenge \Rightarrow^F$) is not needed, and for G3T and G3S4, $\Delta^\natural$ is empty. For G3S5, we admit $\Delta^\natural$ as nonempty but in the definition for it (and for $\Gamma^\star$), we take into account only the first element of the union.

4.3 Cut Elimination

In this section, we will discuss necessary adjustments which must be done in proving cut eliminability for LK or cut admissibility for G3 with added rules for $\Box$. Similar results may be obtained also for systems equipped with rules for both modalities but with several additional minor complications of inessential character, so we omit a presentation.

4.3.1 Mix Elimination for LKT, LKS4

The first proof of cut eliminability for LKT and LKS4 was provided by Ohnishi and Matsumoto as an extension of Gentzen's proof of Mix elimination. It is not necessary to repeat all details, so we will refer to respective points in Gentzen proof from subsection 2.4.1, and demonstrate transformations involving modal rules. Recall that Gentzen's proof eliminates mix.

In case the rank=2 (point 4.1 of Gentzen's proof; point 3.1—the case of an axiom or weakening in at least one premiss is the same as in the classical case), we have two possible situations in LKT according to which modal rule was applied in the right premiss. They look like that

$$(Mix) \cfrac{(\Rightarrow \Box)\cfrac{\Gamma \Rightarrow \varphi}{\Box\Gamma \Rightarrow \Box\varphi} \qquad \cfrac{\varphi, \Delta \Rightarrow \Sigma}{\Box\varphi, \Delta \Rightarrow \Sigma}(\Box \Rightarrow)}{\Box\Gamma, \Delta \Rightarrow \Sigma}$$

or

$$(Mix) \cfrac{(\Rightarrow \Box)\cfrac{\Gamma \Rightarrow \varphi}{\Box\Gamma \Rightarrow \Box\varphi} \qquad \cfrac{\varphi^k, \Delta \Rightarrow \psi}{\Box\varphi^k, \Box\Delta \Rightarrow \Box\psi}(\Rightarrow \Box)}{\Box\Gamma, \Box\Delta \Rightarrow \Box\psi}$$

Note that in the second case, even if rank=2, it is possible that more than one occurrence of the cut-formula is in the right premiss. In both the cases the reduction of complexity is straightforward

$$(\Box \Rightarrow)\cfrac{(Cut)\cfrac{\Gamma \Rightarrow \varphi \qquad \varphi, \Delta \Rightarrow \Sigma}{\Gamma, \Delta \Rightarrow \Sigma}}{\cfrac{\vdots}{\Box\Gamma, \Delta \Rightarrow \Sigma}}$$

or

$$(\Rightarrow \Box)\cfrac{(Mix)\cfrac{\Gamma \Rightarrow \varphi \qquad \varphi^k, \Delta \Rightarrow \psi}{\Gamma, \Delta \Rightarrow \psi}}{\Box\Gamma, \Box\Delta \Rightarrow \Box\psi}$$

In case of LKS4 there is only one case of rank 2

$$(\Rightarrow \Box^4) \quad \cfrac{\cfrac{\Box\Gamma \Rightarrow \varphi}{\Box\Gamma \Rightarrow \Box\varphi} \qquad \cfrac{\varphi, \Delta \Rightarrow \Sigma}{\Box\varphi, \Delta \Rightarrow \Sigma} (\Box \Rightarrow)}{\Box\Gamma, \Delta \Rightarrow \Sigma}$$
$$(Mix)$$

which is transformed accordingly

$$(Mix) \quad \cfrac{\Box\Gamma \Rightarrow \varphi \qquad \varphi, \Delta \Rightarrow \Sigma}{\Box\Gamma, \Delta \Rightarrow \Sigma}$$

The reduction of rank in LKT is not problematic either. First, note that $(\Rightarrow \Box)$ cannot be the last rule in the left premiss since there are no parametric formulae in the succedent, so part B (which was left to the reader) is not involved. Similarly, this rule is not taken into account in proving part A since there are no parametric formulae in the antecedent. We decided to treat them as active since they change their form—it was, in fact, the case considered above. As for $(\Box \Rightarrow)$ in the left or right premiss, there is no problem since it is context insensitive and permutes with other rules. More specifically, in proving part A, we can check that the addition of $(\Box \Rightarrow)$ as suitable subcase in cases A1 and A2 does not do any harm. Only the case A3 calls for adjustment—in subcase A3.1 where the cut-formula is principal we have the following situation:

$$\cfrac{\Gamma \Rightarrow \Delta, \Box\varphi^k \qquad \cfrac{\varphi, \Box\varphi^n, \Pi \Rightarrow \Sigma}{\Box\varphi^{n+1}, \Pi \Rightarrow \Sigma} (\Box \Rightarrow)}{\Gamma, \Pi \Rightarrow \Delta, \Sigma} (Mix)$$

which is changed into

$$\cfrac{\Gamma \Rightarrow \Delta, \Box\varphi^k \qquad \cfrac{\cfrac{\Gamma \Rightarrow \Delta, \Box\varphi^k \qquad \varphi, \Box\varphi^n, \Pi \Rightarrow \Sigma}{\varphi, \Gamma, \Pi \Rightarrow \Delta, \Sigma} (Mix)}{\cfrac{\Box\varphi, \Gamma, \Pi \Rightarrow \Delta, \Sigma}{\Gamma, \Gamma, \Pi \Rightarrow \Delta, \Delta, \Sigma}} \genfrac{}{}{0pt}{}{(\Box \Rightarrow)}{(Mix)}}{\begin{array}{c} \vdots \\ \Gamma, \Pi \Rightarrow \Delta, \Sigma \end{array}} (C)$$

where both applications of mix are of lower rank.

In case of LKS4, things are slightly more complicated, since in $(\Rightarrow \Box^4)$ formulae from the antecedent are treated as parametric. Hence, it is possible that in reducing the rank, a modal formula is involved which is parametric in the right premiss obtained by the application of $(\Rightarrow \Box^4)$

$$\cfrac{\Gamma \Rightarrow \Sigma, \Box\varphi^k \qquad \cfrac{\Box\varphi^n, \Box\Delta \Rightarrow \psi}{\Box\varphi^n, \Box\Delta \Rightarrow \Box\psi} (\Rightarrow \Box^4)}{\Gamma, \Box\Delta \Rightarrow \Sigma, \Box\psi} (Mix)$$

In this case, if the left premiss was obtained by the application of some Boolean rule or $(\Box \Rightarrow)$, we cannot, in general, reduce the right rank since it will lead to the following figure:

$$(Mix) \ \frac{\Gamma \Rightarrow \Sigma, \Box\varphi^k \qquad \Box\varphi^n, \Box\Delta \Rightarrow \psi}{\Gamma, \Box\Delta \Rightarrow \Sigma, \psi}$$

where the application of $(\Rightarrow \Box)$ may not be possible, if formulae in Γ are not modalised, and Σ in the succedent of the left premiss is not empty. But in such a case either the cut-formula in the left premiss is principal or not. If it is parametric, we reduce the left rank instead (so a proof of part B is essentially enriched). On the other hand, if the left premiss was obtained by $(\Rightarrow \Box^4)$ we have the following:

$$\begin{array}{c} (\Rightarrow \Box^4) \\ (Mix) \end{array} \frac{\dfrac{\Box\Gamma \Rightarrow \varphi}{\Box\Gamma \Rightarrow \Box\varphi} \qquad \dfrac{\Box\varphi^n, \Box\Delta \Rightarrow \psi}{\Box\varphi^n, \Box\Delta \Rightarrow \Box\psi} (\Rightarrow \Box^4)}{\Box\Gamma, \Box\Delta \Rightarrow \Box\psi}$$

In this case we reduce the right rank

$$\begin{array}{c} (\Rightarrow \Box^4) \\ (Mix) \\ (\Rightarrow \Box) \end{array} \frac{\dfrac{\dfrac{\Box\Gamma \Rightarrow \varphi}{\Box\Gamma \Rightarrow \Box\varphi} \qquad \Box\varphi^n, \Box\Delta \Rightarrow \psi}{\Box\Gamma, \Box\Delta \Rightarrow \psi}}{\Box\Gamma, \Box\Delta \Rightarrow \Box\psi}$$

This is possible because all formulae in the antecedent of the left premiss are also modalised, and the list of parameters in the succedent is empty.

It is not difficult to apply other strategies of local proofs of cut elimination considered in chapter 2. Girard's proof needs adjustments in the proof of the mix reduction lemma (lemma 2.7). Reductions on the height are performed in the same way as reductions on the rank in both LKT and LKS4. The only case which looks different is when both cut-formulae are principal. In case of $\Box\varphi$ in LKT we have

$$\begin{array}{c} (\Rightarrow \Box) \\ (Mix) \end{array} \frac{\dfrac{\Gamma \Rightarrow \varphi}{\Box\Gamma \Rightarrow \Box\varphi} \qquad \dfrac{\varphi, \Box\varphi^k, \Delta \Rightarrow \Sigma}{\Box\varphi^{k+1}, \Delta \Rightarrow \Sigma} (\Box \Rightarrow)}{\Box\Gamma, \Delta \Rightarrow \Sigma}$$

which is changed into

$$\frac{\Gamma \Rightarrow \varphi \qquad \dfrac{\dfrac{\Box\Gamma \Rightarrow \Box\varphi \qquad \varphi, \Box\varphi^k, \Delta \Rightarrow \Sigma}{\varphi, \Box\Gamma, \Delta \Rightarrow \Sigma} (Mix)}{\Gamma, \Box\Gamma_\varphi, \Delta_\varphi \Rightarrow \Sigma} (\Box \Rightarrow)}{}$$

$$\vdots$$
$$\frac{\Box\Gamma, \Box\Gamma_\varphi, \Delta_\varphi \Rightarrow \Sigma}{} (C \Rightarrow, W \Rightarrow)$$
$$\vdots$$
$$\overline{\Box\Gamma, \Delta \Rightarrow \Sigma}$$

where the first applications of mix is of lower height and the second of lower degree.

Exercise 4.10 *Check the correctness of Girard's proof for LKS4.*

In a similar way, we can extend the reductive proof from section 2.4.3, to cover both LKT and LKS4. As for global proofs; one may find in the last chapter of Curry [56], a sketch of suitable modification for **S4** which works also for **T**. However, Buss' proof cannot be applied to modal logics in a straightforward way.

Exercise 4.11 *Reprove Right reduction lemma (lemma 2.9) for LKT and LKS4.*

Although for LKS5, we have already provided an example of a proof with irreducible cut, after consideration of cut elimination proofs, we can also show why in general such proofs break down in this case. Let us take the situation where the left premiss is obtained by $(\Rightarrow \Box^5)$ but the cut-formula is parametric

$$(\Rightarrow \Box^5) \, \cfrac{\cfrac{\Box\Gamma \Rightarrow \Box\Pi, \psi, \Box\varphi}{\Box\Gamma \Rightarrow \Box\Pi, \Box\psi, \Box\varphi} \quad \cfrac{\varphi, \Delta \Rightarrow \Sigma}{\Box\varphi, \Delta \Rightarrow \Sigma} \, (\Box \Rightarrow)}{\Box\Gamma, \Delta \Rightarrow \Box\Pi, \Box\psi, \Sigma}$$
$$(Cut)$$

in this case after the reduction of the height on the left we get

$$(Cut) \, \cfrac{\Box\Gamma \Rightarrow \Box\Pi, \psi, \Box\varphi \quad \Box\varphi, \Delta \Rightarrow \Sigma}{\Box\Gamma, \Delta \Rightarrow \Box\Pi, \psi, \Sigma}$$

and there is no possibility to apply $(\Rightarrow \Box^5)$ if Δ and Σ are nonempty and not modalised.

4.3.2 Admissibility of Cut in G3T and G3S4

In case of the modal versions of G3, in order to apply any proof of admissibility of cut presented in chapter 3, we must take care of all preliminary results first. We check them one by one.

First, lemma 3.1, i.e. height-preserving admissibility of weakening, is easy to obtain. $(\Box \Rightarrow)$ behaves like classical rules and for $(\Rightarrow \Box)$ and $(\Rightarrow \Box^4)$, we simply add a suitable formula to conclusion with no necessity of using the induction hypothesis. Also, lemma 3.2, concerning the generalisation of axioms holds. $\Gamma, \Box\varphi \Rightarrow \Box\varphi, \Delta$ is provable easily from $\varphi \Rightarrow \varphi$ or $\Box\varphi, \varphi \Rightarrow \varphi$ in all systems.

As for invertibility, $(\Box \Rightarrow)$ is obviously (height-preserving) invertible by (h-p admissible) weakening, but neither $(\Rightarrow \Box)$ nor $(\Rightarrow \Box^4)$ is invertible. Moreover, we can ask if the addition of these rules to G3 does not destroy the result for other rules. Fortunately, it does not. We must check again the case where a suitable Boolean formula is parametric in the last step. Consider a new situation when $(\Rightarrow \Box)$ or $(\Rightarrow \Box^4)$ was applied in this step. The formula in question is not modal, hence it must be an element of Π or Σ in the schema of the application of the modal rule, and it was not present at all in the premiss. So it is enough to change it into suitable subformulae and the induction hypothesis does not apply here. As a result, lemma 3.3, still holds for all Boolean rules and for $(\Box \Rightarrow)$.

Finally, we must prove height-preserving admissibility of contraction in the presence of the new rules. Again, we supplement the proof of lemma 3.4, from subsection 3.2.2. In case of $(\Box \Rightarrow)$ as the last rule, it is straightforward in both situations, i.e. when the contracted formula was parametric and when one occurrence of it was principal. If the latter holds note that two occurrences of $\Box\varphi$ are already in the premiss so one is deleted by the induction hypothesis. If the last rule was $(\Rightarrow \Box)$ or $(\Rightarrow \Box^4)$, then if two occurrences of a formula are in Π or Σ, we simply delete one. Otherwise, the only contraction in the antecedent on some element of $\Box\Gamma$ must be considered since in the succedent we have only one modal formula. In case of $(\Rightarrow \Box)$ or $(\Rightarrow \Box^4)$ it is reduced immediately in the premiss by the induction hypothesis on the height. Thus, contraction is height-preserving admissible for G3T and G3S4.

Incidentally, note that this proof does not hold for G3S5. In the case of $(\Rightarrow \Box^5)$, we have additionally the situation with two occurrences of $\Box\varphi$ in the succedent, where one is parametric and one principal. So we have in the (succedent of the) premiss φ and $\Box\varphi$ and there is no possibility of reduction.

Let us now consider the proof of admissibility of cut in Dragalin's style (subsection 3.3.1). First, we take two cases with the modal cut-formula principal in both premisses. Note that in this case the reduction of the complexity of the cut-formula is connected only with the application of $(\Rightarrow \Box)$ in both premisses in G3T

$$(\Rightarrow \Box)\ \frac{\Gamma \Rightarrow \varphi}{\Pi, \Box\Gamma \Rightarrow \Sigma, \Box\varphi} \qquad \frac{\varphi, \Delta \Rightarrow \psi}{\Box\varphi, \Lambda, \Box\Delta \Rightarrow \Theta, \Box\psi}\ (\Rightarrow \Box)$$
$$(Cut)\ \overline{\Pi, \Lambda, \Box\Gamma, \Box\Delta \Rightarrow \Sigma, \Theta, \Box\psi}$$

which is changed into

$$(Cut)\ \frac{\Gamma \Rightarrow \varphi \qquad \varphi, \Delta \Rightarrow \psi}{\Gamma, \Delta \Rightarrow \psi}$$
$$(\Rightarrow \Box)\ \overline{\Pi, \Lambda, \Box\Gamma, \Box\Delta \Rightarrow \Sigma, \Theta, \Box\psi}$$

In case of $(\Box \Rightarrow)$ on the right we have

$$(\Rightarrow \Box)\ \frac{\Gamma \Rightarrow \varphi}{\Pi, \Box\Gamma \Rightarrow \Sigma, \Box\varphi} \qquad \frac{\Box\varphi, \varphi, \Delta \Rightarrow \Lambda}{\Box\varphi, \Delta \Rightarrow \Lambda}\ (\Box \Rightarrow)$$
$$(Cut)\ \overline{\Pi, \Box\Gamma, \Delta \Rightarrow \Sigma, \Lambda}$$

and provide the following reduction

$$\frac{\dfrac{\Pi, \Box\Gamma \Rightarrow \Sigma, \Box\varphi \qquad \Box\varphi, \varphi, \Delta \Rightarrow \Lambda}{\varphi, \Pi, \Box\Gamma, \Delta \Rightarrow \Sigma, \Lambda}\ (Cut)}{\Gamma \Rightarrow \varphi \qquad \qquad \qquad}$$

$$\frac{}{\Pi, \Gamma, \Box\Gamma, \Delta \Rightarrow \Sigma, \Lambda}\ (Cut)$$
$$\overline{\vdots}\ (\Box \Rightarrow), (C \Rightarrow)$$
$$\overline{\Pi, \Box\Gamma, \Delta \Rightarrow \Sigma, \Lambda}$$

where the first cut is eliminable by the induction hypothesis on the height and the second by the induction hypothesis on the complexity of the cut-formula.

Exercise 4.12 *Check that the reduction of the height (part 3 of the proof) is correct in the presence of modal rules.*

In G3S4, we do not have any such new situation, since in $(\Box \Rightarrow)$ and $(\Rightarrow \Box^4)$ we have the cut-formula also in the premisses, so we must provide a reduction of the height on the right premiss.

Exercise 4.13 *Provide reduction of the height in the right premiss for two cases:*
$(\Rightarrow \Box^4)$ *on the left and on the right.*
$(\Rightarrow \Box^4)$ *on the left and* $(\Box \Rightarrow)$ *on the right.*

Smullyan's proof (subsection 3.3.2), of admissibility of A-cut may be also adapted to G3T and G3S4. We illustrate it with one case for G3T. When both cut-formulae are principal but in the right premiss $(\Box \Rightarrow)$ was applied, we have

$$(\Rightarrow \Box) \cfrac{\cfrac{\Gamma \Rightarrow \varphi}{\Pi, \Box\Gamma \Rightarrow \Sigma, \Box\varphi} \quad \cfrac{\Box\varphi, \varphi, \Pi, \Box\Gamma \Rightarrow \Sigma}{\Box\varphi, \Pi, \Box\Gamma \Rightarrow \Sigma}(\Box \Rightarrow)}{\Pi, \Box\Gamma \Rightarrow \Sigma}(Cut)$$

this is changed into

$$(W) \cfrac{\cfrac{\Gamma \Rightarrow \varphi}{\Pi, \Gamma, \Box\Gamma \Rightarrow \Sigma, \varphi} \quad \cfrac{\cfrac{\varphi, \Pi, \Box\Gamma \Rightarrow \Sigma, \Box\varphi \quad \Box\varphi, \varphi, \Pi, \Box\Gamma \Rightarrow \Sigma}{\varphi, \Pi, \Box\Gamma \Rightarrow \Sigma}(Cut)}{\varphi, \Pi, \Gamma, \Box\Gamma \Rightarrow \Sigma}(W \Rightarrow)}{\cfrac{\Pi, \Gamma, \Box\Gamma \Rightarrow \Sigma}{\vdots}(Cut)}(\Box \Rightarrow), (C \Rightarrow)$$
$$\overline{\Pi, \Box\Gamma \Rightarrow \Sigma}$$

W is h-p admissible which guarantees that the first cut has indeed lower height.

Exercise 4.14 *Complete a proof for G3T and provide a proof for G3S4.*

Schütte's proof (subsection 3.3.3), can be also used but note that now it is not enough to make only one induction on the height of the proof of one premiss which is performed only in the basis of the induction on complexity. Since one rule for $\Box$ is not invertible, we are forced to make a subsidiary induction on the height of the left premiss in the inductive step when cut-formula is modal.

Exercise 4.15 *Provide a proof of admissibility of cut for G3T and G3S4 using Schütte's strategy.*

4.4 Decidability

We already know that proof-search and establishing decidability is simpler in the setting of G3, hence in this section we will be dealing only with systems G3T and G3S4. However, in case of modal logics even in this setting, we can find problems already discussed in chapter 2. Since invertibility of all rules is lost, both calculi are not confluent, and in general, we cannot avoid backtracking or some other techniques which allow for tracking control on all possible choices. In G3S4, things are even worse due to the possibility of infinite branches, but we postpone this problem and first limit considerations to G3T. For both the systems, we modify the notation slightly to make explicit that $(\square \Rightarrow)$ was applied to some $\square$-formula in the root-first proof search and need not be applied again to the repeated principal formula in the premiss, unless some new circumstances take place. We will use $\boxtimes$ instead of $\square$ for making it clear, so in this section $(\square \Rightarrow)$ is represented by means of the following schema:

$$\frac{\boxtimes\varphi, \varphi, \Gamma \Rightarrow \Delta}{\square\varphi, \Gamma \Rightarrow \Delta}$$

4.4.1 Proof Search in G3T

In proof search, we can basically follow a depth-first procedure described in subsection 1.6.3, but with some adjustments. The stage of the application of rules on the chosen branch should be divided into two parts: first apply all boolean rules and $(\square \Rightarrow)$, then $(\Rightarrow \square)$ should be applied (to some chosen boxed formula in the succedent) when all other formulae are either atomic or boxed in the succedent and boxed with $\boxtimes$ in the antecedent. Informally, it corresponds to the situation that we checked all formulae in some state (saturation of a state) before we introduced another, accessible state.

Exercise 4.16 *Generalise the algorithm of proof search provided in section 1.6.3.*

However, it is not enough to provide a fair procedure. Let us consider the thesis $\square(\neg p \wedge q) \to (\neg\square r \to \square(q \vee p))$. Although it is cut-free provable, we can miss the point as the following failed proof search tree shows:

$$
\begin{array}{c}
(\Rightarrow\to) \dfrac{
(\Rightarrow\to) \dfrac{
(\neg \Rightarrow) \dfrac{
(\square \Rightarrow) \dfrac{
(\wedge \Rightarrow) \dfrac{
(\neg \Rightarrow) \dfrac{
(\Rightarrow \square) \dfrac{
(\wedge \Rightarrow) \dfrac{
(\neg \Rightarrow) \dfrac{q \Rightarrow r, p}{\neg p, q \Rightarrow r}}{\neg p \wedge q \Rightarrow r}
}{\boxtimes(\neg p \wedge q), q \Rightarrow \square(q \vee p), \square r, p}
}{\boxtimes(\neg p \wedge q), \neg p, q \Rightarrow \square(q \vee p), \square r}
}{\boxtimes(\neg p \wedge q), \neg p \wedge q \Rightarrow \square(q \vee p), \square r}
}{\square(\neg p \wedge q) \Rightarrow \square(q \vee p), \square r}
}{\square(\neg p \wedge q), \neg\square r \Rightarrow \square(q \vee p)}
}{\square(\neg p \wedge q) \Rightarrow \neg\square r \to \square(q \vee p)}
}{\Rightarrow \square(\neg p \wedge q) \to (\neg\square r \to \square(q \vee p))}
\end{array}
$$

Since G3T is not confluent, we cannot be sure that the root sequent is unprovable if we finish some branch with nonaxiomatic but atomic leaf. Perhaps some other routes may lead to success. The example shows where the problem lies. We can apply $(\Rightarrow \square)$ only to one formula in the succedent, the remaining ones are erased. But one may easily see that the selection of the other $\square$-formula leads to a proof

$$(\Rightarrow \Box) \cfrac{(\land \Rightarrow) \cfrac{(\Rightarrow \lor) \cfrac{\neg p, q \Rightarrow q, p}{\neg p, q \Rightarrow q \lor p}}{\neg p \land q \Rightarrow q \lor p}}{\text{...}}$$

$$
\begin{array}{ll}
(\Rightarrow \Box) & \cfrac{\neg p \land q \Rightarrow q \lor p}{\boxtimes(\neg p \land q), q \Rightarrow \Box(q \lor p), \Box r, p} \\[2mm]
(\neg \Rightarrow) & \cfrac{}{\boxtimes(\neg p \land q), \neg p, q \Rightarrow \Box(q \lor p), \Box r} \\[2mm]
(\land \Rightarrow) & \cfrac{}{\boxtimes(\neg p \land q), \neg p \land q \Rightarrow \Box(q \lor p), \Box r} \\[2mm]
(\Box \Rightarrow) & \cfrac{}{\Box(\neg p \land q) \Rightarrow \Box(q \lor p), \Box r} \\[2mm]
(\neg \Rightarrow) & \cfrac{}{\Box(\neg p \land q), \neg \Box r \Rightarrow \Box(q \lor p)} \\[2mm]
(\Rightarrow\!\rightarrow) & \cfrac{}{\Box(\neg p \land q) \Rightarrow \neg \Box r \rightarrow \Box(q \lor p)} \\[2mm]
(\Rightarrow\!\rightarrow) & \cfrac{}{\Rightarrow \Box(\neg p \land q) \rightarrow (\neg \Box r \rightarrow \Box(q \lor p))}
\end{array}
$$

Incidentally, we can find even shorter proof if we resign from rigid application of our procedure and apply $(\Rightarrow \Box)$ before all possible applications of boolean rules and $(\Box \Rightarrow)$. The resulting proof is

$$
\begin{array}{ll}
& (\Rightarrow \lor) \cfrac{\neg p, q \Rightarrow q, p}{\neg p, q \Rightarrow q \lor p} \\[2mm]
& (\land \Rightarrow) \cfrac{}{\neg p \land q \Rightarrow q \lor p} \\[2mm]
(\Rightarrow \Box) & \cfrac{}{\Box(\neg p \land q), \neg \Box r \Rightarrow \Box(q \lor p)} \\[2mm]
(\Rightarrow\!\rightarrow) & \cfrac{}{\Box(\neg p \land q) \Rightarrow \neg \Box r \rightarrow \Box(q \lor p)} \\[2mm]
(\Rightarrow\!\rightarrow) & \cfrac{}{\Rightarrow \Box(\neg p \land q) \rightarrow (\neg \Box r \rightarrow \Box(q \lor p))}
\end{array}
$$

However, in general, we will follow the usual order of rule's application.

Also, in case of unprovable sequents if we try to obtain a falsifying model for an input, we find that in case of any $(\Rightarrow \Box)$ application all $\Box$-formulae from the succedent must be dealt with. Otherwise, our construction is not sufficient for building a model in which all formulae generated during a proof-search will be satisfied. To avoid such problems, we must have a possibility of going back to this stage of proof search where we have made a selection and try with the next candidate. We have made some remarks on these matters in section 2.6, but postponed a more detailed discussion until now.

Several solutions were proposed to the effect that backtracking is formally secured somehow which may be roughly divided into two main approaches. In the first approach, we are building a collection of proof trees for a sequent and either at least one is a proof or all contain some nonaxiomatic leaves. In the latter case, a falsifying model is constructed on the basis of material provided by all proof trees in the collection. Such an approach is presented in detail by Zeman [276] and Goré [103], for tableau systems and may be applied also to SC. The advantage of this approach is that we do not introduce any new kind of items except proof trees used for proof search in **CPL** and described in section 1.6.

The other approach generalises the notion of a proof search tree in such a way that another kind of branching is introduced. Such trees with two types of branching were introduced first by Beth [28], in tableau system for intuitionistic logic. We have made some remarks on these matters in section 2.6, and proposed a general schema of a rule (DB) of disjunctive branching. In case of G3T, we may simplify (DB) significantly by taking into account only alternative applications of $(\Rightarrow \Box)$. Such a solution was applied, e.g. by Sambin [223]); also a decision algorithm for basic monotonic logics sketched in Indrzejczak [128], is based on such a meta-rule of Subtree Generation (SG). For G3T it has the form

$$(SG) \quad \frac{\varphi_1, ..., \varphi_l \Rightarrow \psi_1 \quad ..... \quad \varphi_1, ..., \varphi_l \Rightarrow \psi_k}{\Gamma, \boxtimes\varphi_1, ..., \boxtimes\varphi_l \Rightarrow \Box\psi_1, ..., \Box\psi_k, \; \Delta}$$

where Γ and Δ contain only atomic formulae.

Notice that in contrast to the very general schema (DB) this schema is strictly restricted. Only applications of $(\Rightarrow \Box)$ are taken into account, all compound boolean formulae and all $\Box$-formulae in the antecedent must be used before its application and all $\Box$-formulae in the succedent must be applied in one step.

Of course, branching occurring in (SG) has a different character than in all branching rules of G3; it is disjunctive in the sense that for provability of the conclusion, we need provability of at least one premiss. In case $k = 1$ (SG) is just an application of $(\Rightarrow \Box)$, otherwise each premiss is a root sequent of a separate subtree.

We can again slightly change our procedure of proof search devised in section 1.6. Moreover, by an atomic sequent we now mean any sequent of the form: $\Gamma, \boxtimes\varphi_1, ..., \boxtimes\varphi_l \Rightarrow \Delta$ with Γ and Δ containing only atomic formulae. Sequents of the form: $\Gamma, \boxtimes\varphi_1, ..., \boxtimes\varphi_l \Rightarrow \Box\psi_1, ..., \Box\psi_k, \; \Delta$ are called m-sequents (from modal).

Notice that in G3T if we run our procedure on any sequent S, and apply first only boolean rules and $(\Box \Rightarrow)$ to all compound formulae, there are three possible outcomes:

1. all leaves axiomatic;

2. at least one leaf nonaxiomatic but atomic;

3. all nonaxiomatic leaves being m-sequents.

Cases 1 and 2 are like in K and lead to termination with a proof or disproof of S. In case 3, we apply our procedure again to every premiss $\varphi_1, ..., \varphi_l \Rightarrow \psi_i$ of the application of (SG) in the leftmost fashion one by one and repeating application of (SG), if case 3 holds for some subtree. If for at least one subtree generated by the application of (SG) case 1 holds, we delete all other subtrees and obtain a proof of S (this way our application of (SG) becomes an application of $(\Rightarrow \Box)$). Otherwise, we continue until all subtrees satisfy case 2. The procedure must terminate since every application of (SG) diminishes the modal depth of formulae in the conclusion sequent and the branching factor is bounded by k (the number of $\Box$-formulae in the succedent).

Exercise 4.17 *Describe formally an algorithm.*

Let us illustrate this approach with an examination of an unprovable sequent

$$(\vee \Rightarrow) \, \dfrac{\dfrac{p \Rightarrow p \qquad q \Rightarrow p}{p \vee q \Rightarrow p} \qquad \dfrac{p \Rightarrow q \qquad q \Rightarrow q}{p \vee q \Rightarrow q}}{}$$

$$(SG)$$
$$(\vee \Rightarrow) \, \dfrac{\boxtimes(p \vee q), p \Rightarrow \Box p, \Box q \qquad\qquad \boxtimes(p \vee q), q \Rightarrow \Box p, \Box q}{}$$

$$(\Box \Rightarrow) \, \dfrac{\boxtimes(p \vee q), p \vee q \Rightarrow \Box p, \Box q}{}$$

$$(\Rightarrow \vee) \, \dfrac{\Box(p \vee q) \Rightarrow \Box p, \Box q}{\Box(p \vee q) \Rightarrow \Box p \vee \Box q}$$

We start with $(\Rightarrow \vee)$, $(\Box \Rightarrow)$ and $(\vee \Rightarrow)$. Since the procedure works in a depth-first fashion, we continue with the leftmost leaf which, by (SG) leads to two subtrees. The first, by $(\vee \Rightarrow)$ yields two atomic leaves with one nonaxiomatic. We continue with the rightmost subtree and obtain the same result. Since both subtrees have one nonaxiomatic branch, we stop and do not continue with the rightmost leaf. The countermodel has three worlds: a root w_0 satisfying p and two worlds accessible from it—w_1 satisfying q (and not p) and w_2 satisfying p (but not q).

Exercise 4.18 *Check that a countermodel satisfies all formulae in antecedents of subsequent sequents and falsifies all formulae from succedents.*

4.4.2 Completeness of G3T

On the basis of this proof search algorithm, we can provide a constructive proof of completeness for G3T. First, we add one more condition to the definition of downward saturated set (see definition 1.10, in section 1.7)

9. if $\Box\varphi \in \Gamma$, then $\varphi \in \Gamma$.

The notion of a Hintikka tuple is now based on the above definition. Moreover, we introduce the notion of a Hintikka system as a (finite) collection HS of Hintikka tuples with a relation $\mathcal{R}$ satisfying two conditions:

(a) If $(\Gamma, \Delta)\mathcal{R}(\Pi, \Sigma)$, then $\Gamma^\Box \subseteq \Pi$.
(b) If $\Box\varphi \in \Delta$, then there is some (Π, Σ) such that $(\Gamma, \Delta)\mathcal{R}(\Pi, \Sigma)$ and $\varphi \in \Sigma$.

where $\Gamma^\Box = \{\varphi : \Box\varphi \in \Gamma\}$

Now on the basis of any Hintikka system, we can define a **T**-model $\mathfrak{M}_{HS}$ where possible worlds are Hintikka tuples and $\mathcal{R}$ is an accessibility relation. Valuation is defined for each atom p: $V(p) = \{(\Gamma, \Delta) : p \in \Gamma\}$. Reflexivity of $\mathcal{R}$ follows from condition 9. One may easily prove

Lemma 4.6 (Truth lemma). *For each (Γ, Δ) in HS*

- *if $\varphi \in \Gamma$, then $\mathfrak{M}_{HS}, (\Gamma, \Delta) \vDash \varphi$;*

- *if $\varphi \in \Delta$, then $\mathfrak{M}_{HS}, (\Gamma, \Delta) \nvDash \varphi$.*

PROOF: As the proof in section 1.7, by induction on the complexity of φ. We consider the case of $\varphi := \Box\psi$. Let $\Box\psi \in \Gamma$. By condition (a) if $(\Gamma, \Delta)\mathcal{R}(\Pi, \Sigma)$, then $\Gamma^\Box \subseteq \Pi$, so $\psi \in \Pi$. By the induction hypothesis $(\Pi, \Sigma) \vDash \psi$, hence $(\Gamma, \Delta) \vDash \varphi$.

Now let $\Box\psi \in \Delta$. By condition (b) there is some (Π, Σ) such that $(\Gamma, \Delta)\mathcal{R}(\Pi, \Sigma)$ and $\psi \in \Sigma$. By the induction hypothesis $(\Pi, \Sigma) \nvDash \psi$, hence $(\Gamma, \Delta) \nvDash \varphi$. $\qquad\square$

Let us say that HS is a Hintikka system for $\Gamma \Rightarrow \Delta$ iff $\Gamma \subseteq \Pi$ and $\Delta \subseteq \Sigma$ for some (Π, Σ) in this HS. Now we are ready to prove

Lemma 4.7 (Saturation). *For any $\Gamma \Rightarrow \Delta$, either $G3T \vdash \Gamma \Rightarrow \Delta$ or there is a (finite) HS for $\Gamma \Rightarrow \Delta$ such that for each (Π, Σ) in HS, $\Pi \cup \Sigma \subseteq SF(\Gamma \cup \Delta)$.*

PROOF: We examine our proof search procedure for some unprovable $\Gamma \Rightarrow \Delta$. If the procedure finishes after the first round of saturation, i.e. with some leaf atomic, then we take as (Π, Σ) the unions of all formulae in the antecedents, and succedents, respectively, of sequents from this branch and define HS as having just this one Hintikka tuple as the only world. The examination that (Π, Σ) is indeed a Hintikka tuple is just like for K (cf. lemma 1.13). The additional condition 9 is satisfied by demanding that all $\Box$-formulae in the antecedents are used by $(\Box \Rightarrow)$. This requirement guarantees also that $(\Pi, \Sigma)\mathcal{R}(\Pi, \Sigma)$ and so $\mathcal{R}$ is reflexive. Otherwise, we had at least one application of (SG) and there are $k \geq 1$ subtrees each of which also finishes with case 2 or 3. Again we define a Hintikka tuple (Π_i, Σ_i) for every $i \leq k$ from unions of formulae from antecedents and succedents of the selected open branch. Also, it is obvious from the schema of (SG) that for each $i \leq k$, $(\Pi, \Sigma)\mathcal{R}(\Pi_i, \Sigma_i)$ since there is $\Box\psi_i \in \Sigma$ such that $\psi_i \in \Sigma_i$ and for every $\boxtimes\varphi \in \Pi$ we have $\varphi \in \Pi_i$. The situation repeats with every application of (SG) so we obtain a HS which satisfies the conditions stated and yields a model (indeed rooted one) where a root-world is just (Π, Σ). $\qquad\square$

This yields

Theorem 4.3 (Completeness). *If $T \models \Gamma \Rightarrow \Delta$, then $G3T \vdash \Gamma \Rightarrow \Delta$.*

PROOF: If $\Gamma \Rightarrow \Delta$ is unprovable, then we have a HS for it by the saturation lemma. By the truth lemma in $\mathfrak{M}_{HS}$ there is a world (Π, Σ) such that $(\Pi, \Sigma) \vDash \varphi$ for every $\varphi \in \Gamma$ and $(\Pi, \Sigma) \nvDash \psi$ for every $\psi \in \Delta$. Hence $\nvDash \Gamma \Rightarrow \Delta$. $\qquad\square$

Exercise 4.19 *Prove completeness by direct strategy also applied in section 1.7 (see lemma 1.14).*

4.4.3   Proof Search in G3S4

Since, in general, the proof search in G3S4 is similar to proof search in G3T, we will be brief and point out only the most interesting things. One must notice that in G3S4 things are worse than in G3T since confluency is not the only thing we have lost. Let us consider the following proof search:

$$\vdots$$

$$\cfrac{\cfrac{\cfrac{\cfrac{\cfrac{\cfrac{\cfrac{\cfrac{\cfrac{\cfrac{\cfrac{\cfrac{\Box p, \Box\neg\Box\neg p \Rightarrow \neg p}{p, \Box p, \Box\neg\Box\neg p \Rightarrow \Box\neg p}\,(\Rightarrow \Box)}{p, \Box p, \neg\Box\neg p, \Box\neg\Box\neg p \Rightarrow}\,(\neg \Rightarrow)}{p, \Box p, \Box\neg\Box\neg p \Rightarrow}\,(\Box \Rightarrow)}{\Box p, \Box\neg\Box\neg p \Rightarrow \neg p}\,(\Rightarrow \neg)}{p, \Box p, \Box\neg\Box\neg p \Rightarrow \Box\neg p}\,(\Rightarrow \Box)}{p, \Box p, \neg\Box\neg p, \Box\neg\Box\neg p \Rightarrow}\,(\neg \Rightarrow)}{p, \Box p, \Box\neg\Box\neg p \Rightarrow}\,(\Box \Rightarrow)}{\Box p, \Box\neg\Box\neg p \Rightarrow}\,(\Box \Rightarrow)}{\Box\neg\Box\neg p \Rightarrow \neg\Box p}\,(\Rightarrow \neg)}{\Box\neg\Box\neg p \Rightarrow \Box\neg\Box p, \Box\neg p}\,(\Rightarrow \Box)}{\neg\Box\neg p, \Box\neg\Box\neg p \Rightarrow \Box\neg\Box p}\,(\neg \Rightarrow)}{\Box\neg\Box\neg p \Rightarrow \Box\neg\Box p}\,(\Box \Rightarrow)}{\Rightarrow \Box\neg\Box\neg p \to \Box\neg\Box p}\,(\Rightarrow\to)$$

Here, we see that our systematic proof-search may lead to the generation of an infinite tree. If we are interested only in proving completeness of our system, this is not a problem. We should take care only of the fairness of our procedure, i.e. we must be sure that every $\Box$-formula in the succedent is eventually used to create a part of the tree corresponding to a new world, and that every $\Box$-formula from the antecedent is transferred to this new world, then the saturation part of our procedure provides sufficient information for the description of the (infinite) model. But we know that **S4** is also decidable and we can falsify each invalid formula by means of a finite model. How to extract a finite model? One possibility is to apply some form of loop-check. Before we state the details, let us introduce suitable modifications. The version of (SG) for G3S4 has the following schema:

$$(SG)\ \frac{\Box\varphi_1, ..., \Box\varphi_l \Rightarrow \psi_1 \quad ..... \quad \Box\varphi_1, ..., \Box\varphi_l \Rightarrow \psi_k}{\Gamma,\ \boxtimes\varphi_1, ..., \boxtimes\varphi_l \Rightarrow \Box\psi_1, ..., \Box\psi_k,\ \Delta}$$

where Γ and Δ contain only atomic formulae.

Here, we can see that, in accordance with the definition of $(\Rightarrow \Box)^4$, we rewrite in the premisses all $\Box$-formulae from the antecedent of the conclusion but we again make them all ready for new applications of $(\Box \Rightarrow)$. Intuitively it corresponds to the fact that these formulae are now in a new world of an attempted model where they were not dealt with so far.

Although in G3S4 we may generate infinite branches, due to the subformula property, the number of different sequents generated from the root sequent is finite and sooner or later we run into cycles. In the example, we may notice that the uppermost sequent is indeed a copy of a sequent generated earlier. There is no need to continue with such branches, hence in the algorithm, we require that after every application of (SG) we check each premiss if it has no earlier appearance. If this is the case, then this premiss is not considered further. Semantically, a world corresponding to this sequent, say w_k, is identified with the earlier world, say w_i, corresponding to the earlier occurrence of this sequent. Thus, we have a loop and in the attempted model we introduce a cluster which contains all worlds occurring between w_i and $w_k(= w_i)$. Hence, the final model is a tree of clusters rather than of single worlds (single worlds are simple clusters). Details of the construction of falsifying models are similar to the case of the completeness proof for G3T but with one difference. Now, for every pair of saturated tuples we define $\mathcal{R}$ by changing a condition (a):

If $(\Gamma, \Delta)\mathcal{R}(\Pi, \Sigma)$, then $\Gamma_\Box \subseteq \Pi$.

where $\Gamma_\Box = \{\Box\varphi : \Box\varphi \in \Gamma\}$.

This allows for proving that $\mathcal{R}$ is transitive.

Exercise 4.20 *Provide a completeness proof for G3S4.*

The loop control mechanism described above has a rather global character. One may also think about introducing some local syntactical constraints blocking unwanted inferences, similar in spirit to the one of using $\boxtimes$ instead of $\Box$. We sketch here one such solution provided by Heuerding, Seyfried and Zimmermann [115]. They generalise the notion of a sequent by addition of a kind of 'history' to both sides of a sequent[8]. Thus, their rules are defined on sequents of the form $[\Gamma] \Delta \Rightarrow \Sigma [\Pi]$, where a history (of modal rules' application) is put in $[\ ]$ and each multiset may be empty. In particular, all formulae in Γ are $\Box$-formulae.

The rules for boolean connectives are standard but performed only on the elements of Δ and Σ. For $\Box$ we have

$$(\Box\Rightarrow)^1 \ \frac{[\Gamma] \ \varphi, \Delta \Rightarrow \Sigma \ [\Pi]}{[\Gamma] \ \Box\varphi, \Delta \Rightarrow \Sigma \ [\Pi]} \qquad (\Box\Rightarrow)^2 \ \frac{[\Box\varphi, \Gamma] \ \varphi, \Delta \Rightarrow \Sigma \ [\ ]}{[\Gamma] \ \Box\varphi, \Delta \Rightarrow \Sigma \ [\Pi]} \qquad (\Rightarrow\Box) \ \frac{[\Gamma] \ \Gamma \Rightarrow \varphi \ [\varphi, \Theta, \Pi]}{[\Gamma] \ \Lambda \Rightarrow \Box\Theta, \Box\varphi \ [\Pi]}$$

All the rules have side conditions

for $(\Box\Rightarrow)^1$, $\Box\varphi \in \Gamma$;
for $(\Box\Rightarrow)^2$, $\Box\varphi \notin \Gamma$;
for $(\Rightarrow\Box)$, $\varphi \notin \Pi$ and there is no box-formula in Λ.

It should be clear that $[\Gamma]$ is just a notationally different way of expressing the same effect which we obtained by using $\boxtimes$. The role of $[\Pi]$ is more complicated. It serves as a record of formulae which are freed of $\Box$ in effect of the applications of $(\Rightarrow \Box)$.

We can show that our example is blocked when applying these rules

---

[8]In fact, they are working with Schütte-style one-sided sequents, so they are just using a kind of tripartite structures $\Gamma \ \| \ \Delta \ | \ \Sigma$. We just translate their rules into standard SC.

$$
\cfrac{
\cfrac{
\cfrac{
\cfrac{
\cfrac{
\cfrac{
\cfrac{
\cfrac{
\cfrac{
\cfrac{
\cfrac{
\cfrac{
\cfrac{
[\Box\neg\Box\neg p,\Box p]\; p \Rightarrow \Box\neg p\; [\neg p]}
{[\Box\neg\Box\neg p,\Box p]\; \neg\Box\neg p, p \Rightarrow\; [\neg p]}(\neg\Rightarrow)}
{[\Box\neg\Box\neg p,\Box p]\; \neg\Box\neg p, p, p \Rightarrow\; [\neg p]}(W\Rightarrow)}
{[\Box\neg\Box\neg p,\Box p]\; \neg\Box\neg p, p \Rightarrow \neg p\; [\neg p]}(\Rightarrow\neg)}
{[\Box\neg\Box\neg p,\Box p]\; \Box\neg\Box\neg p, p \Rightarrow \neg p\; [\neg p]}(\Box\Rightarrow)^1}
{[\Box\neg\Box\neg p,\Box p]\; \Box\neg\Box\neg p, \Box p \Rightarrow \neg p\; [\neg p]}(\Box\Rightarrow)^1}
{[\Box\neg\Box\neg p,\Box p]\; p \Rightarrow \Box\neg p\; [\;]}(\Rightarrow\Box)}
{[\Box\neg\Box\neg p,\Box p]\; \neg\Box\neg p, p \Rightarrow\; [\;]}(\neg\Rightarrow)}
{[\Box\neg\Box\neg p,\Box p]\; \Box\neg\Box\neg p, p \Rightarrow\; [\;]}(\Box\Rightarrow)^1}
{[\Box\neg\Box\neg p]\; \Box\neg\Box\neg p, \Box p \Rightarrow\; [\neg\Box p, \neg p]}(\Box\Rightarrow)^2}
{[\Box\neg\Box\neg p]\; \Box\neg\Box\neg p \Rightarrow \neg\Box p\; [\neg\Box p, \neg p]}(\Rightarrow\neg)}
{[\Box\neg\Box\neg p]\; \Rightarrow \Box\neg\Box p, \Box\neg p\; [\;]}(\Rightarrow\Box)}
{[\Box\neg\Box\neg p]\; \neg\Box\neg p \Rightarrow \Box\neg\Box p\; [\;]}(\neg\Rightarrow)}
{[\;]\; \Box\neg\Box\neg p \Rightarrow \Box\neg\Box p\; [\;]}(\Box\Rightarrow)^2}
{[\;]\; \Rightarrow \Box\neg\Box\neg p \to \Box\neg\Box p\; [\;]}(\Rightarrow\to)
$$

In the leaf $(\Rightarrow\Box)$ cannot be applied since $\neg p$ is in the history on the right. In fact, we can finish the proof search even quicker if we select $\Box\neg p$ as the principal formula in the first (from the bottom) application of $(\Rightarrow\Box)$. We obtain

$$
\cfrac{
\cfrac{
\cfrac{
\cfrac{
\cfrac{
\cfrac{
\cfrac{
[\Box\neg\Box\neg p]\; p \Rightarrow \Box\neg p\; [\neg\Box p, \neg p]}
{[\Box\neg\Box\neg p]\; \neg\Box\neg p, p \Rightarrow\; [\neg\Box p, \neg p]}(\neg\Rightarrow)}
{[\Box\neg\Box\neg p]\; \Box\neg\Box\neg p, p \Rightarrow\; [\neg\Box p, \neg p]}(\Box\Rightarrow)^1}
{[\Box\neg\Box\neg p]\; \Box\neg\Box\neg p \Rightarrow \neg p\; [\neg\Box p, \neg p]}(\Rightarrow\neg)}
{[\Box\neg\Box\neg p]\; \Rightarrow \Box\neg\Box p, \Box\neg p\; [\;]}(\Rightarrow\Box)}
{[\Box\neg\Box\neg p]\; \neg\Box\neg p \Rightarrow \Box\neg\Box p\; [\;]}(\neg\Rightarrow)}
{[\;]\; \Box\neg\Box\neg p \Rightarrow \Box\neg\Box p\; [\;]}(\Box\Rightarrow)^2}
{[\;]\; \Rightarrow \Box\neg\Box\neg p \to \Box\neg\Box p\; [\;]}(\Rightarrow\to)
$$

Again, we cannot apply $(\Rightarrow\Box)$ on $\Box\neg p$ since $\neg p$ is in the history.

In this system, we have an example of generalised SC based on structured sequents. However, it is worth remarking that Heuerding, Seyfried and Zimmermann use the addition of histories to sequents only as a kind of book-keeping device for obtaining a terminating proof-search procedure. They are not interested in developing a full-fledged syntactic system, for example, they did not consider the problem of cut admissibility. More extended theory of such a kind of SC was proposed by Bilkova [30], to prove a generalised (uniform) interpolation theorem. She provided a system only for **T** so only with the left history required, i.e. with sequents of the form $[\Gamma]\,\Delta \Rightarrow \Sigma$ and modal rules

$$(\Box\Rightarrow)\quad \frac{[\Box\varphi,\Gamma]\;\varphi,\Delta\Rightarrow\Sigma}{[\Gamma]\;\Box\varphi,\Delta\Rightarrow\Sigma}\qquad\qquad (\Rightarrow\Box)\quad \frac{[\;]\;\Gamma\Rightarrow\varphi}{[\Box\Gamma]\;\Lambda\Rightarrow\Theta,\Box\varphi}$$

At first sight, it may seem that it is also only a notational device, similar in effect to the introduction of ⊠. But in addition to terminating proof procedure a syntactic proof theory is developed. In particular, proofs of admissibility of structural rules, including cut, are provided. In addition to standard W and C rules, we have two special ones

$$(W\Box\Rightarrow) \quad \frac{[\Gamma]\,\Delta\Rightarrow\Sigma}{[\Box\varphi,\Gamma]\,\Delta\Rightarrow\Sigma} \qquad\qquad (C\Box\Rightarrow) \quad \frac{[\Box\varphi,\Box\varphi,\Gamma]\,\Delta\Rightarrow\Sigma}{[\Box\varphi,\Gamma]\,\Delta\Rightarrow\Sigma}$$

There are also two cut rules of the form

$$(Cut) \quad \frac{[\,]\,\Gamma\Rightarrow\Delta,\varphi \quad [\,]\,\varphi,\Pi\Rightarrow\Sigma}{[\,]\,\Gamma,\Pi\Rightarrow\Delta,\Sigma} \qquad (Cut\Box) \quad \frac{[\Gamma]\,\Sigma\Rightarrow\Delta,\Box\varphi \quad [\Box\varphi,\Pi]\,\Lambda\Rightarrow\Theta}{[\Gamma,\Pi]\,\Sigma,\Lambda\Rightarrow\Delta,\Theta}$$

It is interesting to note that ordinary cut is correct only with an empty history, otherwise we could 'prove' something like $[\Box p]\Rightarrow p$ which is not a provable sequent.

Remark 4.1 Both systems represent a kind of generalised SC based on the use of structured, or higher-arity sequents. It is a family of systems that multiply the number of parts of a sequent. The natural place for such a solution was of course in many-valued logics (cf. [44, 221]), where the number of arguments corresponds to the number of truth values (see section 5.5). But this approach has also some representation in modal logics, where application of more arguments is not always based on such a direct semantical motivation. In the setting of modal logics, such an approach was for the first time proposed by Sato [224], and we briefly describe his approach in the last section of this chapter. But other proposals should also be mentioned briefly. In addition to systems discussed above, where additional arguments in sequents serve only to encode a 'history' of attempted falsifying models, there are systems with more involved motivations, where different parts of sequents allow to distinguish formulae simply true from necessarily true (necessarily true in the past or future in temporal case). Blamey and Humberstone [34] developed such a generalised framework for some modal logics (including all logics discussed here) based on sequents with four parts. Nishimura [188], proposed a similar approach for temporal logics with 6-ary sequents. But in their system, an interpretation of such sequents is different than in SCs described above. Informally, we say that such a sequent is satisfied whenever, if all formulae in two parts of the antecedent are simply true and necessarily true, respectively, then in two parts of the succedent at least one formula is simply true and at least one necessarily true. Hence, additional arguments of a sequent do not serve just as containers for keeping some data for later use but have specific structural and logical rules. Unfortunately, cut (or rather different forms of cut present in this system) is not eliminated in their structured SC. One may find a more detailed description also in Wansing [270] and Poggiolesi [198]. We will devote more attention to this approach in the next chapter, where SC formalisations of many-valued logics will be considered.

4.4.4 Modal Logic with Simplified Proof Search Tree

In general, the shape of rules for the introduction of $\Box$ to the succedent makes it impossible to obtain a confluent procedure of proof search. Even if some of the variants are invertible, the lack of context independence blocks permutability with other rules, and proof search must be always divided into iterated segments: exhaustive proof-search by boolean rules (and $(\Box \Rightarrow)$), then application of $(\Rightarrow \Box)$. Loosely, this corresponds to checking a situation in one state in a model before we go to the next one. In order not to overlook some possibility, we proposed generalisations of $(\Rightarrow \Box)$. It certainly facilitates the control over all possible states we must check but at the price of complication in the structure of the proof-search tree (mixing conjunction and disjunction branching).

It is worthwhile to mention that there are modal logics for which it is possible to avoid such problems. One example of such a logic is **S4.3**, an extension of **S4** axiomatically characterised by the addition of

$$3: \Box(\Box\varphi \to \psi) \vee \Box(\Box\psi \to \varphi)$$

to Hilbert system for **S4**. It is characterised by the class of frames which are reflexive, transitive and connected, i.e. satisfy the condition

$$\forall xyx(Rxy \wedge Rxz \to Ryz \vee Rzy)$$

This in effect makes **S4.3**, the simplest logic of frames with the linear accessibility relation. It is not surprising that a lot of approaches to its formalisation were proposed in the framework of tableau systems and SCs, both standard and generalised. A detailed history and description of several systems may be found in Indrzejczak [130] (see also [136]), here we briefly recall an approach based on standard SC, introduced by Zeman [276], and significantly developed by Shimura [237]. In order to obtain an adequate formalisation it suffices to change the respective rule for $(\Rightarrow \Box)^4$ into:

$$(\Rightarrow \Box)^3 \quad \frac{\Box\Gamma \Rightarrow \varphi_1, \Box\varphi_2, ..., \Box\varphi_k \quad ... \quad \Box\Gamma \Rightarrow \Box\varphi_1, ..., \Box\varphi_{k-1}, \varphi_k}{\Box\Gamma \Rightarrow \Box\varphi_1, ..., \Box\varphi_k}$$

This time the number of premisses is not fixed but depends on the number of $\Box$-formulae in the succedent of the conclusion. A proof of an axiom 3 is straightforward in LK

$$
\cfrac{
  \cfrac{
    \cfrac{
      \cfrac{
        \cfrac{
          \cfrac{
            \cfrac{
              (\Box\Rightarrow)\ \cfrac{p\Rightarrow p}{\Box p\Rightarrow p}
            }{\Box p,\Box q\Rightarrow p}\,(W\Rightarrow)
          }{\Box p\Rightarrow \Box q\to p}\,(\Rightarrow\to)
        }{\Box p\Rightarrow \Box(\Box q\to p)}\,(\Rightarrow\Box)^3
      }{\Box p\Rightarrow q,\Box(\Box q\to p)}\,(\Rightarrow W)
    }{\Rightarrow \Box p\to q,\Box(\Box q\to p)}\,(\Rightarrow\to)
    \qquad
    \cfrac{
      \cfrac{
        \cfrac{
          \cfrac{
            \cfrac{
              \cfrac{q\Rightarrow q}{\Box q\Rightarrow q}\,(\Box\Rightarrow)
            }{\Box p,\Box q\Rightarrow q}\,(W\Rightarrow)
          }{\Box q\Rightarrow \Box p\to q}\,(\Rightarrow\to)
        }{\Box q\Rightarrow \Box(\Box p\to q)}\,(\Rightarrow\Box)^3
      }{\Box q\Rightarrow p,\Box(\Box p\to q)}\,(\Rightarrow W)
    }{\Rightarrow \Box(\Box p\to q),\Box q\to p}\,(\Rightarrow\to)\ (\Rightarrow\Box)^3
  }{\Rightarrow \Box(\Box p\to q),\Box(\Box q\to p)}
}{\Rightarrow \Box(\Box p\to q)\vee\Box(\Box q\to p)}\,(\Rightarrow\vee)
$$

Exercise 4.21 *Prove validity-preservation of $(\Rightarrow\Box)^3$ in connected frames.*

Shimura [237] proved cut elimination for LK formalisation with the above rule. Hence, we can prove also a decidability result based on our proof search procedure. Clearly it is better to base it on G3T with

$$
(\Rightarrow\Box)^3\ \ \cfrac{\Box\Gamma\Rightarrow\varphi_1,\Box\varphi_2,...,\Box\varphi_k\ \ ...\ \ \Box\Gamma\Rightarrow\Box\varphi_1,...,\Box\varphi_{k-1},\varphi_k}{\Delta,\boxtimes\Gamma\Rightarrow\Box\varphi_1,...,\Box\varphi_k,\Sigma}
$$

where Δ and Σ contain only atoms.

Both variants of this rule for $(\Rightarrow\Box)$ look like a collection of all possible premisses generated by $(\Rightarrow\Box)^5$ from the same sequent as the conclusion. But there is one significant difference. Here, we have a conjunctive branching like in other SC rules with more than one premiss. In other words, it means that the conclusion is provable if all premisses are provable (not if at least one of them is provable). In proof search for G3S4.3, we are not forced to backtracking like in G3T or G3S4 with primitive (one-premiss) rule for $(\Rightarrow\Box)$. Again, if we introduce in G3T or G3S4 generalised rules to avoid backtracking, we have an iteration of two kinds of branching in one tree; we are generating not a proof tree but their rooted family. In contrast, for G3S4.3, we avoid backtracking and two kinds of branching from scratch.

All solutions presented in this section have direct character in the sense that proof search is defined directly for SC and to avoid problems, we introduce additional devices which in the end change standard SC into a generalised calculus. It is also possible to develop some procedures for proof search in other frameworks which are free of the problems sketched above and then provide a translation from proofs in this better-behaving system into SC. An interesting example of such indirect approach to proof search in SC for modal logics is presented by Leszczyńska-Jasion [164] where proofs in confluent erotetic calculi for some modal logics are translated into standard SC proofs.

4.5 The Case of S5

Although the rules of standard SC for **S5** are simple and straightforward, we have seen that cut elimination cannot be proved by any method which was introduced. In fact, Lellman and Patinson [159], provided a kind of impossibility result which implies that it is not possible to obtain a standard SC for **S5** that admits standard techniques of proving cut elimination.

So what can we obtain for **S5** in the setting of standard SC? Below we briefly describe some solutions. They may be roughly divided into direct and indirect. In the former we deal with systems for **S5** for which either cut elimination (of some sort) was proved or cut is somewhat restricted. In fact, often such systems are not especially good for practical applications. Indirect solutions are based on results showing that **S5** may be embedded in some other logic for which a decent SC is defined.

4.5.1 Cut-free and Cut-restricted Solutions

Ohnishi and Matsumoto [189, 190] provided two solutions to the problem of obtaining cut-free LKS5, direct and indirect. The first one is based on the idea of using only so-called modal 1M-formulae, i.e. formulae of modal depth 1 (where no box is nested inside another box). There is a well known fact that for **S5** every modal formula may be transformed into an equivalent normal form of such kind (see, e.g. Hughes and Cresswell [124]). Ohnishi and Matsumoto have shown constructively that every application of cut (or rather mix) in the system having only modal 1M-formulae may be eliminated. The bad thing is that there is no procedure based on SC for reducing **S5** M-formulae to their 1M-equivalents. It is just shown that cut is eliminable in the calculus restricted to 1M-formulae[9].

One of the earliest ordinary (but not standard) cut-free SCS5 was provided by Mints [180], who enriched Ohnishi and Matsumoto's system with the following rule:

$$(\Rightarrow \Box^-) \ \frac{\Gamma \Rightarrow \Delta, \Box(\wedge\Pi \rightarrow \vee\Sigma)}{\Gamma, \Pi \Rightarrow \Delta, \Sigma}$$

Adequacy is proved by translation from cut-free SC to a restricted form of first-order logic. Although this rule is far from standard and even the subformula property is lost, it is worth mentioning since many later and more general approaches are somewhat based on a similar idea of separation of some formulae by means of a box for later use (in root-first proof search).

A different solution was proposed by Serebriannikov [236], who proposed to introduce global side conditions for checking the correctness of the application of

---

[9]Recently, Indrzejczak [138], provided a fully formalised account of a proof search procedure in SC based on this solution; however, some standard rules need to be slightly generalised for this aim.

modal rules in a few logics, including **S5**. His proposal is based on the strategy applied by Prawitz in normalisation proofs for natural deduction systems. We can find a similar solution in Bräuner [39], who provided a cut-free SC with simple modal rules on the basis of translation from SC for monadic first-order classical logic. In this approach, the modal rules are very simple and straightforward

$$(\Box\Rightarrow) \quad \frac{\varphi,\Gamma\Rightarrow\Delta}{\Box\varphi,\Gamma\Rightarrow\Delta} \qquad\qquad (\Rightarrow\Box) \quad \frac{\Gamma\Rightarrow\Delta,\varphi}{\Gamma\Rightarrow\Delta,\Box\varphi}$$

Clearly, his ($\Rightarrow \Box$) cannot be applied with no side conditions. Following Prawitz' [203] natural deduction system he stated them in terms of dependency of occurrences of formulae in a proof. ($\Rightarrow \Box$) may be correctly applied if no occurrences of formulae in Γ, Δ depend on the occurrence of φ in this proof. We direct the reader to Bräuner [39], for explication of the notion of dependency in his system. In contrast to other proposals here, we obtain a direct constructive proof of cut elimination. On the other hand, such a system does not allow for defining a proof-search yielding a decision procedure since the control of dependency is possible only for ready proof trees.

Many researchers provided solutions for **S5** (and other modal logics) based on the idea of restricting applications of cut, mostly in tableau framework. Following Sonobe's suggestions Fitting [83] developed an approach where cut-formulae are restricted to subformulae of the root sequent but closed under prefixing of $\Box$. In Goré [103], one may find two other, closely related tableau systems for **S5** with special versions of cut. Since Hintikka-style tableau systems are directly translatable into ordinary SC (see subsection 3.4.2), we can do it in this case with no problem. In both cases we get a version of G3 but with three-premiss rules for boolean connectives replacing two-premiss ones (see subsection 1.4.1), and added some new rules. The first system contains both modal rules for **S4**, and moreover two special rules

$$(Cut\Box) \quad \frac{\Gamma\Rightarrow\Delta,\Box\varphi,\varphi \quad \varphi,\Gamma\Rightarrow\Delta,\Box\varphi}{\Gamma\Rightarrow\Delta,\Box\varphi} \qquad (5) \quad \frac{\Box\neg\Box\varphi,\Gamma\Rightarrow\Delta,\Box\varphi}{\Pi,\Gamma\Rightarrow\Delta,\Box\varphi}$$

The first one is a special form of A-cut whereas the second is a kind of elimination rule. In the second system ($\Rightarrow \Box$)[4] is simply replaced with ($\Rightarrow \Box$)[5] and (5) is deleted. Both systems are proved complete constructively on the basis of saturation (see Goré [103]).

However, in case of **S5** even the usual analytic cut is sufficient. Takano [259] proposed a system of Ohnishi and Matsumoto with an analytic form of cut, where the cut-formula must be a subformula of other formulae in both premises. He proved constructively that every proof in Ohnishi's and Matsumoto's system may be transformed into a proof with only analytic cut applications. In consequence, we obtain the subformula property of the system. Notice that a proof of B in Ohnishi's and Matsumoto's system displayed in subsection 4.2.1, satisfies this restriction since cut is analytic there. In fact, Sato provided a semantic proof of this fact even earlier (for a more general, epistemic multimodal variant of **S5**—but it is possible to extract his result for simpler monomodal case). For the lack of space, we omit the rather elaborated, syntactical proof of Takano and present a simplified version of Sato's result.

He begins with proving a kind of Lindenbaum lemma which is essentially the analytic version of this result which we have proved in subsection 1.11.1 (lemma 1.26). For any set of formulae Θ which is closed under subformulae we consider a set of unprovable sequents $\Gamma \Rightarrow \Delta$ which are maximal in Θ, i.e. $\Gamma \cup \Delta = \Theta$ and call it $Seq\Theta$. An **S5**-model is defined with $W = Seq\Theta$, $V(p) = \{\Gamma \Rightarrow \Delta \in Seq\Theta : p \in \Gamma\}$ and $\Gamma \Rightarrow \Delta \mathcal{R} \Pi \Rightarrow \Sigma$ iff $\Gamma^\square = \Pi^\square$. It is obvious that such relation is an equivalence on $Seq\Theta$, hence we obtain **S5**-model.

Lemma 4.8 (Truth lemma). *For any $\varphi \in \Theta$ and any $\Gamma \Rightarrow \Delta \in Seq\Theta$:*

  if $\varphi \in \Gamma$, then $\Gamma \Rightarrow \Delta \vDash \varphi$

  if $\varphi \in \Delta$, then $\Gamma \Rightarrow \Delta \nvDash \varphi$

PROOF: By induction on the complexity of φ. We provide only the case of $\varphi = \square\psi$.

Let $\square\psi \in \Gamma$ and consider any $\Pi \Rightarrow \Sigma$ which is $\mathcal{R}$-related to $\Gamma \Rightarrow \Delta$. By definition of $\mathcal{R}$ $\square\psi \in \Pi$ and by maximality $\psi \in \Pi \cup \Sigma$. If $\psi \in \Sigma$, then $\Pi \Rightarrow \Sigma$ would be provable, so $\psi \in \Pi$ and by the induction hypothesis $\Pi \Rightarrow \Sigma \vDash \psi$. Since $\Pi \Rightarrow \Sigma$ was arbitrary $\Gamma \Rightarrow \Delta \vDash \square\psi$.

Assume that $\square\psi \in \Delta$, we must show that for some $\Pi \Rightarrow \Sigma$ which is $\mathcal{R}$-related to $\Gamma \Rightarrow \Delta$, $\psi \in \Sigma$, hence by the induction hypothesis $\Pi \Rightarrow \Sigma \nvDash \psi$. Consider $\Gamma_\square \Rightarrow \Delta_\square$; it must be unprovable since otherwise $\Gamma \Rightarrow \Delta$ would be provable by W. Similarly, $\Gamma_\square \Rightarrow \Delta_\square, \psi$ must be unprovable since otherwise $\Gamma_\square \Rightarrow \Delta_\square$ would be provable by ($\Rightarrow \square$). By the Lindenbaum lemma there is a maximal extension $\Pi \Rightarrow \Sigma$ of this sequent. Clearly, $\Gamma^\square \subseteq \Pi^\square$, otherwise for some χ such that $\square\chi \in \Gamma_\square \subseteq \Pi$ we would have by maximality $\chi \in \Sigma$ and $\vdash \Pi \Rightarrow \Sigma$. We must show that $\Pi^\square \subseteq \Gamma^\square$ as well. Assume that there is $\chi \in \Pi^\square$ which is not in $\Gamma^\square$. By maximality $\chi \in \Delta$ and $\square\chi \in \Delta$ but then $\square\chi \in \Sigma$ which yields $\vdash \Pi \Rightarrow \Sigma$. Hence $\Gamma \Rightarrow \Delta \mathcal{R} \Pi \Rightarrow \Sigma$ with $\psi \in \Sigma$ and we are done. $\square$

Completeness follows from this lemma. Moreover, we obtain also decidability; it is enough to take $\Theta = SF(\varphi)$ for any unprovable φ. We can find a model with at most 2^n states, where n is the cardinality of Θ.

Avron [12] presented extension of this result to the strong completeness theorem, i.e. provability from sequents (see section 1.11). Of course, in this case, the cut-formulae may be taken also from the set of formulae occurring in nonaxiomatic leaves.

4.5.2 Indirect Solutions

The second solution of Ohnishi and Matsumoto is indirect in the sense that it is based on the translation of LKS5 into LKS4 which is based on the following:

Theorem 4.4 $\vdash_{S5} \varphi$ *iff* $\vdash_{S4} \Box\Diamond\Box\varphi$

Since **S4** has a cut-free SC we obtain indirectly a decision procedure for **S5**. Note that, in fact, it may be simplified since we know that in **S4** $\vdash \Box\varphi$ iff $\vdash \varphi$. We provide a syntactic proof of this simplified version; the reader may find a semantic proof in Fitting [83].

PROOF: $\Longleftarrow$ is obvious since **S4** is a sublogic of **S5** and $\Diamond\Box\varphi \to \varphi$ is a theorem of **S5** (the converse of B).

$\Longrightarrow$ This direction is more involved. A proof goes by induction on the height of a proof in the H-system for **S5**. We must show that for every axiom φ its counterpart $\Diamond\Box\varphi$ is provable in the H-system for **S4** and that the primitive rules preserve this feature. As for axioms K, T, 4 it is simple since they are axioms of **S4**, so by GR and T' we obtain the respective theses.

Now, we prove that $\vdash_{S4} \Diamond\Box(5')$. By **CPL** $\Box\varphi \to (\Diamond\Box\varphi \to \Box\varphi)$ which by GR, K and 4 yields $\Box\varphi \to \Box(\Diamond\Box\varphi \to \Box\varphi)$. By contraposition and RR, we have $\Box\neg\Box(\Diamond\Box\varphi \to \Box\varphi) \to \Box\neg\Box\varphi$ and by **CPL** inferences we obtain $\Box\neg\Box(\Diamond\Box\varphi \to \Box\varphi) \to (\Diamond\Box\varphi \to \Box\varphi)$ which again by GR, K and 4 yields $\Box\neg\Box(\Diamond\Box\varphi \to \Box\varphi) \to \Box(\Diamond\Box\varphi \to \Box\varphi)$ This, by T' yields $\Box\neg\Box(\Diamond\Box\varphi \to \Box\varphi) \to \Diamond\Box(\Diamond\Box\varphi \to \Box\varphi)$ which implies by **CPL** $\Diamond\Box(\Diamond\Box\varphi \to \Box\varphi)$.

If in **S5** we get $\Box\varphi$ by GR, then by the induction hypothesis we have $\Diamond\Box\varphi$ provable in **S4**. Then by $\vdash \Box\Box\varphi \leftrightarrow \Box\varphi$ and closure of H-**S4** with respect to the extensionality principle, we obtain $\Diamond\Box\Box\varphi$.

The last thing is to show that MP is closed under prefixing $\Diamond\Box$ in **S4**. By K and T', we have $\Box(\varphi \to \psi) \wedge \Box\varphi \to \Diamond\Box\psi$. This is in **CPL** equivalent to $\Box(\varphi \to \psi) \to \neg\Box\varphi \vee \Diamond\Box\psi$. Since $\Diamond$ is distributive with respect to $\vee$ we obtain $\Box(\varphi \to \psi) \to \Diamond(\neg\varphi \vee \Box\psi)$ By RR' we infer $\Diamond\Box(\varphi \to \psi) \to \Diamond\Diamond(\neg\varphi \vee \Box\psi)$ which by the induction hypothesis yields $\Diamond\Diamond(\neg\varphi \vee \Box\psi)$. This implies by 4' $\Diamond(\neg\varphi \vee \Box\psi)$ and by distributivity again $\Diamond\neg\varphi \vee \Diamond\Box\psi$ which is equivalent to $\Box\varphi \to \Diamond\Box\psi$. Again by RR' we obtain $\Diamond\Box\varphi \to \Diamond\Diamond\Box\psi$ which by the induction hypothesis and 4 yields $\Diamond\Box\psi$. $\Box$

On the basis of this relationship, to provide a proof for φ in **S5** it is enough to find a cut-free proof of $\Box\neg\Box\varphi \Rightarrow$ in SC for **S4**. For the sake of illustration, we provide a cut-free proof of B (or rather $\Diamond\Box(B)$):

$$
\begin{array}{ll}
(\neg \Rightarrow) & \dfrac{p \Rightarrow p}{p, \neg p \Rightarrow} \\[4pt]
(\Rightarrow W) & \dfrac{}{p, \neg p \Rightarrow \Box\neg\Box p} \\[4pt]
(\Rightarrow\to) & \dfrac{}{p \Rightarrow \neg p \to \Box\neg\Box p} \\[4pt]
(\Box \Rightarrow) & \dfrac{}{\Box p \Rightarrow \neg p \to \Box\neg\Box p} \\[4pt]
(\Rightarrow \Box) & \dfrac{}{\Box p \Rightarrow \Box(\neg p \to \Box\neg\Box p)} \\[4pt]
(\neg \Rightarrow) & \dfrac{}{\neg\Box(\neg p \to \Box\neg\Box p), \Box p \Rightarrow} \\[4pt]
(\Box \Rightarrow) & \dfrac{}{\Box\neg\Box(\neg p \to \Box\neg\Box p), \Box p \Rightarrow} \\[4pt]
(\Rightarrow \neg) & \dfrac{}{\Box\neg\Box(\neg p \to \Box\neg\Box p) \Rightarrow \neg\Box p} \\[4pt]
(\Rightarrow \Box) & \dfrac{}{\Box\neg\Box(\neg p \to \Box\neg\Box p) \Rightarrow \Box\neg\Box p} \\[4pt]
(W \Rightarrow) & \dfrac{}{\Box\neg\Box(\neg p \to \Box\neg\Box p), \neg p \Rightarrow \Box\neg\Box p} \\[4pt]
(\Rightarrow\to) & \dfrac{}{\Box\neg\Box(\neg p \to \Box\neg\Box p) \Rightarrow \neg p \to \Box\neg\Box p} \\[4pt]
(\Rightarrow \Box) & \dfrac{}{\Box\neg\Box(\neg p \to \Box\neg\Box p) \Rightarrow \Box(\neg p \to \Box\neg\Box p)} \\[4pt]
(\neg \Rightarrow) & \dfrac{}{\neg\Box(\neg p \to \Box\neg\Box p), \Box\neg\Box(\neg p \to \Box\neg\Box p) \Rightarrow} \\[4pt]
(\Box \Rightarrow) & \dfrac{}{\Box\neg\Box(\neg p \to \Box\neg\Box p), \Box\neg\Box(\neg p \to \Box\neg\Box p) \Rightarrow} \\[4pt]
(C \Rightarrow) & \dfrac{}{\Box\neg\Box(\neg p \to \Box\neg\Box p) \Rightarrow}
\end{array}
$$

Ohnishi and Matsumoto's result may be reasonably improved by replacing **S4** with **K45** which is axiomatically **S5** but without axiom T, or, in semantic terms, the logic of transitive and euclidean (but not reflexive) frames. It may be easily proved that

Theorem 4.5 $\vdash_{S5} \varphi$ *iff* $\vdash_{K45} \Box\varphi$

PROOF: $\Longleftarrow$ Again simple, by containment of **K45** in **S5** and T.

$\Longrightarrow$ This is harder and goes like a proof of theorem 4.4, by induction on the height of the proof in H-**S5**. Axioms K, 4, 5 and GR hold just by the application of GR and MP by K since both $\Box(\varphi \to \psi)$ and $\Box\varphi$ are provable in **K45** by the induction hypothesis. It remains to prove $\Box(\Box\varphi \to \psi)$ in **K45**. Note that both $\varphi \to (\Box\varphi \to \varphi)$ and $\neg\Box\varphi \to (\Box\varphi \to \varphi)$ are **CPL** theses. By GR and K we get $\Box\varphi \to \Box(\Box\varphi \to \varphi)$ and $\Box\neg\Box\varphi \to \Box(\Box\varphi \to \varphi)$ The latter by 5 reduces to $\neg\Box\varphi \to \Box(\Box\varphi \to \varphi)$, hence by **CPL** we obtain $\Box(\Box\varphi \to \varphi)$. $\square$

But what is good for **S5** in SC setting from this fact? Shvarts [240] provided LK for **K45** which is cut-free. It contains only one modal rule of the form

$$(\Rightarrow\Box)^{45} \quad \dfrac{\Gamma, \Box\Delta \Rightarrow \Box\Sigma, \varphi}{\Box\Gamma, \Box\Delta \Rightarrow \Box\Sigma, \Box\varphi}$$

with side condition that $\Box\Delta$ is nonempty. One may easily show either its derivability in H-**K45** (via I_G) or its validity-preservation. We take the second option

Lemma 4.9 $(\Rightarrow\Box)^{45}$ *is validity-preserving in transitive and euclidean models.*

PROOF: Assume that the premiss is valid but in some model and state w_1 all formulae in $\Box\Gamma$ and $\Box\Delta$ are true whereas $\Box\varphi$ and all formulae in $\Box\Sigma$ are false. So in some accessible w_2 φ is false and all members of Γ and $\Box\Delta$ true. Since the premiss is valid, there must be at least one formula in $\Box\Sigma$ which is true in w_2. Let it be $\Box\psi$; since it was false in w_1 there is some accessible state w_3 where ψ is false. Since by euclideaness w_2 is accessible from w_3, then $\Box\psi$ must be false in w_2 and we have a contradiction. $\Box$

Hence soundness follows. Completeness may be easily demonstrated by proving the axioms, which we leave to the reader.

Exercise 4.22 *Provide a proof of derivability of Shvarts' rule in H-**K45***.*
Prove K, 4 and 5 in LKK45. Note that using contraction may be necessary.

Goré [103] provided in tableau framework a simplification of Shvart's rule of the form

$$(\Rightarrow\Box)^{45}, \quad \frac{\Gamma, \Box\Gamma \Rightarrow \Box\Delta, \Box\varphi, \varphi}{\Box\Gamma \Rightarrow \Box\Delta, \Box\varphi}$$

now $\Box\Delta$ is not required to be nonempty, since $\Box\varphi$ is repeated in the premiss.

Exercise 4.23 *Prove syntactically or semantically that this rule is correct for **K45**.*

Shvarts is using LK and provides constructive proof of cut elimination for his SC. Goré instead provides a constructive completeness proof by saturation. We will provide a syntactic proof of cut elimination but for LK with Goré's rule since this simplifies a proof significantly. Only the case of reduction of the complexity of the cut-formula in case the cut-formula is boxed will be considered. The reader is invited to complete the proof. We apply the strategy Girard used in the proof of the mix reduction lemma (see subsection 2.4.2, and remark 2.2 in particular).

Let us consider the following application of Mix

$$(\Rightarrow\Box)\ \frac{\Gamma, \Box\Gamma \Rightarrow \Box\Delta, \Box\varphi^i, \varphi \qquad \varphi^k, \Box\varphi^k, \Pi, \Box\Pi \Rightarrow \Box\Sigma, \Box\psi, \psi}{(Mix)\ \frac{\Box\Gamma \Rightarrow \Box\Delta, \Box\varphi^i \qquad\qquad\qquad \Box\varphi^k, \Box\Pi \Rightarrow \Box\Sigma, \Box\psi}{\Box\Gamma, \Box\Pi \Rightarrow \Box\Delta, \Box\Sigma, \Box\psi}}\ (\Rightarrow\Box)$$

and this is replaced with

$$\frac{\dfrac{\dfrac{\Gamma, \Box\Gamma \Rightarrow \Box\Delta, \Box\varphi^i, \varphi \qquad \Box\varphi^k, \Box\Pi \Rightarrow \Box\Sigma, \Box\psi}{\Gamma, \Box\Gamma, \Box\Pi \Rightarrow \Box\Delta, \Box\Sigma, \Box\psi, \varphi} \qquad \dfrac{\Box\Gamma \Rightarrow \Box\Delta, \Box\varphi^i \qquad \varphi^k, \Box\varphi^k, \Pi, \Box\Pi \Rightarrow \Box\Sigma, \Box\psi, \psi}{\varphi, \Box\Gamma, \Pi, \Box\Pi \Rightarrow \Box\Delta, \Box\Sigma, \Box\psi, \psi}(Mix)}{\dfrac{\Gamma, \Box\Gamma, \Box\Pi, \Box\Gamma_\varphi, \Pi_\varphi, \Box\Pi_\varphi \Rightarrow \Box\Delta_\varphi, \Box\Sigma_\varphi, \Box\Delta, \Box\Sigma, \Box\psi, \Box\psi, \psi}{\dfrac{\Gamma, \Box\Gamma, \Pi, \Box\Pi \Rightarrow \Box\Delta, \Box\Sigma, \Box\psi, \psi}{\Box\Gamma, \Box\Pi \Rightarrow \Box\Delta, \Box\Sigma, \Box\psi}(\Rightarrow\Box)}(C)(P)(W)}(Mix)}{}$$

where two uppermost mixes are eliminable by the induction hypothesis. The last one is not eliminable but it has lower degree.

Exercise 4.24 *Complete the proof of cut elimination. Try different strategies.*

What we gain now is the possibility of devising a decision procedure for **S5** based on cut-free proof system for **K45** by using G3 with Goré's rule (possibly in the weakening-absorbing version) as the only modal rule and starting always with $\Box\varphi$ at the root. But it appears that we can do even better. We finish this section with, perhaps a bit surprising result concerning the eliminability of TE (see subsection 3.7.1), in a variant of standard SC. Fitting on the ground of tableau calculus provided an even simpler solution which may be easily transferred to SC. In [83] he introduced a tableau system for **S5** with an extra rule for the addition of $\Box$ to formulae on the branch. It is a consequence of his completeness proof that it is sufficient to apply such a rule only once, at the very beginning. This observation is then exploited in [86], for two variants of this system differing in the definition of modalised formula. In both the cases, there is no extra rule for addition of $\Box$ but only a requirement that when building a tableau for φ we start with $\Box\varphi$ (assumed to be false). Fitting's solution may be easily simulated in G3S5. One may use the following contraction-absorbing versions (and weakening-absorbing in the second case) versions of Ohnishi's and Matsumoto's rules:

$$(\Box\Rightarrow)\ \ \frac{\varphi,\Box\varphi,\Gamma\Rightarrow\Delta}{\Box\varphi,\Gamma\Rightarrow\Delta} \qquad\qquad (\Rightarrow\Box)\ \ \frac{\Box\Gamma\Rightarrow\Box\Delta,\Box\varphi,\varphi}{\Pi,\Box\Gamma\Rightarrow\Box\Delta,\Sigma,\Box\varphi}$$

For such a system, we can prove the following weak adequacy result

Theorem 4.6 $\vdash\varphi$ *iff* $\vdash\Rightarrow\Box\varphi$

Soundness is obvious. Two different semantic proofs of completeness may be found in Fitting and in Goré [103]. We provide a different route to show its adequacy. First, note that all preliminary results concerning invertibility of rules and admissibility of weakening and contraction hold also for this version of G3S5. One may easily check that the addition of modal rules in this form does not destroy the proofs of any of these results.

As for completeness, one may rigorously demonstrate that every tableau proof of $\Box\varphi$ (i.e. a closed tableau starting with $\neg\Box\varphi$) may be translated into a proof of $\Rightarrow\Box\varphi$ in our system. Hence, the semantic completeness proof of Fitting applies to this sequent system as well. This will show that we obtain a cut-free SC for **S5** but we want more: a constructive syntactical proof of its admissibility. Hence, for the moment, we assume that cut is admissible and we prove syntactically

Theorem 4.7 $\vdash_H\varphi$ *then* $\vdash\Rightarrow\Box\varphi$

where $\vdash_H$ denotes provability in axiomatic system for **S5**.

PROOF: It is easy to provide proofs for all (boxed) axioms, in particular a proof of (boxed) axiom B looks like that

$$
\begin{array}{c}
(\neg \Rightarrow) \dfrac{p, \Box p \Rightarrow p, \Box(\neg p \to \Box \neg \Box p), \Box \neg \Box p}{p, \neg p, \Box p \Rightarrow \Box(\neg p \to \Box \neg \Box p), \Box \neg \Box p, \Box \neg \Box p} \\[4pt]
(\Rightarrow \to) \dfrac{\rule{0pt}{0pt}}{p, \Box p \Rightarrow \Box(\neg p \to \Box \neg \Box p), \Box \neg \Box p, \neg p \to \Box \neg \Box p} \\[4pt]
(\Box \Rightarrow) \dfrac{\rule{0pt}{0pt}}{\Box p \Rightarrow \Box(\neg p \to \Box \neg \Box p), \Box \neg \Box p, \neg p \to \Box \neg \Box p} \\[4pt]
(\Rightarrow \Box) \dfrac{\rule{0pt}{0pt}}{\Box p \Rightarrow \Box(\neg p \to \Box \neg \Box p), \Box \neg \Box p} \\[4pt]
(\Rightarrow \neg) \dfrac{\rule{0pt}{0pt}}{\Rightarrow \Box(\neg p \to \Box \neg \Box p), \Box \neg \Box p, \neg p} \\[4pt]
(\Rightarrow \Box) \dfrac{\rule{0pt}{0pt}}{\neg p \Rightarrow \Box(\neg p \to \Box \neg \Box p), \Box \neg \Box p} \\[4pt]
(\Rightarrow \to) \dfrac{\rule{0pt}{0pt}}{\Rightarrow \Box(\neg p \to \Box \neg \Box p), \neg p \to \Box \neg \Box p} \\[4pt]
(\Rightarrow \Box) \dfrac{\rule{0pt}{0pt}}{\Rightarrow \Box(\neg p \to \Box \neg \Box p)}
\end{array}
$$

We must of course also demonstrate how applications of GR and MP are simulated on (boxed) theses. For GR, we have

$$
(\Rightarrow W) \dfrac{\Rightarrow \Box \varphi}{\Rightarrow \Box \varphi, \Box \Box \varphi} \\[4pt]
(\Rightarrow \Box) \dfrac{\rule{0pt}{0pt}}{\Rightarrow \Box \Box \varphi}
$$

For MP we have:

$$
\dfrac{\dfrac{\Box(\varphi \to \psi), \Box \varphi, \varphi \Rightarrow \varphi, \Box \psi \qquad \Box(\varphi \to \psi), \Box \varphi, \psi \Rightarrow \psi, \Box \psi}{\dfrac{\varphi \to \psi, \Box(\varphi \to \psi), \Box \varphi, \varphi \Rightarrow \psi, \Box \psi}{\dfrac{\Box(\varphi \to \psi), \Box \varphi, \varphi \Rightarrow \psi, \Box \psi}{\dfrac{\Box(\varphi \to \psi), \Box \varphi \Rightarrow \psi, \Box \psi}{\Box(\varphi \to \psi), \Box \varphi \Rightarrow \Box \psi} \, (\Rightarrow \Box)} \, (\Box \Rightarrow)} \, (\Box \Rightarrow)} \, (\to \Rightarrow)}{}
$$

$$
\dfrac{\Rightarrow \Box(\varphi \to \psi) \qquad \dfrac{\Box(\varphi \to \psi), \Box \varphi \Rightarrow \Box \psi}{\Box \varphi \Rightarrow \Box \psi} \, (Cut)}{\Rightarrow \Box \psi}
$$

Hence, the theorem follows by induction on the height of axiomatic proofs.
□

Now we consider this SC without cut. Although there are many similarities with the preceding result concerning the equivalence with **K45**, there is one important difference. In SC for **K45** there was no $(\Box \Rightarrow)$, and as we noticed cut is not eliminable if the left premiss is generated by $(\Rightarrow \Box)$ whereas the right premiss is obtained by $(\Box \Rightarrow)$. Now both rules are again present and we cannot proceed as in the proof for Shvarts' system. However, for this system we can prove

Theorem 4.8 (Admissibility of TE). *For any φ, Γ, Δ, if $\vdash \varphi \to \varphi, \Gamma \Rightarrow \Delta$, then $\vdash \Gamma \Rightarrow \Delta$.*

PROOF: The structure of the proof is the same as in subsection 3.7.1. Again, we prove that $\vdash \Gamma \Rightarrow \Delta$ by induction on the complexity of φ and utilising the fact that by invertibility from (a) $\vdash \varphi \to \varphi, \Gamma \Rightarrow \Delta$ we get (b) $\vdash \Gamma \Rightarrow \Delta, \varphi$ and (c) $\vdash \varphi, \Gamma \Rightarrow \Delta$. Also, on this basis, we perform subsidiary induction on the height of $\vdash \varphi, \Gamma \Rightarrow \Delta$. Now additionally, in the induction step we must consider modal rules. For $(\Box \Rightarrow)$ the proof is trivial like for Boolean rules. In case of $(\Rightarrow \Box)$ φ may be introduced only as a part of weakening of the antecedent, i.e. we have

$\varphi, \Gamma', \Box\Pi \Rightarrow \Delta', \Box\Sigma, \Box\psi$ (where $\Gamma = \Gamma', \Box\Pi$ and $\Delta = \Delta', \Box\Sigma, \Box\psi$) deduced from $\Box\Pi \Rightarrow \Box\Sigma, \Box\psi, \psi$ and it is enough to deduce $\Gamma', \Box\Pi \Rightarrow \Delta', \Box\Sigma, \Box\psi$ by the same rule.

Induction step: Again we assume as the induction hypothesis that the lemma holds for all formulae of lower complexity than φ. The proof goes by cases. For all boolean formulae, it is the same (i.e. based on the invertibility of respective rules). In case of $\varphi := \Box\psi$ we prove the claim: If $\vdash \varphi, \Gamma \Rightarrow \Delta$, then $\vdash \Gamma \Rightarrow \Delta$ by a subsidiary induction on the height of (c).

The basis is trivial since, if $\vdash \varphi, \Gamma \Rightarrow \Delta$ is an axiom, then $\vdash \Gamma \Rightarrow \Delta$ is an axiom too. In the inductive step, the cases of parametric φ in the application of boolean rules and both $(\Rightarrow \Box)$ rules are also trivial—we just delete φ by the induction hypothesis and apply a suitable rule. The only interesting case is when φ is principal in the application of $(\Box \Rightarrow)$. The premiss is $\Box\psi, \psi, \Gamma \Rightarrow \Delta$ and by the induction hypothesis, we obtain (d) $\psi, \Gamma \Rightarrow \Delta$. In order to deduce $\Gamma \Rightarrow \Delta$ we must additionally consider (b). Now we prove the claim: If $\vdash \Gamma \Rightarrow \Delta, \varphi$, then $\vdash \Gamma \Rightarrow \Delta$, by subsidiary induction on the height of (b).

Again it is straightforward for the case of an axiom and φ parametric in boolean and all modal rules. We must consider the case of φ principal in both $(\Rightarrow \Box)$-rules. We have $\Gamma', \Box\Pi \Rightarrow \Box\Sigma, \Delta', \Box\psi$ (where $\Gamma = \Gamma', \Box\Pi$ and $\Delta = \Box\Sigma, \Delta'$) deduced from $\Box\Pi \Rightarrow \Box\Sigma, \Box\psi, \psi$. We use this premiss and (d) to obtain the following proof with IH denoting the application of the induction hypothesis

$$IH \cfrac{\cfrac{\Box\Pi \Rightarrow \Box\Sigma, \Box\psi, \psi}{\Gamma \Rightarrow \Delta, \psi} \qquad \psi, \Gamma \Rightarrow \Delta}{(\rightarrow\Rightarrow) \cfrac{\psi \rightarrow \psi, \Gamma \Rightarrow \Delta}{IH \cfrac{}{\Gamma \Rightarrow \Delta}}}$$

This completes the proof. □

A specific feature of this proof, when compared with the proof for classical logic, is the fact that we must perform two subsidiary inductions on the height of both (b) and (c) in modal case. By the extension of TE admissibility to G3 for **S5**, we obtain (indirect) proof of cut admissibility for this calculus.

4.6 Generalised SC

As we could see, standard SC is too weak for providing cut-free, or at least analytic formalisation of **S5** without bringing some additional features into the market. Even if we can find some (maybe even all) solutions described in the previous section quite satisfactory, it appeared that many other modal logics which are semantically characterised by frames with symmetric or euclidean (or other troublesome) accessibility relations are hard to deal with in such a way. This was the main motivation for the generalisation of the standard SC. A complete picture of the proposed systems is far beyond this survey. In what follows we focus only on a few approaches illustrated with some systems developed for **S5**. A more detailed description of other generalised SC and for other modal logics one should consult Wansing [270] and Poggiolesi [198].

4.6.1 Labelled Sequent Calculi

One of the most important generalised approaches to formalisation of modal logics is based on the application of labels. In fact, this technique is not only connected with modal logics but has a really wide scope of application in several branches of logic. Generally, labels are very handy if we deal with information that has complex structure, especially when different sorts of data require different forms of processing. For example, labels may be used to represent: the set of assumptions for a formula (e.g. Anderson and Belnap [4], ND-systems for relevant logics), truth values or the sets of truth values for a formula (e.g. Carnielli [44] or Hähnle [106], tableau systems for many-valued logics), fuzzy reliability value n ($0 \leq n \leq 1$) used mainly in expert systems, and many more[10].

In modal logic labels extend a language with a representation of states in a model and their addition strengthens considerably the flexibility of expression. To understand why, notice that there is an asymmetry between the local perspective of relational semantics and the global perspective of standard modal language. States in a model, essential in relational semantics, are not represented in modal syntax. In particular, standard modal languages have no mechanisms for naming particular states in a model, asserting or denying the equality of states, talking about accessibility of one state from another. In consequence, they do not provide an adequate representation of many semantic features and this generates problems with developing a suitable modal proof theory.

Labels provide a remedy for the problem of discrepancy between a syntax and a semantics but there are a lot of possible solutions. Blackburn [32] distinguishes between the internalised and the external approach. The former consists of an enrichment of the object language obtained via sorting (of the atoms) and addition of the new operators and/or modalities. It is the way of doing hybrid logic, where labels are part of the language[11].

In the external approach, labels are just an additional technical apparatus. Even in this group, we can find a variety of different solutions, according to the strength of semantical commitment expressed by labels. We can talk about

1. Weak labelling—labels as a very limited technical device supporting proof construction.

2. Strong labelling—system of labels as an exact representation of an attempted falsifying model.

---

[10]Dov Gabbay [91] considered several applications on different fields in his general theory of LDS's (labelled deductive systems).

[11]Logics of this sort and some of their SC formalisations will be dealt with in the second volume.

3. Medium labelling—with no special calculus for labels but still sufficient for the construction of falsifying modelss, e.g. Fitting's [83] prefixed tableau calculi for modal logics or single-step tableaux of Massacci [173, 174], explored by Goré [103] under the name explicit systems.

There are a lot of deductive systems where labels of several sorts are applied in a very limited way. Probably the earliest one is SC of Kanger [146] for **S5**, where labels are linked only to propositional variables. For temporal logics tableau systems of Rescher and Urquhart [213] and of Marx, Mikulas, Reynolds [171], Mints' [182], indexed SC for modal logics, as well as multisequent calculi for temporal logics of Indrzejczak [127], belong to this group. Despite different rules and interpretations of labels in all these systems labels play only a supporting role in a deduction, separating some parts of a derivation. Their motivation is in fact semantical but only a very modest part of interpretation is involved. The apparatus of labels in itself is too weak in these approaches to help building a falsifying model, in contrast to medium and strong labelling, where we can directly extract a model from labels.

The opposite solutions are represented by strongly labelled systems in Gabbay's tradition [91]. In such an approach labels save as much as possible from suitable semantics, but in contrast to the internalised approach, labels are not part of a language. Instead, we have a composition of two languages: an object language of a logic and a language of the algebra of labels. In deductive systems of this sort in addition to rules for logical constants we have rules governing the behaviour of labels, and usually, some rules which correlate both levels.

Such an approach is very popular. One may find several simplified variants of easy to use tableau systems of this kind for many non-classical logics, including modal ones, in several textbooks (cf. e.g. books of Girle [100] and of Priest [205]). There are also more theoretically oriented works investigating labelled ND-systems and SCs for modal logics, e.g. Russo [222] and Basin, Mathews and Vigano [21] (also for other non-classical logics in [22] and [274])[12], Castellini and Smaill [46, 47], Negri [187] and Negri and von Plato [186].

The last solution, i.e the addition of the machinery of prefixes to formulae, due to Fitting[13] [83], is situated between the extrema of weak and strong labelling and has a lot of advantages. It is quite simple and natural since the technical apparatus is kept in reasonable bounds. Fitting's labels (prefixes) are not as direct a way of encoding semantics as strong labels or the internalised approach, but they still may be easily used for the construction of falsifying models. Moreover, this approach is quite extensive—Fitting's original systems cover a lot of modal logics, Massacci's version formalises even more. So, in contrast to strongly labelled systems, we obtain quite satisfactory results with the help of a relatively modest apparatus. In this

---

[12]One may mention also ND-systems with strong labelling for other kinds of temporal logics not discussed in this book, like **PLTL** or **CTL** provided by Renteria and Hausler [211] or by Bolotov, Basukoski, Grigoriev and Shangin in [35, 36].

[13]Fitting [82], in fact, refers to earlier note of Fitch [80], as a source of inspiration.

approach, there are no operations on labels made in an extra language like in Gabbay style systems—labels are always connected with formulae of the object language. Intuitively $\sigma : \varphi$ means that φ is satisfied at a state of a model denoted by prefix σ. It is important that Fitting-style labels have a structure of their own, which helps to build a model using them as building blocks. It is possible because each label is not only the name of a state in a model, but its structure encodes the place of this state (via an accessibility relation) in a falsifying model we are attempting to build. In general, prefixes are finite sequences of natural numbers, separated with dots, with 1 as the first digit. So 1 is the name of the root of a model, 1.1, 1.2, are names of two worlds accessible from 1, etc. Such prefixes can take more complex shape and contain additional semantic information, as for example, in Indrzejczak [130].

4.6.2   Simple Labelled System

On the other hand, in case of **S5**, the notion of a prefix may be simplified remarkably, if we refer to semantics without the accessibility relation. It is enough to use just natural numbers as names of the worlds. This solution was first applied by Kanger [146], but with prefixes attached only to atomic formulae. Tableaux or SC of this kind can be found in many places, e.g. Fitting [83] or Wansing [269].

The construction of such a system is extremely easy. Just add prefixes (natural numbers) to every formula. In case of boolean rules, it is important that side and principal formulae must have the same label which means that this inference is carried in one state denoted by this label. For example, the rules for conjunction are

$$(\wedge\Rightarrow) \ \frac{k : \varphi, k : \psi, \Gamma \Rightarrow \Delta}{k : \varphi \wedge \psi, \Gamma \Rightarrow \Delta} \qquad\qquad (\Rightarrow\wedge) \ \frac{\Gamma \Rightarrow \Delta, k : \varphi \quad \Gamma \Rightarrow \Delta, k : \psi}{\Gamma \Rightarrow \Delta, k : \varphi \wedge \psi}$$

Of course, Γ, Δ consist of labelled formulae and labels may be different than label k of active formulae. Rules for modal formulae are

$$(\Box\Rightarrow) \ \frac{i : \Box\varphi, k : \varphi, \Gamma \Rightarrow \Delta}{i : \Box\varphi, \Gamma \Rightarrow \Delta} \qquad\qquad (\Rightarrow\Box) \ \frac{\Box\Gamma \Rightarrow k : \varphi}{\Pi, \Box\Gamma \Rightarrow \Sigma, i : \Box\varphi}$$

where k is new in $(\Rightarrow\Box)$ but in $(\Box\Rightarrow)$ it may be any prefix, including $i = k$. Informally, in terms of root-first proof search, it means that if $\Box\varphi$ is false in i, then there is some k where φ is false. We must take a new one since we cannot assume that it is one of the states we considered so far. On the other hand, if $\Box\varphi$ is true in i it must be true in any state of the attempted model, including i. A proof of B is straightforward

$$(\neg \Rightarrow)\, \dfrac{2:\Box p, 1:p \Rightarrow 1:p}{}$$
$$(\Box \Rightarrow)\, \dfrac{2:\Box p, 1:p, 1:\neg p \Rightarrow}{2:\Box p, 1:\neg p \Rightarrow}$$
$$(\Rightarrow \neg)\, \dfrac{}{1:\neg p \Rightarrow 2:\neg\Box p}$$
$$(\Rightarrow \Box)\, \dfrac{}{1:\neg p \Rightarrow 1:\Box\neg\Box p}$$
$$(\Rightarrow\rightarrow)\, \dfrac{}{\Rightarrow 1:\neg p \rightarrow \Box\neg\Box p}$$

The soundness proof is easy. Let us say that a labelled sequent is valid iff for every **S5** model, we can provide an interpretation such that labels are mapped onto states and at least one formula in the antecedent is false in a state correlated with its label, or at least one in the succedent is true in its correlated state (under this mapping). Accordingly, a sequent is falsified iff there is a model and a mapping of labels such that all formulae in the antecedent are true in corresponding states, and all formulae in the succedent are false in corresponding states.

Exercise 4.25 *Show that modal rules are validity-preserving.*

We can provide a proof of the admissibility of cut for LK or G3 with labels (see e.g. Wansing [269]), but it needs some additional technical results concerning substitution (of labels) which it is more natural to discuss first in the context of first-order logic. So we only provide a sketch of the completeness proof based on our proof search procedure. This time we define a set of labelled saturated pairs in the following way:

Definition 4.9 *Let* (Γ, Δ) *be an ordered pair of labelled formulae and* X *the set of all occuring labels. It is downward saturated iff for every* k *in* X *it satisfies the following conditions:*

1. if $k : \neg\varphi \in \Gamma$, then $k : \varphi \in \Delta$
2. if $k : \neg\varphi \in \Delta$, then $k : \varphi \in \Gamma$
3. if $k : \varphi \wedge \psi \in \Gamma$, then $k : \varphi \in \Gamma$ and $k : \psi \in \Gamma$
4. if $k : \varphi \wedge \psi \in \Delta$, then $k : \varphi \in \Delta$ or $k : \psi \in \Delta$
5. if $k : \varphi \vee \psi \in \Gamma$, then $k : \varphi \in \Gamma$ or $k : \psi \in \Gamma$
6. if $k : \varphi \vee \psi \in \Delta$, then $k : \varphi \in \Delta$ and $k : \psi \in \Delta$
7. if $k : \varphi \rightarrow \psi \in \Gamma$, then $k : \varphi \in \Delta$ or $k : \psi \in \Gamma$
8. if $k : \varphi \rightarrow \psi \in \Delta$, then $k : \varphi \in \Gamma$ and $k : \psi \in \Delta$
9. if $k : \Box\varphi \in \Delta$, then $i : \varphi \in \Delta$ for some $i \in X$
10. if $k : \Box\varphi \in \Gamma$, then $i : \varphi \in \Gamma$ for all $i \in X$.

Moreover, it is a Hintikka labelled pair iff it is consistent, i.e. there is no $k : \varphi \in \Gamma \cap \Delta$.

As one may expect we define a model for labelled Hintikka pairs taking X to be its domain and $V(p) = \{k \in X : k : p \in \Gamma\}$ for any atom.

Exercise 4.26 *Prove the truth lemma for this construction.*

Proving completeness is routine if we can show that for every unprovable sequent $\Gamma \Rightarrow \Delta$ we can obtain a Hintikka pair (Π, Σ) such that $\Gamma \subseteq \Pi$ and $\Delta \subseteq \Sigma$. Clearly, $\Gamma \Rightarrow \Delta$ is falsified in state 1. It may be done by defining an exhaustive and fair procedure of proof search or by the direct method (see section 1.7), Let us consider the former approach in order to compare it with the proof-search procedures for G3T and G3S4 described above. We can apply our proof-search procedure basically as in the case of nonlabelled G3, by taking care of the saturation of the set of formulae with the same label before we apply $(\Rightarrow \Box)$ and introduce a new one. But now immediately after the introduction of a new label to the sequent, we must apply $(\Box \Rightarrow)$ to all $\Box$-formulae in the antecedent with respect to this new label. Of course, the same formula $\Box\varphi$ may occur with different labels in the antecedent so it is enough to apply the respective rule only to one of them to avoid multiple occurrences. Moreover, by the end of saturation of this new label, we must apply $(\Box \Rightarrow)$ to new box-formulae with this label also with respect to all labels already present on the branch. This, of course, forces us to check again if respective sets of formulae are saturated. We omit the details of description of a fair proof-search procedure; one can consult Fitting [83] or Goré [103].

Note that, similarly as in the case of G3S4, we can run into an infinite branch while applying this procedure. If we only want to prove completeness it is not a problem, but if we want to have decidability, we must do a little more. Let us analyse the source of the problem. Due to the subformula property of all boolean rules and $(\Box \Rightarrow)$ with respect to single label the set of formulae with the same label must be finite. But because of the necessity of application of $(\Box \Rightarrow)$ with respect to every introduced label it is possible that new boxed formulae in the succedent will appear and generate new labels. Since the number of distinct sets of formulae generated from the root sequent is finite, we run a loop. Hence, if we want to have a terminating procedure to obtain finite models we must apply loop control as in the case of **S4**. But things here are harder. In case of **S4**, it would be enough to stop once and for all if the set of saturated formulae with prefix k is included in the set of saturated formulae with prefix i which was saturated earlier. In case of **S5**, however, such simple form of loop control does not work.

First, it is not sufficient that the set of formulae with label k is a subset of some earlier set labelled by i—they must contain the same set of formulae. Then we say that k is blocked by i and we can identify both sets. Second, because of the possibility of enlargement of some (blocked or blocking) set in later stages of proof search, we must admit that blocking will be broken. So, in contrast to G3S4, here we must assume that blocking is provisory and may be broken if one of the blocking set changes. This solution is similar to the technique of 'dynamic blocking', devised by Horrocks [121], for the implementation of tableau system for some versions of description logics.

Despite the complications that must be taken into account for the construction of a procedure of exhaustive proof search this kind of SC is very handy in practice. In particular, the nice thing is that we do not need to introduce a generalisation of ($\Rightarrow \Box$) and disjunctive branching since all boxed formulae in the succedent are examined one by one on the same branch.

4.7 Hypersequent Calculus

Hypersequent calculi (HC) are amongst the most interesting ones in the family of generalised SC which we called the many-sequent approach in subsection 3.5.2. The name derives from the fact that rules are defined on items called hypersequents by Avron [7]. In general, hypersequents are some finite collections of ordinary sequents. As in the case of ordinary SC, some way of understanding this collection is significant. Commonly hypersequents are treated as sets or multisets of sequents, but it is also possible to interpret them as sequences of sequents. All these choices have a great influence on the character of structural rules necessary for the development of the system but also on the scope of its application and possibility of using several proof strategies, in particular, for proving admissibility of cut and other rules.

It is commonly believed that HC was originally introduced for few modal logics in a short abstract by Pottinger [202]. However, this information should be revised since a similar idea was earlier introduced by Mints in [177] and [179], to formalise **S5**. Unfortunately, these papers were written in Russian and unknown to the wider community. Even much later, when an English translation of [179], was presented in Mints [181], he did not care to underline his priority in this respect[14]. But it was Avron [7], who not only independently introduced such kind of SC but developed its theory, first for relevant, then for many other non-classical logics (see, e.g. [10]).

In fact, HC may be seen as a special simplified case of Došen's [60], a more general framework where one is dealing with a hierarchy of sequents of order $n+1$ with arguments being finite sets of sequents of order n. In particular, sequents of order 2 consist of finite sets of ordinary sequents (of order 1) on both sides, where elements of the antecedent are treated conjunctively, and elements of the succedent disjunctively. In this perspective, hypersequents are just sequents of order 2 with empty antecedents.

Došen's general approach may be seen also as a general conceptual framework for other cases of many sequent systems. In particular, all calculi based on embedding of sequents inside other sequents which were later exploited under different names (e.g. deep inference, nested sequents, tree hypersequents) by Bull [42], Kashima [147], Stouppa [246], Brünnler [41], Poggiolesi [198]. This shows a deep relationship between these approaches. In particular, if hypersequents are defined not as sets or multisets of sequents, but rather as their sequences, then

---

[14]In fact both the notation and applied terminology is different; hypersequents are called tableaux by Mints in [181], and he rather stressed the similarity of his system to Kripke's tableaux.

HC may be interpreted as a restricted version of nested sequent calculi, called by Lellmann [161] linear nested SC and by Indrzejczak [136] non-commutative HC.

HC is also related to other kinds of generalised SC. In particular, it may be seen as a special kind of display calculus (DC), introduced by Belnap [25] (see also [53]). While in DC a family of structural connectives of fixed arity is introduced, in HC a separator of sequents may be treated as the only added structural connective of nonfixed arity. One may find results concerning embeddings of HC in DC in Wansing [269] and Ramanayake [209]. It is also easy to establish a similar relationship of HC to the much stronger framework of labelled sequent calculi (see Negri [187]).

It seems that HC, when compared to other generalised SC, has rather simple form, anyway it increases significantly the expressive power of ordinary SC by allowing the additional transfer of information between different sequents. It proved to be very useful for the construction of cut-free formalisations of many non-classical logics including modal, temporal, many-valued, relevant, paraconsistent and fuzzy logics (see for example Avron [7, 10], Baaz, Ciabattoni and Fermüller [17], Metcalfe, Olivetti and Gabbay [176]). In the field of modal logics there are surprisingly many different cut-free systems for **S5** (Mints [177], Pottinger [202], Avron [10], Restall [215], Poggiolesi [198], Lahav [156], Kurokawa [154], Bednarska, and Indrzejczak [24]). For other modal logics the situation is worse. One can find case studies of some logics of linear frames; there are HC for **S4.3** (Indrzejczak [132], Kurokawa [154]), later generalised to **K4.3** and **KD4.3** (Indrzejczak [135]). Kurokawa [154] provided also HC for **K4.2**, many temporal logics of linear frames were formalised in Indrzejczak [136]. Recently some more general approaches were provided: Lahav [156] proposed a uniform treatment of various normal modal logics based on the translation of semantic conditions. Some general approach of a different character is developed by Lellmann [160].

In what follows we focus only on solutions provided for **S5**. Despite the variety of systems, we can divide them into two types. An approach of Mints and Pottinger, characteristic also for Restall's and Poggiolesi's system, consists in providing special modal rules introducing □ and characterising **S5** on the basis of HC for **CPL**. On the other hand, the approach of Avron, characteristic also for Kurokawa and Lahav, builds the system for **S5** by means of a special quasi-structural rule added to HC system which is already equipped with modal rules introducing □ that are adequate for **S4**. In general, we could say that the former approach is more semantically oriented, whereas the latter approach is more syntactically oriented, but this is rather an oversimplification. For example, Restall's system has also syntactical character manifested in using a special structural rule, and Poggiolesi's system, despite semantic motivation for modal rules, is presented in strictly proof-theoretic style. On the other hand, Lahav's characteristic rule is obtained by means of translation from semantic condition and only semantic proofs of completeness for cut-free HC are provided.

4.7.1 The Basic HC

Before we present a survey of these approaches and several methods of proving cut admissibility for them, we need some conventions and a description of the basic HC for **CPL**. We define hypersequents as finite multisets of ordinary Gentzen sequents. The following notation will be applied:

- $\Gamma_1 \Rightarrow \Delta_1 \mid ... \mid \Gamma_n \Rightarrow \Delta_n$ stand for (n-element) hypersequents.

- $S \mid G$ (or $\Gamma \Rightarrow \Delta \mid G$) stand for hypersequents with displayed sequent S ($\Gamma \Rightarrow \Delta$); hence, in particular, in the schemata of rules, G, H are used to denote the (external) context, i.e. the remaining (possibly empty) multisets of sequents.

- G, H alone can be also used for representing hypersequents; in this case, it is assumed that they are nonempty.

- $S \in G$ means that S is an element (possibly the only one) of G.

How to interpret hypersequents? In general, hypersequents are interpreted as metalevel disjunctions but details vary depending on the kind of logics which are formalised this way. In syntactical terms we can express the meaning of a hypersequent in **S5** (and other modal logics) by means of the following translation:

$$\Im(\Gamma_1 \Rightarrow \Delta_1 \mid ... \mid \Gamma_n \Rightarrow \Delta_n) = \Box I(\Gamma_1 \Rightarrow \Delta_1) \vee ... \vee \Box I(\Gamma_n \Rightarrow \Delta_n)$$

where I may be any of the possible translations for sequents considered in section 3.4. Alternatively, we can extend suitable semantic notions to hypersequents in the following way:,

- $\models G$ (G is valid) iff $\mathfrak{M} \models G$ for all models $\mathfrak{M}$

- $\mathfrak{M} \models G$ iff there is some $S \in G$ such that $\mathfrak{M} \models S$

- $\mathfrak{M} \models S$ iff $\mathfrak{M}, w \models S$, for all w in the domain of $\mathfrak{M}$

- $\mathfrak{M}, w \models \Gamma \Rightarrow \Delta$ iff $\mathfrak{M}, w \models \wedge\Gamma \rightarrow \vee\Delta$.

Note, that as a consequence we have

$\not\models G$ iff there is some $\mathfrak{M}$ such that $\mathfrak{M} \not\models G$, and $\mathfrak{M} \not\models G$ iff for all $S \in G, \mathfrak{M} \not\models S$ which eventually means that for every $S \in G$ there is some w such that $\mathfrak{M}, w \not\models S$.

HC for **CPL** can be introduced as a generalisation of LK or G3. In the latter option the basic HG3 consists of generalised atomic axioms $p, \Gamma \Rightarrow \Delta, p \mid G$ and the following logical rules:

$$(\neg\Rightarrow) \quad \frac{\Gamma \Rightarrow \Delta, \varphi \mid G}{\neg\varphi, \Gamma \Rightarrow \Delta \mid G} \qquad\qquad (\Rightarrow\neg) \quad \frac{\varphi, \Gamma \Rightarrow \Delta \mid G}{\Gamma \Rightarrow \Delta, \neg\varphi \mid G}$$

$$(\wedge\Rightarrow) \quad \frac{\varphi, \psi, \Gamma \Rightarrow \Delta \mid G}{\varphi \wedge \psi, \Gamma \Rightarrow \Delta \mid G} \qquad\qquad (\Rightarrow\wedge) \quad \frac{\Gamma \Rightarrow \Delta, \varphi \mid G \qquad \Gamma \Rightarrow \Delta, \psi \mid G}{\Gamma \Rightarrow \Delta, \varphi \wedge \psi \mid G}$$

$$(\Rightarrow\vee) \quad \frac{\Gamma \Rightarrow \Delta, \varphi, \psi \mid G}{\Gamma \Rightarrow \Delta, \varphi \vee \psi \mid G} \qquad\qquad (\vee\Rightarrow) \quad \frac{\varphi, \Gamma \Rightarrow \Delta \mid G \qquad \psi, \Gamma \Rightarrow \Delta \mid G}{\varphi \vee \psi, \Gamma \Rightarrow \Delta \mid G}$$

$$(\Rightarrow\rightarrow) \quad \frac{\varphi, \Gamma \Rightarrow \Delta, \psi \mid G}{\Gamma \Rightarrow \Delta, \varphi \rightarrow \psi \mid G} \qquad\qquad (\rightarrow\Rightarrow) \quad \frac{\Gamma \Rightarrow \Delta, \varphi \mid G \qquad \psi, \Gamma \Rightarrow \Delta \mid G}{\varphi \rightarrow \psi, \Gamma \Rightarrow \Delta \mid G}$$

A proof of a hypersequent G in HC is defined in the usual way as a tree of hypersequents with G as the root and axioms as leaves. On the basis of the specified interpretation, one may easily establish

Lemma 4.10 *All rules of HC are validity-preserving in* **CPL**.

Exercise 4.27 *Prove the above lemma and soundness for HG3 either directly in semantical terms or by using translation* $\mathfrak{S}$.

All rules satisfy the subformula property as in the standard SC. In some cases (e.g. Poggiolesi [198]) this basis, after addition of logical rules for $\Box$, is sufficient for adequacy but in general structural rules are required, in particular if the basic HC is a version of LK (call it HLK). In this case we can use axioms of the form $\varphi \Rightarrow \varphi \mid G$, hypersequent versions of logical rules as stated for LK, and hypersequent generalisations of ordinary W and C

$$(IW \Rightarrow) \quad \frac{\Gamma \Rightarrow \Delta \mid G}{\varphi, \Gamma \Rightarrow \Delta \mid G} \qquad (\Rightarrow IW) \quad \frac{\Gamma \Rightarrow \Delta \mid G}{\Gamma \Rightarrow \Delta, \varphi \mid G}$$

$$(IC \Rightarrow) \quad \frac{\varphi, \varphi, \Gamma \Rightarrow \Delta \mid G}{\varphi, \Gamma \Rightarrow \Delta \mid G} \qquad (\Rightarrow IC) \quad \frac{\Gamma \Rightarrow \Delta, \varphi, \varphi \mid G}{\Gamma \Rightarrow \Delta, \varphi \mid G}$$

But this is not the whole story. The above rules may be called internal structural rules (hence names IW, IC) but we can also consider external structural rules operating on the whole sequents. The most important external structural rules are generalisations of C and W

$$(EC) \quad \frac{\Gamma \Rightarrow \Delta \mid \Gamma \Rightarrow \Delta \mid G}{\Gamma \Rightarrow \Delta \mid G} \qquad (EW) \quad \frac{G}{\Gamma \Rightarrow \Delta \mid G}$$

Note that if EW is primitive in HLK we can use simple axioms $\varphi \Rightarrow \varphi$.

In fact, other solutions are also possible. In particular, Restall's system [215] differs slightly in the selection of structural rules. Instead of (EC) he applies the special rule of merging and both external and internal weakenings are combined into a specific pair of rules

$$(Merge) \quad \frac{\Gamma \Rightarrow \Delta \mid \Pi \Rightarrow \Sigma \mid G}{\Gamma, \Pi \Rightarrow \Delta, \Sigma \mid G} \quad (\Rightarrow WE) \quad \frac{G}{\Rightarrow \varphi \mid G} \quad (WE \Rightarrow) \quad \frac{G}{\varphi \Rightarrow \mid G}$$

It is clear that these special weakening rules allow the derivability of the usual IW and EW rules with the help of $(Merge)$. EC is also derivable by means of $(Merge)$ and IC. On the other hand, Restall's weakening rules are just special instances of EW, and $(Merge)$ is derivable by IW and EC. Hence, Restall's set of structural rules is equivalent to the more commonly adapted set but often allows for simpler proofs of many hypersequents. In particular, $(Merge)$ is very handy for constructing more compact proofs as it combines applications of IW and EC. It should be noticed that $(Merge)$ is based on the similar idea as the special modal rule of Mints [180], (see subsection 4.5.1), but whereas the latter is defined for standard SC and destroys subformula property, $(Merge)$, due to some extra machinery of HC, allows for the subformula property to be preserved.

What about cut? In fact, different forms of cut were encountered in the framework of HC. The most direct adaptation of a standard (multiplicative) cut is $(H - Cut)$

$$\frac{\Gamma \Rightarrow \Delta, \varphi \mid G \qquad \varphi, \Sigma \Rightarrow \Pi \mid H}{\Gamma, \Sigma \Rightarrow \Delta, \Pi \mid G \mid H}$$

where $G \mid H$ denotes a concatenation of possibly different external contexts. An additive version is also possible in the weaker (G and H admitted as different contexts) or stronger (the same external context in both premises) but we limit our interest only to multiplicative $(H - Cut)$. A hypersequential counterpart of Mix is $(H - Mix)$

$$\frac{\Gamma \Rightarrow \Delta, \varphi^i \mid G \qquad \varphi^j, \Sigma \Rightarrow \Pi \mid H}{\Gamma, \Sigma \Rightarrow \Delta, \Pi \mid G \mid H}$$

where we tacitly assume that there are no other occurrences of φ in Δ and Σ.

Similarly, as in the standard sequent calculus, both rules are equivalent in the presence of weakening and contraction, we thus have

Lemma 4.11 *G is provable in HC with $(H - Cut)$ iff G is provable in HC with $(H - Mix)$.*

Since in HC we can have also EC or $(Merge)$ as primitive rules, this introduces additional complications analogous to those with C in ordinary SC, but now concerning different sequents. In order to deal with them, Avron introduced a yet more general version of $(H - Mix)$ which we call here $(SH - Mix)$ (S for strong):

$$\frac{\Gamma_1 \Rightarrow \Delta_1, \varphi^i \mid ... \mid \Gamma_n \Rightarrow \Delta_n, \varphi^j \mid G \qquad \varphi^k, \Sigma_1 \Rightarrow \Pi_1 \mid ... \mid \varphi^n, \Sigma_k \Rightarrow \Pi_k \mid H}{\Gamma_1, ..., \Gamma_n, \Sigma_1, ..., \Sigma_k \Rightarrow \Delta_1, ..., \Delta_n, \Pi_1, ..., \Pi_k \mid G \mid H}$$

In this way, we can cut not only multiple occurrences of a formula in one sequent of a premise, but in many sequents in one step. Note, however, that in case of this rule the situation is a bit different with respect to its strength; namely, the following holds:

Lemma 4.12 *If G is provable in HC with $(SH - Mix)$, then G is provable in HC with $(H - Cut)$.*

PROOF: In order to simulate an application of $(SH - Mix)$ with $(H - Cut)$ it is enough to apply successively IW, EC and IC to each premiss. This way from the left premiss we obtain $\Gamma_1, ..., \Gamma_n \Rightarrow \Delta_1, ...\Delta_n, \varphi$, and similarly for the right premiss. From these hypersequents, the result of the application of $(SH - Mix)$ follows by $(H - Cut)$. $\qquad\qquad\square$

However, not every application of $(H - Cut)$ may be simulated by $(SH - Mix)$. Consider the following instance (with no other occurrences of φ):

$$\frac{\Gamma \Rightarrow \Delta, \varphi \mid G \qquad \varphi, \Sigma \Rightarrow \Pi \mid \varphi, \Lambda \Rightarrow \Theta \mid H}{\Gamma, \Sigma \Rightarrow \Delta, \Pi \mid \varphi, \Lambda \Rightarrow \Theta \mid G \mid H}$$

If we apply $(SH - Mix)$, we obtain $\Gamma, \Sigma, \Lambda \Rightarrow \Delta, \Pi, \Theta \mid G \mid H$ and there is no possibility to derive the original conclusion (we can restore $\varphi, \Lambda \Rightarrow \Theta$ by EW but we cannot delete Λ and Θ in the active sequent). This is rather unfortunate, since even if we can prove elimination of $(SH - Mix)$, it does not guarantee that all cuts may be eliminable. In order to have equal strength of rules, we must rather use some selective version of Mix which allows for deleting some but not necessarily all occurrences of φ in both premisses. Let us call such a variant $(SH - Multicut)$; the above schema is sufficient for its expression, we just do not assume that all occurrences of φ are displayed. Since in this case $(H - Cut)$ is just a special case of $(SH - Multicut)$ we obtain

Lemma 4.13 *G is provable in HC with $(H - Cut)$ iff G is provable in HC with $(SH - Multicut)$.*

Before we focus on systems characterising **S5** it should be noted that the HC framework may help to make life easier also in other logics which apparently do not need generalised SC. In case of **T** and **S4** we focused on problems with proof search requiring either backtracking or introducing more general proof-search trees with additional disjunctive branching regulated by meta-rules like (DB) in section 2.7 or more specialised (SG) rules from section 4.4. But this layer of branching is implicit in hypersequents so instead, we can formulate suitable rules of $\square$ introduction into the succedent. For example, in case of **S4**, we can use the following rule:

$$\frac{\square\Gamma \Rightarrow \varphi_1 \mid ... \mid \square\Gamma \Rightarrow \varphi_k \mid G}{\Delta, \square\Gamma \Rightarrow \square\varphi_1, ..., \square\varphi_k, \Sigma \mid G}$$

Additionally, for the need of the proof search procedure, we may add side condition that Δ, Σ consist of atomic formulae only. One may also overcome the aforementioned difficulties in a different way and use a rule (introduced by Pottinger [202]) which deals only with one selected box-formula but keeps the rest for later use

$$\frac{\Delta, \Box\Gamma \Rightarrow \Sigma \mid \Box\Gamma \Rightarrow \varphi \mid G}{\Delta, \Box\Gamma \Rightarrow \Box\varphi, \Sigma \mid G}$$

If we delete $\Delta, \Box\Gamma \Rightarrow \Sigma$ in the premiss we obtain just simple (weakening-absorbing) hypersequent version of $(\Rightarrow \Box)$ for **S4**. The reader may easily provide suitable versions of all such rules for **T**.

Exercise 4.28 *Prove* **S4**-*validity-preservation of both rules.*

In fact, a kind of generalisation imposed by hypersequents may be treated on a par with the step which is already done in ordinary SC for **CPL** if we take sequents with at most one formula in the succedent as basic items. Avron [11] points out that if we consider LK rule $(\Rightarrow \lor)$ in such framework (natural for intuitionistic logic—see section 5.1), introducing multiplicative disjunction requires admissibility of more formulae in the succedent (G3 rule for $(\Rightarrow \lor)$) but for additive disjunction instead of two rules we may use just the hypersequent rule

$$\frac{\Gamma \Rightarrow \varphi \mid \Gamma \Rightarrow \psi \mid G}{\Gamma \Rightarrow \varphi \lor \psi \mid G}$$

4.7.2 Systems for S5

We start a description of HCs for **S5** with an approach which was provisionally called syntactical and initiated with Avron [10]. He proposed HLK with simple axioms $\varphi \Rightarrow \varphi$ and primitive structural rules of both kinds. The modal part is based on Ohnishi and Matsumoto's rules for **S4** but in the hypersequent shape, i.e.

$$(\Box\Rightarrow^T) \ \frac{\varphi, \Gamma \Rightarrow \Delta \mid G}{\Box\varphi, \Gamma \Rightarrow \Delta \mid G} \qquad (\Rightarrow\Box^{S4}) \ \frac{\Box\Gamma \Rightarrow \varphi \mid G}{\Box\Gamma \Rightarrow \Box\varphi \mid G}$$

To obtain HC for **S5** Avron introduced a special rule (MS) of modal splitting which has a combined character; it is partly structural but with displayed multisets of modal formulae:

$$(MS) \ \frac{\Box\Gamma, \Pi \Rightarrow \Box\Delta, \Sigma \mid G}{\Box\Gamma \Rightarrow \Box\Delta \mid \Pi \Rightarrow \Sigma \mid G}$$

Lemma 4.14 *All rules of Avron's system are validity-preserving.*

PROOF: We take (MS) as an example: Assume that $\models \Box\Gamma, \Pi \Rightarrow \Box\Delta, \Sigma \mid G$ and $\not\models \Box\Gamma \Rightarrow \Box\Delta \mid \Pi \Rightarrow \Sigma \mid G$ so we know that there are w_1, w_2 such that $w_1 \not\models \Box\Gamma \Rightarrow \Box\Delta$, $w_2 \not\models \Pi \Rightarrow \Sigma$ and that $\not\models G$. According to the definition of validity we know that for all $\Box\psi \in \Box\Gamma, w_1 \models \Box\psi$ and for all $\Box\chi \in \Box\Delta, w_1 \not\models \Box\chi$ which implies that it holds for each w_i, in particular for w_2 but this falsifies $\Box\Gamma, \Pi \Rightarrow \Box\Delta, \Sigma \mid G$, contrary to the assumption. $\qquad\square$

In fact, (MS) is so strong that the system may be modified in two ways either by changing a rule for $(\Rightarrow \Box)$ or by weakening of (MS). In the first case, $(\Rightarrow \Box^{S4})$ may be replaced with any of the following:

$$(\Rightarrow \Box^{S5}) \quad \frac{\Box \Gamma \Rightarrow \Box \Delta, \varphi \mid G}{\Box \Gamma \Rightarrow \Box \Delta, \Box \varphi \mid G} \qquad (\Rightarrow \Box^{T}) \quad \frac{\Gamma \Rightarrow \varphi \mid G}{\Box \Gamma \Rightarrow \Box \varphi \mid G} \qquad (\Rightarrow \Box^{G}) \quad \frac{\Rightarrow \varphi \mid G}{\Rightarrow \Box \varphi \mid G}$$

The first one is stronger since it is a hypersequent version of LK rule for **S5**. But the other two are weaker as they are hypersequent counterparts of the LK rule for **T** (and **K** of course) and of Gödel's rule. To show that it is sufficient we can show that by means of (MS) and $(\Rightarrow \Box^{G})$ we can prove the derivability of the primitive rule of Avron

$$
\begin{array}{c}
(MS) \ \dfrac{\Box \Gamma \Rightarrow \varphi \mid G}{\Box \Gamma \Rightarrow \mid \Rightarrow \varphi \mid G} \\[4pt]
(\Rightarrow \Box^{G}) \ \dfrac{}{\Box \Gamma \Rightarrow \mid \Rightarrow \Box \varphi \mid G} \\[4pt]
IW \ \dfrac{}{\Box \Gamma \Rightarrow \Box \varphi \mid \Box \Gamma \Rightarrow \Box \varphi \mid G} \\[4pt]
(EC) \ \dfrac{}{\Box \Gamma \Rightarrow \Box \varphi \mid G}
\end{array}
$$

In general, Avron's system strongly depends on the application of structural rules of contraction both in internal (standard) and external versions. The proof of 5 is a good example

$$
\begin{array}{c}
(\neg \Rightarrow) \ \dfrac{\Box p \Rightarrow \Box p}{\Box p, \neg \Box p \Rightarrow} \\[4pt]
(MS) \ \dfrac{}{\Box p \Rightarrow \mid \neg \Box p \Rightarrow} \\[4pt]
(\Rightarrow \neg) \ \dfrac{}{\Rightarrow \neg \Box p \mid \neg \Box p \Rightarrow} \\[4pt]
(\Rightarrow \Box) \ \dfrac{}{\Rightarrow \Box \neg \Box p \mid \neg \Box p \Rightarrow} \\[4pt]
IW \times 2 \ \dfrac{}{\neg \Box p \Rightarrow \Box \neg \Box p \mid \neg \Box p \Rightarrow \Box \neg \Box p} \\[4pt]
(EC) \ \dfrac{}{\neg \Box p \Rightarrow \Box \neg \Box p} \\[4pt]
(\Rightarrow \rightarrow) \ \dfrac{}{\Rightarrow \neg \Box p \rightarrow \Box \neg \Box p}
\end{array}
$$

Exercise 4.29 *Prove K, 4 and B in Avron's system. To simplify use $(Merge)$ instead of EC.*

Lahav [156] and Kurokawa's [154] systems may be seen as refinements of Avron's approach obtained by special rules which are weaker than (MS). Kurokawa shows HC for some extensions of **S4** including **S5**. His basic system is almost exactly like Avron's calculus but instead of (MS) he is using its weaker version:

$$(MS^{K}) \quad \frac{\Box \Gamma, \Pi \Rightarrow \Sigma \mid G}{\Box \Gamma \Rightarrow \mid \Pi \Rightarrow \Sigma \mid G}$$

It is weaker in the sense that only boxed formulae from the antecedent are taken into account. One should check again the proof of 5 in Avron's system to see that the omission of boxed formulae in the succedent of the premiss and (one of the) conclusion sequent makes no harm. Below we put a proof of B to show that such a solution works

$$
\begin{array}{c}
(\square \Rightarrow) \dfrac{p \Rightarrow p}{\square p \Rightarrow p} \\[4pt]
(\neg \Rightarrow) \dfrac{}{\square p, \neg p \Rightarrow} \\[4pt]
(MS^K) \dfrac{}{\square p \Rightarrow \mid \neg p \Rightarrow} \\[4pt]
(\Rightarrow \neg) \dfrac{}{\Rightarrow \neg \square p \mid \neg p \Rightarrow} \\[4pt]
(\Rightarrow \square) \dfrac{}{\Rightarrow \square \neg \square p \mid \neg p \Rightarrow} \\[4pt]
IW \times 2 \ \dfrac{}{\neg p \Rightarrow \square \neg \square p \mid \neg p \Rightarrow \square \neg \square p} \\[4pt]
(EC) \dfrac{}{\neg p \Rightarrow \square \neg \square p} \\[4pt]
(\Rightarrow \rightarrow) \dfrac{}{\Rightarrow \neg p \rightarrow \square \neg \square p}
\end{array}
$$

Lahav [156] presents a general method for generating hypersequent rules from some frame. His basic system for **K** is defined on sequents built from sets so contraction is implicit but both IW and EW are primitive. The only modal rule is $(\Rightarrow \square^T)$ and his solution for **S5** is based on the addition of the following rule encoding the property of universality:

$$
(U) \ \dfrac{\Gamma, \Pi \Rightarrow \Delta \mid G}{\Lambda, \square \Pi \Rightarrow \Sigma \mid \Gamma, \square \Xi \Rightarrow \Delta \mid G}
$$

Closer inspection shows that Lahav's specific rule for **S5** may be seen as a (weaker) variant of (MS) with additional deletion of $\square$ in elements of Π in the premiss. In fact, (U) may be easily derived in Avron's system

$$
\begin{array}{c}
(\square \Rightarrow) \dfrac{\Gamma, \Pi \Rightarrow \Delta \mid G}{\Gamma, \square \Pi \Rightarrow \Delta \mid G} \\[4pt]
(MS) \dfrac{}{\square \Pi \Rightarrow \mid \Gamma \Rightarrow \Delta \mid G} \\[4pt]
IW \ \dfrac{}{\Lambda, \square \Pi \Rightarrow \Sigma \mid \Gamma, \square \Xi \Rightarrow \Delta \mid G}
\end{array}
$$

To derive Avron's rules in Lahav's system we need to use cut, which is interpreted by the author as showing that his system is in a sense stronger as it implies the admissibility of cut in Avron's calculus. Note that the qualification of this rule as a weaker version of (MS) is not connected with the lack of $\square$ in front of elements of Π in the premiss but with the fact that only boxed formulae from the antecedent (of one of the sequents in the conclusion) are put in the antecedent of the premiss (without boxes, however).

Exercise 4.30 *Show that rules of Kurokawa and Lahav are validity-preserving in* **S5**.

Prove the axioms of **S5** *in both systems.*

The approach which we provisionally called semantic was initiated by Mints in [177] and independently by Pottinger [202]. The characteristic feature of it is the application of rules for $\Box$ which are specific for **S5** (of course not all but at least one of them). Mints [177] is using HLK for **CPL** with addition of EW and the following rules:

$$(\Rightarrow\Box^K) \; \frac{\Gamma \Rightarrow \Delta \mid \Rightarrow \varphi \mid G}{\Gamma \Rightarrow \Delta, \Box\varphi \mid G} \qquad (\Rightarrow\Box^G) \; \frac{\Rightarrow \varphi \mid G}{\Rightarrow \Box\varphi \mid G}$$

$$(\Box\Rightarrow^T) \; \frac{\varphi, \Gamma \Rightarrow \Delta \mid G}{\Box\varphi, \Gamma \Rightarrow \Delta \mid G} \qquad (\Box\Rightarrow^5) \; \frac{\Gamma \Rightarrow \Delta \mid \varphi, \Sigma \Rightarrow \Theta \mid G}{\Box\varphi, \Gamma \Rightarrow \Delta \mid \Sigma \Rightarrow \Theta \mid G}$$

Two of them are already known and $(\Rightarrow\Box^K)$ is, similarly like $(\Rightarrow \Box^G)$ validity-preserving for **K** (hence the superscript). The only rule which is specific for **S5** is $(\Box\Rightarrow^5)$, hence the superscript 5.

Exercise 4.31 *Prove that* $(\Rightarrow\Box^K)$ *is validity-preserving for any frame and that* $(\Box\Rightarrow^5)$ *is validity-preserving in* **S5**.

Mints' rules are canonical (see subsection 3.5.5), although $(\Rightarrow \Box^G)$ is in a sense not context independent. We can define also suitable dual rules for $\Diamond$

$$(\Diamond\Rightarrow^K) \; \frac{\Gamma \Rightarrow \Delta \mid \varphi \Rightarrow \mid G}{\Diamond\varphi, \Gamma \Rightarrow \Delta \mid G} \qquad (\Diamond\Rightarrow^G) \; \frac{\varphi \Rightarrow \mid G}{\Diamond\varphi \Rightarrow \mid G}$$

$$(\Rightarrow\Diamond^T) \; \frac{\Gamma \Rightarrow \Delta, \varphi \mid G}{\Gamma \Rightarrow \Delta, \Diamond\varphi \mid G} \qquad (\Rightarrow\Diamond^5) \; \frac{\Gamma \Rightarrow \Delta \mid \Sigma \Rightarrow \Theta, \varphi \mid G}{\Gamma \Rightarrow \Delta, \Diamond\varphi \mid \Sigma \Rightarrow \Theta \mid G}$$

Here is an example of a half of the proof showing interdefinability of $\Box$ and $\Diamond$

$$
\begin{array}{l}
(\neg\Rightarrow) \; \dfrac{\Rightarrow\mid p \Rightarrow p}{\Rightarrow\mid p, \neg p \Rightarrow} \\[4pt]
(\Box\Rightarrow^5) \; \dfrac{}{\Box p \Rightarrow\mid \neg p \Rightarrow} \\[4pt]
(\Diamond\Rightarrow^K) \; \dfrac{}{\Diamond\neg p, \Box p \Rightarrow} \\[4pt]
(\Rightarrow\neg) \; \dfrac{}{\Box p \Rightarrow \neg\Diamond\neg p}
\end{array}
$$

Exercise 4.32 *Prove other definitional implications for* $\Box$ *and* $\Diamond$.

On the other hand, this set of rules is redundant. $(\Rightarrow\Box^G)$ is derivable by means of *EW* and $(\Rightarrow\Box^K)$ in the following way:

$$
\begin{array}{l}
(EW) \; \dfrac{\Rightarrow \varphi \mid G}{\Rightarrow\mid\Rightarrow \varphi \mid G} \\[4pt]
(\Rightarrow\Box^G) \; \dfrac{}{\Rightarrow \Box\varphi \mid G}
\end{array}
$$

In fact, also $(\Rightarrow\Box^K)$ is derivable by means of $(\Rightarrow\Box^G)$ and $(Merge)$ (or IW and EC) but Mints did not introduce these rules. Similarly, $(\Box\Rightarrow^T)$ is derivable by means of EW, $(\Box\Rightarrow^5)$ and either IW and EC, or $(Merge)$ and IC. Later Mints [179], reduced the set of rules by deleting $(\Rightarrow\Box^G)$.

Exercise 4.33 *Prove the derivability of the rules as stated above.*

Restall [215] also uses HLK but with specific structural rules described above and a nonredundant set of two rules for introduction of $\Box$

$$(\Box\Rightarrow^R) \quad \frac{\varphi, \Gamma \Rightarrow \Delta \mid G}{\Box\varphi \Rightarrow\mid \Gamma \Rightarrow \Delta \mid G} \qquad (\Rightarrow\Box^G) \quad \frac{\Rightarrow \varphi \mid G}{\Rightarrow \Box\varphi \mid G}$$

The first is a simpler version of $(\Box\Rightarrow^5)$ and the second is the same as in Mints' set. Note that Restall's $(\Box\Rightarrow^R)$ looks also like an extremely simplified version of (MS), so his system may as well be seen as a drastic simplification of Avron's system in that it keeps only the minimal resources necessary for an adequate characterisation of **S5**. Indeed both modal rules (in the presence of $(Merge)$) are sufficient to obtain an adequate system for **S5**. Below we display a proof of K in his system:

$$
\begin{array}{l}
(\to\Rightarrow) \dfrac{p \Rightarrow p \qquad q \Rightarrow q}{p \to q, p \Rightarrow q} \\[4pt]
(\Box\Rightarrow^R) \dfrac{}{\Box(p \to q) \Rightarrow\mid p \Rightarrow q} \\[4pt]
(\Box\Rightarrow^R) \dfrac{}{\Box(p \to q) \Rightarrow\mid \Box p \Rightarrow\mid\Rightarrow q} \\[4pt]
(\Rightarrow\Box^G) \dfrac{}{\Box(p \to q) \Rightarrow\mid \Box p \Rightarrow\mid\Rightarrow \Box q} \\[4pt]
(Merge) \dfrac{}{\Box(p \to q) \Rightarrow\mid \Box p \Rightarrow \Box q} \\[4pt]
(\Rightarrow\to) \dfrac{}{\Box(p \to q) \Rightarrow\mid\Rightarrow \Box p \to \Box q} \\[4pt]
(Merge) \dfrac{}{\Box(p \to q) \Rightarrow \Box p \to \Box q} \\[4pt]
(\Rightarrow\to) \dfrac{}{\Rightarrow \Box(p \to q) \to (\Box p \to \Box q)}
\end{array}
$$

Exercise 4.34 *Prove other axioms and B.*

The soundness of this system is straightforward. It is enough to demonstrate that

Lemma 4.15 $(\Box\Rightarrow^R)$ *is validity-preserving in* **S5**.

PROOF: Assume that $\models \varphi, \Gamma \Rightarrow \Delta \mid G$ and $\not\models \Box\varphi \Rightarrow\mid \Gamma \Rightarrow \Delta \mid G$. Hence we obtain that there exists some $\mathfrak{M}$ with $w_1 \not\models \Box\varphi \Rightarrow$, $w_2 \not\models \Gamma \Rightarrow \Delta$ and $\not\models G$. So $w_1 \models \Box\varphi$ which means that $w_2 \models \varphi$ which falsifies the premiss, but this is contrary to assumption. $\qquad\Box$

Restall also provided a variant of his system in the spirit of Kleene's solution for a constructive semantic proof of completeness. In the case of rules for boolean constants, it consists in repeating the principal formulae in the premises; in the case of modals we must repeat the whole sequents which yields the following rules:

$$(\Box\Rightarrow^{5'}) \; \frac{\Box\varphi, \Pi \Rightarrow \Sigma \mid \varphi, \Gamma \Rightarrow \Delta \mid G}{\Box\varphi, \Pi \Rightarrow \Sigma \mid \Gamma \Rightarrow \Delta \mid G} \qquad (\Rightarrow\Box^{K'}) \; \frac{\Gamma \Rightarrow \Delta, \Box\varphi \mid \Rightarrow \varphi \mid G}{\Gamma \Rightarrow \Delta, \Box\varphi \mid G}$$

which are in fact contraction-absorbing versions of $(\Box\Rightarrow^{5})$ and $(\Rightarrow\Box^{K})$.

Poggiolesi [198] is using HG3, hence only logical rules are taken as primitive. In fact, she needs structural rules like $(Merge)$ for providing a syntactical proof of cut elimination; however, these are not primitive but admissible, in contrast to Restall's system. The price for that is the presence of two rules for introducing $\Box$ in the antecedent. The only rule for introducing $\Box$ in the succedent is $(\Rightarrow\Box^{K})$; the remaining two rules are contraction-absorbing versions of Mints' rules:

$$(\Box\Rightarrow^{T'}) \; \frac{\varphi, \Box\varphi, \Gamma \Rightarrow \Delta \mid G}{\Box\varphi, \Gamma \Rightarrow \Delta \mid G} \qquad (\Box\Rightarrow^{5'}) \; \frac{\Box\varphi, \Pi \Rightarrow \Sigma \mid \varphi, \Gamma \Rightarrow \Delta \mid G}{\Box\varphi, \Pi \Rightarrow \Sigma \mid \Gamma \Rightarrow \Delta \mid G}$$

The soundness of this system is immediate and the proof of the axioms of **S5** are simple.

Exercise 4.35 *Prove axioms of* **S5**.

Although Poggiolesi provides a very elegant proof of cut admissibility which implies the adequacy of her system, we provide a different proof here which is based on the possibility of the simulation of the labelled system presented in section 4.6. It is easy since her system in the most direct way encodes semantic features of **S5** in terms of syntactical rules for modals. First, note that every sequent in the labelled system may be represented in the following way: $\Gamma_1, ..., \Gamma_n \Rightarrow \Delta_1, ..., \Delta_n$ where n is a maximal label occurring in the sequent and each $\Gamma_i(\Delta_i)$ is the multiset (possibly empty) of all formulae labelled with i. Define

$$\Im(\Gamma_1, ..., \Gamma_n \Rightarrow \Delta_1, ..., \Delta_n) \; = \Im(\Gamma_1) \Rightarrow \Im(\Delta_1) \mid ... \mid \Im(\Gamma_n) \Rightarrow \Im(\Delta_n)$$

where $\Im(\Gamma_i)$ is the same multiset with deleted labels. It is straightforward to observe that every application of $(\Rightarrow\Box)$ in the labelled system is simulated by $(\Rightarrow\Box^{K})$, whereas every application of $(\Box\Rightarrow)$ is simulated either by $(\Box\Rightarrow^{T'})$ (if the same label is preserved) or by $(\Box\Rightarrow^{5'})$. Thus, it is routine to show that every proof in labelled SC is stepwise simulated in Poggiolesi's system. It follows that this cut-free system is complete.

The comparison of the second variant of Restall's system with Poggiolesi's system is also very instructive. At first, one might suspect that Restall's rules are insufficient but this is not true. One can derive $(\Box\Rightarrow^{T'})$ in his system in the following way:

$$
\begin{array}{c}
(\Box\Rightarrow^{R}) \dfrac{\varphi,\Box\varphi,\Gamma\Rightarrow\Delta\mid G}{\Box\varphi\Rightarrow\mid\Box\varphi,\Gamma\Rightarrow\Delta\mid G} \\[2mm]
(Merge) \dfrac{}{\Box\varphi,\Box\varphi,\Gamma\Rightarrow\Delta\mid G} \\[2mm]
(C\Rightarrow) \dfrac{}{\Box\varphi,\Gamma\Rightarrow\Delta\mid G}
\end{array}
$$

This may lead to the suspicion that Poggiolesi's system is redundant. No, it is not since it has no external structural rules as primitive. One can easily notice that in Restall's system we cannot prove axiom T without using (*Merge*), hence this rule must be primitive. On the contrary, in Poggiolesi's system, the presence of a special rule corresponding to T makes (*Merge*) redundant.

One may consider if we can avoid the apparent inelegance of having two rules for introducing $\Box$ into the antecedent in the system but without the need for such strong structural rules like (*Merge*) or EC. It seems that Pottinger's system provides a solution. It was stated very briefly in the half-page long abstract [202], and as far as we know, was never presented in its full version. The abstract contains rules of HC for modal logics **T**, **S4** and **S5**. Essentially, it is also HG3 although with some peculiarities which may be disregard here. Two modal rules for introduction of $\Box$ are of the form

$$
(\Box\Rightarrow^{P})\ \dfrac{\varphi,\Box\varphi,\Gamma\Rightarrow\Delta\mid\Box\varphi,\Gamma_{1}\Rightarrow\Delta_{1}\mid\cdots\mid\Box\varphi,\Gamma_{n}\Rightarrow\Delta_{n}}{\Box\varphi,\Gamma\Rightarrow\Delta\mid\Gamma_{1}\Rightarrow\Delta_{1}\mid\cdots\mid\Gamma_{n}\Rightarrow\Delta_{n}} \qquad (\Rightarrow\Box^{4})\ \dfrac{\Delta,\Box\Gamma\Rightarrow\Sigma\mid\Box\Gamma\Rightarrow\varphi\mid G}{\Delta,\Box\Gamma\Rightarrow\Box\varphi,\Sigma\mid G}
$$

Both the rules are rather semantically oriented on the actual search of either a proof or a falsifying model. In contrast to several modal rules presented above these are rather global and in effect quite redundant. Also, construction of proofs may be rather involved; let us see a proof of K as an example

$$
\begin{array}{c}
(\rightarrow\Rightarrow) \dfrac{S\mid\Box(p\rightarrow q),\Box p,p\Rightarrow q,p \qquad S\mid\Box(p\rightarrow q),\Box p,p,q\Rightarrow q}{\Box(p\rightarrow q),\Box(p\rightarrow q),\Box p,\Box p\Rightarrow\mid\Box(p\rightarrow q),\Box p,p\rightarrow q,p\Rightarrow q} \\[2mm]
(\Box\Rightarrow^{P}) \dfrac{\Box(p\rightarrow q),\Box p,\Box p\Rightarrow\mid\Box(p\rightarrow q),\Box p,p\Rightarrow q}{} \\[2mm]
(\Box\Rightarrow^{P}) \dfrac{\Box(p\rightarrow q),\Box p\Rightarrow\mid\Box(p\rightarrow q),\Box p\Rightarrow q}{} \\[2mm]
(\Rightarrow\Box^{4}) \dfrac{\Box(p\rightarrow q),\Box p\Rightarrow\Box q}{} \\[2mm]
(\Rightarrow\rightarrow) \dfrac{\Box(p\rightarrow q)\Rightarrow\Box p\rightarrow\Box q}{} \\[2mm]
(\Rightarrow\rightarrow) \dfrac{\Box(p\rightarrow q)\rightarrow(\Box p\rightarrow\Box q)}{}
\end{array}
$$

where S is the sequent $\Box(p\rightarrow q),\Box(p\rightarrow q),\Box p,\Box p\Rightarrow$.

Exercise 4.36 *Prove the remaining axioms of* **S5**.

The abstract of Pottinger does not contain any proof; the system is only claimed to be adequate in cut-free form but it is not known if it was proved syntactically or semantically. It is easy to prove soundness; it is enough to demonstrate:

Lemma 4.16 $(\Box\Rightarrow^{P})$ *is validity-preserving in* **S5**.

PROOF: Assume that $\models \varphi, \Box\varphi, \Gamma \Rightarrow \Delta \mid \Box\varphi, \Gamma_1 \Rightarrow \Delta_1 \mid \cdots \mid \Box\varphi, \Gamma_n \Rightarrow \Delta_n$ and $\not\models \Box\varphi, \Gamma \Rightarrow \Delta \mid \Gamma_1 \Rightarrow \Delta_1 \mid \cdots \mid \Gamma_n \Rightarrow \Delta_n$. By definition we have that there is $\mathfrak{M}$ and $w_1, \ldots, w_n$ such that $w_1 \not\models \Box\varphi, \Gamma \Rightarrow \Delta$, $w_2 \not\models \Gamma_1 \Rightarrow \Delta_1 \ldots w_n \not\models \Gamma_n \Rightarrow \Delta_n$ thus $w_1 \models \Box\varphi$ which implies that for all $w_i \in \mathcal{W}_{\mathfrak{M}}, w_i \models \Box\varphi$ but this falsifies every sequent in the premiss and contradicts our assumption. $\Box$

As for completeness, it is easier to prove it semantically by a Hintikka-style argument or by translation from the labelled system like we did for Poggiolesi's system.

Exercise 4.37 *Prove completeness of Pottinger's system.*

The structure of modal rules of Pottinger make them not very suitable for constructing a syntactic proof of cut elimination. In Bednarska and Indrzejczak [24], a refined version of Pottinger's HG3 was proposed which admits a proof of cut admissibility but at the cost of having primitive IC. As a result a proof is provided directly for $(H - Mix)$. However, it may be further refined in such a way that a fully logical system may be obtained exactly as in Poggiolesi and with only two modal rules:

$$(\Box\Rightarrow^{P'}) \ \frac{\varphi, \Box\varphi, \Gamma \Rightarrow \Delta \mid \varphi, \Gamma_1 \Rightarrow \Delta_1 \mid \cdots \mid \varphi, \Gamma_n \Rightarrow \Delta_n}{\Box\varphi, \Gamma \Rightarrow \Delta \mid \Gamma_1 \Rightarrow \Delta_1 \mid \cdots \mid \Gamma_n \Rightarrow \Delta_n} \qquad (\Rightarrow\Box^{K'}) \ \frac{\Gamma \Rightarrow \Delta, \Box\varphi \mid \Rightarrow \varphi \mid G}{\Gamma \Rightarrow \Delta, \Box\varphi \mid G}$$

The version from Bednarska and Indrzejczak [24], used just $(\Rightarrow \Box^K)$ and this was the reason for keeping IC as a primitive rule.

4.7.3   Admissibility of Cut

A lot of different methods for proving cut elimination/admissibility in the framework of HC have been offered so far. Some syntactical proofs of cut admissibility (elimination) for HC are performed by means of a suitable technique for tracing the cut-formula through a proof (see, e.g. the 'history technique' of Avron [7] or the 'decoration technique' of Baaz and Ciabattoni [16]). There were also some applications of a Schütte-style proof to HC by Baaz and Ciabattoni [16] and by Indrzejczak [140]. These approaches were applied to logics not dealt with in this book, so we will not discuss them below. Instead, we will focus on three strategies which were actually applied to HC for **S5**.

The first strategy is based on the application of a special multicut version suitable for hypersequents and it was proposed by Avron [10]. The proof was only sketched in Avron [10], but it was presented in detail in Bednarska and Indrzejczak [24]. It is rather complicated since it requires two different forms of the cut (or rather multicut) rule. Strictly speaking, Avron did not introduce this rule (which we call $(BSH - Mix)$—see the next page), or even $(SH - Mix)$) as a special rule but rather demonstrated the admissibility of $(H - Cut)$ by means of more general theorem where both forms of Mix are involved in the induction hypothesis. However, it seems to be most transparent to define a special HC calculus with both rules explicitly formulated. We follow in this respect the form of presentation of HC for Gödel logics in Baaz, Ciabattoni and Fermüller [17]. Application of two versions of Multicut leads in consequence to the multiplication of subcases which must be considered. So it is not a perfect solution but may serve as an example of the original idea peculiar to HC. Below we only briefly characterise its specific features.

Why are two forms of multicut necessary for this proof? We already pointed out that in order to deal with EC Avron was forced to generalise an ordinary mix. The most interesting point with Avron's proof is that despite the generalisation of $(H - Mix)$ to $(SH - Mix)$ (or rather $(SH - Multicut)$) to deal with EC he was still unable to prove cut elimination syntactically. Let us look at the following figure:

$$(SH\text{-}Multicut) \cfrac{(MS) \cfrac{G \mid \Gamma, \Box\Pi \Rightarrow \Delta, \Box\Sigma, \Box\varphi}{G \mid \Gamma \Rightarrow \Delta \mid \Box\Pi \Rightarrow \Box\Sigma, \Box\varphi} \qquad H \mid \Box\varphi, \Lambda \Rightarrow \Theta}{G \mid H \mid \Gamma \Rightarrow \Delta \mid \Box\Pi, \Lambda \Rightarrow \Box\Sigma, \Theta}$$

where we assume that Λ, Θ consist of nonmodal formulae and that $\Box\varphi$ does not belong to Δ. If we want to reduce the height of (SH-Multicut), we obtain

$$(SH\text{-}Multicut) \cfrac{G \mid \Gamma, \Box\Pi \Rightarrow \Delta, \Box\Sigma, \Box\varphi \qquad H \mid \Box\varphi, \Lambda \Rightarrow \Theta}{(MS) \cfrac{G \mid H \mid \Gamma, \Box\Pi, \Lambda \Rightarrow \Delta, \Box\Sigma, \Theta}{G \mid H \mid \Gamma, \Lambda \Rightarrow \Delta, \Theta \mid \Box\Pi \Rightarrow \Box\Sigma}}$$

From the last hypersequent, we have no way to obtain the last sequent of the original proof. In order to deal with the problem Avron restricted the application of $(SH - Mix)$ (or rather $(SH - Multicut)$) to nonmodal formulae and introduced one more special form of mix for cutting boxed formulae which we call $(BSH - Mix), (BSH - Multicut)$ (B - for boxed)

$$\cfrac{G \mid \Gamma_1 \Rightarrow \Delta_1, \Box\varphi^i \mid ... \mid \Gamma_n \Rightarrow \Delta_n, \Box\varphi^j \qquad H \mid \Box\varphi^k, \Sigma_1 \Rightarrow \Pi_1 \mid ... \mid \Box\varphi^l, \Sigma_k \Rightarrow \Pi_k}{G \mid H \mid \Gamma_1 \Rightarrow \Delta_1 \mid ... \mid \Gamma_n \Rightarrow \Delta_n \mid \Sigma_1 \Rightarrow \Pi_1 \mid ... \mid \Sigma_k \Rightarrow \Pi_k}$$

One may easily check that this rule is validity-preserving only for **S5**, hence it cannot be used in general as an admissible form of cut for HC. It may be easily demonstrated that a system with both forms of Multicut, i.e. $(SH - Multicut)$ and $(BSH - Multicut)$ is equivalent to the primary system with nonrestricted applications of $(SH - Multicut)$

Theorem 4.9 *G is provable in the original HLKS5 iff G is provable in the modified system.*

PROOF: From left to right, it is enough to show that any application of $(SH - Multicut)$ with modal formulae as cut-formulae may be simulated by $(BSH - Multicut)$. It works like that

$$(BSHM) \cfrac{G \mid \Gamma_1 \Rightarrow \Delta_1, \Box\varphi^i \mid ... \mid \Gamma_n \Rightarrow \Delta_n, \Box\varphi^j \qquad H \mid \Box\varphi^k, \Sigma_1 \Rightarrow \Pi_1 \mid ... \mid \Box\varphi^l, \Sigma_k \Rightarrow \Pi_k}{IW, EC \cfrac{G \mid H \mid \Gamma_1 \Rightarrow \Delta_1 \mid ... \mid \Gamma_n \Rightarrow \Delta_n \mid \Sigma_1 \Rightarrow \Pi_1 \mid ... \mid \Sigma_k \Rightarrow \Pi_k}{G \mid H \mid \Gamma_1, ..., \Gamma_n, \Sigma_1, ..., \Sigma_k \Rightarrow \Delta_1, ..., \Delta_n, \Pi_1, ..., \Pi_k}}$$

From right to left, it is sufficient to show that every application of $(BSH - Multicut)$ may be simulated by $(SH - Multicut)$. It looks like that

$$(MS)\ \frac{G \mid \Gamma_1 \Rightarrow \Delta_1, \Box\varphi^i \mid ... \mid \Gamma_n \Rightarrow \Delta_n, \Box\varphi^j}{G \mid \Gamma_1 \Rightarrow \Delta_1 \mid ... \mid \Gamma_n \Rightarrow \Delta_n \mid \Rightarrow \Box\varphi^i \mid ... \mid \Rightarrow \Box\alpha^j} \qquad (?)\ \frac{H \mid \Box\varphi^k, \Sigma_1 \Rightarrow \Pi_1 \mid ... \mid \Box\varphi^l, \Sigma_k \Rightarrow \Pi_k}{H \mid \Box\alpha^k \Rightarrow \mid ... \mid \Box\alpha^l \Rightarrow \mid \Sigma_1 \Rightarrow \Pi_1 \mid ... \mid \Sigma_k \Rightarrow \Pi_k}$$

$$(SHM)\ \frac{}{G \mid H \mid \Gamma_1 \Rightarrow \Delta_1 \mid ... \mid \Gamma_n \Rightarrow \Delta_n \mid \Rightarrow \mid \Sigma_1 \Rightarrow \Pi_1 \mid ... \mid \Sigma_k \Rightarrow \Pi_k}$$

$$(IW)\ \frac{}{G \mid H \mid \Gamma_1 \Rightarrow \Delta_1 \mid ... \mid \Gamma_n \Rightarrow \Delta_n \mid \Gamma_n \Rightarrow \Delta_n \mid \Sigma_1 \Rightarrow \Pi_1 \mid ... \mid \Sigma_k \Rightarrow \Pi_k}$$

$$(EC)\ \frac{}{G \mid H \mid \Gamma_1 \Rightarrow \Delta_1 \mid ... \mid \Gamma_n \Rightarrow \Delta_n \mid \Sigma_1 \Rightarrow \Pi_1 \mid ... \mid \Sigma_k \Rightarrow \Pi_k}$$

$\Box$

One can prove the cut elimination theorem for HSC with two multicuts. In order to carry out a subsidiary induction on the height, not on the rank, in Bednarska and Indrzejczak [24], a slightly generalised Girard-strategy based on cross-cuts was applied. It has the advantage that the rank is a more complicated measure already for ordinary SC and its adaptation to hypersequent calculi encounter further difficulties whereas such a measure like height is simpler for control.

One may ask: 1) if it is possible to deal with only one version of cut?; 2) if it is possible to obtain a proof showing directly elimination or admissibility of $(H - Cut)$? In fact, both eventualities are realisable and below we provide two proofs of this kind.

As for the possibility of reduction to the system having only one rule, the method of Metcalfe, Olivetti and Gabbay [176], which was presented already in subsection 2.4.3, may be used. In fact, they introduced this strategy in the framework of HC as a very general and elegant method for proving cut elimination in the presence of contraction for numerous fuzzy logics. Let us recall that this method, called reductive by us, is in some sense a half-way between Gentzen's original proof of cut elimination and proofs based on global transformations of derivations like in Curry [56]. In contrast to Gentzen's proof, we eliminate the cut rule, not multicut, (or Mix) which is its generalisation absorbing contraction. But the original proof of Metcalfe, Olivetti and Gabbay [176], is strongly based on the predefined notion of 'substitutivity' of rules where the result of multiple applications of cut is absorbed. The adaptation of this method of proving cut elimination to extensions of t-norm logic **MTL** and related fuzzy logics with truth stresser modalities was provided by Ciabattoni, Metcalfe and Montagna [52]. This proof is particularly important since it deals with rules which are not 'substitutive' in the sense of [176], and may be applied here although not to the original system of Avron. The key point is that the reduction step for (SHM) fails because (MS) introduces boxed formula on both sides of the new sequent in the conclusion. But we noticed that we do not need such a strong form of (MS). In Kurokawa [154] (and similarly in Lahav [156]), a weaker version is used which operates only on the antecedent. For a system with such a rule, a special $(BSHM)$ is not required. The modal rules of Kurokawa enable an application of this strategy to his system for **S5**. His original proof is performed for $(WH - Multicut)$ according to the lines of proof from [52]. However, instead of reproducing here Kurokawa's proof we will adapt this kind of proof to Restall's system.

Restall's [215] original proof applies the global strategy of elimination of cuts in the proof, introduced by Curry [56] and refined by Belnap [25], in the context of display calculus. His solution is based on the fact that all rules of the system (including modal ones and $(Merge)$) are regular in the sense of allowing unrestricted permutation with cuts performed on parametric formulae. It is an elegant solution but shown in a very sketchy way which leaves some essential points of necessary transformations open. It seems, however, that the application of Curry's solution based on an inductive definition of the set of ancestors of the respective sequent may be adapted here.

In order to apply the reductive proof of cut elimination (see subsection 2.4.3), let us recall that we defined *the cut-degree $d\varphi$* as the complexity of the cut-formula φ and let the proof-degree dD be defined as the complexity of the most complex cut-formula plus 1. Hence, a cut-free derivation $\mathcal{D}$ has $d\mathcal{D} = 0$.

As we stated in section 2.4.3, the proof is based on two reduction lemmata. The first of them in HC framework is:

Lemma 4.17 *Let $\mathcal{D}_1$ and $\mathcal{D}_2$ be derivations such that*

1. *$\mathcal{D}_1$ is a derivation of $G \mid \varphi^i, \Gamma_1 \Rightarrow \Delta_1 \mid \ldots \mid \varphi^k, \Gamma_n \Rightarrow \Delta_n$;*

2. *$\mathcal{D}_2$ is a derivation of $H \mid \Pi \Rightarrow \Sigma, \varphi$;*

3. *$d\mathcal{D}_1 \leq d\varphi$ and $d\mathcal{D}_2 \leq d\varphi$;*

4. *φ is a principal formula of a logical rule in $\mathcal{D}_2$.*

Then a derivation $\mathcal{D}$ can be constructed of $H' = G \mid H \mid \Pi^i, \Gamma_1 \Rightarrow \Delta_1, \Sigma^i \mid \ldots \mid \Pi^k, \Gamma_n \Rightarrow \Delta_n, \Sigma^k$; with $d\mathcal{D} \leq d\varphi$.

PROOF: The proof is by induction on the height of $\mathcal{D}_1$. If it is an axiom, then we are done since either H' is also an axiom or is derivable from $H \mid \Pi \Rightarrow \Sigma, \varphi$ by external and internal weakenings. Otherwise, we must analyse the last rule applied in $\mathcal{D}_1$. In case the rule is applied to some element of G, the claim follows by the induction hypothesis and the application of the rule. If it is some nonmodal logical rule or $(\Box \Rightarrow)$ applied to some of the displayed sequents but with all occurrences of φ parametric, then the claim follows by the context independence of the rules, the induction hypothesis, and the application of the rule, or with the application of IW and IC in case φ is an auxiliary formula in the premise(s). Similarly, in case of structural rules; it makes no difference if W or $(Merge)$ applies.

In the case of nonmodal logical rules with φ principal, we must use additionally one or two cuts with premiss(es) of $H \mid \Pi \Rightarrow \Sigma, \varphi$; here the assumption 4 of the lemma is essential. Let us consider as an example the case of $\varphi = \psi \vee \chi$. We have two premises:

$$G \mid \psi \vee \chi^{i-1}, \psi, \Gamma_1 \Rightarrow \Delta_1 \mid \ldots \mid \psi \vee \chi^k, \Gamma_n \Rightarrow \Delta_n \text{ and}$$

$$G \mid \psi \vee \chi^{i-1}, \chi, \Gamma_1 \Rightarrow \Delta_1 \mid \ldots \mid \psi \vee \chi^k, \Gamma_n \Rightarrow \Delta_n.$$

By the induction hypothesis, we obtain derivations of

$$G \mid H \mid \Pi^{i-1}, \psi, \Gamma_1 \Rightarrow \Delta_1, \Sigma^{i-1} \mid \dots \mid \Pi^k, \Gamma_n \Rightarrow \Delta_n, \Sigma^k \text{ and}$$

$$G \mid H \mid \Pi^{i-1}, \chi, \Gamma_1 \Rightarrow \Delta_1, \Sigma^{i-1} \mid \dots \mid \Pi^k, \Gamma_n \Rightarrow \Delta_n, \Sigma^k$$

both of cut-degree $\leq d\varphi$. By two cuts with $H \mid \Pi \Rightarrow \Sigma, \psi, \chi$, followed by applications of IC and EC we obtain H'; the derivation has cut-degree $\leq d\varphi$ too.

The application of $(Merge)$ is not a problem; we obtain a result by the induction hypothesis and $(Merge)$. Similarly, with $(\Rightarrow \Box)$.

If the last rule is $(\Box \Rightarrow)$ with $\varphi = \Box\psi$ principal, we have

$$\frac{G \mid \Box\psi^{i-1}, \psi, \Gamma_1 \Rightarrow \Delta_1 \mid \dots \mid \Box\psi^k, \Gamma_n \Rightarrow \Delta_n}{G \mid \Box\psi \Rightarrow \mid \Box\psi^{i-1}, \psi, \Gamma_1 \Rightarrow \Delta_1 \mid \dots \mid \Box\psi^k, \Gamma_n \Rightarrow \Delta_n}$$

By the induction hypothesis, we obtain $H \mid G \mid \psi, \Gamma_1 \Rightarrow \Delta_1 \mid \dots \mid \Gamma_n \Rightarrow \Delta_n$. This by cut with $H \mid\Rightarrow \psi$ (which by the assumption 4 of the lemma must be the premiss of $H \mid\Rightarrow \varphi$) yields $H \mid H \mid G \mid \Gamma_1 \Rightarrow \Delta_1 \mid \dots \mid \Gamma_n \Rightarrow \Delta_n$ which after application of EC gives us the required result. Note that resulting derivation has cut-degree $\leq d\varphi$.

$\square$

Lemma 4.18 *Let $\mathcal{D}_1$ and $\mathcal{D}_2$ be derivations such that*

1. $\mathcal{D}_1$ *is a derivation of* $G \mid \Gamma_1 \Rightarrow \Delta_1, \varphi^i \mid \dots \mid \Gamma_n \Rightarrow \Delta_n, \varphi^k$;

2. $\mathcal{D}_2$ *is a derivation of* $H \mid \varphi, \Pi \Rightarrow \Sigma$;

3. $d\mathcal{D}_1 \leq d\varphi$ *and* $d\mathcal{D}_2 \leq d\varphi$;

Then a derivation $\mathcal{D}$ can be constructed of $H' = G \mid H \mid \Pi^i, \Gamma_1 \Rightarrow \Delta_1, \Sigma^i \mid \dots \mid \Pi^k, \Gamma_n \Rightarrow \Delta_n, \Sigma^k$; with $d\mathcal{D} \leq d\varphi$.

PROOF: The proof is by induction on the height of $\mathcal{D}_1$ and is very similar to the proof of the preceding lemma. The main difference is that when the last rule is logical with φ principal we refer not only to the induction hypothesis but also to lemma 4.17. Let us illustrate the point with $(\Rightarrow \Box)$ as the last rule. Then $\varphi = \Box\psi$ and it looks like that

$$\frac{G \mid \Gamma_1 \Rightarrow \Delta_1, \varphi^i \mid \dots \mid\Rightarrow \psi}{G \mid \Gamma_1 \Rightarrow \Delta_1, \varphi^i \mid \dots \mid\Rightarrow \varphi}$$

By the induction hypothesis and $(\Rightarrow \Box)$ we obtain a derivation of $G \mid H \mid \Pi^i, \Gamma_1 \Rightarrow \Delta_1, \Sigma^i \mid \dots \mid\Rightarrow \Box\psi$. Since $\Box\psi$ is principal the claim follows by lemma 4.17.

$\square$

Theorem 4.10 (Cut) *is eliminable.*

PROOF: Consider any derivation $\mathcal{D}$ with $d\mathcal{D} > 0$, The proof proceeds by a double induction on $d\mathcal{D}$ and $nd\mathcal{D}$—the number of applications of cut with cut-degree

$d\mathcal{D}$. We start with an uppermost application of cut with cut-degree $d\mathcal{D}$ and apply lemma 4.18 to its premises. As a result, we decrease either $d\mathcal{D}$ or $nd\mathcal{D}$. $\qquad\square$

Let us notice that from this proof of cut elimination yet another rule (or rather a pair of rules) may be extracted which we call $(WH - Mix)$ (W for weak)

$$\frac{G \mid \Gamma \Rightarrow \Delta, \varphi \qquad H \mid \varphi^i, \Sigma_1 \Rightarrow \Pi_1 \mid ... \mid \varphi^j, \Sigma_k \Rightarrow \Pi_k}{G \mid H \mid \Gamma^i, \Sigma_1 \Rightarrow \Pi_1, \Delta^i \mid ... \mid \Gamma^j, \Sigma_k \Rightarrow \Pi_k, \Delta^j}$$

and

$$\frac{G \mid \Gamma_1 \Rightarrow \Delta_1, \varphi^i \mid ... \mid \Gamma_n \Rightarrow \Delta_n, \varphi^j \qquad H \mid \varphi, \Sigma \Rightarrow \Pi}{G \mid H \mid \Sigma^i, \Gamma_1 \Rightarrow \Delta_1, \Pi^i \mid ... \mid \Sigma^j, \Gamma_n \Rightarrow \Delta_n, \Pi^j}$$

and again the version is possible which does not require deletion of all occurrences of φ in all sequents of one of the premises, which we can call $(WH - Multicut)$. The relations between these rules and $(H - Cut)$ are the same as in the case of $(SH - Mix)$ and $(SH - Multicut)$.

Eventually, we provide a proof of admissibility of cut for HG3 with refined Pottinger's rules in Dragalin-style. We follow Poggiolesi's proof as closely as possible but there are some significant differences, which will be commented on after presentation. Despite its semantical motivation, Poggiolesi's [198], system allows for very elegant syntactic proof of the admissibility of cut. It is constructed along the lines of Dragalin's proof for G3 and avoids many complications of the two proofs presented above. In her system $(Merge)$ is height-preserving admissible, which simplifies further steps in an essential way. We do not enter into details since the proof is described in [198] in an exact way, so we only sketch it. First of all, Poggiolesi must prove that axioms in atomic form may be generalised to arbitrary formula φ on both sides. The next step is the proof that $(Merge)$ is height-preserving admissible. From this follows height-preserving admissibility of IW and EW and height-preserving invertibility of logical rules. The additional machinery of admissible tools allows for a smooth proof of $(H - Cut)$.

As we noted the price for having a fully logical system (i.e. with no structural primitive rules) admitting a simple proof of admissibility of $(H - Cut)$ is a certain inelegance of having two rules for box introduction in the antecedent. We will finish this survey of proof methods with Dragalin-style proof for such a fully logical system but with only one rule for $(\square \Rightarrow)$.

One can easily check that for Pottinger's system with refined rules (see above) we obtain the usual preliminary results

Lemma 4.19 $\vdash \varphi, \Gamma \Rightarrow \Delta, \varphi$

PROOF: as usual by induction on the complexity of φ; left to the reader. $\qquad\square$

Lemma 4.20 *IW and EW are height-preserving admissible.*

PROOF: By induction on the height of the proof of the premise. Note that first we must prove both forms of IW since it is necessary for proving h-p amissibility of

EW. The former is straightforward since all rules, including both modal ones, are context independent. In proving EW, it is similar but the case when the last applied rule was $(\square \Rightarrow^{P'})$ needs IW. In this case, our G is of the form $\square\varphi, \Gamma_1 \Rightarrow \Delta_1 \mid \dots \mid \Gamma_n \Rightarrow \Delta_n$ inferred from $\square\varphi, \varphi, \Gamma_1 \Rightarrow \Delta_1 \mid \dots \mid \varphi, \Gamma_n \Rightarrow \Delta_n$. By the induction hypothesis, we obtain $\square\varphi, \varphi, \Gamma_1 \Rightarrow \Delta_1 \mid \dots \mid \varphi, \Gamma_n \Rightarrow \Delta_n \mid \Sigma \Rightarrow \Theta$ and then by h-p admissibility of IW we get $\square\varphi, \varphi, \Gamma_1 \Rightarrow \Delta_1 \mid \dots \mid \varphi, \Gamma_n \Rightarrow \Delta_n \mid \varphi, \Sigma \Rightarrow \Theta$ from which $\square\varphi, \Gamma_1 \Rightarrow \Delta_1 \mid \dots \mid \Gamma_n \Rightarrow \Delta_n \mid \Sigma \Rightarrow \Theta$ follows by $(\square \Rightarrow^{P'})$ as required. $\square$

Lemma 4.21 *All logical rules are height-preserving invertible.*

PROOF: By straightforward induction on the height of the proof of the premises. In case of boolean rules, it goes exactly as for ordinary G3; of course, we must take into account that modal rules are applied in the last step but it does not encounter any complications. Both modal rules are h-p invertible just by h-p admissibility of IW and EW. $\square$

Having all that we can prove:

Lemma 4.22 *Both forms of IC and EC is height-preserving admissible.*

PROOF: The proof is by induction on the height of the proof, and in general, is similar to the proof for G3. In case of IC when the contracted formula is parametric in both occurrences, we just apply the induction hypothesis and perform a suitable rule on the contracted premiss(es). Again, if the last rule is modal, this does not create any problem. When one of the occurrences of the contracted formula was principal, then in case of boolean formulae we apply h-p invertibility. In case of modal formulae, we need only to apply the induction hypothesis to the premiss since both rules are contraction-absorbing. For EC the situation is similar. Again in the case where the last rule was modal, the induction hypothesis justifies the claim. For example

$$(\square \Rightarrow^{P'}) \; \frac{\square\varphi, \varphi, \Gamma_1 \Rightarrow \Delta_1 \mid \square\varphi, \varphi, \Gamma_1 \Rightarrow \Delta_1 \mid \dots \mid \varphi, \Gamma_n \Rightarrow \Delta_n}{\square\varphi, \Gamma_1 \Rightarrow \Delta_1 \mid \square\varphi, \Gamma_1 \Rightarrow \Delta_1 \mid \dots \mid \Gamma_n \Rightarrow \Delta_n}$$

by the induction hypothesis and $(\square \Rightarrow^{P'})$ yields

$$(\square \Rightarrow^{P'}) \; \frac{\square\varphi, \varphi, \Gamma_1 \Rightarrow \Delta_1 \mid \dots \mid \varphi, \Gamma_n \Rightarrow \Delta_n}{\square\varphi, \Gamma_1 \Rightarrow \Delta_1 \mid \dots \mid \Gamma_n \Rightarrow \Delta_n}$$

In case of boolean rules, we must again apply h-p invertibility of rules. Let us illustrate this point with the case of $(\Rightarrow \vee)$ as the last applied rule

$$\frac{G \mid \Gamma \Rightarrow \Delta, \varphi \vee \psi \mid \Gamma \Rightarrow \Delta, \varphi, \psi}{G \mid \Gamma \Rightarrow \Delta, \varphi \vee \psi \mid \Gamma \Rightarrow \Delta, \varphi \vee \psi}$$

this is transformed into

$$\frac{G \mid \Gamma \Rightarrow \Delta, \varphi, \psi \mid \Gamma \Rightarrow \Delta, \varphi, \psi}{\dfrac{G \mid \Gamma \Rightarrow \Delta, \varphi, \psi}{G \mid \Gamma \Rightarrow \Delta, \varphi \vee \psi}}$$

where the first line is by height-preserving invertibility of $(\Rightarrow \vee)$ and the second by the induction hypothesis, so the height of the last line is the same as in the first proof.

Now we are ready to prove the admissibility of (H-Cut).

Theorem 4.11 *(H-Cut) is admissible.*

PROOF: It is similar to the Dragalin-style proof for G3, by induction on the complexity of the cut-formula and a subsidiary induction on the height of the proof. In particular, the cases where one premiss is an axiom or where the cut-formula is parametric are straightforward. The reader should check that reduction of the height of the premiss does not make any problems when the last rule is modal. As for the cases where in both premisses the cut-formula is principal the only difference is that we must apply EC after making cut on subformulae of cut-formula. The essentially different point is

$$(H\text{-}Cut) \ \frac{\dfrac{G \mid \Gamma \Rightarrow \Delta, \Box\varphi \mid \Rightarrow \varphi}{G \mid \Gamma \Rightarrow \Delta, \Box\varphi} \quad \dfrac{\Box\varphi, \varphi, \Sigma_1 \Rightarrow \Pi_1 \mid ... \mid \varphi, \Sigma_n \Rightarrow \Pi_n}{\Box\varphi, \Sigma_1 \Rightarrow \Pi_1 \mid ... \mid \Sigma_n \Rightarrow \Pi_n}}{G \mid \Gamma, \Sigma_1 \Rightarrow \Delta, \Pi_1 \mid ... \mid \Sigma_n \Rightarrow \Pi_n}$$

First, we must perform two cross-cuts on $\Box\varphi$ reducing the height in both cases

$$(H\text{-}Cut) \ \frac{G \mid \Gamma \Rightarrow \Delta, \Box\varphi \quad \Box\varphi, \varphi, \Sigma_1 \Rightarrow \Pi_1 \mid ... \mid \varphi, \Sigma_n \Rightarrow \Pi_n}{G \mid \varphi, \Gamma, \Sigma_1 \Rightarrow \Delta, \Pi_1 \mid ... \mid \varphi, \Sigma_n \Rightarrow \Pi_n}$$

$$(H\text{-}Cut) \ \frac{G \mid \Gamma \Rightarrow \Delta, \Box\varphi \mid \Rightarrow \varphi \quad \Box\varphi, \Sigma_1 \Rightarrow \Pi_1 \mid ... \mid \Sigma_n \Rightarrow \Pi_n}{G \mid \Gamma, \Sigma_1 \Rightarrow \Delta, \Pi_1 \mid ... \mid \Sigma_n \Rightarrow \Pi_n \mid \Rightarrow \varphi}$$

Now, we must apply $(H\text{-}Cut)$ on φ to both hypersequents derived above

$$(H\text{-}Cut) \ \frac{G \mid \Gamma, \Sigma_1 \Rightarrow \Delta, \Pi_1 \mid ... \mid \Sigma_n \Rightarrow \Pi_n \mid \Rightarrow \varphi \quad G \mid \varphi, \Gamma, \Sigma_1 \Rightarrow \Delta, \Pi_1 \mid ... \mid \varphi, \Sigma_n \Rightarrow \Pi_n}{G \mid G \mid \Gamma, \Sigma_1 \Rightarrow \Delta, \Pi_1 \mid ... \mid \Sigma_n \Rightarrow \Pi_n \mid \Gamma, \Sigma_1 \Rightarrow \Delta, \Pi_1 \mid ... \mid \varphi, \Sigma_n \Rightarrow \Pi_n}$$

We repeat $(H-Cut)$ on this sequent again with the conclusion of the previous $(H-Cut)$ systematically $n-1$ times to all displayed occurrences of φ. Since we are cutting φ, all performed cuts are eliminable by the induction hypothesis on the complexity. Finally, by EC, we obtain the desired result. $\qquad\square$

By the way, one may notice why Pottinger's original rules do not work for such a proof. First when applying $(\Rightarrow \Box^P)$ if some boxed formulae are in Γ we will have $\Gamma^\Box \Rightarrow \varphi$ instead of $\Rightarrow \varphi$ and this sequent when mixed with some $\varphi, \Sigma_i \Rightarrow \Pi_i$ yields $\Gamma^\Box, \Sigma_i \Rightarrow \Pi_i$. There seems to be no way to get rid of $\Gamma^\Box$ in any such case to obtain the desired result. Second, in the application of $(\Box \Rightarrow^P)$ we have $\Box\varphi$ instead of φ added to every $\Sigma_i \Rightarrow \Pi_i$ and it is impossible to perform a reduction on cut-formula complexity in the series of steps described above.

We finish with a comparison of this proof with a proof of Poggiolesi. She is proving h-p admissibility of $(Merge)$ from scratch and then applies this rule for obtaining other results. For the modified Pottinger's rule, it is not possible. Consider the case when the last applied rule was $(\Box \Rightarrow^{P'})$, so we have

$$(\Box \Rightarrow^{P'}) \ \frac{\Box\varphi, \varphi, \Gamma_1 \Rightarrow \Delta_1 \mid \varphi, \Gamma_2 \Rightarrow \Delta_2 \mid ... \mid \varphi, \Gamma_n \Rightarrow \Delta_n}{\Box\varphi, \Gamma_1 \Rightarrow \Delta_1 \mid \Gamma_2 \Rightarrow \Delta_2 \mid ... \mid \Gamma_n \Rightarrow \Delta_n}$$

and we want to show that $\Box\varphi, \Gamma_1, \Gamma_2 \Rightarrow \Delta_1, \Delta_2 \mid ... \mid \Gamma_n \Rightarrow \Delta_n$ is provable. But by the induction hypothesis we obtain $\Box\varphi, \varphi, \varphi, \Gamma_1, \Gamma_2 \Rightarrow \Delta_1, \Delta_2 \mid ... \mid \varphi, \Gamma_n \Rightarrow \Delta_n$ and there is no way to derive the desired result without using IC.

4.8   Pairs of Sequents are Sufficient

We emphasised that HC is a rather weak generalisation of standard SC but still sufficient to increase the expressive power of the ordinary Gentzen apparatus. But for many modal logics (for example symmetric ones like **B**), and in general, many non-classical logics, this is not sufficient. Nested sequents (or tree hypersequents) introduced first by Bull [42] and Kashima [147], provide an even more general framework but also a more complicated one. We do not aim to introduce this approach since one may find an excellent presentation in Poggiolesi [198][15]. The general idea is that other sequents may be elements of sequents, so that such an object corresponds to a tree of sequents. Lellman [161], noticed that in many cases we can obtain a simpler form by allowing only one sequent to be nested. In this way, we obtain something which may be interpreted either as a restricted form of nested sequents or something like hypersequents with a linear ordering of sequents. Hence, they may be called linear nested sequents (Lellmann [161]) or non-commutative hypersequents (Indrzejczak [136]). Even this half-way solution (between hypersequents and nested sequents) offers greater flexibility. For example, Indrzejczak [136] provided cut-free systems for many temporal logics in this framework. In general, providing constructive proofs of cut admissibility is difficult for such systems, but not impossible (see e.g. Indrzejczak [140, 143], Kuznets and Lellmann [155]). On the other hand, semantical proofs of completeness may be easily obtained at least for some such systems since every linear nested sequent, in fact, simulates a stage in proof search along the lines of section 4.4. (see Lellmann [161]). Further generalisations obtained by Baelde, Lick and Schmitz [20], where clusters of sequents are admitted as components of linear nested sequents, lead to

---

[15]Other variants of such systems were developed, for example, by Stouppa [246], on the basis of earlier systems of deep inference of slightly different character

decision procedures with optimal efficiency for temporal logics. All these versions of many sequent calculi may be seen as specialised forms of the general approach of Došen [60], and it seems that the potential applications of this framework are far from being fully examined.

However, taking into account our restricted interests in **S5** only we may ask if in this case we cannot find an even simpler solution. In the framework of labelled calculi, the notion of a label was extremely simplified in case of **S5**. Perhaps for other approaches, the situation is similar. Below we present two systems that may be seen either as hypersequent calculi, or alternatively as nested sequent calculi of a very modest kind, using only (ordered) pairs of sequents, which we call bisequents. Both bisequent calculi are cut-free, moreover for one of them cut elimination is proved constructively. To show the expressive power of this simplified version of HC we first recall two calculi belonging to different families of generalised SCs but we will eventually show that both may be translated into bisequents. The first system is due to Sato and provides another example of structured SC operating on 4-argument sequents which we already introduced in section 4.4. The second is due to Indrzejczak and belongs to the family of multisequent calculi, where different kinds of sequents are used in one system.

4.8.1 Sato's Structured System

The original system of Sato is formulated by means of 4-argument sequents, similarly as the system of Heuerding, Seyfried and Zimmermann [115]. However, the interpretation is different, and to make differences more visible we put additional arguments of a sequent inside, i.e. immediately on two sides of $\Rightarrow$ (in fact, in accordance with Sato's original convention). So we are using sequents of the form $\Gamma [\Pi] \Rightarrow [\Sigma] \Delta$ where arguments are defined as sets of formulae. We will call them external (for Γ, Δ) and internal (for Π, Σ) arguments or contexts. The language contains only $\bot, \rightarrow$ and $\Box$. Most of the rules are defined only for the external arguments and they are, in fact, performed also when the internal arguments are empty sets. So following Sato, we omit [ ] on both sides of an arrow in these cases, i.e. instead of writing $\Gamma [\,] \Rightarrow [\,] \Delta$ we will just write $\Gamma \Rightarrow \Delta$. Hence, sequents with empty internal arguments are just standard sequents.

Sato's system consists of axioms $\varphi \Rightarrow \varphi$ and $\bot \Rightarrow$ and the following rules:

$$(EW) \ \frac{\Gamma \Rightarrow \Delta}{\Pi, \Gamma \Rightarrow \Delta, \Sigma} \qquad (IW) \ \frac{\Gamma [\Pi] \Rightarrow [\Sigma] \Delta}{\Gamma [\Lambda, \Pi] \Rightarrow [\Sigma, \Theta] \Delta}$$

$$(E \rightarrow \Rightarrow) \ \frac{\Gamma \Rightarrow \Delta, \varphi, \psi \qquad \varphi, \Pi \Rightarrow \Sigma, \psi \qquad \varphi, \psi, \Lambda \Rightarrow \Theta}{\varphi \rightarrow \psi, \Gamma, \Pi, \Lambda \Rightarrow \Delta, \Sigma, \Theta}$$

$$(E \Rightarrow \rightarrow) \ \frac{\varphi, \Gamma \Rightarrow \Delta, \psi}{\Gamma \Rightarrow \Delta, \varphi \rightarrow \psi} \qquad (I \Rightarrow \rightarrow) \ \frac{\Gamma [\varphi, \Pi] \Rightarrow [\Sigma, \psi] \Delta}{\Gamma [\Pi] \Rightarrow [\Sigma, \varphi \rightarrow \psi] \Delta}$$

$$(I\rightarrow\Rightarrow)\ \frac{\Gamma\ [\Pi]\Rightarrow[\Sigma,\varphi,\psi]\ \Delta\qquad\Gamma\ [\varphi,\Lambda]\Rightarrow[\Theta,\psi]\ \Delta\qquad\Gamma\ [\varphi,\psi,\Xi]\Rightarrow[\Omega]\ \Delta}{\Gamma\ [\varphi\rightarrow\psi,\Pi,\Lambda,\Xi]\Rightarrow[\Sigma,\Theta,\Omega]\ \Delta}$$

$$(\Box\Rightarrow)\ \frac{\varphi,\Gamma\Rightarrow\Delta}{\Box\varphi,\Gamma\Rightarrow\Delta}\qquad\qquad(\Rightarrow\Box)\ \frac{\Box\Gamma\Rightarrow\Box\Delta,\varphi}{\Box\Gamma\Rightarrow\Box\Delta,\Box\varphi}$$

$$(Enter\Rightarrow)\ \frac{\Gamma,\Box\varphi\ [\Pi]\Rightarrow[\Sigma]\ \Delta}{\Gamma\ [\Box\varphi,\Pi]\Rightarrow[\Sigma]\ \Delta}\qquad(\Rightarrow Enter)\ \frac{\Gamma\ [\Pi]\Rightarrow[\Sigma]\ \Box\varphi,\Delta}{\Gamma\ [\Pi]\Rightarrow[\Sigma,\Box\varphi]\ \Delta}$$

$$(\Rightarrow[\Box])\ \frac{\Gamma\ [\ ]\Rightarrow[\varphi]\ \Delta}{\Gamma\ [\ ]\Rightarrow[\ ]\ \Box\varphi,\Delta}\qquad(Cut)\ \frac{\Gamma\Rightarrow\Delta,\varphi\qquad\varphi,\Pi\Rightarrow\Sigma}{\Gamma,\Pi\Rightarrow\Delta,\Sigma}$$

A proof is defined in a standard way but only for standard sequents as possible roots of the proof tree. Note that the treatment of external and internal (i.e. inside [ ]) arguments is not the same. There are no rules for $\Box$ inside [ ], only the rules allowing for the transfer to external arguments, also axioms are defined only on external arguments. So external arguments are privileged and the role of internal context is, in fact, quite similar to the role of the respective components in Heuerding, Seyfried and Zimmermann's approach. However, they are not just 'history'-boxes; we can apply in them weakening and Boolean rules but only until we do not obtain $\Box$-formulae which may be transported back to an external context. The specific form of the rules for $\rightarrow\Rightarrow$ is not accidental, which will be evident in the proof of completeness.

Since the rules of external context are just Ohnishi and Matsumoto's rules of LKS5 (in restricted boolean language) there is no problem in proving that

Lemma 4.23 *If* $\vdash\Gamma\Rightarrow\Delta$ *in LKS5, then* $\vdash\Gamma\Rightarrow\Delta$ *in Sato's system.*

PROOF: By induction on the height of a proof. Since all rules of LKS5 are directly simulated it is enough to show that ordinary $(\rightarrow\Rightarrow)$ is provable in Sato's system which is left to the reader (cf. a consideration on 3−premiss variants of standard rules in section 1.4). $\qquad\Box$

So the system is complete. But the addition of internal context allows to build cut-free proofs of sequents (in fact, always) which in Ohnishi and Matsumoto's system are not cut-free provable. For example, a cut-free proof of B is straightforward

$$(EW) \frac{\dfrac{p \Rightarrow p}{p \Rightarrow p, \bot} \qquad \dfrac{p \Rightarrow p}{p, \bot \Rightarrow p} \qquad \dfrac{\bot \Rightarrow}{p, p, \bot \Rightarrow}}{}$$

$$(\square \Rightarrow) \frac{p, p \to \bot \Rightarrow}{p, \square(p \to \bot) \Rightarrow}$$

$$(IW) \frac{}{p, \square(p \to \bot) [\,] \Rightarrow [\bot]}$$

$$(Enter \Rightarrow) \frac{}{p \, [\square(p \to \bot)] \Rightarrow [\bot]}$$

$$(\Rightarrow \to) \frac{}{p \, [\,] \Rightarrow [\square(p \to \bot) \to \bot]}$$

$$(\Rightarrow \square) \frac{}{p \Rightarrow \square(\square(p \to \bot) \to \bot)}$$

$$(\Rightarrow \to) \frac{}{\Rightarrow p \to \square(\square(p \to \bot) \to \bot)}$$

Sato provides the following interpretation of his sequents:

$$\Im(\Gamma \, [\Pi] \Rightarrow [\Sigma] \, \Delta) = \Gamma \Rightarrow \Delta, \square(\wedge\Pi \to \vee\Sigma)$$

with $\top$ representing empty Π and $\bot$ empty Σ.

On the basis of this translation, we can either prove the validity-preservation of all rules and demonstrate the soundness of the system, or provide a syntactical proof of the converse of lemma 4.23. Following Sato, we choose the latter option

Lemma 4.24 *If $\vdash \Gamma \Rightarrow \Delta$ in Sato's system, then $\vdash \Gamma \Rightarrow \Delta$ in LKS5.*

PROOF: Again by induction on the height of a proof. We must prove that all Sato's rules are derivable in Ohnishi and Matsumoto's system under translation. Let us show the case of $(Enter \Rightarrow)$ which under translation is

$$\frac{\Gamma, \square\varphi \Rightarrow \Delta, \square(\pi \to \sigma)}{\Gamma \Rightarrow \Delta, \square(\square\varphi \wedge \pi \to \sigma)}$$

where π represents $\wedge\Pi$ and σ $\vee\Sigma$.

$$
\frac{
  \dfrac{
    \dfrac{
      \dfrac{
        \dfrac{
          \dfrac{\Gamma, \square\varphi \Rightarrow \Delta, \square(\pi \to \sigma)}{\Gamma \Rightarrow \Delta, \square\varphi \to \square(\pi \to \sigma)}
        }{}
      }{}
    }{}
  }{}
  \qquad
  \dfrac{
    \dfrac{
      \dfrac{
        \dfrac{
          \dfrac{
            \dfrac{
              \dfrac{
                \dfrac{\square\varphi \Rightarrow \square\varphi}{\square\varphi \Rightarrow \square\varphi, \sigma} (\Rightarrow W)
              }{\square\varphi \wedge \pi \Rightarrow \square\varphi, \sigma} (\wedge \Rightarrow)
            }{\Rightarrow \square\varphi, \square\varphi \wedge \pi \to \sigma} (\Rightarrow \to)
          }{\Rightarrow \square\varphi, \square(\square\varphi \wedge \pi \to \sigma)} (\Rightarrow \square)
          \qquad
          \dfrac{
            \dfrac{
              \dfrac{
                \dfrac{\dfrac{\pi \Rightarrow \pi \qquad \sigma \Rightarrow \sigma}{\pi, \pi \to \sigma \Rightarrow \sigma} (\to \Rightarrow)}{\square\varphi \wedge \pi, \pi \to \sigma \Rightarrow \sigma} (\wedge \Rightarrow)
              }{\pi \to \sigma \Rightarrow \square\varphi \wedge \pi \to \sigma} (\Rightarrow \to)
            }{\square(\pi \to \sigma) \Rightarrow \square\varphi \wedge \pi \to \sigma} (\square \Rightarrow)
          }{\square(\pi \to \sigma) \Rightarrow \square(\square\varphi \wedge \pi \to \sigma)} (\Rightarrow \square)
        }{\square\varphi \to \square(\pi \to \sigma) \Rightarrow \square(\square\varphi \wedge \pi \to \sigma), \square(\square\varphi \wedge \pi \to \sigma)} (\to \Rightarrow)
      }{\square\varphi \to \square(\pi \to \sigma) \Rightarrow \square(\square\varphi \wedge \pi \to \sigma)} (\to C)
    }{}
  }{}
}{\Gamma \Rightarrow \Delta, \square(\square\varphi \wedge \pi \to \sigma)} (Cut)
$$

$\square$

Exercise 4.38 *Complete the proof of lemma 4.24.*
Show validity-preservation of Sato's rules.

Sato did not provide a syntactic proof of cut elimination for this system (he did it for SC presented in [225], instead) and it is not clear if we can provide such a proof. Therefore, we present (essentially the original Sato's) proof of the completeness of the cut-free version of his system.

The definition of saturated sequent is a bit different. $\Gamma \Rightarrow \Delta$ is saturated iff

1. $\Gamma \Rightarrow \Delta$ is unprovable (in the system without cut)

2. if $\varphi \to \psi \in \Gamma$ or $\varphi \Rightarrow \psi \in \Delta$, then $\{\varphi, \psi\} \subseteq \Gamma \cup \Delta$

3. if $\Box\varphi \in \Gamma$, then $\varphi \in \Gamma$

4. if $\Box\varphi \in \Delta$, then $SF_\Box(\varphi) \subseteq \Gamma \cup \Delta$

   where $SF_\Box(\varphi) = \{\Box\psi : \Box\psi \in SF(\varphi)\}$

Lemma 4.25 *If $\Gamma \Rightarrow \Delta$ is unprovable, then there is saturated $\Pi \Rightarrow \Sigma$ such that $\Gamma \subseteq \Pi$ and $\Delta \subseteq \Sigma$.*

PROOF: We construct a finite sequence of unprovable sequents built from $SF(\Gamma \cup \Delta)$ in the same way as in the proof of lemma 1.26 with $\Gamma_1 \Rightarrow \Delta_1 = \Gamma \Rightarrow \Delta$.

   If at the stage n a sequent $\Gamma_n \Rightarrow \Delta_n$ is defined and it is saturated, then $\Gamma_n \Rightarrow \Delta_n = \Pi \Rightarrow \Sigma$. Otherwise, some of the conditions 2-4 is not satisfied. If condition 2 is not satisfied then, if $\varphi \to \psi$ is in Δ_n, we take $\varphi, \Gamma_n \Rightarrow \Delta_n, \psi = \Gamma_{n+1} \Rightarrow \Delta_{n+1}$. It must be unprovable, otherwise $\Gamma_n \Rightarrow \Delta_n$ would be provable by $(\Rightarrow \to)$. If it is in Γ, then for the same reason at least one of the $\Gamma_n \Rightarrow \Delta_n, \varphi, \psi$ or $\psi, \Gamma_n \Rightarrow \Delta_n, \varphi$ or $\varphi, \psi, \Gamma_n \Rightarrow \Delta_n$ must be unprovable and we take it as $\Gamma_{n+1} \Rightarrow \Delta_{n+1}$. Condition 3 is obvious so it remains to consider condition 4.

   Assume that $\Box\varphi \in \Delta_n$ but it does not hold that $SF_\Box(\varphi) \subseteq \Gamma_n \cup \Delta_n$. Let $\Box\psi$ be a formula of maximal complexity from $SF_\Box(\varphi) - \Gamma_n \cup \Delta_n$. Since $\Box\varphi \in \Delta_n$ and $\Box\psi \in SF(\varphi)$ there must be at least one $\Box\chi \in \Gamma_n \cup \Delta_n$ such that $\Box\psi \in SF(\chi)$ ($\Box\varphi$ is such if no other). If there is more than one, take a formula $\Box\chi$ with minimal complexity. If $\Box\chi \in \Gamma_n$, then $\chi, \Gamma_n \Rightarrow \Delta_n$ is unprovable, otherwise $\Gamma_n \Rightarrow \Delta_n$ would be provable by $(\Box \Rightarrow)$. Since $\Box\psi \in SF(\chi)$ but not in the scope of any other $\Box$ then if we apply (root-first) rules for implication and EW to $\chi, \Gamma_n \Rightarrow \Delta_n$ we obtain eventually either $\Box\psi, \Gamma_n \Rightarrow \Delta_n$ or $\Gamma_n \Rightarrow \Delta_n, \Box\psi$. But both cannot be provable, otherwise $\chi, \Gamma_n \Rightarrow \Delta_n$ would be provable. So one of these sequents is taken as $\Gamma_{n+1} \Rightarrow \Delta_{n+1}$. If $\Box\chi \in \Delta_n$, then $\Gamma_n \, [\,] \Rightarrow [\chi] \, \Delta_n$ is unprovable, otherwise $\Gamma_n \Rightarrow \Delta_n$ would be provable by $(\Rightarrow \Box)$. Now by the internal rules for implication and IW, we obtain eventually either $\Gamma_n \, [\Box\psi] \Rightarrow [\,] \, \Delta_n$ or $\Gamma_n \, [\,] \Rightarrow [\Box\psi] \, \Delta_n$. In both the cases by Enter-rules either $\Box\psi, \Gamma_n \Rightarrow \Delta_n$ or $\Gamma_n \Rightarrow \Delta_n, \Box\psi$ must be unprovable and is taken as $\Gamma_{n+1} \Rightarrow \Delta_{n+1}$. $\Box$

Remark 4.2 *Careful analysis of the proof of the last condition shows why Sato used a $3-$premiss variant of $(\to\Rightarrow)$. For example, let $\chi := \Box\psi \to \varphi$, then applying standard $(\to\Rightarrow)$ we obtain only $\Gamma_n \Rightarrow \Delta_n, \Box\psi$ and $\varphi, \Gamma_n \Rightarrow \Delta_n$. It may be the case that the latter is unprovable and condition 4 is still not satisfied.*

Note that the model is infinite and we do not obtain a decision procedure for **S5** in this way, in contrast to the construction with analytic cut described in section 4.5.

4.8.2 Double Sequent Calculus

The system of Indrzejczak [125], uses two kinds of sequents. In addition to ordinary sequents, there are modal ones of the form $\Gamma \,\Box{\Rightarrow}\Delta$. If it is inessential whether a standard or a modal sequent is applied both kinds are denoted as $\Gamma \,(\Box){\Rightarrow} \Delta$. The idea of using special kinds of sequents is due to Curry [56], and it was also used by Zeman [276]. In both the cases, additional sequents were introduced to express the modal character of suitable operations. In fact, its use in Curry's formulation of **S4** is not necessary; in Zeman it is essential for obtaining a modal rule characterising **S4.2**, a logic not dealt with in this book. Two kinds of sequents were applied also in Avron, Honsell, Miculan and Paravano [13], but in a totally different character. In their system, two kinds of sequents correspond to two different deducibility relation (see section 4.1). Indrzejczak introduced a general construction where several types of modal [126], and temporal sequents [127] were applied in one SC but in case of **S5** a considerable reduction is possible to the effect that only one type of modal sequent is required. Below we briefly describe this system; in addition to [125] one may find a fuller account and comparison with other approaches in Poggiolesi [198] and Wansing [270].

In addition to modal sequents, the language is enriched with a special structural operation of transition (from one argument of a sequent to another). It is unary like negation but cannot be iterated; it is only allowed to be added in front of a formula or to be deleted. We will use a sign $-$ for it, so any formula φ may be transformed into a shifted formula $-\varphi$. In the schemata we will use a convention φ^* in the sense that for ordinary formula φ, $\varphi^* = -\varphi$ and $(-\varphi)^* = \varphi$. Also $\Gamma^* = \{\varphi^* : \varphi \in \Gamma\}$.

Most rules are standard and work the same way on both kinds of sequents. However, in order to block the uncontrolled transition from one side of a sequent to the other for negation and implication, we have symmetric variants (like in SC used for interpolation proofs in section 3.7)

$$(\neg{\Rightarrow}) \quad \frac{-\varphi, \Gamma(\Box){\Rightarrow} \Delta}{\neg\varphi, \Gamma(\Box){\Rightarrow} \Delta} \qquad ({\Rightarrow}\neg) \quad \frac{\Gamma(\Box){\Rightarrow} \Delta, -\varphi}{\Gamma(\Box){\Rightarrow} \Delta, \neg\varphi}$$

$$({\Rightarrow}{\rightarrow}) \quad \frac{\Gamma(\Box){\Rightarrow} \Delta, -\varphi, \psi}{\Gamma(\Box){\Rightarrow} \Delta, \varphi \rightarrow \psi} \qquad ({\rightarrow}{\Rightarrow}) \quad \frac{-\varphi, \Gamma(\Box){\Rightarrow} \Delta \quad \psi, \Gamma(\Box){\Rightarrow} \Delta}{\varphi \rightarrow \psi, \Gamma(\Box){\Rightarrow} \Delta}$$

Clearly, Γ and Δ may contain ordinary formulae, as well as shifted formulae; the same remark applies to further rules. We need special rules for transition of the form

$$({\Rightarrow}*) \quad \frac{\varphi, \Gamma \Rightarrow \Delta}{\Gamma \Rightarrow \Delta, \varphi^*} \qquad (*{\Rightarrow}) \quad \frac{\Gamma \Rightarrow \Delta, \varphi}{\varphi^*, \Gamma \Rightarrow \Delta} \qquad (TR) \quad \frac{\Gamma \,\Box{\Rightarrow} \Delta}{\Delta^* \,\Box{\Rightarrow} \Gamma^*}$$

and modal rules

$$(\Box \Rightarrow) \ \frac{\varphi, \Gamma(\Box) \Rightarrow \Delta}{\Box\varphi, \Gamma(\Box) \Rightarrow \Delta} \qquad (\Rightarrow \Box) \ \frac{\Gamma \Box \Rightarrow M\Delta, \varphi}{\Gamma \Rightarrow M\Delta, \Box\varphi} \qquad (NC) \ \frac{\Gamma \Rightarrow \Delta}{\Gamma \Box \Rightarrow \Delta}$$

where $M\Delta$ contains only M-formulae of the form $\Box\psi, -\Box\psi$ and in (NC) one of the Γ, Δ is either empty or contains only M-formulae. If we admit $\Diamond$ as a primitive operator, we have dual rules for it and the notion of M-formula is extended to include $\Diamond\psi, -\Diamond\psi$.

Similarly to Sato's system, only standard sequents are to be proved and modal sequents may occur only as nodes of proof trees. Here is an example of a proof

$$
\begin{array}{l}
(\Box \Rightarrow) \ \dfrac{p \Rightarrow p}{\Box p \Rightarrow p} \\[4pt]
(NC) \ \dfrac{}{\Box p \Box \Rightarrow p} \\[4pt]
(TR) \ \dfrac{}{-p \Box \Rightarrow \ -\Box p} \\[4pt]
(\Rightarrow \neg) \ \dfrac{}{-p \Box \Rightarrow \neg\Box p} \\[4pt]
(\Rightarrow \Box) \ \dfrac{}{-p \Rightarrow \Box\neg\Box p} \\[4pt]
(\neg \Rightarrow) \ \dfrac{}{\neg p \Rightarrow \Box\neg\Box p} \\[4pt]
(\Rightarrow *) \ \dfrac{}{\Rightarrow -\neg p, \Box\neg\Box p} \\[4pt]
(\Rightarrow \to) \ \dfrac{}{\Rightarrow \neg p \to \Box\neg\Box p}
\end{array}
$$

Exercise 4.39 *Prove axioms K, 4 and 5.*

It is easy to prove soundness under syntactic translation where standard sequents are dealt with as Gentzen transforms with the addition that shift formulae are translated as negations. Modal sequents are translated as $\wedge\Gamma \to \Box(\vee\Delta)$ with the same proviso for shift formulae.

Exercise 4.40 *Prove validity-preservation of modal rules and (NC).*

This system is cut-free and has a generalised subformula property in the sense that the only formulae which must occur in any proof of $\Gamma \Rightarrow \Delta$ are of the form $\varphi, -\varphi$ for every $\varphi \in SF(\Gamma \cup \Delta)$. Completeness and decidability is proved by Hintikka-style argument. We omit the details and only point out that for the need of exhaustive proof search procedure (NC) is replaced with two disjunctive branching rules (see subsection 4.4.1)

$$(NCG) \ \frac{\Gamma \Rightarrow \Sigma \qquad \Rightarrow \Delta, \Sigma}{\Gamma \Box \Rightarrow \Delta, \Sigma} \qquad \frac{\Gamma, \Sigma \Rightarrow \qquad \Sigma \Rightarrow \Delta}{\Gamma, \Sigma \Box \Rightarrow \Delta}$$

where Σ contains only M-formulae and Γ, Δ only atomic ones (including shift formulae). Also, some other minor changes are needed like introducing a contraction-absorbing version of $(\Box \Rightarrow)$ and the elimination of weakening.

4.8.3 Bisequent Calculi

Although the two systems presented above in this section differ significantly, we can find a uniform perspective which shows hidden similarities between them. In both the cases, we can operate not only on single but also on pairs of states in one sequent. It will be more transparent if we just use hypersequents but limited to two sequents only and provide a translation. We will call such hypersequents as just bisequents. The obvious translation to this setting is the following: Sato's structured sequent $\Gamma\ [\Pi] \Rightarrow [\Sigma]\ \Delta$ is represented as $\Gamma \Rightarrow \Delta \mid \Pi \Rightarrow \Sigma$ and the latter is equivalent to modal sequent $\Gamma, -\Delta\ \Box\!\Rightarrow -\Pi, \Sigma$.

Exercise 4.41 *Provide a translation of transitional modal rules from both SC into bisequent framework.*

In this way, we can observe an inherent similarity of these generalised calculi to the hypersequent approach and, moreover, it shows that in case of **S5** the framework of hypersequents may be simplified in a similar way as it was done in the labelled framework. This time there is a similarity with respect to the cardinality of the multiset of admissible elements of any hypersequent. Below, we provide yet another set (or rather two) of rules which are extremely simple and easy to use.

In general, we are using bisequents but in case one component is empty we can omit it, and a bisequent with single nonempty sequent is just a standard sequent. But in contrast to Sato's system in both components, we have the same set of rules. For simplicity, but also because of some further changes in the system, in the schemata of the rules we will state the active components always in the left sequent but in the course of the proof, they are allowed in both sequents. For the classical basis, we just take LK but in case of two-premiss rules, we keep the second, non-active component the same in both premisses. On the contrary for cut, we admit different sequents which are mixed in the conclusion, hence we have

$$(\rightarrow\Rightarrow)\ \frac{\Gamma \Rightarrow \Delta, \varphi \mid \Pi \Rightarrow \Sigma \qquad \psi, \Lambda \Rightarrow \Theta \mid \Pi \Rightarrow \Sigma}{\varphi \rightarrow \psi, \Gamma, \Lambda \Rightarrow \Delta, \Theta \mid \Pi \Rightarrow \Sigma}$$

but

$$(Cut)\ \frac{\Gamma \Rightarrow \Delta, \varphi \mid \Lambda \Rightarrow \Theta \qquad \varphi, \Pi \Rightarrow \Sigma \mid \Xi \Rightarrow \Omega}{\Gamma, \Pi \Rightarrow \Delta, \Sigma \mid \Lambda, \Xi \Rightarrow \Theta, \Omega}$$

Clearly, in both cases the parametric (non-active) component of a bisequent may be put on the left or on the right; the latter is only a convention. In addition to ordinary structural rules W and C, we have two transitional structural rules

$$(TR\Rightarrow)\ \frac{\Box\varphi, \Gamma \Rightarrow \Delta \mid \Pi \Rightarrow \Sigma}{\Gamma \Rightarrow \Delta \mid \Box\varphi, \Pi \Rightarrow \Sigma} \qquad (\Rightarrow TR)\ \frac{\Gamma \Rightarrow \Delta, \Box\varphi \mid \Pi \Rightarrow \Sigma}{\Gamma \Rightarrow \Delta \mid \Pi \Rightarrow \Sigma, \Box\varphi}$$

and two modal (static) rules

$$(\Box\Rightarrow)\ \frac{\varphi, \Gamma \Rightarrow \Delta \mid \Pi \Rightarrow \Sigma}{\Box\varphi, \Gamma \Rightarrow \Delta \mid, \Pi \Rightarrow \Sigma} \qquad (\Rightarrow\Box)\ \frac{\Rightarrow \varphi \mid \Gamma \Rightarrow \Delta}{\Rightarrow \Box\varphi \mid \Gamma \Rightarrow \Delta}$$

This system will be called BSC1. The proof is defined as a tree of bisequents. As we can see the only transitional rules are just the enter-rules of Sato, but in contrast to his system both sequents in the present system are treated on a par, hence we have an admissible transition in both directions and all rules allowed in both components. It is interesting to compare this pair of modal rules with those of Sato and with hypersequent rules of Mints, Poggiolesi and Restall. In Sato's system $(\Rightarrow [\Box])$ is transitional but $(\Rightarrow \Box)$ is static whereas in Restall's system it is just the contrary. In Poggiolesi's system, both rules are transitional (and the former also static—we have two rules of $(\Box \Rightarrow))$, whereas in this system both rules are static. It makes the system more uniform and elegant; all logical rules are static and the only transitional rules are structural. But it also has a more important aspect—such a set of rules allows, as we will see, to obtain a cut-free system.

As we can see all rules are symmetric, explicit and separate. We can easily add rules for $\Diamond$ which are just duals of those for $\Box$ (and of course suitable transitional rules). So we have

$$(\Rightarrow\Diamond) \ \frac{\Gamma \Rightarrow \Delta, \varphi \mid \Pi \Rightarrow \Sigma}{\Gamma \Rightarrow \Delta, \Diamond\varphi, \mid, \Pi \Rightarrow \Sigma} \qquad (\Diamond\Rightarrow) \ \frac{\varphi \Rightarrow \mid \Gamma \Rightarrow \Delta}{\Diamond\varphi \Rightarrow \mid \Gamma \Rightarrow \Delta}$$

$$(TR\Rightarrow) \ \frac{\Diamond\varphi, \Gamma \Rightarrow \Delta \mid \Pi \Rightarrow \Sigma}{\Gamma \Rightarrow \Delta \mid \Diamond\varphi, \Pi \Rightarrow \Sigma} \qquad (\Rightarrow TR) \ \frac{\Gamma \Rightarrow \Delta, \Diamond\varphi \mid \Pi \Rightarrow \Sigma}{\Gamma \Rightarrow \Delta \mid \Pi \Rightarrow \Sigma, \Diamond\varphi}$$

Soundness is easy to prove. We proceed as in the case of hypersequents and apply the same translation.

Exercise 4.42 *Prove validity-preservation or derivability of translation of rules in* H-**S5**

Proofs of the axioms are very easy, for example 5

$$
(TR \Rightarrow) \ \frac{\Box p \Rightarrow \Box p \mid \Rightarrow}{
(\neg \Rightarrow) \ \frac{\Rightarrow \Box p \mid \Box p \Rightarrow}{
(\Rightarrow \neg) \ \frac{\neg\Box p \Rightarrow \mid \Box p \Rightarrow}{
(\Rightarrow \Box) \ \frac{\neg\Box p \Rightarrow \mid \Rightarrow \neg\Box p}{
(\Rightarrow TR) \ \frac{\neg\Box p \Rightarrow \mid \Rightarrow \Box\neg\Box p}{
(\Rightarrow\rightarrow) \ \frac{\neg\Box p \Rightarrow \Box\neg\Box p}{
\Rightarrow \neg\Box p \rightarrow \Box\neg\Box p}}}}}}
$$

Exercise 4.43 *Provide proofs of the remaining axioms and B.*

Since (GR) is just $(\Rightarrow \Box)$ and MP is simulated by cut, we easily obtain completeness. What with cut elimination? One may also prove that the rules of

Indrzejczak's system are derivable in BSC1 without cut under the translation we have provided[16].

Exercise 4.44 *Show derivability of all rules of double sequent calculus in cut-free BSC1.*

Since both systems are cut-free we obtain indirectly a proof that the cut-free version of the present calculus is also complete. Hence, we have

Theorem 4.12 $\Gamma \vdash_{S5} \varphi$ *iff* $BSC1 \vdash \Gamma \Rightarrow \varphi$ *iff* $BSC1\text{-}(Cut) \vdash \Gamma \Rightarrow \varphi$.

We may do even better and prove elimination of cut constructively but first we must modify the calculus slightly to obtain its variant BSC2. First of all, we restrict the application of all static rules to left sequents only. So what in BSC1 was only a convention for schemata of rules, now is a rigid requirement. Note that in consequence of this restriction the right sequent is either empty or modal and plays only an auxiliary role similarly to Heuerding, Seyfried and Zimmermann's approach; it serves for storing modal data. To simplify things we restrict the language to $\square$ only, but the proof works also in the presence of $\lozenge$.

We also introduce (Mix) instead of (Cut) to deal with C. It is obvious that the system with mix is equivalent to the system with cut by exactly the same argument as stated for LK.

To deal with transitional rules we must add the second form of cut or rather mix, similarly to the case of Avron's HSC. Thus, let (Mix') denote (Mix) restricted to nonmodal cut-formulae and $(MMix)$ denote the following rule:

$$(MMix) \quad \frac{\Gamma \Rightarrow \Delta, \square\varphi^i \mid \Lambda \Rightarrow \Theta, \square\varphi^j \qquad \square\varphi^k, \Pi \Rightarrow \Sigma \mid \square\varphi^n, \Xi \Rightarrow \Omega}{\Gamma, \Pi \Rightarrow \Delta, \Sigma \mid \Lambda, \Xi \Rightarrow \Theta, \Omega}$$

with $i + j \geq 1$ and $k + n \geq 1$.

Note that $(MMix)$ similarly like TR-rules works also on the right sequents, even if $i = k = 0$; if $j = n = 0$ it works like (Mix).

Let us call the system with these two variants of mix BSC2'. One may easily prove that

Lemma 4.26 $BSC2 \vdash \Gamma \Rightarrow \Delta$ *iff* $BSC2' \vdash \Gamma \Rightarrow \Delta$

PROOF: From left to right it is enough to show that the application of (Mix) on modal formula is derivable by $(MMix)$. If $j = n = 0$ it is the same. Otherwise, after the application of $(MMix)$ we must introduce the missing number of occurrences of the cut-formula by W to the left sequent and then by TR move them to the right sequent to restore its full shape.

From right to left it is enough to show that $(MMix)$ is derivable by (Mix). Again only the case with $j \geq 1$ or $n \geq 1$ needs to be considered. We apply (TR)

---

[16]Note that in the case od Sato's system a demonstration of provability of the 3-premiss rules for implication in BSC1 requires application of cut.

to such occurrences of the cut-formula to move them to the left sequent in both premises, then we apply (Mix) so all these occurrences are deleted from resulting bisequent. □

Before we prove the elimination of cut for BSC2 one important thing should be noted. Clearly, with cut BSC1 and BSC2 are equivalent. It is also easy to observe that without cut everything provable in BSC2 must be provable in BSC1 since the former is just a restricted form of the latter. But is BSC2 without cut equivalent to BSC1? No, just try to prove B in BSC2. In BSC1 it is simple

$$
\begin{array}{ll}
(\Box \Rightarrow) & \dfrac{p \Rightarrow p}{\Box p \Rightarrow p} \\[6pt]
(TR \Rightarrow) & \dfrac{}{\Rightarrow p \mid \Box p \Rightarrow} \\[6pt]
(\neg \Rightarrow) & \dfrac{}{\neg p \Rightarrow \mid \Box p \Rightarrow} \\[6pt]
(\Rightarrow \neg) & \dfrac{}{\neg p \Rightarrow \mid \Rightarrow \neg \Box p} \\[6pt]
(\Rightarrow \Box) & \dfrac{}{\neg p \Rightarrow \mid \Rightarrow \Box \neg \Box p} \\[6pt]
(\Rightarrow TR) & \dfrac{}{\neg p \Rightarrow \Box \neg \Box p} \\[6pt]
(\Rightarrow \rightarrow) & \dfrac{}{\Rightarrow \neg p \rightarrow \Box \neg \Box p}
\end{array}
$$

But in BSC2 the application of static rules in the right sequent is forbidden and without cut we are not able to prove B. We are in need of the solution due to Fitting and applied in subsection 4.5.2, but now it may be done simpler due to the use of bisequents. Note first that

Lemma 4.27 *In BSC2 (without cut)* $\vdash \Rightarrow \Box\varphi$ *iff* $\vdash \Rightarrow \varphi \mid \Rightarrow \Box\varphi$

PROOF: From left to right we just apply $(\Rightarrow TR)$ and $(\Rightarrow W)$; conversely we apply $(\Rightarrow \Box)$, then $(\Rightarrow TR)$ (but to the right sequent) and $(\Rightarrow C)$. □

Exercise 4.45 *Prove in BSC2 without cut* $\Rightarrow \neg p \rightarrow \Box \neg \Box p \mid \Rightarrow \Box(\neg p \rightarrow \Box \neg \Box p)$

Now we can prove

Theorem 4.13 *If BSC2* $\vdash \Rightarrow \varphi \mid \Rightarrow \Box\varphi$, *then BSC2-cut* $\vdash \Rightarrow \varphi \mid \Rightarrow \Box\varphi$

PROOF: We will use the method of Girard (see section 2.4), based on the application of cross-cuts. But we apply Gentzen's overall strategy, i.e we will prove the result for the case where both premises of (Mix') or $(MMix)$ are cut-free.

The case where one premiss, say the left one, is axiomatic is simple; we show it only for $(MMix)$:

$$
(MMix)\quad \dfrac{\Box\varphi \Rightarrow \Box\varphi \qquad \Box\varphi^i, \Gamma \Rightarrow \Delta \mid \Box\varphi^j, \Pi \Rightarrow \Sigma}{\Box\varphi, \Gamma \Rightarrow \Delta \mid \Pi \Rightarrow \Sigma}
$$

is replaced by:

$$(TR \Rightarrow) \frac{\Box\varphi^i, \Gamma \Rightarrow \Delta \mid \Box\varphi^j, \Pi \Rightarrow \Sigma}{\dfrac{\Box\varphi^{i+j}, \Gamma \Rightarrow \Delta \mid \Pi \Rightarrow \Sigma}{\Box\varphi, \Gamma \Rightarrow \Delta \mid \Pi \Rightarrow \Sigma}} (C \Rightarrow)$$

The cases where one cut-formula in one premiss is parametric in all occurrences are similar to reductions in standard SC. For illustration, we consider the case of $(MMix)$ when the left premiss is obtained by $(\rightarrow\Rightarrow)$

$$(\rightarrow\Rightarrow) \frac{\Gamma \Rightarrow \Delta, \varphi, \Box\chi^i \mid \Pi \Rightarrow \Sigma, \Box\chi^j \qquad \psi, \Gamma \Rightarrow \Delta, \Box\chi^i \mid \Pi \Rightarrow \Sigma, \Box\chi^j}{(MMix) \dfrac{\varphi \rightarrow \psi, \Gamma \Rightarrow \Delta, \Box\chi^i \mid \Pi \Rightarrow \Sigma, \Box\chi^j \qquad \Box\chi^k, \Lambda \Rightarrow \Theta \mid \Box\chi^n, \Xi \Rightarrow \Upsilon}{\varphi \rightarrow \psi, \Gamma, \Lambda \Rightarrow \Delta, \Theta \mid \Pi, \Xi \Rightarrow \Sigma, \Upsilon}}$$

is transformed into

$$(MMix) \frac{\dfrac{\Gamma \Rightarrow \Delta, \varphi, \Box\chi^i \mid \Pi \Rightarrow \Sigma, \Box\chi^j \qquad \Box\chi^k, \Lambda \Rightarrow \Theta \mid \Box\chi^n, \Xi \Rightarrow \Upsilon}{\Gamma, \Lambda \Rightarrow \Delta, \Theta, \varphi \mid \Pi, \Xi \Rightarrow \Sigma, \Upsilon} \quad D}{(\rightarrow\Rightarrow) \;\; \varphi \rightarrow \psi, \Gamma, \Lambda \Rightarrow \Delta, \Theta \mid \Pi, \Xi \Rightarrow \Sigma, \Upsilon}$$

where D is

$$\frac{\psi, \Gamma \Rightarrow \Delta, \Box\chi^i \mid \Pi \Rightarrow \Sigma, \Box\chi^j \qquad \Box\chi^k, \Lambda \Rightarrow \Theta \mid \Box\chi^n, \Xi \Rightarrow \Upsilon}{\psi, \Gamma, \Lambda \Rightarrow \Delta, \Theta \mid \Pi, \Xi \Rightarrow \Sigma, \Upsilon} (MMix)$$

Note that in case $\varphi = \Box\chi$ we must additionally restore φ by $(\Rightarrow W)$ to be able to derive the last sequent by $(\rightarrow\Rightarrow)$.

It should be noted that when TR is performed we can always reduce the height even if the left sequents are inactive.

The most troublesome cases are where cut-formulae are principal in both premisses. Let us consider the case of $\Box\varphi$

$$(MMix) \frac{(\Rightarrow \Box) \dfrac{\Rightarrow \varphi \mid \Gamma \Rightarrow \Delta, \Box\varphi^i}{\Rightarrow \Box\varphi \mid \Gamma \Rightarrow \Delta, \Box\varphi^i} \qquad \dfrac{\Box\varphi^j, \varphi, \Lambda \Rightarrow \Theta \mid \Box\varphi^k, \Pi \Rightarrow \Sigma}{\Box\varphi^{j+1}, \Lambda \Rightarrow \Theta \mid \Box\varphi^k, \Pi \Rightarrow \Sigma} (\Box \Rightarrow)}{\Lambda \Rightarrow \Theta \mid \Gamma, \Pi \Rightarrow \Delta, \Sigma}$$

if $i = j = k = 0$ it is enough to perform (Mix') on φ and then possibly restore by $(W \Rightarrow)$ some occurences of φ in Λ. Moreover, if $\varphi = \Box\psi$ and there are some occurrences of it in Δ or Π we actually perform $(MMix)$ and must also restore deleted occurrences in these multisets by W. In case some of $i, j, k \geq 0$ we must first make cross-cuts to delete occurrences of $\Box\varphi$ first. Of course, the most difficult situation is when all of $i, j, k \geq 1$; we perform

$$\frac{\dfrac{\Rightarrow \varphi \mid \Gamma \Rightarrow \Delta, \Box\varphi^i \qquad \Box\varphi^{j+1}, \Lambda \Rightarrow \Theta \mid \Box\varphi^k, \Pi \Rightarrow \Sigma}{\Lambda \Rightarrow \Theta, \varphi \mid \Gamma, \Pi \Rightarrow \Delta, \Sigma} \qquad \dfrac{\Rightarrow \Box\varphi \mid \Gamma \Rightarrow \Delta, \Box\varphi^i \qquad \Box\varphi^j, \varphi, \Lambda \Rightarrow \Theta \mid \Box\varphi^k, \Pi \Rightarrow \Sigma}{\varphi, \Lambda \Rightarrow \Theta \mid \Gamma, \Pi \Rightarrow \Delta, \Sigma} (Mix')}{\dfrac{\Lambda, \Lambda_\varphi \Rightarrow \Theta_\varphi, \Theta \mid \Gamma, \Gamma, \Pi, \Pi \Rightarrow \Delta, \Delta, \Sigma, \Sigma}{\Lambda, \Rightarrow \Theta \mid \Gamma, \Pi \Rightarrow \Delta, \Sigma} (IC)}$$

where both applications of $(MMix)$ have lower height and the application of (Mix') is of lower complexity. Again, if $\varphi = \Box\psi$ and there are some occurrences of it in Δ, Λ, Π we perform rather $(MMix)$ and restore by W deleted occurrences of φ. $\Box$

Exercise 4.46 *Prove the remaining cases in the above proof.*

Chapter 5

Alternatives to CPL

In the preceding chapter, we have illustrated the problems with the extension of standard SC techniques to non-classical logics which are stronger than **CPL**. But there is a huge number of non-classical logics which are weaker than **CPL** like intuitionistic and intermediate (between intuitionistic and classical) logics, many-valued logics or logics of relevant implication. Since our aim is an analysis of applications of SC to such logics, we are not going to present them in detail, in particular, a discussion of the reasons for deviations from classical logics will be avoided[1]. As in the case of modal logic, we focus on the technical problems connected with the adaptation of SC toolbox to this field. There is also one interesting question closely related to our aim. It appeared that many non-classical logics, originally constructed on the basis of totally different considerations, later have shown surprising connections with the framework of SC. In fact, the search for SC formalisations of non-classical logics being weaker than classical logic, like logics of relevant implication or many-valued logics demonstrated serious limitations of the original Gentzen's framework, significantly different from those described in the last chapter. In particular, for many logics, some structural rules were shown invalid in general (e.g. weakening for relevant logic, contraction for many-valued ones), at least in the framework of ordinary SC. This route led eventually to the development of extensive research on these logics which can be formalised as SCs obtained by deletion of some (or all) structural rules from the standard SC framework. Nowadays these logics are commonly called *substructural logics*[2].

In section 5.1, we will discuss briefly the earliest, and probably the most important, weakening of **CPL**, namely intuitionistic logic **INT**. After a survey of its basic features, we introduce some variants of standard SC for **INT** and prove the most important results focusing mainly on differences with SC for **CPL**. Next,

---

[1] There are many good introductions to non-classical logics and some concerned with special kinds of them will be mentioned later. For general considerations of motives which led to their construction, one should consult Priest [205] or Beall and van Fraassen [23].

[2] This name was introduced first by Došen and Schroeder-Heister [62]

© Springer Nature Switzerland AG 2021
A. Indrzejczak, *Sequents and Trees*, Studies in Universal Logic,
https://doi.org/10.1007/978-3-030-57145-0_5

we take a look at substructural logics. After the presentation of Girard's linear logic, some attention will be paid to a few logics of relevant implication. In the last two sections, we present some important many-valued logics and examine the most popular approaches to their formalisation in terms of SC. At first, we present some ordinary SCs. In section 5.5, we introduce two kinds of generalised SCs for them. Although originally they were presented mainly in tableau setting and in different forms (e.g. as labelled systems), we will present them uniformly as built from structured 4-argument sequents.

5.1   Intuitionistic Logic

The intuitionistic logic **INT** was proposed by Brouwer as a constructivist alternative to classical logic. We are not going to present philosophical and mathematical motivations underlying intuitionism and constructivism in general[3] We will be concerned as usual with SCs for **INT** but some introductory information on the axiomatic and semantical formulation is briefly recalled first. The syntax of the propositional part of **INT** is the same as for **CPL**, so most of the terminology from chapter 1 applies with no changes. **INT** was first presented in axiomatic form by Heyting[4]. In fact, one may use the same axiomatization as stated in subsection 1.1.3, for **CPL** but with one different axiom; instead of axiom 7 we must use a weaker form of the transposition law

$$(\varphi \to \psi) \to (\neg\psi \to \neg\varphi)$$

The remaining axioms yield a formalisation of positive (intuitionistic[5]) logic. The only rule of inference is MP and the notion of proof/provability relation and consistency are defined as in section 1.1.

Intuitionistic negation and implication are essentially non-classical, as a result, conjunction and disjunction, when mixed with these connectives, also lack many classical properties. Not only such classical laws like the law of excluded middle or the double negation elimination $\neg\neg\varphi \to \varphi$ fail to hold. Also, the classical definitional equivalences do not hold in general.

Example 5.1. The following implications are not theses of **INT** (although their converses are provable)

$$\neg(\varphi \wedge \psi) \to \neg\varphi \vee \neg\psi \qquad \neg(\varphi \to \neg\psi) \to \varphi \wedge \psi \qquad (\varphi \to \psi) \to \neg\varphi \vee \psi$$

---

[3]One can find many good introductions, for example in Dummett [65], Mints [183], Dragalin [64]. More technical presentation of the results obtained for propositional **INT** and many of its extensions is provided in Chagrov and Zakharyaschev [48]. Fitting's first part of [81] and the last chapter of [83], provide good introductions to proof methods for **INT**, with special attention paid to tableau methods. Troelstra and Schwichtenberg [264], Waaler and Wallen [268] and Negri and von Plato [185],, provide accounts of different SCs for **INT**.

[4]A slightly different formalisation of Brouwer's ideas were proposed by Kolmogorov.

[5]In order to get classical positive logic one must add Peirce's law $((\varphi \to \psi) \to \varphi) \to \varphi$ to this basis.

This suggests that, in contrast to **CPL**, intuitionistic connectives are fully independent. Of course, the fact that known classical equivalences do not hold does not preclude the possibility that other, more complicated, ways of proving interdefinability work. However, it was shown independently by Wajsberg and McKinsey that intuitionistic connectives are indeed independent. An accessible semantical proof of that fact is provided by Fitting [83].

Since intuitionistic negation is not involutive (a failure of double elimination), one may have objections that this connective is too weak. There were proposals to introduce a stronger negation but without destroying the constructive character of the logic. The most important systems of this kind are Nelson's logics of constructive negation **N3** and **N4**. One may obtain their axiomatization by the addition of the following axioms to positive **INT**:

$$\neg\neg\varphi \leftrightarrow \varphi$$
$$\neg(\varphi \wedge \psi) \leftrightarrow \neg\varphi \vee \neg\psi$$
$$\neg(p \vee q) \leftrightarrow \neg p \wedge \neg q$$
$$\neg(\varphi \rightarrow \psi) \leftrightarrow \varphi \wedge \neg\psi$$

This extension gives H-system for **N4** which is an example of (constructive) paraconsistent logic[6]. To obtain **N3** we must add Duns law: $\varphi \wedge \neg\varphi \rightarrow \psi$.

5.1.1 Relational Semantics

The first semantic characterisations of **INT** were provided in terms of infinite matrices by Gödel and Jaśkowski already in 1930s, there are also topological and algebraic semantics of several kinds (see, e.g. Chagrov and Zakharyashev [48]). However, the best recognised semantics for **INT** was introduced by Kripke in the late 1950s (Beth [28], provided a slightly different variant). It is very similar to the relational semantics for modal logics. A frame is a pair $\mathfrak{F} = \langle \mathcal{S}, \leq \rangle$ where:

- the domain $\mathcal{S} \neq \varnothing$ is usually interpreted as a set of states of knowledge;

- $\leq$ is a binary relation on $\mathcal{S}$ called an accessibility or a hereditary relation, and satisfying conditions of reflexivity and transitivity.

A model on a frame $\mathfrak{F}$ is a pair $\mathfrak{M} = \langle \mathfrak{F}, V \rangle$ with V being a valuation function $V : PROP \longrightarrow \mathcal{P}(\mathcal{S})$ which satisfies the heredity condition:

if $s \in V(p)$ and $s \leq s'$, then $s' \in V(p)$.

A satisfiability of a formula φ in a state s of a model $\mathfrak{M}$ ($\mathfrak{M}, s \vDash \varphi$) is defined in the standard way:

---

[6]Paraconsistent logics are one of the most important and rich class of non-classical logics where the explosive effects of Duns law are under control. In fact, many logics introduced later in this chapter will be also paraconsistent. An extensive treatment, covering also SC characterisations of many logics, is provided by Avron, Arieli and Zamansky [15].

$$
\begin{array}{lll}
\mathfrak{M}, s \vDash \varphi & \text{iff} & s \in V(\varphi) \text{ for any } \varphi \in PROP \\
\mathfrak{M}, s \vDash \neg\varphi & \text{iff} & \mathfrak{M}, s' \nvDash \varphi \text{ for any } s' \text{ such that } s \le s' \\
\mathfrak{M}, s \vDash \varphi \wedge \psi & \text{iff} & \mathfrak{M}, s \vDash \varphi \text{ and } \mathfrak{M}, s \vDash \psi \\
\mathfrak{M}, s \vDash \varphi \vee \psi & \text{iff} & \mathfrak{M}, s \vDash \varphi \text{ or } \mathfrak{M}, s \vDash \psi \\
\mathfrak{M}, s \vDash \varphi \to \psi & \text{iff} & \mathfrak{M}, s' \nvDash \varphi \text{ or } \mathfrak{M}, s' \vDash \psi \text{ for any } s' \ge s
\end{array}
$$

It holds:

Claim 5.1. *If $s \vDash \varphi$ and $s \le s'$, then $s' \vDash \varphi$, for any φ.*

As before we write $\mathfrak{M}, s \vDash \Gamma$ iff $\mathfrak{M}, s \vDash \psi$ for all $\psi \in \Gamma$. $\mathfrak{M}, s \nvDash \varphi$ denotes falsity of φ in s; $\mathfrak{M}, s \nvDash \Gamma$ denotes a falsity of at least one formula from Γ in s. In case of the previously established model, we simply write $s \vDash \varphi$ $(s \nvDash \varphi)$ or $s \vDash \Gamma$ $(s \nvDash \Gamma)$ for a set of formulae. The set of all states satisfying a formula (a set) is denoted as

$$
\begin{aligned}
&\|\varphi\|_{\mathfrak{M}} = \{s \in \mathcal{S}_{\mathfrak{M}} : s \vDash \varphi\}; \\
&\|\Gamma\|_{\mathfrak{M}} = \bigcap \|\psi\|_{\mathfrak{M}} \text{ for all } \psi \in \Gamma
\end{aligned}
$$

Again we usually use an abbreviated form $\|\varphi\|$ $(\|\Gamma\|)$ if $\mathfrak{M}$ is default or fixed. The most important semantical notions are displayed below

Definition 5.1.   • Γ *is satisfiable in* $\mathfrak{M}$ *iff* $\|\Gamma\|_{\mathfrak{M}} \ne \varnothing$.

- Γ *is satisfiable iff there exists a model in which it is satisfied.*

- Γ *is falsifiable iff there exists a model* $\mathfrak{M}$ *in which it is false* $(\|\Gamma\|_{\mathfrak{M}} \ne \mathcal{S}_{\mathfrak{M}})$.

- Γ *is unsatisfiable iff no model satisfies it.*

- φ *is true (globally) in a model* $(\mathfrak{M} \vDash \varphi)$ *iff for all* $s \in \mathcal{S}_{\mathfrak{M}}$, $\mathfrak{M}, s \vDash \varphi$ *(or* $\|\varphi\|_{\mathfrak{M}} = \mathcal{S}_{\mathfrak{M}}$, *analogously for* Γ, $\mathfrak{M} \vDash \Gamma$ *iff* $\|\Gamma\|_{\mathfrak{M}} = \mathcal{S}_{\mathfrak{M}}$*).*

- φ *is valid (intuitionistic tautology)* $(\vDash \varphi)$ *iff in all* $\mathfrak{M}$, $\mathfrak{M} \vDash \varphi$.

- φ *follows from* Γ $(\Gamma \vDash \varphi)$ *iff* $\|\Gamma\|_{\mathfrak{M}} \subseteq \|\varphi\|_{\mathfrak{M}}$ *for all* $\mathfrak{M}$ *(or: for all* $\mathfrak{M}$ *and all* $s \in \mathcal{S}_{\mathfrak{M}}$, *if* $\mathfrak{M}, s \vDash \Gamma$, *then* $\mathfrak{M}, s \vDash \varphi$*).*

$\nvDash \varphi$ is used to denote invalidity of a formula and $\Gamma \nvDash \varphi$ means that φ does not follow from Γ.

Note that similarly as in the case of modal logics, we may introduce also the global notion of consequence relation but it will be of no use for us.

We display without a proof some important features of **INT**

Theorem 5.1. *The following holds:*

1. **INT** $\subset$ **CPL**

2. $\varphi \in$ **CPL** iff $\neg\neg\varphi \in$ **INT**

3. $\varphi \vee \psi \in$ **INT** iff $\varphi \in$ **INT** or $\psi \in$ **INT**

4. $\Gamma, \varphi \models \psi$ iff $\Gamma \models \varphi \to \psi$

5. $\Gamma, \varphi \models \bot$ iff $\Gamma \models \neg\varphi$

6. if $\Gamma \models \varphi$, then $\Gamma, \neg\varphi \models \bot$

The second point in theorem 5.1, due to Glivenko, shows that **CPL** may be embedded in **INT**. In fact, it may be demonstrated in many different ways; possibly the best known translation was provided by Gödel. We will show a result of this kind below in the framework of SC. An analysis of relational semantics suggests other possible embeddings of **INT**. One may easily notice that **INT** is semantically characterised by **S4**-frames. This connection between the two logics may be precisely described as an embedding of **INT** into **S4**. Let $\Im$ be a translation function from $FOR(\textbf{INT})$ into $FOR(\textbf{S4})$ defined in the following way:

$$
\begin{aligned}
\Im(p) &= \Box p \\
\Im(\varphi \wedge \psi) &= \Im(\varphi) \wedge \Im(\psi) \\
\Im(\varphi \vee \psi) &= \Im(\varphi) \vee \Im(\psi) \\
\Im(\varphi \to \psi) &= \Box(\Im(\varphi) \to \Im(\psi)) \\
\Im(\neg\varphi) &= \Box\neg\Im(\varphi)
\end{aligned}
$$

The following holds:

Theorem 5.2. $\models_{INT} \varphi$ *iff* $\models_{S4} \Im(\varphi)$

Troelstra and Schwichtenberg [264] provide syntactical proofs concerning this translation (and some other as well).

We finish this brief presentation of propositional **INT** with the profound result that the semantical characterisation coincides with the Hilbert system

Theorem 5.3. (Adequacy) $\Gamma \models \varphi$ *iff* $\Gamma' \vdash \varphi$ *where* $\Gamma' \subseteq \Gamma$

5.1.2 Gentzen's Characterisation of INT

In fact, a sequent formalisation of **INT** was introduced more than 20 years before the advent of relational semantics. Gentzen [95] provided both natural deduction and sequent calculus for **INT**, the latter called LJ. The main difference with LK is in a sense structural; sequents are restricted in succedents to at most one formula. This is the reason for using the name intuitionistic sequents for such single-succedent sequents. Due to this restriction in LJ, we can dispense with structural rules $(\Rightarrow P)$ and $(\Rightarrow C)$ and the logical rules must be suitably redefined. Axioms are like in LK. Here is the list of all rules of LJ with Δ empty or with one formula only

Structural rules

$$
(Cut) \ \frac{\Gamma \Rightarrow \varphi \quad \varphi, \Pi \Rightarrow \Delta}{\Gamma, \Pi \Rightarrow \Delta} \qquad (P\Rightarrow) \ \frac{\Pi, \varphi, \psi, \Gamma \Rightarrow \Delta}{\Pi, \psi, \varphi, \Gamma \Rightarrow \Delta}
$$

$(W\Rightarrow) \quad \dfrac{\Gamma\Rightarrow\Delta}{\varphi,\Gamma\Rightarrow\Delta}$
$\qquad\qquad\qquad$
$(C\Rightarrow) \quad \dfrac{\varphi,\varphi,\Gamma\Rightarrow\Delta}{\varphi,\Gamma\Rightarrow\Delta}$

Logical rules

$(\neg\Rightarrow) \quad \dfrac{\Gamma\Rightarrow\varphi}{\neg\varphi,\Gamma\Rightarrow}$
$\qquad\qquad\qquad$
$(\Rightarrow\neg) \quad \dfrac{\varphi,\Gamma\Rightarrow}{\Gamma\Rightarrow\neg\varphi}$

$(\wedge\Rightarrow) \quad \dfrac{\varphi,\Gamma\Rightarrow\Delta}{\varphi\wedge\psi,\Gamma\Rightarrow\Delta}$
$\qquad\qquad\qquad$
$(\wedge\Rightarrow) \quad \dfrac{\psi,\Gamma\Rightarrow\Delta}{\varphi\wedge\psi,\Gamma\Rightarrow\Delta}$

$(\Rightarrow\wedge) \quad \dfrac{\Gamma\Rightarrow\varphi \quad\cdot\quad \Gamma\Rightarrow\psi}{\Gamma\Rightarrow\varphi\wedge\psi}$
$\qquad\qquad$
$(\vee\Rightarrow) \quad \dfrac{\varphi,\Gamma\Rightarrow\Delta \quad \psi,\Gamma\Rightarrow\Delta}{\varphi\vee\psi,\Gamma\Rightarrow\Delta}$

$(\Rightarrow\vee) \quad \dfrac{\Gamma\Rightarrow\varphi}{\Gamma\Rightarrow\varphi\vee\psi}$
$\qquad\qquad\qquad$
$(\Rightarrow\vee) \quad \dfrac{\Gamma\Rightarrow\psi}{\Gamma\Rightarrow\varphi\vee\psi}$

$(\to\Rightarrow) \quad \dfrac{\Gamma\Rightarrow\varphi \quad \psi,\Pi\Rightarrow\Delta}{\varphi\to\psi,\Gamma,\Pi\Rightarrow\Delta}$
$\qquad\qquad$
$(\Rightarrow\to) \quad \dfrac{\varphi,\Gamma\Rightarrow\psi}{\Gamma\Rightarrow\varphi\to\psi}$

Now we can understand what was Gentzen's reason for using the additive form of rules $(\Rightarrow\vee)$ (and $(\wedge\Rightarrow)$ by duality) instead of a more convenient multiplicative form. Due to the restriction on the succedent, only one part of a disjunction may appear in the premiss.

The notion of a proof is the same as for LK.

Exercise 5.1. *Prove the converses of nonvalid implications from example 5.1. Try to prove the implications and look why it is not possible. Prove* $\neg(p\vee q)\leftrightarrow\neg p\wedge\neg q$, $\neg(p\to\neg q)\to\neg\neg p\wedge\neg q$ *and* $\neg p\to q\leftrightarrow p\vee q$.

With cut as a primitive rule, one can easily prove the equivalence of LJ with Hilbert system for **INT**. What is even more interesting is the affinity of LJ with ND formulation of **INT** introduced by Gentzen. Although the original form of Gentzen's ND (called NJ for **INT**) was different, we can refer to the sequent form of ND described in subsection 3.5.4. It is enough to delete one rule for negation, namely $(\Rightarrow\neg E)$ to obtain ND in a sequent form for **INT**:

Theorem 5.4. *If* $\Gamma\vdash_{NJ}\varphi$, *then* $\vdash_{LJ}\Gamma\Rightarrow\varphi$

PROOF: Formally, it is an induction on the height of a proof in NJ. It is enough to demonstrate that every rule of NJ is simulated in LJ. Note that the introduction of assumptions corresponds to axiomatic sequents. All introduction rules are already present in LJ as rules of introduction to the succedent. There is only a slight difference between $(\wedge I)$ and $(\Rightarrow\wedge)$, since the former has a multiplicative form, so we must first apply W to both premises to unify the antecedents. In case of elimination rules, we must use suitable rules of introduction to the antecedent and cut. As an example, we demonstrate the case of MP (i.e. $(\to E)$)

$$\dfrac{\Delta \Rightarrow \varphi \rightarrow \psi \quad \dfrac{\Gamma \Rightarrow \varphi \quad \dfrac{\dfrac{\varphi \Rightarrow \varphi \quad \psi \Rightarrow \psi}{\varphi, \varphi \rightarrow \psi \Rightarrow \psi}\ (\rightarrow \Rightarrow)}{\varphi \rightarrow \psi, \Gamma \Rightarrow \psi}\ (Cut)}{\Gamma, \Delta \Rightarrow \psi}\ (Cut)$$

□

Exercise 5.2. *Prove the remaining cases.*

Instead of proving the converse of theorem 5.4, we prove

Theorem 5.5. *If $\vdash_{LJ} \Gamma \Rightarrow \varphi$ then $\Gamma \models \varphi$.*

PROOF: It is enough to prove validity-preservation of all rules. The only cases which are different from classical counterparts involve implication and negation. Let us analyse $(\Rightarrow\rightarrow)$. If the conclusion is not valid, then in some state s of some model all formulae from Γ are satisfied but $s \not\Vdash \varphi \rightarrow \psi$. Hence in some accessible s', φ is satisfied but ψ is not. However, Γ is also satisfied in s' hence s' falsified the premiss, contrary to our assumption. □

Exercise 5.3. *Prove the remaining cases.*

It is rather not surprising that the cut elimination theorem holds for LJ and the proof is similar to the classical case. Gentzen in fact proved his Hauptsatz simultaneously for LJ and LK. We leave the reader to check the details.

Exercise 5.4. *Prove mix elimination for LJ using Gentzen's original strategy. Prove cut elimination for LJ by means of other strategies from chapter 2.*

Things stated so far suggest that LJ, being only a restricted form of LK, preserves almost all features of LK and the results which were previously proved for it. For example, a proof of cut elimination for LJ does not require new techniques and is in fact a bit simpler. To some extent it is true but there are also significant differences. Consistency follows immediately and decidability of **INT** may be proved by exactly the same method which was presented in chapter 2 for LK. In fact, Gentzen provided the first proof of decidability of propositional **INT** by means of LJ, before any semantic method was accessible for that. Also, the interpolation theorem may be proved constructively in a similar way as for **CPL** when using Maehara's method. But if we try to apply Smullyan's strategy we encounter problems with defining a symmetric version of SC for **INT** (although Fitting [83] provides a solution based on some exceptions to the general format of symmetric rules).

Exercise 5.5. *Prove the interpolation theorem for **INT** by means of Maehara's method. Note that it is simpler than for **CPL** due to single-succedent sequents.*

Even more significant differences appear when we consider the problem of permutability of rules. Kleene's results concerning permutability of rules have shown the problem in the early 1950s [151]. One may easily check that, in contrast to

LK, not all rules are permutable in LJ[7]. For example, although $\varphi \vee \psi \Rightarrow \psi \vee \varphi$ is provable, we can do it only if we apply root-first $(\vee \Rightarrow)$ and then $(\Rightarrow \vee)$ in each branch. Changing the order leads to unprovable sequent $\varphi \vee \psi \Rightarrow \varphi$ or $\varphi \vee \psi \Rightarrow \psi$. In case of implication rules, we have the opposite situation: one must first apply $(\Rightarrow \rightarrow)$, then $(\rightarrow \Rightarrow)$. Consider the following proof (we disregard applications of P):

$$(W \Rightarrow) \dfrac{q \Rightarrow q}{q, p \Rightarrow q} \qquad \dfrac{\dfrac{\dfrac{p \Rightarrow p}{\neg p, p \Rightarrow} (\neg \Rightarrow)}{r, q, p \Rightarrow \neg\neg p} (\Rightarrow \neg), (W \Rightarrow)}{q, p, q \rightarrow r \Rightarrow \neg\neg p} (\rightarrow \Rightarrow)}{p, q \rightarrow r \Rightarrow q \rightarrow \neg\neg p} (\Rightarrow \rightarrow)$$

In fact, we could even forget about the application of $(\rightarrow \Rightarrow)$ and obtain a simpler proof just by the application of W and negation rules. However, if we apply first $(\rightarrow \Rightarrow)$, we obtain the following:

$$\dfrac{p \Rightarrow q \qquad \dfrac{\dfrac{\dfrac{p \Rightarrow p}{\neg p, p \Rightarrow} (\neg \Rightarrow)}{q, r, p \Rightarrow \neg\neg p} (\Rightarrow \neg), (W \Rightarrow)}{r, p \Rightarrow q \rightarrow \neg\neg p} (\Rightarrow \rightarrow)}{p, q \rightarrow r \Rightarrow q \rightarrow \neg\neg p} (\rightarrow \Rightarrow)$$

where the left leaf is not axiomatic.

Exercise 5.6. *Prove the contraction law* $((p \rightarrow (p \rightarrow q)) \rightarrow (p \rightarrow q))$*. Try to do it with* $(\rightarrow \Rightarrow)$ *applied (root-)first as soon as possible.*

The above considerations show that rules for disjunction and implication do not permute. This has certainly an impact on proof search, since in particular, we must take into account that $(\vee \Rightarrow)$ must be applied before $(\Rightarrow \vee)$ and $(\Rightarrow \rightarrow)$ before $(\rightarrow \Rightarrow)$. Also $(\Rightarrow \neg)$ must be applied before $(\neg \Rightarrow)$ for obvious reason. However, we know already that LK is not good for proof search, so perhaps it is not surprising that LJ is even worse. Perhaps with some other versions of SC for **INT** we can improve the situation. Before checking this we consider one more question: how to use LJ to prove the embedding of **CPL** into **INT**? A solution provided below is based on Kolmogorov's and Glivenko's idea. As indicated in the point 2 of theorem 5.1, every classical thesis, when doubly negated, becomes a thesis of **INT**.

Example 5.2. Let us consider a proof of $\neg\neg(p \vee \neg p)$ in LJ

---

[7]Actually Kleene was using a bit different versions of SC for **CPL** and **INT** but the problem remains the same even if in some variants of SC we are able to improve the results—see Waaler and Wallen [268].

$$(\Rightarrow \vee) \dfrac{\dfrac{p \Rightarrow p}{p \Rightarrow p \vee \neg p}}{\,}$$

$$(\neg \Rightarrow) \dfrac{p \Rightarrow p \vee \neg p}{\neg(p \vee \neg p), p \Rightarrow}$$

$$(P \Rightarrow) \dfrac{\neg(p \vee \neg p), p \Rightarrow}{p, \neg(p \vee \neg p) \Rightarrow}$$

$$(\Rightarrow \neg) \dfrac{p, \neg(p \vee \neg p) \Rightarrow}{\neg(p \vee \neg p) \Rightarrow \neg p}$$

$$(\Rightarrow \vee) \dfrac{\neg(p \vee \neg p) \Rightarrow \neg p}{\neg(p \vee \neg p) \Rightarrow p \vee \neg p}$$

$$(\neg \Rightarrow) \dfrac{\neg(p \vee \neg p) \Rightarrow p \vee \neg p}{\neg(p \vee \neg p), \neg(p \vee \neg p) \Rightarrow}$$

$$(C \Rightarrow) \dfrac{\neg(p \vee \neg p), \neg(p \vee \neg p) \Rightarrow}{\neg(p \vee \neg p) \Rightarrow}$$

$$(\Rightarrow \neg) \dfrac{\neg(p \vee \neg p) \Rightarrow}{\Rightarrow \neg\neg(p \vee \neg p)}$$

In fact, for any classical thesis φ, its double negation can be provable in LJ in a similar way, i.e. by using C immediately after $(\Rightarrow \neg)$. However, we provide below a formal proof of embedding of **CPL** into **INT**, by means of the translation enabling a step-by-step simulation of a proof in LK by a proof of the translated sequent in LJ. This will serve as an illustration of one more technique applied successfully in the framework of SC. Let us consider the following translation:

$$
\begin{aligned}
\Im(p) &= p \\
\Im(\neg\varphi) &= \neg\Im(\varphi) \\
\Im(\varphi \wedge \psi) &= \neg\neg\Im(\varphi) \wedge \neg\neg\Im(\psi) \\
\Im(\varphi \vee \psi) &= \neg\neg\Im(\varphi) \vee \neg\neg\Im(\psi) \\
\Im(\varphi \to \psi) &= \neg\neg\Im(\varphi) \to \neg\neg\Im(\psi)
\end{aligned}
$$

It naturally extends to sequences in the sense that $\Im(\Gamma)$ is the sequence of translations of all its elements. It holds that

Theorem 5.6. *If $\vdash_{LK} \Gamma \Rightarrow \Delta$, then $\vdash_{LJ} \neg\neg\Im(\Gamma), \neg\Im(\Delta) \Rightarrow$*

PROOF: By induction on the height of a proof in LK (with applications of P disregarded). For axioms we have the following LJ proof:

$$(\neg \Rightarrow) \dfrac{\Im(\varphi) \Rightarrow \Im(\varphi)}{\neg\Im(\varphi), \Im(\varphi) \Rightarrow}$$

$$(\Rightarrow \neg) \dfrac{\neg\Im(\varphi), \Im(\varphi) \Rightarrow}{\neg\Im(\varphi) \Rightarrow \neg\Im(\varphi)}$$

$$(\neg \Rightarrow) \dfrac{\neg\Im(\varphi) \Rightarrow \neg\Im(\varphi)}{\neg\neg\Im(\varphi), \neg\Im(\varphi) \Rightarrow}$$

Applications of all structural rules are straightforward to simulate. For cut, we have

$$(\Rightarrow \neg) \dfrac{\neg\neg\Im(\Gamma), \neg\Im(\varphi) \Rightarrow}{\neg\neg\Im(\Gamma) \Rightarrow \neg\neg\Im(\varphi)} \qquad \neg\neg\Im(\varphi), \neg\neg\Im(\Pi), \neg\Im(\Delta) \Rightarrow$$

$$(Cut) \dfrac{}{\neg\neg\Im(\Gamma), \neg\neg\Im(\Pi), \neg\Im(\Delta) \Rightarrow}$$

where both leaves are provable by the induction hypothesis.

For $(\neg \Rightarrow)$, we get

$$(\Rightarrow \neg) \; \frac{\neg\neg\Im(\Gamma), \neg\Im(\varphi) \; [= \Im\neg(\varphi)], \neg\Im(\Delta) \Rightarrow}{\neg\neg\Im(\Gamma), \neg\Im(\Delta) \Rightarrow \neg\Im\neg(\varphi)}$$
$$(\neg \Rightarrow) \; \frac{}{\neg\neg\Im\neg(\varphi), \neg\neg\Im(\Gamma), \neg\Im(\Delta) \Rightarrow}$$

For $(\Rightarrow \neg)$ there is nothing to prove, since $\neg\neg\Im(\varphi) = \neg\Im\neg(\varphi)$. Both rules for implication are simulated as follows:

$$(\Rightarrow \neg) \; \frac{\neg\neg\Im(\varphi), \neg\neg\Im(\Gamma), \neg\Im(\psi), \neg\Im(\Delta) \Rightarrow}{\neg\neg\Im(\varphi), \neg\neg\Im(\Gamma), \neg\Im(\Delta) \Rightarrow \neg\neg\Im(\psi)}$$
$$(\Rightarrow\rightarrow) \; \frac{}{\neg\neg\Im(\Gamma), \neg\Im(\Delta) \Rightarrow \neg\neg\Im(\varphi) \rightarrow \neg\neg\Im(\psi) \; [= \Im(\varphi \rightarrow \psi)]}$$
$$(\neg \Rightarrow) \; \frac{}{\neg\neg\Im(\Gamma), \neg\Im(\varphi \rightarrow \psi), \neg\Im(\Delta) \Rightarrow}$$

$$(\Rightarrow \neg) \; \frac{\neg\Im(\varphi), \neg\neg\Im(\Gamma), \neg\Im(\Delta) \Rightarrow}{\neg\neg\Im(\Gamma), \neg\Im(\Delta) \Rightarrow \neg\neg\Im(\varphi)} \qquad \neg\neg\Im(\psi), \neg\neg\Im(\Pi), \neg\Im(\Sigma) \Rightarrow$$
$$(\rightarrow\Rightarrow) \; \frac{}{\neg\neg\Im(\varphi) \rightarrow \neg\neg\Im(\psi) \; [= \Im(\varphi \rightarrow \psi)], \neg\neg\Im(\Gamma), \neg\neg\Im(\Pi), \neg\Im(\Delta), \neg\Im(\Sigma) \Rightarrow}$$
$$(\Rightarrow \neg) \; \frac{}{\neg\neg\Im(\Gamma), \neg\neg\Im(\Pi), \neg\Im(\Delta), \neg\Im(\Sigma) \Rightarrow \neg\Im(\varphi \rightarrow \psi)}$$
$$(\neg \Rightarrow) \; \frac{}{\neg\neg\Im(\varphi \rightarrow \psi), \neg\neg\Im(\Gamma), \neg\neg\Im(\Pi), \neg\Im(\Delta), \neg\Im(\Sigma) \Rightarrow}$$

$\square$

Exercise 5.7. *Prove translations of the applications of rules for $\wedge$ and $\vee$.*

Goubault-Larrecq and Mackie [104] present a formal proof of embedding by means of a very similar translation. The only difference is that the clause for implication is $\Im(\varphi \rightarrow \psi) = \neg\Im(\psi) \rightarrow \neg\Im(\varphi)$ and they prove that if $\vdash_{LK} \Gamma \Rightarrow \Delta$, then $\vdash_{LJ} \Im(\Gamma), \neg\Im(\Delta) \Rightarrow$. Other examples of translations are considered by Bimbo [31].

5.1.3 Other Variants of SC

One may characterise **INT** also in Ketonen's style and obtain G3i which consists of logical rules only. Such a calculus was introduced by Troelstra and Schwichtenberg [264], under influence of Dragalin's calculus [64], which we characterise below. At first, it may seem that to obtain purely logical SC for **INT** it is sufficient to put the same restriction on the succedents of G3 rules as it was done on sequents of LK to obtain LJ. Clearly, two rules of $(\Rightarrow \vee)$ are needed due to this restriction but the rest should work. However, it is not enough to dispense with contraction. Let us look at the example 5.2, where the application of $(C \Rightarrow)$ was essential. The same problem appears also in G3i with negation as primitive and other rules unmodified. One may overcome a difficulty by introducing $\perp$ as primitive instead of $\neg$ which is now definable: $\neg\varphi := \varphi \rightarrow \perp$. Hence for $\perp$ we have additional axiom: $\perp, \Gamma \Rightarrow \Delta$. We also restrict axioms to atomic ones, i.e. of the form $\Gamma, p \Rightarrow p$. Note that when proving that $\Gamma, \varphi \Rightarrow \varphi$ is provable for any φ, we must take care of the order of applied rules. In this respect, G3i is not so much different from LJ.

Exercise 5.8. *Prove that $\vdash_{G3i} \Gamma, \varphi \Rightarrow \varphi$*

But this is still not enough and we must also modify $(\to\Rightarrow)$ in the contraction-absorbing style

$$(\to\Rightarrow) \quad \frac{\varphi \to \psi, \Gamma \Rightarrow \varphi \quad \psi, \Gamma \Rightarrow \Delta}{\varphi \to \psi, \Gamma \Rightarrow \Delta}$$

Note that we need to repeat the principal formula only in the left premiss.

In such a calculus the law of doubly negated excluded middle may be proved in the following way:

$$
\begin{array}{l}
(\Rightarrow \vee) \ \dfrac{p \Rightarrow p}{p \Rightarrow p \vee (p \to \bot)} \qquad \bot \Rightarrow \bot \\[2pt]
(\to\Rightarrow) \ \dfrac{}{p, p \vee (p \to \bot) \to \bot \Rightarrow \bot} \\[2pt]
(\Rightarrow\to) \ \dfrac{}{p \vee (p \to \bot) \to \bot \Rightarrow p \to \bot} \\[2pt]
(\Rightarrow \vee) \ \dfrac{}{p \vee (p \to \bot) \to \bot \Rightarrow p \vee (p \to \bot)} \qquad \bot \Rightarrow \bot \\[2pt]
(\to\Rightarrow) \ \dfrac{}{p \vee (p \to \bot) \to \bot \Rightarrow \bot} \\[2pt]
(\Rightarrow\to) \ \dfrac{}{\Rightarrow (p \vee (p \to \bot) \to \bot) \to \bot}
\end{array}
$$

For such a version of G3i, we can provide Dragalin-style proof of cut admissibility, of course, on the basis of the previously demonstrated auxiliary results concerning the admissibility of other structural rules and invertibility of logical rules. Note that the last result does not hold for all cases but only for $(\wedge \Rightarrow), (\Rightarrow \wedge), (\vee \Rightarrow), (\Rightarrow\to)$ and for $(\to\Rightarrow)$ but only with respect to the right premiss, i.e: if $\vdash \varphi \to \psi, \Gamma \Rightarrow \Delta$, then $\vdash \psi, \Gamma \Rightarrow \Delta$

Exercise 5.9. *Show semantically that $(\to\Rightarrow)$ is not invertible with respect to its left premiss.*

However, even this limited set of invertible rules allows for proving admissibility of $(C \Rightarrow)$ (in fact, invertibility of $(\Rightarrow \wedge)$ and $(\Rightarrow\to)$ are dispensable) and then for the proof of admissibility of cut. We do not provide this proof; one can try to do it alone or consult Negri and von Plato [185] or Troelstra and Schwichtenberg [264].

Exercise 5.10. *Prove admissibility of cut. Take care of the case of implication being the principal formula in both premisses of cut.*

Having shown that cut is admissible for G3i we can obtain also some original features of **INT** which do not hold for **CPL**. For LJ it is also possible but in some cases not so straightforward due to contraction. First of all one may constructively show in G3i the underivability of classical theses which are not intuitionistically valid. For example, to get a proof of $\Rightarrow p \vee \neg p$ in LK, one must first apply $(\Rightarrow C)$, but already in LJ this is impossible and neither $\Rightarrow p$ nor $\Rightarrow \neg p$ are provable.

Exercise 5.11. *Show that the implications from example 5.1. are underivable in G3i.*

Such considerations allow for demonstration of one of the interesting features of **INT** which is called the *disjunction property* and does not hold for **CPL** (provablity of the excluded middle law in the above form provides a counterexample)

Lemma 5.1. $\vdash_{INT} \varphi \vee \psi$ *iff* $\vdash_{INT} \varphi$ *or* $\vdash_{INT} \psi$

PROOF $\Rightarrow$: Without cut and ($\Rightarrow C$) the only way of proving a disjunction is to apply ($\Rightarrow W$) or ($\Rightarrow \vee$). The former is impossible since **INT** is consistent and $\Rightarrow$ is not provable. Hence, either $\Rightarrow \varphi$ or $\Rightarrow \psi$ must be provable. $\Longleftarrow$: By ($\Rightarrow \vee$) from any of the disjuncts. $\square$

Let us focus briefly on the problem of decidability and proof search in G3i. Since we know that G3 is a better tool for defining proof search procedures than LK we may try with G3i as well. However, things are still more complicated here. G3i is not confluent which is evident because of the lack of invertibility of all rules. In particular, we cannot avoid backtracking even in G3i, since ($\Rightarrow \vee$) is additive. Moreover, the shape of the contraction-absorbing ($\rightarrow\Rightarrow$) leads to loops in proof search, similarly as in SC for modal logics. Due to these obstacles (backtracking, restricted permutation) even in G3i a proof search is harder than in G3 and decidability proof is more complicated.

One may consider if all rules really need restriction to at most one formula in the succedent. After all, conjunction and disjunction are characterised classically in the semantics. In fact, only two rules: ($\Rightarrow \neg$) and ($\Rightarrow\rightarrow$) need such restriction which is not surprising if we again take a look at satisfiability conditions. Such a multisuccedent version of LJ was introduced by Maehara [168]. Only ($\Rightarrow\rightarrow$) and ($\Rightarrow \neg$) (if negation is primitive) are restricted to one formula in the succedent. This improves permutability features, since we have classical rules for disjunction. But still, not all problems are resolved. In particular, structural rules are primitive and ($C \Rightarrow$) must be applied to implications in the antecedent before ($\rightarrow\Rightarrow$) will be used, and the same is needed for negations in antecedents, if negation is primitive. This reflects the semantical fact that all true formulae must be kept for use in all accessible states. In case of conjunction and disjunction, side formulae of ($\wedge \Rightarrow$) and ($\vee \Rightarrow$) are also in the antecedent so repetition of principal formulae is not needed. In case of implication and negation, the situation is different, since at least one side formula is transferred to the succedent. Also, rules for implication (and negation) are still not permutable which suggests that ($\Rightarrow\rightarrow$) has a priority. But from the standpoint of proof search, it is not so obvious, since before that, other formulae in the succedent must be deleted by ($\Rightarrow W$) and sometimes their presence may lead to finding a proof. So backtracking may be unavoidable anyway. A discussion of several subtleties connected with such kind of SC (called there LB from Beth) may be found in Waaler and Wallen [268].

The same relaxation of rules may be applied to G3i. In fact, such calculi were first provided by Beth [28] and Dragalin [64], and the latter was then reformulated by Troelstra and Schwichtenberg [264], as a single-succedent calculus. Dragalin's

SC has two axioms: $\Gamma, p \Rightarrow p, \Delta$ and $\perp, \Gamma \Rightarrow \Delta$ and standard G3 rules for $\vee$ and $\wedge$. For implication we have (negation is not concerned)

$$(\rightarrow\Rightarrow) \quad \frac{\varphi \rightarrow \psi, \Gamma \Rightarrow \varphi \quad \psi, \Gamma \Rightarrow \Delta}{\varphi \rightarrow \psi, \Gamma \Rightarrow \Delta} \qquad\qquad (\Rightarrow\rightarrow) \quad \frac{\varphi, \Gamma \Rightarrow \psi}{\Gamma \Rightarrow \Delta, \varphi \rightarrow \psi}$$

One of the rules is a contraction-(in the left premiss) and the second is a weakening-absorbing rule. Thus, we are putting into the format of logical rules all necessary applications of structural rules. Note that the repetition of implication in the right premiss of $(\rightarrow\Rightarrow)$ is not necessary since one of its side formula is kept in the antecedent. It is also important to note that the weakening-absorbing version of $(\Rightarrow\rightarrow)$ is not invertible.

Exercise 5.12. *Prove* $\vdash \varphi, \Gamma \Rightarrow \Delta, \varphi$ *in Dragalin's system. Note that in case of implication the order of the application of rules is still essential.*

Prove invertibility of all rules, except $(\rightarrow\Rightarrow)$ *with respect to the left premiss, and* $(\Rightarrow\rightarrow)$.

Prove admissibility of W, C and cut.

Dragalin provides only syntactic account of the features of his calculus and does not consider proof search. It seems that in this respect this calculus may be improved. Let us call it GD and add rules for negation. Since we discuss it as a system suitable for better proof search it is enough to take axioms $\varphi, \Gamma \Rightarrow \Delta, \varphi$. Rules for $\vee, \wedge$ are taken from G3, and for $\neg$ and $\rightarrow$ we have

$$(\neg\Rightarrow) \quad \frac{\neg\varphi, \Gamma \Rightarrow \Delta, \varphi}{\neg\varphi, \Gamma \Rightarrow \Delta} \qquad\qquad (\Rightarrow\neg) \quad \frac{\varphi, \Gamma \Rightarrow}{\Gamma \Rightarrow \neg\varphi, \Delta}$$

$$(\rightarrow\Rightarrow) \quad \frac{\varphi \rightarrow \psi, \Gamma \Rightarrow \Delta, \varphi \quad \psi, \Gamma \Rightarrow \Delta}{\varphi \rightarrow \psi, \Gamma \Rightarrow \Delta} \qquad (\Rightarrow\rightarrow) \quad \frac{\varphi, \Gamma \Rightarrow \psi}{\Gamma \Rightarrow \Delta, \varphi \rightarrow \psi}$$

GD corresponds to one of the tableau calculus of Fitting [81] based on Beth rules [28]. Note that all the rules are invertible except weakening-absorbing versions of $(\Rightarrow \neg)$ and $(\Rightarrow\rightarrow)$. $(\rightarrow\Rightarrow)$ is invertible with respect to the left premiss because Δ is present in the succedent. From all versions of SC defined so far GD is the simplest system for proving Hintikka-style constructive completeness proof.

Exercise 5.13. *Provide Hintikka-style completeness proof for GD.*

For proof search and decidability, not all discussed problems can be removed. Due to $(\Rightarrow \neg)$ and $(\Rightarrow\rightarrow)$, backtracking is still unavoidable. Due to $(\neg \Rightarrow)$ and $(\rightarrow\Rightarrow)$, a loop may be generated. We can try to dissolve backtracking in the way similar to solutions presented in chapter 3 and 4, by the addition of some disjunctive-branching rule of the form

$$(SG) \quad \frac{\Gamma, \varphi_1 \Rightarrow, ..., \Gamma, \varphi_l \Rightarrow, \ \Gamma, \psi_1 \Rightarrow \chi_1 ..... \Gamma, \psi_k \Rightarrow \chi_k}{\Gamma \Rightarrow \neg\varphi_1, ..., \neg\varphi_l, \psi_1 \rightarrow \chi_1, ..., \psi_k \rightarrow \chi_k, \ \Delta}$$

where Γ and Δ contain only atomic formulae, $l + k \geq 1$.

Clearly, such a solution may be developed in the framework of hypersequent calculi in the way analogous to our treatment of modal logics in section 4.4. Then (SG) is rewritten as one-premiss rule operating on hypersequents. In fact, Mints [183], developed such a system for **INT** but instead he introduced the following rule:

$$(\Rightarrow\rightarrow^{M}) \quad \frac{G \mid \Gamma \Rightarrow \Delta, \varphi \rightarrow \psi \mid \Gamma, \varphi \Rightarrow \psi}{G \mid \Gamma \Rightarrow \Delta, \varphi \rightarrow \psi}$$

In his system, $\bot$ is primitive, hence this is the only rule introducing additional component in the premiss; otherwise we need an analogous rule for negated formula in the succedent.

Now, we can define an algorithm which exploits all other rules before we apply $(\Rightarrow\rightarrow)$ and $(\Rightarrow\neg)$. Informally, it corresponds to the situation where we decompose all formulae in one state before we proceed with generation of all new states forced by the presence of falsified implications and negations in the current state. Note that this solution does not eliminate loops since they are connected with the application of contraction-absorbing versions of $(\rightarrow\Rightarrow)$ and $(\neg\Rightarrow)$. Hence, to obtain termination we must introduce some loop check or to propose a refinement of contraction-absorbing rules.

In order to provide SC for **INT** which is terminating without the need for loop control, Vorobeev [275] introduced a system where instead of one $(\rightarrow\Rightarrow)$ we have a collection of rules dealing with implication in the antecedent of a sequent, depending on the shape of its antecedent. Similar idea was independently developed in sequent systems of Dyckhoff [70] and Hudelmaier [123]. All these calculi have interesting properties but the special rules for implication lack the subformula property. They are not standard although they are ordinary variants of SC. We omit the presentation of these approaches; one may find a description of Dyckhoff's solution in many places, e.g. in Waaler and Wallen [268], Troelstra and Schwichtenberg [264] and Negri and von Plato [185].

Many proposals defined particularly for better proof search avoiding loops are based on the use of structured sequents with the addition of several forms of history (Heuerding, Seyfried and Zimmermann [115], Dyckhoff and Pinto [72]) or some focused formulae (Herbelin [110]). A comparison (from the point of view of their efficiency) of Dyckhoff's and Pinto's with Heuerding, Seyfried and Zimmermann's approach, is provided by Howe [122], who uses Herbelin's calculus as the basis. Recently Ferrari, Fiorentini and Fiorino [75] provided an interesting and very efficient SC with the additional set of formulae in the antecedent which does not work as a history.

One may easily provide SC for **N3** and **N4**; we will examine only the case of LJ. It is enough to get rid of rules for negation but instead to add the following negated rules:

$$(\neg\neg\Rightarrow) \quad \frac{\varphi, \Gamma \Rightarrow \Delta}{\neg\neg\varphi, \Gamma \Rightarrow \Delta} \qquad\qquad (\Rightarrow\neg\neg) \quad \frac{\Gamma \Rightarrow \varphi}{\Gamma \Rightarrow \neg\neg\varphi}$$

$$(\Rightarrow\neg\wedge) \quad \frac{\Gamma \Rightarrow \neg\varphi}{\Gamma \Rightarrow \neg(\varphi\wedge\psi)} \qquad (\Rightarrow\neg\wedge) \quad \frac{\Gamma \Rightarrow \neg\psi}{\Gamma \Rightarrow \neg(\varphi\wedge\psi)} \qquad (\neg\wedge\Rightarrow) \quad \frac{\neg\varphi, \Gamma \Rightarrow \Delta \quad \neg\psi, \Gamma \Rightarrow \Delta}{\neg(\varphi\wedge\psi), \Gamma \Rightarrow \Delta}$$

$$(\neg\vee\Rightarrow) \quad \frac{\neg\varphi, \neg\psi, \Gamma \Rightarrow \Delta}{\neg(\varphi\vee\psi), \Gamma \Rightarrow \Delta} \qquad\qquad (\Rightarrow\neg\vee) \quad \frac{\Gamma \Rightarrow \neg\varphi \quad \Gamma \Rightarrow \neg\psi}{\Gamma \Rightarrow \neg(\varphi\vee\psi)}$$

$$(\neg\rightarrow\Rightarrow) \quad \frac{\varphi, \neg\psi, \Gamma \Rightarrow \Delta}{\neg(\varphi\rightarrow\psi), \Gamma \Rightarrow \Delta} \qquad\qquad (\Rightarrow\neg\rightarrow) \quad \frac{\Gamma \Rightarrow \varphi \quad \Gamma \Rightarrow \neg\psi}{\Gamma \Rightarrow \neg(\varphi\rightarrow\psi)}$$

This is sufficient for **N4**; to get **N3** one must add axioms of the form $\varphi, \neg\varphi \Rightarrow$. It is also important to note that if we consider a version of **N4** based on G3i with atomic axioms, we must add also axioms of the form $\Gamma, \neg p \Rightarrow \neg p$, otherwise, we are unable to prove that the general form of axiomatic sequent is provable (for **N3** $\Gamma, p, \neg p \Rightarrow \Delta$ is enough).

Exercise 5.14. *Prove axioms of* **N3** *and* **N4** *in this system.*

It is clear that such a system is just a weaker version of Smullyan's symmetric SC presented in section 3.6. However, two differences are important: Smullyan's system was equivalent to standard SCs for **CPL** and completely symmetric, whereas the present one provides a formalisation of logics stronger than **INT** and treats implication in the standard way, by nonsymmetric rules.

Kamide and Wansing [145] provide an indirect proof of cut elimination for such system by embedding into ordinary cut-free system for **INT**. But it is possible also to prove cut elimination for these calculi in a similar way as for LJ. The only difference is in the complexity reduction step when negated formulae must be additionally dealt with. We illustrate this with one example and leave the rest to the reader. Let us suppose that we apply Girard's strategy and consider the following application of mix:

$$(Mix) \quad \frac{(\Rightarrow\neg\wedge) \dfrac{\Gamma \Rightarrow \neg\varphi}{\Gamma \Rightarrow \neg(\varphi\wedge\psi)} \qquad \dfrac{\neg\varphi, \neg(\varphi\wedge\psi)^i, \Sigma \Rightarrow \Delta \qquad \neg\psi, \neg(\varphi\wedge\psi)^i, \Sigma \Rightarrow \Delta}{\neg(\varphi\wedge\psi)^{i+1}, \Sigma \Rightarrow \Delta} (\neg\wedge\Rightarrow)}{\Gamma, \Sigma \Rightarrow \Delta}$$

we proceed as follows:

$$\frac{\Gamma \Rightarrow \neg\varphi \qquad \dfrac{\Gamma \Rightarrow \neg(\varphi\wedge\psi) \qquad \neg\varphi, \neg(\varphi\wedge\psi)^i, \Sigma \Rightarrow \Delta}{\neg\varphi, \Gamma, \Sigma \Rightarrow \Delta} (Mix)}{\dfrac{\Gamma, \Gamma_{\neg\varphi}, \Sigma_{\neg\varphi} \Rightarrow \Delta}{\Gamma, \Sigma \Rightarrow \Delta} C, W} (Mix)$$

where first the height is reduced and then the complexity.

Exercise 5.15. *Provide a proof of mix elimination for LJN3 and LJN4. Try to use different strategies.*

5.2   Substructural Logics

What do we obtain if we get rid of all, or some, of the structural rules? Of course, we think about such form of SC in which these rules are not even admissible. Clearly, such restrictions must lead to some weaker logics; our considerations concerning SCs for **INT** may serve as an example. But can we expect that such experiments performed on LK or LJ will lead to some reasonable logics? The answer is positive and the family of logics which may be obtained by restricting, or modifying the set of structural rules is commonly called substructural logics. The systematic investigations on such kind of logics started not so long ago; in fact, the term was invented by Došen and Schroeder-Heister in 1993 [62]. This approach certainly offers a fruitful research perspective which is evident from the fact that some new interesting logics were invented in this way. Perhaps even more interesting is the fact that a lot of important non-classical logics which were developed much earlier, like several relevant or many-valued logics, appeared to be substructural, at least if formalised in the framework of standard SC. It is a nice feature of SC that it provides a general syntactic framework for capturing several logics which come from different sources and traditions; briefly—they are very different in many respects, yet SC introduces a new perspective for their comparison. On the other hand, the overall picture is not so clear as we would like to see. In well developed classes of logics mentioned above, i.e. in relevant and many-valued logics, many important systems are hardly dealt with in standard SC just by taming structural rules. And sometimes even if such substructural formalisation is possible it behaves badly, in particular cut may be not eliminable. These problems naturally lead to developing generalised SCs but in many cases, generalised calculi allow for reintroduction of some structural rules which were forbidden in standard SC. This makes the notion of substructural logic somewhat relative to the kind of applied formal framework. We will provide several examples of this irregular behaviour in this and the next sections. In fact, this is the main reason that (some) relevant logics and many-valued logics will be treated separately and not in this section devoted to substructural logics in the strict sense. However, we will point out in suitable places which formalisations have such a substructural character.

As usual, we are not going to provide a fuller description of substructural logics. In particular, we omit a discussion of several semantics provided for these logics, since for us they are interesting mainly as an illustration of the explanative power of standard SC. There is a lot of comprehensive books and papers which may be consulted for more information. In particular, Došen and Schroeder-Heister [62] and two monographs due to Restall [214] and Paoli [193]. Girard [98], Troelstra [263] and Roorda [220], offer excellent presentations of linear logics. For relevant logics, the two-volume massive work of Anderson and Belnap [4] (with many other contributors), may be still treated as the fundamental, and very readable, source. If we are concerned mainly with the proof theory of these logics, the good starting point is Ono [191], concise but very informative, and the comprehensive treatment in Bimbo [31].

5.2.1 The Significance of Structural Rules

To understand what is the sense and power of substructural logics let us consider first LK or LJ with all structural rules deleted. We already introduced (following Girard) the names: additive for rules characterising $\wedge$, and $\vee$ and multiplicative for rules characterising $\rightarrow$ in LK. The characteristic feature of the first type of rules is that many-premiss rules are context-sharing but one-premiss rule(s) display(s) only one argument of the principal formula in the premiss. In the second type, many-premiss rules are context-free, whereas one-premiss rules display both arguments of the principal formula in the premiss. The additive rules for $(\Rightarrow\rightarrow)$ in LK have the form (in LJ Δ is empty)

$$(\Rightarrow\rightarrow) \quad \frac{\varphi, \Gamma \Rightarrow \Delta}{\Gamma \Rightarrow \Delta, \varphi \rightarrow \psi} \qquad\qquad \frac{\Gamma \Rightarrow \Delta, \psi}{\Gamma \Rightarrow \Delta, \varphi \rightarrow \psi}$$

Note that Ketonen-style systems, like K or G3, mix both kinds of rules (many-premiss additive, and one-premiss multiplicative). In the context of SC for **CPL** or **INT** it does not matter which kind of rules we use since any combination defines the same connective. It may be seen by showing that additive rules are provable by means of multiplicative ones and vice versa. We will show it for respective rules for $\vee$ in the context of LK.

Example 5.3. (a) Additive $\Longrightarrow$ multiplicative

$$\begin{array}{c} (\Rightarrow \vee) \dfrac{\Gamma \Rightarrow \Delta, \varphi, \psi}{(\Rightarrow P) \dfrac{\Gamma \Rightarrow \Delta, \varphi, \varphi \vee \psi}{(\Rightarrow \vee) \dfrac{\Gamma \Rightarrow \Delta, \varphi \vee \psi, \varphi}{(\Rightarrow C) \dfrac{\Gamma \Rightarrow \Delta, \varphi \vee \psi, \varphi \vee \psi}{\Gamma \Rightarrow \Delta, \varphi \vee \psi}}}} \end{array}$$

$$\begin{array}{c} (W),(P) \\ (\vee \Rightarrow) \end{array} \dfrac{\dfrac{\varphi, \Gamma \Rightarrow \Delta}{\varphi, \Gamma, \Pi \Rightarrow \Delta, \Sigma} \quad \dfrac{\psi, \Pi \Rightarrow \Sigma}{\psi, \Gamma, \Pi \Rightarrow \Delta, \Sigma} (W),(P)}{\varphi \vee \psi, \Gamma, \Pi \Rightarrow \Delta, \Sigma}$$

(b) Multiplicative $\Longrightarrow$ additive

$$\begin{array}{c} (\Rightarrow W) \dfrac{\Gamma \Rightarrow \Delta, \psi}{(\Rightarrow P) \dfrac{\Gamma \Rightarrow \Delta, \psi, \varphi}{(\Rightarrow \vee) \dfrac{\Gamma \Rightarrow \Delta, \varphi, \psi}{\Gamma \Rightarrow \Delta, \varphi \vee \psi}}} \end{array}$$

the second one similarly (but without $(\Rightarrow P)$)

$$(\vee \Rightarrow) \, \frac{\varphi, \Gamma \Rightarrow \Delta \qquad \psi, \Gamma \Rightarrow \Delta}{\varphi \vee \psi, \Gamma, \Gamma \Rightarrow \Delta, \Delta}$$
$$(P), (C) \, \frac{}{\varphi \vee \psi, \Gamma \Rightarrow \Delta}$$

Exercise 5.16. *Prove that the rules for $\wedge$ and $\rightarrow$ are also interderivable.*

It is evident that structural rules play a crucial role in these proofs. In fact, in the absence of weakening and contraction, we obtain a characterisation of two different connectives. The additive ones are also called extensional (in particular in the context of relevant logics) or lattice-theoretical, since their behaviour is similar to the behaviour of lattice operations of join and meet. The multiplicative rules are also called intensional or group-theoretical.

Let us keep $\wedge$ and $\vee$ for additive conjunction and disjunction, and introduce $\otimes$ and $\oplus$ for their multiplicative counterparts (we do not consider the additive implication). The latter are in some sense better-behaving then the former since they still correspond to the role of a comma in the antecedent and the succedent. To be more precise, one can still prove (in **LK** without contraction and weakening) that

$$\vdash \varphi_1, ..., \varphi_k \Rightarrow \psi_1, ..., \psi_n \text{ iff } \vdash \varphi_1 \otimes ... \otimes \varphi_k \Rightarrow \psi_1 \oplus ... \oplus \psi_n$$

whereas

$$\vdash \varphi_1, ..., \varphi_k \Rightarrow \psi_1, ..., \psi_n \text{ iff } \varphi_1 \wedge ... \wedge \varphi_k \Rightarrow \psi_1 \vee ... \vee \psi_n \text{ is not provable.}$$

Exercise 5.17. *Prove the first equivalence above.*

In the framework of substructural logics one more extension of terminology is useful. In addition to $\top$ and $\bot$ (lattice-theoretical constants) which are characterised by axioms $\bot, \Gamma \Rightarrow \Delta$ and $\Gamma \Rightarrow \Delta, \top$, one may need more 'sensitive' constants. In **LK**, we can without problems prove that $\top \leftrightarrow \varphi$, for any provable φ but in the absence of $(W \Rightarrow)$ we cannot prove $\top \rightarrow \varphi$. To repair the problem we add new (group-theoretical) constants t and f which denote something like the weakest provable and the strongest contradictory formula. Thus, we add new axioms $\Rightarrow t$ and $f \Rightarrow$ and two rules:

$$(t \Rightarrow) \, \frac{\Gamma \Rightarrow \Delta}{t, \Gamma \Rightarrow \Delta} \qquad\qquad (\Rightarrow f) \, \frac{\Gamma \Rightarrow \Delta}{\Gamma \Rightarrow \Delta, f}$$

These rules have the effect of a very restricted weakening.

Exercise 5.18. *Prove in LK by means of weakening $\Rightarrow \top \leftrightarrow t$ and $\Rightarrow \bot \leftrightarrow f$.*

5.2.2   Linear Logic

If we take **LJ** without structural rules but with added axioms and rules for $\top, \bot, t, f$ and $\otimes$[8] we obtain a calculus **FL** characterising so-called Full Lambek Calculus

---

[8] We cannot add rules for $\oplus$ due to the restriction upon succedents in **LJ**.

which is often treated as the basic substructural logic[9] If we use LK instead of
LJ, we get a calculus CFL, where we can add also rules for $\oplus$. However, the
most popular substructural logic is undoubtedly *Linear Logic* LL[10] introduced by
Girard [98], which is just CFL with added permutation (but without t and f).
Since most of non-classical logics which were discovered earlier and then shown to
be in some sense substructural admit also permutation, linear logic LL (except its
own advantages) seems to be the most convenient basis. Hence, in the remaining
sections, we will be dealing mostly with calculi using sequents built from finite
multisets (or even sets) for convenience.

Before we will go to some well known logics of this sort, we also consider
some further extensions of LL. Formulae in LL are treated as concrete pieces
of information hence neither contraction nor weakening is reasonable as such.
However, one may recover the application of these rules in a controlled way by
introducing a special kind of modal functors called exponentials. We use standard
notation $\Box$ and $\Diamond$ instead of Girard's ! and ?; informally we can read them as 'of
course' and 'why not'. They are characterised by the following rules:

$$(\Box \Rightarrow) \quad \frac{\varphi, \Gamma \Rightarrow \Delta}{\Box\varphi, \Gamma \Rightarrow \Delta} \qquad\qquad (\Rightarrow \Box) \quad \frac{\Box\Gamma \Rightarrow \Diamond\Delta, \varphi}{\Box\Gamma \Rightarrow \Diamond\Delta, \Box\varphi}$$

$$(\Diamond \Rightarrow) \quad \frac{\varphi, \Box\Gamma \Rightarrow \Diamond\Delta}{\Diamond\varphi, \Box\Gamma \Rightarrow \Diamond\Delta} \qquad\qquad (\Rightarrow \Diamond) \quad \frac{\Gamma \Rightarrow \Delta, \varphi}{\Gamma \Rightarrow \Delta, \Diamond\varphi}$$

$$(W \Rightarrow) \quad \frac{\Gamma \Rightarrow \Delta}{\Box\varphi, \Gamma \Rightarrow \Delta} \qquad\qquad (\Rightarrow W) \quad \frac{\Gamma \Rightarrow \Delta}{\Gamma \Rightarrow \Delta, \Diamond\varphi}$$

$$(C \Rightarrow) \quad \frac{\Box\varphi, \Box\varphi, \Gamma \Rightarrow \Delta}{\Box\varphi\Gamma \Rightarrow \Delta} \qquad\qquad (\Rightarrow C) \quad \frac{\Gamma \Rightarrow \Delta, \Diamond\varphi, \Diamond\varphi}{\Gamma \Rightarrow \Delta, \Diamond\varphi}$$

One can easily recognise that we obtain **S4** modalities plus structural rules in
restricted form. Such extension is connected with the original idea of Girard that
LL is not just a new alternative to classical logic but a system allowing for better
analysis of proofs by a stricter control over data.

What happens to cut admissibility in LL? We can expect that removing C
may only improve the matters. But on the other hand, we may be afraid that
the lack of W leads to problems. However, it is not so, at least in LL without
modalities. One may directly prove

Theorem 5.7. *Cut is admissible in LL.*

PROOF: It may be carried in Dragalin-style but without proving auxiliary results
first. The reader should check that the cases with axioms (including the new ones)
and with reduction of the height are performed in the same manner as in G3 for

---

[9]Not necessarily the weakest one. One may consider nonassociative Lambek Calculus which
requires even more sensitive data structures as arguments of sequents, since lists are associative.

[10]In this case, we simply do not distinguish between a calculus and a logic it defines, hence we
write LL not **LL**.

CPL. To show that the lack of W and C does not lead to problems in cases with both principal cut-formulae, we compare the transformation for $\vee$ and $\oplus$. For the former, we have

$$(\Rightarrow \vee) \dfrac{\dfrac{\Gamma \Rightarrow \Delta, \varphi}{\Gamma \Rightarrow \Delta, \varphi \vee \psi} \qquad \dfrac{\varphi, \Pi \Rightarrow \Sigma \qquad \psi, \Pi \Rightarrow \Sigma}{\varphi \vee \psi, \Pi \Rightarrow \Sigma} (\vee \Rightarrow)}{\Gamma, \Pi \Rightarrow \Delta, \Sigma} (Cut)$$

which is replaced with:

$$\dfrac{\Gamma \Rightarrow \Delta, \varphi \qquad \varphi, \Pi \Rightarrow \Sigma}{\Gamma, \Pi \Rightarrow \Delta, \Sigma} (Cut)$$

That's all. For the latter, we have

$$(\Rightarrow \oplus) \dfrac{\dfrac{\Gamma \Rightarrow \Delta, \varphi, \psi}{\Gamma \Rightarrow \Delta, \varphi \oplus \psi} \qquad \dfrac{\varphi, \Pi \Rightarrow \Sigma \qquad \psi, \Lambda \Rightarrow \Theta}{\varphi \oplus \psi, \Pi, \Lambda \Rightarrow \Sigma, \Theta} (\oplus \Rightarrow)}{\Gamma, \Pi, \Lambda \Rightarrow \Delta, \Sigma, \Theta} (Cut)$$

which is replaced with

$$(Cut) \dfrac{(Cut) \dfrac{\Gamma \Rightarrow \Delta, \varphi, \psi \qquad \varphi, \Pi \Rightarrow \Sigma}{\Gamma, \Pi \Rightarrow \Delta, \Sigma, \psi} \qquad \psi, \Lambda \Rightarrow \Theta}{\Gamma, \Pi, \Lambda \Rightarrow \Delta, \Sigma, \Theta}$$

As we can see there are no problems with the lack of structural rules. In the additive case, only one premiss of the two-premiss rule is used after transformation but no W is needed, since all parametric formulae are in both premisses. On the other hand, in the multiplicative case, we must use both premisses of the two-premiss rule but no C is needed, since both have independent multisets of parameters. $\square$

Exercise 5.19. *Complete the proof.*

If we consider additional modal operators and restricted structural rules, we must again take care of C and use one of the strategies from chapter 2. We direct the reader to Troelstra [263], for full account of such a proof.

5.3   Relevant Logics

One of the most important class of non-classical logics which may be naturally placed in the family of substructural logics is the class of relevant logics. The point of departure from classical logic is the notion of material implication. Consider the following theses of **CPL**: $p \rightarrow (q \rightarrow p)$, $p \wedge \neg p \rightarrow q$ or $p \rightarrow (q \rightarrow q)$

In all of them, the condition of relevancy between the antecedent and the succedent of implication is somewhat broken. It may be stated as the requirement that for every provable implication at least one propositional symbol is common to the antecedent and the succedent, although other formulations are also proposed.

The prehistory of such logics is connected with Orlov's logic from 1928, and Parry's analytic implication from 1932. In 1950s, Church and Ackermann constructed axiomatic systems of so-called strong implication. However, the great project of the development of such kind of logics started in 1960s, thanks to Anderson and Belnap. They first developed several systems of relevant implication and (the weaker) connective of entailment. In 1970s, Routley and Meyer provided a kind of relational semantics for such logics. A detailed treatment of relevant logics may be found in the works mentioned in the introductory paragraph of the previous section. We restrict our consideration only to the basic systems **R** (of relevant implication) and **E** (of entailment) and their so- called mingle-versions. A presentation of some weaker relevant logics which were investigated by the Australian team of researchers will be omitted.

Let us start with implicational language only. Such restricted system of relevant implication $\mathbf{R}^\rightarrow$ may be axiomatised by means of the following axioms:

- $\varphi \rightarrow \varphi$

- $(\varphi \rightarrow \psi) \rightarrow ((\psi \rightarrow \chi) \rightarrow (\varphi \rightarrow \chi))$

- $(\varphi \rightarrow (\psi \rightarrow \chi)) \rightarrow (\psi \rightarrow (\varphi \rightarrow \chi))$

- $(\varphi \rightarrow (\varphi \rightarrow \psi)) \rightarrow (\varphi \rightarrow \psi)$

The only rule is MP.

Other connectives may be characterised by means of the following axioms:

- $\neg\neg\varphi \rightarrow \varphi \qquad (\varphi \rightarrow \neg\psi) \rightarrow (\psi \rightarrow \neg\varphi)$

- $\varphi \wedge \psi \rightarrow \varphi \qquad \varphi \wedge \psi \rightarrow \psi \qquad (\varphi \rightarrow \chi) \wedge (\psi \rightarrow \chi) \rightarrow (\varphi \wedge \psi \rightarrow \chi)$

- $\varphi \rightarrow \varphi \vee \psi \qquad \psi \rightarrow \varphi \vee \psi \qquad (\varphi \rightarrow \chi) \wedge (\psi \rightarrow \chi) \rightarrow (\varphi \vee \psi \rightarrow \chi)$

- $\varphi \oplus \psi \rightarrow (\neg\varphi \rightarrow \psi) \qquad (\neg\varphi \rightarrow \psi) \rightarrow \varphi \oplus \psi$

- $\varphi \otimes \psi \rightarrow \neg(\neg\varphi \oplus \neg\psi) \qquad \neg(\neg\varphi \oplus \neg\psi) \rightarrow \varphi \otimes \psi$

- $\varphi \wedge (\psi \vee \chi) \rightarrow (\varphi \wedge \psi) \vee (\varphi \vee \chi)$ (or in simpler form $\varphi \wedge (\psi \vee \chi) \rightarrow (\varphi \wedge \psi) \vee \varphi$)

Note that in case of addition of $\wedge$ we need also the rule of adjunction

$\varphi, \ \psi \ / \ \varphi \wedge \psi$

in this way, we obtain the Hilbert system for **R**. It is quite characteristic that we need the last axiom (of distribution) to full characterisation of $\vee$ and $\wedge$. Otherwise, we are unable to prove the laws of distribution. We will see that this is not only an inelegance of axiomatisation; it has also a strong impact on SC for relevant logics and even deeper consequences concerning decidability. In what follows $\mathbf{R}^-$ and $\mathbf{R}^+$ will be applied to denote **R** without distribution axiom and positive variant of **R**, i.e. without negation.

System **R** (in different language variants) may be weakened by the replacement of relevant implication with the connective of entailment which is both relevant and strict, like strict implication in modal logics. The basic system of entailment **E** may be axiomatised by replacement of the third axiom (the permutation of antecedents) with its weaker form

$$((\varphi \to \varphi) \to \psi) \to \psi$$

Both **R** and **E** may be also strengthtened by the addition of the axiom Mingle:

$$\varphi \to (\varphi \to \varphi)$$

In this way, we obtain two mingle-logics **RM**, **EM**.

5.3.1   SC for Relevant Logics

SCs for these relevant logics may be obtained by selection of suitable rules from LL (or FLC) with the addition of contraction rules. We can easily characterise $\mathbf{R}^{\to}$ by using appropriate rules from LJ, namely both rules for $\to$, $(C \Rightarrow)$, (Cut) and axiom $\varphi \Rightarrow \varphi$. For $\mathbf{E}^{\to}$ we must replace $(\Rightarrow\to)$ with the weaker (context-sensitive) rule:

$$(\Rightarrow\to)' \quad \frac{\varphi, \Gamma^{\to} \Rightarrow \psi}{\Gamma^{\to} \Rightarrow \varphi \to \psi}$$

where $\Gamma^{\to}$ contains only implicational formulae.

SC for mingle systems requires the addition of the expansion rule which is the converse of contraction (and in fact a restricted form of weakening).

$$(E \Rightarrow) \quad \frac{\varphi, \Gamma \Rightarrow \Delta}{\varphi, \varphi, \Gamma \Rightarrow \Delta}$$

If we add $\neg$ to implicational versions of these logics, we must admit classical sequents, i.e. change the rules taken from LJ into their counterparts from LK. The extension to other connectives is not problematic. They are just like rules for LL, we also assume that cut is a primitive rule. The most troublesome thing with this LK-like ordinary SC for relevant logic is that it is not complete, i.e. it is not sufficiently strong to cover full **R** characterised as Hilbert system. The above SC is adequate with respect to the distributionless relevant logic $\mathbf{R}^-$, i.e. the system without the last axiom. For this variant a demonstration of the equivalence with Hilbert systems is easy.

Exercise 5.20. *Prove axioms of H-$\mathbf{R}^-$.*

What about cut elimination? It is not so unproblematic as in the case of LL, since we have again C as the primitive rule. Recall that the original Gentzen's proof for LK and LJ to avoid problems with contraction introduced mix instead of cut. But in case of relevant systems, such a rule is too strong since we do not have weakening to recover missing occurrences of deleted formulae. In order to avoid problems, we must use the rule of multicut, often called a selective mix, which deletes at least one occurrence of cut-formula from both premisses but not necessarily all. Recall that similar solution was examined also in case of HSC for **S5** in the previous chapter (subsection 4.7.3). The reader is invited to examine again Gentzen's proof of mix elimination in chapter 2 and check in which cases after reduction we must restrict an application of mix to only one formula displayed in the schema. This is rather straightforward and we omit the details.

It was proved by Urquhart [265], in 1984, that both **E** and **R** are undecidable. However, the restricted versions of these logics are decidable. Decidability of $\mathbf{R}^{\rightarrow}$ was proved already by Kripke in 1950s, then other systems were proved decidable by a similar method. Since one may find numerous descriptions of these proofs in many places[11] we do not reproduce them here.

Difficulties with relevant logics based on the language with disjunction and negation destroy the clear proof-theoretic picture of this family of logics. Despite that, one should remember that stronger languages allow for obtaining many important results not accessible in weaker versions. We finish this subsection with a brief presentation of one such result. It is quite interesting that **CPL** is contained in a rather straightforward way in **E**. To prove an embedding consider the version of **CPL** in the language without (primitive) implication and a Schütte-style system S which is expressed by means of the following rules (where Γ, Δ denote disjunctions):

$$(P) \ \frac{\Gamma \vee \varphi \vee \psi \vee \Delta}{\Gamma \vee \psi \vee \varphi \vee \Delta} \qquad (C) \ \frac{\Gamma \vee \varphi \vee \varphi}{\Gamma \vee \varphi} \qquad (W) \ \frac{\Gamma}{\Gamma \vee \varphi}$$

$$(\neg\vee) \ \frac{\Gamma \vee \neg\varphi \quad \Delta \vee \neg\psi}{\Gamma \vee \Delta \vee \neg(\varphi\vee\psi)} \qquad (\neg\neg) \ \frac{\Gamma \vee \varphi}{\Gamma \vee \neg\neg\varphi}$$

$$(\wedge) \ \frac{\Gamma \vee \varphi \quad \Delta \vee \psi}{\Gamma \vee \Delta \vee (\varphi \wedge \psi)} \qquad (\neg\wedge) \ \frac{\Gamma \vee \neg\varphi \vee \neg\psi}{\Gamma \vee \neg(\varphi \wedge \psi)}$$

Proofs are defined as trees with leaves of the form $\varphi \vee \neg\varphi$. This system is adequate for **CPL** in the sense that $\models \varphi$ iff $\vdash_S \varphi$. We can prove:

Theorem 5.8. *If $\vdash_S \varphi$, then $\vdash_E \Rightarrow \varphi$.*

PROOF: It is just a consequence of the fact that the law of excluded middle is a thesis of **E** and all rules of calculus S correspond to sequents provable in **E**. □

Exercise 5.21. *Complete the proof of theorem 5.8*

---

[11]For example, in Anderson and Belnap [4], Dunn [68], Paoli [193], Ono [191]. A particularly detailed treatment of Kripke's proof is provided by Bimbo [31].

5.3.2   Generalised Sequent Calculi for Relevant Logics

In order to get an adequate formalisation of **R** or **RM** in the full language, without
ad hoc rules expressing distributivity, we must use some generalised version of SC.
As for **RM** and some of its extensions, Avron [7], showed how to apply hypersequent
calculus. The version of HC suitable for this task has simple axioms $\varphi \Rightarrow \varphi$ and
the following logical rules:

$$(\neg\Rightarrow) \quad \frac{\Gamma \Rightarrow \Delta, \varphi \mid G}{\neg\varphi, \Gamma \Rightarrow \Delta \mid G} \qquad\qquad (\Rightarrow\neg) \quad \frac{\varphi, \Gamma \Rightarrow \Delta \mid G}{\Gamma \Rightarrow \Delta, \neg\varphi \mid G}$$

$$(\wedge\Rightarrow1) \quad \frac{\varphi, \Gamma \Rightarrow \Delta \mid G}{\varphi \wedge \psi, \Gamma \Rightarrow \Delta \mid G} \qquad\qquad (\wedge\Rightarrow2) \quad \frac{\psi, \Gamma \Rightarrow \Delta \mid G}{\varphi \wedge \psi, \Gamma \Rightarrow \Delta \mid G}$$

$$(\Rightarrow\wedge) \quad \frac{\Gamma \Rightarrow \Delta, \varphi \mid G \qquad \Gamma \Rightarrow \Delta, \psi \mid G}{\Gamma \Rightarrow \Delta, \varphi \wedge \psi \mid G} \qquad (\vee\Rightarrow) \quad \frac{\varphi, \Gamma \Rightarrow \Delta \mid G \qquad \psi, \Gamma \Rightarrow \Delta \mid G}{\varphi \vee \psi, \Gamma \Rightarrow \Delta \mid G}$$

$$(\Rightarrow\vee1) \quad \frac{\Gamma \Rightarrow \Delta, \varphi \mid G}{\Gamma \Rightarrow \Delta, \varphi \vee \psi \mid G} \qquad\qquad (\Rightarrow\vee2) \quad \frac{\Gamma \Rightarrow \Delta, \psi \mid G}{\Gamma \Rightarrow \Delta, \varphi \vee \psi \mid G}$$

$$(\Rightarrow\rightarrow) \quad \frac{\varphi, \Gamma \Rightarrow \Delta, \psi \mid G}{\Gamma \Rightarrow \Delta, \varphi \rightarrow \psi \mid G} \qquad (\rightarrow\Rightarrow) \quad \frac{\Gamma \Rightarrow \Delta, \varphi \mid G \qquad \psi, \Gamma \Rightarrow \Delta \mid G}{\varphi \rightarrow \psi, \Gamma \Rightarrow \Delta \mid G}$$

$$(\otimes\Rightarrow) \quad \frac{\varphi, \psi, \Gamma \Rightarrow \Delta \mid G}{\varphi \otimes \psi, \Gamma \Rightarrow \Delta \mid G} \qquad (\Rightarrow\otimes) \quad \frac{\Gamma \Rightarrow \Delta, \varphi \mid G \qquad \Gamma \Rightarrow \Delta, \psi \mid G}{\Gamma \Rightarrow \Delta, \varphi \otimes \psi \mid G}$$

$$(\Rightarrow\oplus) \quad \frac{\Gamma \Rightarrow \Delta, \varphi, \psi \mid G}{\Gamma \Rightarrow \Delta, \varphi \oplus \psi \mid G} \qquad (\oplus\Rightarrow) \quad \frac{\varphi, \Gamma \Rightarrow \Delta \mid G \qquad \psi, \Gamma \Rightarrow \Delta \mid G}{\varphi \oplus \psi, \Gamma \Rightarrow \Delta \mid G}$$

One may easily observe that this is just a hypersequential version of LK, or
rather LL since richer language is considered. As for structural rules cut and both
EW and EC (i.e. external versions of weakening and contraction—see section (4.7),
are present but only internal contraction is primitive. Internal weakening is not a
sound rule in this system so Avron's HSC is indeed a substructural hypersequent
system. Only two additional structural rules of splitting and combining are needed
to get an adequate formalisation of **RM**:

$$(Split) \quad \frac{\Gamma, \Pi \Rightarrow \Delta, \Sigma \mid G}{\Gamma \Rightarrow \Delta \mid \Pi \Rightarrow \Sigma \mid G} \qquad (Com) \quad \frac{G \mid \Gamma \Rightarrow \Delta \qquad \Pi \Rightarrow \Sigma \mid H}{\Gamma, \Pi \Rightarrow \Delta, \Sigma \mid G \mid H}$$

The following interpretation of hypersequents is used to prove equivalency
with axiomatic system:

$$\Im(\Gamma_1 \Rightarrow \Delta_1 \mid ... \mid \Gamma_n \Rightarrow \Delta_n) = (\otimes\Gamma_1 \rightarrow \oplus\Delta_1) \vee ... \vee (\otimes\Gamma_n \rightarrow \oplus\Delta_n),$$

where $\otimes\Gamma$ and $\oplus\Delta$ are intensional conjunction and disjunction of all elements
of Γ, Δ, respectively. Cut is proved eliminable but by means of rather complicated
history method which we do not consider here.

On the other hand, the problem of distributivity is still not dealt within the framework of HSC. There is some specific nonstandard form of SC devised particularly for relevant logics which provides a solution. It was developed independently by Mints [178] and Dunn [67], and often called the consecutive calculus. We present rules for $\mathbf{R}^+$. The basic idea is to distinguish between two kinds of composing formulae in the antecedent (in the succedent only one formula is present). In ordinary sequents formulae are just separated by a comma, now we add the second separator ';'. Formally, one introduces the notion of a structure which will be denoted by means of X, Y, Z, W and is defined as follows:

- Every formula is a structure;

- the empty antecedent is a structure;

- if X and Y are structures, then (X, Y) is an extensional structure;

- if X and Y are structures, then $(X; Y)$ is an intensional structure;

- nothing more is a structure.

Now, sequents are of the form $X \Rightarrow \varphi$, and X may represent a complicated nested structure. For example, $((p, (q \wedge r; s)); ((p \to s, r); q) \Rightarrow p$ is a sequent where X is an intensional structure whose left argument is extensional (with the right argument again intensional) and the right is intensional (with the left extensional). We will use a convention $X[Y] \Rightarrow \varphi$ to represent that Y is a substructure of the structure X (which may be empty as well) and in general we omit outer parentheses. $X \circ Y$ means that both separators may be used in this rule or, what comes to the same, that we have a pair of rules for both kinds of a structure. As can be expected, the set of structural rules is much richer. As axioms we have $\varphi \Rightarrow \varphi$ and $\Rightarrow t$ and the following basic set of structural rules

$$(P) \ \frac{X[Y \circ Z] \Rightarrow \varphi}{X[Z \circ Y] \Rightarrow \varphi} \qquad (A) \ \frac{X[(Y \circ Z) \circ W] \Rightarrow \varphi}{X[Y \circ (Z \circ W)] \Rightarrow \varphi} \qquad (C) \ \frac{X[Y \circ Y] \Rightarrow \varphi}{X[Y] \Rightarrow \varphi}$$

$$(W) \ \frac{X[Y] \Rightarrow \varphi}{X[Y, Z] \Rightarrow \varphi} \qquad (Cut) \ \frac{X \Rightarrow \varphi \quad Y[\varphi] \Rightarrow \psi}{Y[X] \Rightarrow \psi}$$

The last two rules are with elaborate side conditions. (W) is admitted only for making extensional structure; moreover, Y must be nonempty. This restricted use of (W) is sufficient for proving distributivity but (in connection with suitably defined logical rules) still blocks derivation of unwanted sequents like, e.g. $\varphi \Rightarrow \psi \to \varphi$.

In the case of cut, if X is empty we substitute φ with t in the conclusion, otherwise we could 'prove', e.g. $\varphi \Rightarrow \psi \to \psi$. In fact, without this side condition, the rule is not correct. The set of logical rules contains

$$(\Rightarrow\rightarrow) \ \frac{X;\varphi \Rightarrow \psi}{X \Rightarrow \varphi \rightarrow \psi} \qquad (\rightarrow\Rightarrow) \ \frac{X \Rightarrow \varphi \qquad Y[\psi] \Rightarrow \chi}{Y[X;\varphi \rightarrow \psi] \Rightarrow \chi}$$

$$(\otimes\Rightarrow) \ \frac{X[\varphi;\psi] \Rightarrow \chi}{X[\varphi \otimes \psi] \Rightarrow \chi} \qquad (\Rightarrow\otimes) \ \frac{X \Rightarrow \varphi \qquad Y \Rightarrow \psi}{X;Y \Rightarrow \varphi \otimes \psi}$$

$$(\wedge\Rightarrow) \ \frac{X[\varphi,\psi] \Rightarrow \chi}{X[\varphi \wedge \psi] \Rightarrow \chi} \qquad (\Rightarrow\wedge) \ \frac{X \Rightarrow \varphi \qquad Y \Rightarrow \psi}{X,Y \Rightarrow \varphi \wedge \psi}$$

$$(\Rightarrow\vee 1) \ \frac{X \Rightarrow \varphi}{X \Rightarrow \varphi \vee \psi} \qquad (\Rightarrow\vee 2) \ \frac{X \Rightarrow \psi}{X \Rightarrow \varphi \vee \psi}$$

$$(t\Rightarrow) \ \frac{X[\varphi] \Rightarrow \psi}{X[\varphi;t] \Rightarrow \psi} \qquad (\vee\Rightarrow) \ \frac{X[\varphi] \Rightarrow \chi \qquad X[\psi] \Rightarrow \chi}{X[\varphi \vee \psi] \Rightarrow \chi}$$

We provide a proof of the distribution law as an example (an analogous proof of the rightmost branch is omited)

$$
\begin{array}{c}
\dfrac{
(W)\ \dfrac{\varphi \Rightarrow \varphi}{\varphi,\psi \Rightarrow \varphi} \atop (\wedge\Rightarrow)\ \dfrac{}{\varphi \wedge \psi \Rightarrow \varphi} }{(\Rightarrow\wedge)}
\quad
\dfrac{\dfrac{\dfrac{\dfrac{\psi \Rightarrow \psi}{\psi \Rightarrow \psi \vee \chi}\,(\Rightarrow\vee 1)}{\psi,\varphi \Rightarrow \psi \vee \chi}\,(W)}{\varphi,\psi \Rightarrow \psi \vee \chi}\,(P)}{\varphi \wedge \psi \Rightarrow \psi \vee \chi}\,(\wedge\Rightarrow)
}{(V\Rightarrow)\ \dfrac{\varphi \wedge \psi \Rightarrow \varphi \wedge (\psi \vee \chi) \qquad\qquad \vdots \qquad \varphi \wedge \chi \Rightarrow \varphi \wedge (\psi \vee \chi)}{(\varphi \wedge \psi) \vee (\varphi \wedge \chi) \Rightarrow \varphi \wedge (\psi \vee \chi)}}
\end{array}
$$

Dunn [67] proved cut elimination for this calculus. Consecution calculus provided a solution to the problem of controlled use of weakening in relevant logic. Still in this calculus, a proper treatment of negation is problematic. To deal with this question a more refined structure was needed. The solution was provided by Belnap [25], who introduced display calculi (often, rather mistakenly, called display logics). This is a class of very general SCs where the whole family of structural operations is introduced to characterise precisely different classes of logics. We omit the presentation of this calculi and direct the reader to works containing a detailed treatment of these calculi for several non-classical logics. One should consult, in particular, Belnap [26], Wansing [269] or Bimbo [31].

5.3.3   First-Degree Entailment

We finish this section with the presentation of nonstandard SC for a common fragment of both **E** and **R** which is called **FDE**—*first-degree entailment*. It will be evident soon that it may be treated as a kind of transition to the next sections devoted to many-valued logics. But for now, we stay in the realm of relevant logics. **FDE** is the set of theses of **E** of the form $\varphi \rightarrow \psi$ where in φ and ψ there is no occurrence of $\to$. These fragments of **R** and **E** coincide and the resulting logic is a very interesting system which has connections to many other logics, in particular to **CPL**.

In his first version, **FDE** was presented by Anderson and Belnap [4], as a kind of axiom system but if we use $\Rightarrow$ instead of $\to$ it may be just treated as a kind of ordinary SC of mixed type with sequents restricted to one formula on both sides. Such a system consists of the following axioms and rules:

- $\varphi \Rightarrow \neg\neg\varphi \qquad \neg\neg\varphi \Rightarrow \varphi$

- $\varphi \wedge \psi \Rightarrow \varphi \quad \varphi \wedge \psi \Rightarrow \psi$

- $\varphi \Rightarrow \varphi \vee \psi \quad \psi \Rightarrow \varphi \vee \psi$

- $\varphi \wedge (\psi \vee \chi) \Rightarrow (\varphi \wedge \psi) \vee (\varphi \vee \chi)$ (or simply $\varphi \wedge (\psi \vee \chi) \Rightarrow (\varphi \wedge \psi) \vee \chi$)

$$(TR) \quad \frac{\varphi \Rightarrow \psi \quad \psi \Rightarrow \chi}{\varphi \Rightarrow \chi} \qquad\qquad (CTP) \quad \frac{\varphi \Rightarrow \psi}{\neg\psi \Rightarrow \neg\varphi}$$

$$(\Rightarrow\wedge) \quad \frac{\chi \Rightarrow \varphi \quad \chi \Rightarrow \psi}{\chi \Rightarrow \varphi \wedge \psi} \qquad\qquad (\vee \Rightarrow) \quad \frac{\varphi \Rightarrow \chi \quad \psi \Rightarrow \chi}{\varphi \vee \psi \Rightarrow \chi}$$

The notion of proof may be defined in the standard way as a tree starting with the instances of axioms. Note that $\varphi \Rightarrow \varphi$ is not an axiom but it may be easily derived from $\varphi \Rightarrow \neg\neg\varphi$ and $\neg\neg\varphi \Rightarrow \varphi$ by (TR) which is just a simplified form of cut. In order to grasp some familiarity with this system, we suggest the reader to:

Exercise 5.22. *Prove sequents expressing laws of commutativity, association, idempotency.*

For the sake of illustration, we provide a schema of the proof of one of the remaining laws of distribution

$$(\Rightarrow \wedge) \quad \frac{\dfrac{\varphi \wedge \psi \Rightarrow \varphi \quad \dfrac{\psi \Rightarrow \psi \vee \chi}{\varphi \wedge \psi \Rightarrow \psi \vee \chi}}{(\vee \Rightarrow) \quad \dfrac{\varphi \wedge \psi \Rightarrow \varphi \wedge (\psi \vee \chi)}{(\varphi \wedge \psi) \vee (\varphi \wedge \chi) \Rightarrow \varphi \wedge (\psi \vee \chi)} \quad \dfrac{\varphi \wedge \chi \Rightarrow \varphi \quad \dfrac{\chi \Rightarrow \psi \vee \chi}{\varphi \wedge \chi \Rightarrow \psi \vee \chi}}{\varphi \wedge \chi \Rightarrow \varphi \wedge (\psi \vee \chi)}} (\Rightarrow \wedge)$$

Note that in two places we used lemma 2.3, and replaced axiomatic sequent $\varphi \wedge \psi \Rightarrow \psi$ ($\varphi \wedge \chi \Rightarrow \chi$, respectively) and (TR) with derived rule to provide more compact tree. We will be using such shortcuts also below.

Several variants of this calculus were proposed which change the balance between the number of axioms and rules. In particular

Lemma 5.2. *SC for* **FDE** *is equivalent to the system where the contraposition rule is replaced with four axioms expressing De Morgan laws.*

PROOF: $\Longrightarrow$ this is easy; we display the proof of one de Morgan law as an example

$$(CTP) \; \cfrac{\cfrac{\varphi \wedge \psi \Rightarrow \varphi}{\neg \varphi \Rightarrow \neg(\varphi \wedge \psi)} \quad \cfrac{\varphi \wedge \psi \Rightarrow \psi}{\neg \psi \Rightarrow \neg(\varphi \wedge \psi)} \; (CTP)}{\neg \varphi \vee \neg \psi \Rightarrow \neg(\varphi \wedge \psi)} \; (\vee \Rightarrow)$$

Exercise 5.23. *Prove the remaining De Morgan laws.*

$\Longleftarrow$ We must prove, by induction on the height of proof, that contraposition rule is admissible in the presence of De Morgan laws. First, we prove that it holds for each schema of an axiom. In case of double negation axioms, it is straightforward. For those concerning $\wedge$ and $\vee$ it is easy.

The only difficult task is with $\varphi \wedge (\psi \vee \chi) \Rightarrow (\varphi \wedge \psi) \vee (\varphi \wedge \chi)$. We prove admissibility of it. First note that the following distributivity law is provable (in the simplest way by means of De Morgan laws)

$$(\varphi \vee \psi) \wedge (\varphi \vee \chi) \Rightarrow \varphi \vee (\psi \wedge \chi);$$

It will be used in the proof, together with other abbreviations implied by lemma 2.3, to shorten the proof. By De Morgan laws and the instance of the above stated distributivity law we have

$$(\Rightarrow \wedge) \; \cfrac{\cfrac{\cfrac{\neg((\varphi \wedge \psi) \vee (\varphi \wedge \chi)) \Rightarrow \neg(\varphi \wedge \psi) \wedge \neg(\varphi \wedge \chi)}{\neg((\varphi \wedge \psi) \vee (\varphi \wedge \chi)) \Rightarrow \neg(\varphi \wedge \psi)}}{\neg((\varphi \wedge \psi) \vee (\varphi \wedge \chi)) \Rightarrow \neg \varphi \vee \neg \psi} \quad \cfrac{\cfrac{\neg((\varphi \wedge \psi) \vee (\varphi \wedge \chi)) \Rightarrow \neg(\varphi \wedge \psi) \wedge \neg(\varphi \wedge \chi)}{\neg((\varphi \wedge \psi) \vee (\varphi \wedge \chi)) \Rightarrow \neg(\varphi \wedge \psi)}}{\neg((\varphi \wedge \psi) \vee (\varphi \wedge \chi)) \Rightarrow \neg \varphi \vee \neg \psi}}{\cfrac{\neg((\varphi \wedge \psi) \vee (\varphi \wedge \chi)) \Rightarrow (\neg \varphi \vee \neg \psi) \wedge (\neg \varphi \vee \neg \chi)}{\neg((\varphi \wedge \psi) \vee (\varphi \wedge \chi)) \Rightarrow \neg \varphi \vee (\neg \psi \wedge \neg \chi)}}$$

By the following:

$$(\vee \Rightarrow) \; \cfrac{\cfrac{\neg \varphi \Rightarrow \neg \varphi \vee \neg(\psi \vee \chi) \quad \cfrac{\neg(\psi \vee \chi) \Rightarrow \neg \varphi \vee \neg(\psi \vee \chi)}{\neg \psi \wedge \neg \chi \Rightarrow \neg \varphi \vee \neg(\psi \vee \chi)}}{\neg \varphi \vee (\neg \psi \wedge \neg \chi) \Rightarrow \neg \varphi \vee \neg(\psi \vee \chi)}}{\neg \varphi \vee (\neg \psi \wedge \neg \chi) \Rightarrow \neg(\varphi \wedge (\psi \vee \chi))}$$

and (TR) we obtain a desired result. $\square$

FDE is related to **CPL** in several interesting ways. In particular, in order to provide a decision procedure, we do not need more than is necessary for **CPL**. In fact, Anderson and Belnap introduced this system to formalise the notion of so-called tautological implication. Roughly speaking these are classically valid implications $\varphi \to \psi$ satisfying relevancy condition and with no implication in φ and ψ. Anderson and Belnap proved

Theorem 5.9. $\models \varphi \to \psi$ *iff* $\vdash_{FDE} \varphi \Rightarrow \psi$.

But what exactly are tautological implications in the sense of Anderson and Belnap? Let us start with the simplest cases built from atoms. We remember that in **CPL** the atomic sequents are valid just in case they have a common atom in antecedent and succedent. In the present context, they are represented by sequents $\varphi \Rightarrow \psi$ where φ is an elementary conjunction and ψ is a clause. Of course, we get rid of all such sequents which do not have a common literal. Thus, although in **CPL** $p, q \Rightarrow p, r$ as well as $p, \neg p, q \Rightarrow r$ and $q \Rightarrow p, \neg p, r$ are all valid (and equivalent), in **FDE** only $p \wedge q \Rightarrow p \vee r$ is valid and not equivalent to the remaining two due to the lack of suitable negation rules. Using only such primitive tautological implications and rules $(\vee \Rightarrow)$ and $(\Rightarrow \wedge)$ we can deduce from them more complex sequents where the antecedent is in DNF and the succedent is in CNF, i.e. of the form $\varphi_1 \vee \ldots \vee \varphi_k \Rightarrow \psi_1 \wedge \ldots \wedge \psi_n$ where each φ_i is an elementary conjunction and each ψ_j is a clause. It is obvious that for sequents in such normal form it holds

Theorem 5.10. $\models \varphi_1 \vee \ldots \vee \varphi_k \Rightarrow \psi_1 \wedge \ldots \wedge \psi_n$ *iff* $\models \varphi_i \Rightarrow \psi_j$ *for each* $i \leq k, j \leq n$.

Since, as we proved, **FDE** has all necessary resources for changing any formula into CNF- or DNF-form we can use it to transform every sequent into equivalent sequent in such normal form. It follows that every tautological sequent may be equivalently expressed as a sequent of the form $\varphi_1 \vee \ldots \vee \varphi_k \Rightarrow \psi_1 \wedge \ldots \wedge \psi_n$ where each φ_i is an elementary conjunction and each ψ_j is a clause. Since for every sequent of the form $\varphi_i \Rightarrow \psi_j$ it is enough to have one common literal in φ_i and ψ_j we obtain decidability for **FDE** by purely classical resources.

Although the system provided above is in the natural correspondence with the notion of tautological implication, it is not very good for finding proofs. One may ask if it is not possible to provide a standard SC for **FDE**. In fact, already in Anderson and Belnap, we can find even two such systems. We briefly present the first, and the second will be mentioned in the next section. The system is using standard multisuccedent sequents and LK rules for $\wedge$ and $\vee$. Moreover, standard structural rules of contraction and weakening are primitive. In particular, the presence of W shows that at least **FDE** can be hardly treated as a substructural logic even in the framework of standard SC. For $\wedge$ and $\vee$ it is just a suitable part of LK, hence all forms of distributivity are provable in this calculus without the need of introducing special axioms or rules.

Exercise 5.24. *Prove all forms of distribution.*

As for negation we have three rules of the form:

$$(*) \quad \frac{\Gamma \Rightarrow \Delta}{*\Delta \Rightarrow *\Gamma} \qquad (\Rightarrow \neg) \quad \frac{\varphi, \Gamma \Rightarrow \Delta}{*\Delta \Rightarrow *\Gamma, \neg\varphi} \qquad (\neg \Rightarrow) \quad \frac{\Gamma \Rightarrow \Delta, \varphi}{\neg\varphi, *\Delta \Rightarrow *\Gamma}$$

where $*$ is an operation of addition of negation to each formula with zero or even number of negations on the left, or deletion of negation before each formula with an odd number of negations on the left. Finally, for $\rightarrow$ there is only one rule

$$(\Rightarrow \rightarrow) \quad \frac{\varphi \Rightarrow \psi}{\Rightarrow \varphi \rightarrow \psi}$$

Thus, this is standard but not canonical SC for **FDE** which is just a proper part of LK with only one specific rule for implication and three special rules for negation. The subformula property occurs in a generalised form, i.e. with the closure under single negations, like in the symmetric SC from section 3.6. Despite the fact that contraction and weakening are primitive, it works much better as a tool of actual root-first proof search since it may be proved complete in cut-free version. In practice, one should just resign from the applications of W and restrict the applications of C to principal formulae of applications of $(\wedge \Rightarrow)$ and $(\Rightarrow \vee)$. Clearly, axioms are taken in the generalised form and the notion of an atomic sequent is applied to sequents built from literals on both sides. In such form, it may be also potentially used for proving decidability. We leave to the reader the task of describing a suitable algorithm. Instead, we point out that we can alternatively apply the strategy of coupled trees applied by Dunn [68], on the basis of Jeffrey's coupled trees for **CPL**. Roughly speaking, if we want to check whether $\varphi \to \psi$ is valid tautological implication, we build in a root-first fashion separate completed proof trees for $\varphi \Rightarrow$ and $\psi \Rightarrow$. Then we compare all leaves in both trees (including axiomatic ones). Let us say that an atomic sequent S_1 covers an atomic sequent S_2 iff it is included in it. Now, if for each leaf in the proof tree for $\varphi \Rightarrow$ there is at least one covered leaf in the proof tree for $\psi \Rightarrow$, then $\varphi \to \psi$ is valid. Note a similarity of this technique to the proofs of the interpolation theorem by splitting method (see section 3.6). In fact, the latter was used by Muskens and Wintein [271] to prove interpolation for **FDE**.

In Anderson and Belnap [4], one more SC is provided where instead of non-standard rules for negation we have rules for negated formulae like in the symmetric SC of Smullyan [245]. However, such kind of a calculus will be examined in the next section.

We finish this section with another interesting remark. It may also be proved that **FDE** corresponds to **N4** in the following sense:

Theorem 5.11. *Let* $\varphi_1, ..., \varphi_n, \psi$ *contain no implication, then* $\vdash_{FDE} \varphi_1 \wedge ... \wedge \varphi_n \Rightarrow \psi$ *iff* $\vdash_{N4} \varphi_1, ..., \varphi_n \Rightarrow \psi$.

The proof may be found in Kamide and Wansing [145].

5.4 Many-Valued Logics

One of the oldest and well known families of non-classical logics being weaker than **CPL** is a family of many-valued logics. They were invented by Jan Łukasiewicz and independently by Emil Post in 1920s. Since then, many-valued logics were actively investigated and found numerous applications in many fields (see, e.g. Malinowski [170], Rescher [212], Urquhart [266]).

The point of departure for this enterprise was the rejection of the principle of bivalence. There is a variety of reasons for this step and we are not going to present them here—the reader may find an interesting discussion in Priest [205]. Such a choice leads to the introduction of more than two truth values—even infinitely many. Semantically many-valued logics are usually characterised by means of many-valued matrices, and we will also use this characterisation. One should notice, however, that the very application of many-valued matrices is not a good criterion for describing the family of many-valued logics and it may be a matter of discussion of what class of logics is properly denoted by this term. Wójcicki [273] has proved that all propositional logics may be characterised by means of infinitely-valued matrices. Hence, **INT** may be treated as a multi-valued logic as well, the same applies to modal logics and other logics described earlier. On the other hand, thanks to so-called Suszko thesis [255], all logics are in a deeper sense two-valued, since despite the number of values used in the characteristic matrix (Suszko called them algebraic values), we can always divide them into two groups: values which are accepted and values which are rejected. Actually, his views and a way of its interpretation is a matter of philosophical dispute[12] but technically, Suszko's thesis may serve as a convenient basis for one of the possible introduction of SC into the realm of many-valued logics.

Of course, it is important to restrict the term in some way to cover just this class of logics which is usually recognised as comprising many-valued logics in the exact sense. However, we are not concerned with general considerations but rather with the case studies of some characteristic logics, hence for us it is not essential to go into details of this dispute (see, e.g. Avron [12]). In what follows we focus on a few well known three-valued logics, including Łukasiewicz's $\mathbf{L}_3$, Kleene's $\mathbf{K}_3$, and on the one important four-valued logic of Belnap and Dunn.

5.4.1 The Basic Logics

Many-valued logics were originally presented in semantic terms and they are still usually characterised by means of matrices being structures of the form $\mathfrak{M} = \langle A, O, D \rangle$, where

- A is a nonempty set of logical values.

- O is a nonempty set of operations on A which correspond to connectives; we will use $\neg, \rightarrow$, etc for denoting suitable operations.

- $D \subset A$ is a nonempty set of designated values (not necessarily $\{1\}$).

A pair $\mathcal{A} = \langle A, O \rangle$ included in a matrix and containing operations corresponding to all connectives of some fixed language is an algebra similar to this language. Instead of valuations on $PROP$ extended by means of suitably defined satisfaction relation to all formulae, we will use mappings defined on FOR in a structure-preserving manner, i.e. homomorphisms. Technically, a homomorphism $h : FOR \longrightarrow \mathcal{A}$ is an interpretation of a language in a matrix satisfying the following conditions:

- $h(\neg\varphi) = \neg(h(\varphi))$

---

[12]One of the recent summary of different positions towards Suszko's thesis is provided by Shramko and Wansing [239].

- $h(\varphi \wedge \psi) \;=\; h(\varphi) \underline{\wedge} h(\psi)$
- $h(\varphi \vee \psi) \;=\; h(\varphi) \underline{\vee} h(\psi)$
- $h(\varphi \rightarrow \psi) \;=\; h(\varphi) \underline{\rightarrow} h(\psi)$

In what follows, we will be using interchangeably terms homomorphism and valuation. The content of a matrix $E(\mathfrak{M})$ is defined in the following way:

$$E(\mathfrak{M}) = \{\varphi : h(\varphi) \in D, \text{ for any homomorphism } h\}$$

If the content of a matrix $\mathfrak{M}$ is identical with the set of tautologies of a logic $\mathbf{L}$, then we say that this matrix determines a logic $\mathbf{L}$ (is adequate for $\mathbf{L}$ or is the characteristic matrix for $\mathbf{L}$). For any matrix, we define a relation of matrix consequence in the following way:

$$\Gamma \models_M \varphi \text{ iff for any homomorphism } h \text{ if } h(\Gamma) \subseteq D, \text{ then } h(\varphi) \in D$$

Let us consider a few matrices determining well known logics. For example, the matrix for $\mathbf{K}_3$ is $\mathfrak{M}_1^3$, where $A = \{0, 1/2, 1\}$, $D = \{1\}$, and O contains an unary operation $\neg : A \longrightarrow A$ and binary operations $\odot : A \times A \longrightarrow A$, where $\odot \in \{\wedge, \vee, \rightarrow\}$. All operations are defined by the following truth tables:

$\underline{\wedge}$	1	1/2	0	$\underline{\vee}$	1	1/2	0
1	1	1/2	0	1	1	1	1
1/2	1/2	1/2	0	1/2	1	1/2	1/2
0	0	0	0	0	1	1/2	0

$\underline{\rightarrow}$	1	1/2	0		$\underline{\neg}$
1	1	1/2	0	1	0
1/2	1	1/2	1/2	1/2	1/2
0	1	1	1	0	1

The matrix for 3-valued logic of Łukasiewicz is different only with respect to the definition of operation $\underline{\rightarrow}$. We add also a characterisation of additive conjunction and disjunction since they are sometimes considered for Łukasiewicz logics:

$\underline{\rightarrow}$	1	1/2	0	$\underline{\otimes}$	1	1/2	0	$\underline{\oplus}$	1	1/2	0
1	1	1/2	0	1	1	1/2	0	1	1	1	1
1/2	1	1	1/2	1/2	1/2	0	0	1/2	1	1	1/2
0	1	1	1	0	0	0	0	0	1	1/2	0

The only difference between Kleene's and Łukasiewicz's implication is for both arguments being $1/2$, but it is critical. For example, $p \rightarrow p$ is a tautology of $\mathbf{Ł}_3$ but not of $\mathbf{K}_3$. In fact, the latter is a logic with the empty set of tautologies since for any formula φ we may define a homomorphism such that $h(\varphi) \notin D$.

If we assume a natural order on logical values corresponding to the order of rational numbers, then operations in a matrix for $\mathbf{L}_3$ may be defined in the following way:

$$\neg x = 1 - x$$

$$x \underline{\wedge} y = min(x, y)$$

$$x \underline{\vee} y = max(x, y)$$

$$x \underline{\rightarrow} y = min(1, 1 - x + y)$$

It is an important fact in the case of a generalisation to a bigger number of logical values, since the only thing which must be changed is A. For example, for any n-valued logic with finite number of values $A = \{0, 1/n - 1, 2/n - 1, ..., 1\}$.

In both logics the third value is informally understood as a truth-value gap; a sentence evaluated as $1/2$ is treated as neither true nor false. If we decide to interpret $1/2$ as a truth-value glut—both true and false, we open the way for expressing paraconsistency in many-valued setting. Now, if truth-preservation is treated as the principal feature of logical consequence it is natural to change the set of designated values for $D = \{1, 1/2\}$. In particular, if we change the set of designated values for Kleene's logic we obtain **LP**—the logic of paradox of Asenjo and Priest. In this logic, the set of tautologies is nonempty due to the extension of the set of designated values; for instance, $\models_{LP} p \vee \neg p$. In fact, the set of tautologies of **LP** is precisely the set of tautologies of **CPL**; the two logics differ with respect to their consequence relation. In this respect, **LP** is a rather strange logic—neither MP nor hypothetical syllogism are validity-preserving. However, if we additionally change the definition of implication into the one introduced by Sobociński

$\underline{\rightarrow}$	1	1/2	0
1	1	0	0
1/2	1	1/2	0
0	1	1	1

we obtain a more interesting (still paraconsistent) system called **RM**$_3$ which is a sublogic of the relevant logic **RM** presented in the previous section.

One of the most interesting four-valued logics may be obtained if we consider both truth-value gaps and gluts as additional values. Such a logic called **B**$_4$ was proposed by Belnap and Dunn [13]. Let us use $\perp$ as a symbol for a gap and $\top$ for a glut. Designated values are 1 and $\top$ (again we keep truth-preservation as decisive factor). **B**$_4$ is characterised by the matrix $\mathfrak{M}_2^4$, where $A = \{0, \perp, \top, 1\}$, $D = \{1, \top\}$, O contains operations defined in the following way:

---

[13]In fact, such 'four corners of truth' were already considered in Indian logic, see Beall and van Fraassen [23].

∧	1	T	⊥	0	∨	1	T	⊥	0
1	1	T	⊥	0	1	1	1	1	1
T	T	T	0	0	T	1	T	1	T
⊥	⊥	0	⊥	0	⊥	1	1	⊥	⊥
0	0	0	0	0	0	1	T	⊥	0

→	1	T	⊥	0		¬
1	1	T	⊥	0	1	0
T	1	T	⊥	0	T	T
⊥	1	1	1	1	⊥	⊥
0	1	1	1	1	0	1

These four values form two lattices: a lattice of truth with the order $\leq_t$ (1 as the supremum and 0 as the infimum), and a lattice of knowledge with the order $\leq_k$ (T—the supremum and ⊥—the infimum). The former looks like this

The consequence relation and the set of tautologies are defined like for other logics. Note that an interesting connection holds between $\mathbf{B}_4$ and one of the relevant logics considered in the previous section

Theorem 5.12. *Let φ, ψ represent formulae with no occurrence of $\rightarrow$, then: $\varphi \models_{\mathbf{B}_4} \psi$ iff $\varphi \rightarrow \psi$ is a thesis of* **FDE**

Thus, in $\mathbf{B}_4$ we obtain another characterisation of **FDE** enriched with syntactically nonrestricted implication. There is also an interesting connection of $\mathbf{B}_4$ with $\mathbf{K}_3$ and **LP**. One can easily check that the three-valued matrices characterising connectives of both logics may be obtained from Dunn-Belnap's four-valued matrices just by deleting one of the values T or ⊥. Thus, $\mathbf{B}_4$ is the intersection of both three-valued logics. We will see that syntactically each of them may be obtained just by the addition of one axiom to SC for $\mathbf{B}_4$.

5.4.2 Proof Theory

As we mentioned above, many-valued logics were originally characterised semantically; the syntactical characterisation was provided later, first in the form of Hilbert calculi. $\mathbf{L}_3$ may be axiomatised just by taking the same axioms for $\vee$ and $\wedge$ as we did for $\mathbf{R}$ (but without distributivity since now it is provable). For $\rightarrow$ and $\neg$ we can use the more economical basis due to Wajsberg

- $(\varphi \rightarrow \psi) \rightarrow ((\psi \rightarrow \chi) \rightarrow (\varphi \rightarrow \chi))$
- $(\neg\varphi \rightarrow \neg\psi) \rightarrow (\psi \rightarrow \varphi)$

- $((\varphi \to \neg\varphi) \to \varphi) \to \varphi$

The only rule is MP.

Of course, $\mathbf{K}_3$ may be only characterised by means of rules. For example, one may add an axiom $\varphi \wedge \neg\varphi \vdash \psi$ to our primary formalisation of **FDE** and obtain a system for $\mathbf{K}_3$ (in the language without $\to$, but this is definable in the standard way) but there is no reason for such artificial restrictions on the side of the number of premisses. In fact, it is not hard to find suitable ordinary SC of any type described in chapter 3. For example, Urquhart [266], provides a Hertz' type formalisation of $\mathbf{K}_3$ which operates on multisuccedent sequents and has all structural rules (cut, C, W). All logical constants are characterised by means of axioms. It is enough to add the following sequents to obtain a characterisation of $\wedge$ and $\neg$:

- $\varphi, \neg\varphi \Rightarrow \qquad \neg\neg\varphi \Rightarrow \varphi \qquad \varphi \Rightarrow \neg\neg\varphi$

- $\varphi \wedge \psi \Rightarrow \varphi \qquad \varphi \wedge \psi \Rightarrow \psi \qquad \varphi, \psi \Rightarrow \varphi \wedge \psi$

- $\neg\varphi \Rightarrow \neg(\varphi \wedge \psi) \quad \neg\psi \Rightarrow \neg(\varphi \wedge \psi) \quad \neg(\varphi \wedge \psi) \Rightarrow \neg\varphi, \neg\psi$

Exercise 5.25. *Provide sequents characterising $\vee$ and $\to$; remember that they are defined in $\mathbf{K}_3$ in the standard way by means of $\neg$ and $\wedge$.*

Note that deletion of the first axiom yields a system for $\mathbf{B}_4$, whereas a replacement of it with $\Rightarrow \varphi, \neg\varphi$ yields **LP**. Such characterisations are elegant but not very practical for proof search. We need rather something more close to standard SC. Note also that the presence of all structural rules show, similarly as in the case of **FDE**, that belonging to the family of substructural logics is in some respects a matter of the chosen framework.

In fact, the first attempt to find other kinds of calculi, in particular, sequent calculi, was far from being ordinary. Schröter [231] and later, independently, Takahashi [258] and Rousseau [221], introduced the notion of n-sided sequent for n-valued logic. Thus, in case of three values, a sequent consists of three parts, and—in general—in case of n values a sequent is built from n parts corresponding to all values. The most developed study of such kind of SCs with proved cut elimination theorems may be found in Baaz, Fermüller and Zach [19]. Very often such a solution was later successfully applied to tableau systems by introducing n labels added to formulae (see, e.g. Surma [251], Suchoń [247], Carnielli [44]). Hence, we can naturally treat such systems as belonging also to the category of labelled calculi. Hähnle [107] pointed out that unfortunately this approach was independently introduced many times by many authors without pushing further the basic research. There is also a significant difference between syntactical treatment (in terms of n-sided sequents) and more semantically oriented one (in terms of n labels attached to formulae), concerning their interpretation. Loosely speaking, the former approach is usually based on the disjunctive (verificationist) interpretation, whereas the latter on the conjunctive (falsificationist) one. We will see below that—in contrast to classical logic—in many-valued setting the choice of one, or the other interpretation, leads to the construction of different calculi. Although such systems have a very natural motivation, they have also serious drawbacks when applied to proof search. In particular, we can have a proliferation of proof search trees. For example, imagine that we are dealing with n-valued logic having only 1 as a designated value and we test a formula which is actually a thesis of this logic in a labelled tableau system based on the conjunctive interpretation. To show that it is a thesis we must check $n-1$ cases, i.e. to built $n-1$ trees before we can conclude that it is a thesis indeed. Clearly, in case of non-thesis, we can stop after the first open tree if we are lucky enough. In the disjunctive interpretation, we can avoid this problem introducing instead axioms of the form $\varphi \mid \varphi \mid ... \mid \varphi$ where we have n-times repeated φ. But still, such calculi are rather inefficient for actual proof search. We will illustrate this problem in subsection 5.5.1.

One possible solution leading to significant improvement of this approach was proposed independently by Doherty [59] and Hähnle [106]. Roughly, the idea is to introduce labels which correspond not to single values but to their (selected) sets. At first, it may seem worse, since now in the case of n-valued logic we are dealing with 2^n possible labels or 2^n-sided sequents. However, not all sets of values are required, and even in case where the number of required sets exceeds the number of values, the systems tend to be simpler. In particular, only one proof search tree is necessary in all cases and the same rules are required independently of the kind of interpretation. We will present generalised sequent calculi for some logics based on this idea and designed specifically for more efficient proof search in subsection 5.5.2

In fact, other approaches to generalised SC, like hypersequent calculi (Avron [10]) or SCs with constants denoting values (e.g. Fitting [84]), were also applied. In what follows we focus only on the two approaches described above but first let us see in what way standard SC may be applied in this field.

5.4.3 Standard SC for Many-Valued Logics

One possible way is to treat many-valued logics as a kind of substructural logics. Prijatelj [207] provided a uniform formalisation of finitely-valued Łukasiewicz logics as systems with restricted contraction. Connectives are characterised by means of respective rules of LL but with added bounded contraction rules of the form:

$$(C^n \Rightarrow) \quad \frac{\varphi^{n+1}, \Gamma \Rightarrow \Delta}{\varphi^n, \Gamma \Rightarrow \Delta} \qquad\qquad (\Rightarrow C^n) \quad \frac{\Gamma \Rightarrow \Delta, \varphi^{n+1}}{\Gamma \Rightarrow \Delta, \varphi^n}$$

where n denotes the number of occurrences of φ. In case of $\mathbf{L}_3$ $n = 2$. This approach is elegant but has one essential drawback—cut is not eliminable.

On the other hand, it is possible to find out SC defined on standard sequents and allowing cut elimination at the expense of using rules which are not canonical. Beziau [29] provided such SC for $\mathbf{L}_3$ in the restricted language with $\neg$ and $\rightarrow$ only.

His approach is an example of formal realisation of Suszko's thesis, in the sense that instead of 3-valued matrix he introduces a 2-valued semantics in terms of bivaluations. This is a reformulation of the original Suszko semantics for $\mathbf{L}_3$. Such an approach is quite general but has some drawbacks; usually, a semantics of this kind is not truth-functional, i.e. bivaluations need not be homomorphisms. This is also the case of this bivaluational semantics for $\mathbf{L}_3$. Logical rules are modelled on semantical conditions and look like this:

$$(\neg\neg\Rightarrow)\ \frac{\varphi,\Gamma\Rightarrow\Delta}{\neg\neg\varphi,\Gamma\Rightarrow\Delta} \qquad (\Rightarrow\neg\neg)\ \frac{\Gamma\Rightarrow\Delta,\varphi}{\Gamma\Rightarrow\Delta,\neg\neg\varphi} \qquad (\neg\Rightarrow)\ \frac{\Gamma\Rightarrow\Delta,\varphi}{\neg\varphi,\Gamma\Rightarrow\Delta}$$

$$(\neg\vee\Rightarrow)\ \frac{\neg\varphi,\neg\psi,\Gamma\Rightarrow\Delta}{\neg(\varphi\vee\psi),\Gamma\Rightarrow\Delta} \qquad (\Rightarrow\neg\vee)\ \frac{\Gamma\Rightarrow\Delta,\neg\varphi \quad \Gamma\Rightarrow\Delta,\neg\psi}{\Gamma\Rightarrow\Delta,\neg(\varphi\vee\psi)}$$

$$(\Rightarrow\to 1)\ \frac{\Gamma\Rightarrow\Delta,\neg\varphi,\psi}{\Gamma\Rightarrow\Delta,\varphi\to\psi} \qquad (\Rightarrow\to 2)\ \frac{\varphi,\Gamma\Rightarrow\Delta \quad \neg\psi,\Gamma\Rightarrow\Delta}{\Gamma\Rightarrow\Delta,\varphi\to\psi}$$

$$(\to\Rightarrow 1)\ \frac{\neg\varphi,\Gamma\Rightarrow\Delta \quad \Gamma\Rightarrow\Delta,\neg\psi}{\varphi\to\psi,\Gamma\Rightarrow\Delta} \qquad (\to\Rightarrow 2)\ \frac{\Gamma\Rightarrow\Delta,\varphi \quad \psi,\Gamma\Rightarrow\Delta}{\varphi\to\psi,\Gamma\Rightarrow\Delta}$$

$$(\Rightarrow\neg\to)\ \frac{\varphi\to\psi,\Gamma\Rightarrow\Delta \quad \Gamma\Rightarrow\Delta,\varphi,\neg\varphi \quad \Gamma\Rightarrow\Delta,\psi,\neg\psi}{\Gamma\Rightarrow\Delta,\neg(\varphi\to\psi)}$$

$$(\neg\to\Rightarrow 1)\ \frac{\varphi\to\psi,\Gamma\Rightarrow\Delta \quad \varphi,\Gamma\Rightarrow\Delta \quad \neg\varphi,\Gamma\Rightarrow\Delta}{\neg(\varphi\to\psi),\Gamma\Rightarrow\Delta}$$

$$(\neg\to\Rightarrow 2)\ \frac{\varphi\to\psi,\Gamma\Rightarrow\Delta \quad \psi,\Gamma\Rightarrow\Delta \quad \neg\psi,\Gamma\Rightarrow\Delta}{\neg(\varphi\to\psi),\Gamma\Rightarrow\Delta}$$

In consequence of a non truth-functional character of bivaluations, we have a proliferation of rules for the introduction of the same kind of formula. However, one can prove adequacy and cut elimination for this SC.

A different and quite general approach was developed by Avron [12]. In order to provide (an almost) standard characterisation for all three-valued logics with standard negation[14] Avron introduced two background 3-valued logics which are functionally complete; one for the matrix with $D = \{1\}$ and the second for the matrix with $D = \{1, 1/2\}$. All logics discussed above are sublogics of one or the other. In both the cases, to obtain functional completeness the language must be enriched with constants $\bot$ and (the new one) I; the latter denotes $1/2$. $\neg, \wedge, \vee$ are defined as above but implication is generally characterised in the following way:

$$h(\varphi\to\psi) = \begin{cases} h(\psi) & \text{if } h(\varphi)\in D \\ 1 & \text{if } h(\varphi)\notin D \end{cases}$$

This implication (called natural by Avron) in case of $D = \{1\}$ obtains the following definition:

---

[14]It excludes logics with Post negation.

$\rightarrow$	1	1/2	0
1	1	1/2	0
1/2	1	1	1
0	1	1	1

and in case of $D = \{1, 1/2\}$ has the following:

$\rightarrow$	1	1/2	0
1	1	1/2	0
1/2	1	1/2	0
0	1	1	1

Both implications were independently introduced earlier. The former by Słupecki and the latter by D'Ottaviano and DaCosta in order to obtain three-valued paraconsistent logic $\mathbf{J_3}$ being a formalisation of Jaśkowski's discussive logic.

In both the cases, we have the same set of rules: cut, weakening, standard G3 rules of introduction for $\wedge, \vee, \rightarrow$ and the following ones for negation:

$$(\neg\neg\Rightarrow) \quad \frac{\varphi, \Gamma \Rightarrow \Delta}{\neg\neg\varphi, \Gamma \Rightarrow \Delta} \qquad\qquad (\Rightarrow\neg\neg) \quad \frac{\Gamma \Rightarrow \Delta, \varphi}{\Gamma \Rightarrow \Delta, \neg\neg\varphi}$$

$$(\neg\vee\Rightarrow) \quad \frac{\neg\varphi, \neg\psi, \Gamma \Rightarrow \Delta}{\neg(\varphi\vee\psi), \Gamma \Rightarrow \Delta} \qquad (\Rightarrow\neg\vee) \quad \frac{\Gamma \Rightarrow \Delta, \neg\varphi \quad \Gamma \Rightarrow \Delta, \neg\psi}{\Gamma \Rightarrow \Delta, \neg(\varphi\vee\psi)}$$

$$(\Rightarrow\neg\wedge) \quad \frac{\Gamma \Rightarrow \Delta, \neg\varphi, \neg\psi}{\Gamma \Rightarrow \Delta, \neg(\varphi\wedge\psi)} \qquad (\neg\wedge\Rightarrow) \quad \frac{\neg\varphi, \Gamma \Rightarrow \Delta \quad \neg\psi, \Gamma \Rightarrow \Delta}{\neg(\varphi\wedge\psi), \Gamma \Rightarrow \Delta}$$

$$(\neg\rightarrow\Rightarrow) \quad \frac{\varphi, \neg\psi, \Gamma \Rightarrow \Delta}{\neg(\varphi\rightarrow\psi), \Gamma \Rightarrow \Delta} \qquad (\Rightarrow\neg\rightarrow) \quad \frac{\Gamma \Rightarrow \Delta, \varphi \quad \Gamma \Rightarrow \Delta, \neg\psi}{\Gamma \Rightarrow \Delta, \neg(\varphi\rightarrow\psi)}$$

As axioms we have sequents of the following form: $\varphi \Rightarrow \varphi$, $\perp \Rightarrow$, $\Rightarrow \neg\perp$

One may easily check that all these rules are validity-preserving independently of the choice of D. This core SC is called GBS.

Exercise 5.26. *Prove validity-preservation of rules of GBS with respect to either selection of D.*

The calculus $GM_3^{\{1\}}$, adequate for the matrix with $D = \{1\}$, is obtained by the addition to GBS the following axioms: $\neg\varphi, \varphi \Rightarrow$; $\quad I \Rightarrow$; $\quad \neg I \Rightarrow$.

The calculus $GM_3^{\{1,1/2\}}$, adequate for the matrix with two designated values, is obtained by addition to GBS the following axioms: $\Rightarrow \neg\varphi, \varphi$; $\quad \Rightarrow I$; $\quad \Rightarrow \neg I$.

Avron proves adequacy and cut elimination for these calculi semantically. However, it is not difficult to provide a purely syntactical proof of cut elimination in the way which was briefly described in section 5.1. (for **N4**).

Exercise 5.27. *Provide a proof of cut elimination for Avron's SCs.*

Avron shows how SCs can be obtained for well known 3-valued logics on this basis by changing (some) rules for $\rightarrow$ and deleting axioms for $\bot$ and I. For example to get SC for $\mathbf{L}_3$ we must change two rules

$$(\Rightarrow\rightarrow) \quad \frac{\varphi, \Gamma \Rightarrow \Delta, \psi \qquad \neg\psi, \Gamma \Rightarrow \Delta, \neg\varphi}{\Gamma \Rightarrow \Delta, \varphi \rightarrow \psi}$$

$$(\rightarrow\Rightarrow) \quad \frac{\psi, \Gamma \Rightarrow \Delta, \varphi \qquad \psi, \Gamma \Rightarrow \Delta \qquad \neg\varphi, \Gamma \Rightarrow \Delta}{\varphi \rightarrow \psi, \Gamma \Rightarrow \Delta}$$

For $\mathbf{RM}_3$ we use the same rules except the last which is now

$$(\rightarrow\Rightarrow) \quad \frac{\neg\varphi, \psi, \Gamma \Rightarrow \Delta \qquad \Gamma \Rightarrow \Delta, \neg\psi \qquad \Gamma \Rightarrow \Delta, \varphi}{\varphi \rightarrow \psi, \Gamma \Rightarrow \Delta}$$

In contrast to Prijatelj's formalisation, this one rather hardly can be treated as substructural but is cut-free instead. Avron, demonstrates also a similar approach to four-valued logics but in this case, much more complicated rules are required.

5.5 Generalised Sequent Calculi

As we remarked above, the earliest versions of SCs for many-valued logics were generalised systems. It is the most popular approach to construction of many-valued sequent or tableau systems, based on the idea of syntactic representation of n values either by means of n-sided sequents or by n labels attached to formulae or sets of formulae. This solution was presented by many authors apparently without recognition of earlier works as was pointed out by Hähnle [107]. Below, we illustrate the basic idea of this approach using $\mathbf{L}_3$ as an example. But instead of using 3-sided sequents of the form $\Gamma \mid \Delta \mid \Sigma$, as it is usually done in papers where generalised SC is used as a framework, we prefer to apply again structured sequents, i.e. with the multiplication of arguments of ordinary sequent. This accords well with Suszko's thesis, and allows for a better comparison with classical sequents—with the minor change that more structure is now used to represent accepted and rejected formulae. Since we do not consider logics with more than four values, we do not need sequents with more than three or four components exactly as in the case of modal logics.

Recall that in addition to Gentzen's original translation there are two possible interpretations of a sequent described in section 3.4, and called I_D and I_C. The former goes in terms of verification of a sequent and applied to $\mathbf{L}_3$ leads to sequents of the form $\Gamma\ [\Delta] \Rightarrow \Sigma$ which are satisfied in a matrix iff some $\varphi \in \Gamma$ is 0, or some $\psi \in \Delta$ is 1/2, or some $\chi \in \Sigma$ is 1, under some h. The latter goes in terms of falsification and may be expressed by means of sequents of the form $\Sigma \Rightarrow [\Delta]\ \Gamma$ which are falsified in a matrix iff all formulae in Σ are 1, all in Δ are 1/2 and all in Γ are 0, under some h. In the classical case, the choice of interpretation has no impact on the shape of rules but now the situation is different. What is really interesting is the fact that in many-valued setting the choice of one of these interpretations leads to different calculi. Clearly, despite the type of interpretation we can use in both cases simply objects of the form $\Gamma \mid \Delta \mid \Sigma$ as is usually done. But even if the unified notation for n-sequents is used for both interpretation, the rules are different. In our opinion it is better to express a difference between both interpretation already on the level of the notion of a sequent, hence we will be using two variants: $\Gamma\ [\Delta] \Rightarrow \Sigma$ and $\Sigma \Rightarrow [\Delta]\ \Gamma$.

5.5.1   N-sided Systems as Structured Sequents

Let us start with the first (verificationist) approach to $\mathbf{L}_3$. Axioms are all sequents with $\Gamma \cap \Delta \cap \Sigma$ nonempty and we have the following rules:

$$(\neg \Rightarrow)\ \frac{\Gamma,\ [\Delta] \Rightarrow \Sigma, \varphi}{\neg\varphi, \Gamma,\ [\Delta] \Rightarrow \Sigma} \qquad ([\neg] \Rightarrow)\ \frac{\Gamma,\ [\varphi, \Delta] \Rightarrow \Sigma}{\Gamma,\ [\neg\varphi, \Delta] \Rightarrow \Sigma} \qquad (\Rightarrow \neg)\ \frac{\varphi, \Gamma,\ [\Delta] \Rightarrow \Sigma}{\Gamma,\ [\Delta] \Rightarrow \Sigma, \neg\varphi}$$

$$(\wedge \Rightarrow)\ \frac{\varphi, \psi, \Gamma,\ [\Delta] \Rightarrow \Sigma}{\varphi \wedge \psi, \Gamma,\ [\Delta] \Rightarrow \Sigma} \qquad (\Rightarrow \wedge)\ \frac{\Gamma,\ [\Delta] \Rightarrow \Sigma, \varphi \quad \Gamma,\ [\Delta] \Rightarrow \Sigma, \psi}{\Gamma,\ [\Delta] \Rightarrow \Sigma, \varphi \wedge \psi}$$

$$([\wedge] \Rightarrow)\ \frac{\Gamma,\ [\varphi, \Delta] \Rightarrow \Sigma, \varphi \quad \Gamma,\ [\varphi, \psi, \Delta] \Rightarrow \Sigma \quad \Gamma,\ [\psi, \Delta] \Rightarrow \Sigma, \psi}{\Gamma,\ [\varphi \wedge \psi, \Delta] \Rightarrow \Sigma}$$

$$(\vee \Rightarrow)\ \frac{\varphi, \Gamma,\ [\Delta] \Rightarrow \Sigma \quad \psi, \Gamma,\ [\Delta] \Rightarrow \Sigma}{\varphi \vee \psi, \Gamma,\ [\Delta] \Rightarrow \Sigma} \qquad (\Rightarrow \vee)\ \frac{\Gamma,\ [\Delta] \Rightarrow \Sigma, \varphi, \psi}{\Gamma,\ [\Delta] \Rightarrow \Sigma, \varphi \vee \psi}$$

$$([\vee] \Rightarrow)\ \frac{\varphi, \Gamma,\ [\varphi, \Delta] \Rightarrow \Sigma \quad \Gamma,\ [\varphi, \psi, \Delta] \Rightarrow \Sigma \quad \psi, \Gamma,\ [\psi, \Delta] \Rightarrow \Sigma}{\Gamma,\ [\varphi \vee \psi, \Delta] \Rightarrow \Sigma}$$

$$(\rightarrow \Rightarrow)\ \frac{\Gamma,\ [\Delta] \Rightarrow \Sigma, \varphi \quad \psi, \Gamma,\ [\Delta] \Rightarrow \Sigma}{\varphi \rightarrow \psi, \Gamma,\ [\Delta] \Rightarrow \Sigma} \qquad ([\rightarrow] \Rightarrow)\ \frac{\Gamma,\ [\varphi, \psi, \Delta] \Rightarrow \Sigma \quad \psi, \Gamma,\ [\Delta] \Rightarrow \Sigma, \varphi}{\Gamma,\ [\varphi \rightarrow \psi, \Delta] \Rightarrow \Sigma}$$

$$(\Rightarrow \rightarrow)\ \frac{\varphi, \Gamma,\ [\varphi, \Delta] \Rightarrow \Sigma, \psi \quad \varphi, \Gamma,\ [\psi, \Delta] \Rightarrow \Sigma, \psi}{\Gamma,\ [\Delta] \Rightarrow \Sigma, \varphi \rightarrow \psi}$$

The definition of proof is standard and we can allow as root of the tree any sequent although natural reading in terms of theses admits only sequents of the form $\Rightarrow \varphi$. In the former case we say that a provable sequent $\Gamma\ [\Delta] \Rightarrow \Sigma$ is 3-valid in the sense that for every h at least one formula φ is such that

$$h(\varphi) = \begin{cases} 0 & \text{if } \varphi \in \Gamma \\ 1/2 & \text{if } \varphi \in \Delta \\ 1 & \text{if } \varphi \in \Sigma \end{cases}$$

In order to prove that proofs may be simulated in the axiomatic system, one may use the following translation. For each sequent

$$\Im(\varphi_1, ..., \varphi_i, [\psi_1, ..., \psi_j] \Rightarrow \chi_1, ..., \chi_k) =$$

$\neg\varphi_1\vee...\vee\neg\varphi_i\vee(\psi_1\rightarrow\neg\psi_1)\wedge(\neg\psi_1\rightarrow\psi_1)\vee...\vee(\psi_j\rightarrow\neg\psi_j)\wedge(\neg\psi_j\rightarrow\psi_j)\vee\chi_1\vee...\vee\chi_k$

Under this translation, one can prove admissibility of SC rules in H-system. Instead, we provide a much simpler semantic treatment. Soundness of this calculus is a simple consequence of the following:

Lemma 5.3. *All rules are validity-preserving (normal).*

PROOF: We check the rules for $\wedge$. Consider $(\wedge\Rightarrow)$ and assume that the premiss is valid. Then for an arbitrary valuation, the premiss is satisfied which means that at least one formula in the (nonbracketed) antecedent is false or in $\Delta-1/2$, or is true in Σ. If it is the member of $\Gamma\cup\Delta\cup\Sigma$, then the conclusion is satisfied as well. Otherwise, φ or ψ is false, but then $\varphi\wedge\psi$ is false in both cases.

For $(\Rightarrow\wedge)$, again only the values of φ,ψ are crucial, since for any valuation where some member of $\Gamma\cup\Delta\cup\Sigma$ obtains the respective value it is transmitted to the conclusion. Now, if this is not the case, then both φ and ψ must be true, hence $\varphi\wedge\psi$ is true as well and the conclusion is satisfied.

The case of $([\wedge]\Rightarrow)$ is more complicated. If we compare all three premisses, we see that for having them satisfied together, we must have that: either (1) $h(\varphi)=h(\psi)=1/2$, or (2) $h(\varphi)=1/2$ but $h(\psi)=1$, or (3) $h(\varphi)=1$ and $h(\psi)=1/2$. The reader can easily check that these are all the situations yielding $h(\varphi\wedge\psi)=1/2$.   □

Exercise 5.28. *Complete the proof of validity-preservation for other rules.*

It is essentially the calculus of Rousseau [221] or Takahashi [258], who provided a general theory of such calculi for several finitely-valued logics. If we prefer LK-like version, we can follow Baaz, Fermüller and Zach [19] and admit axioms of the form $\varphi\,[\varphi]\Rightarrow\varphi$ and the following structural rules:

$$(C\Rightarrow)\;\frac{\varphi,\varphi,\Gamma,\,[\Delta]\Rightarrow\Sigma}{\varphi,\Gamma,\,[\Delta]\Rightarrow\Sigma}\qquad([C]\Rightarrow)\;\frac{\Gamma,\,[\varphi,\varphi,\Delta]\Rightarrow\Sigma}{\Gamma,\,[\varphi,\Delta]\Rightarrow\Sigma}\qquad(\Rightarrow C)\;\frac{\Gamma,\,[\Delta]\Rightarrow\Sigma,\varphi,\varphi}{\Gamma,\,[\Delta]\Rightarrow\Sigma,\varphi}$$

$$(W\Rightarrow)\;\frac{\Gamma,\,[\Delta]\Rightarrow\Sigma}{\varphi,\Gamma,\,[\Delta]\Rightarrow\Sigma}\qquad([W]\Rightarrow)\;\frac{\Gamma,\,[\Delta]\Rightarrow\Sigma}{\Gamma,\,[\varphi,\Delta]\Rightarrow\Sigma}\qquad(\Rightarrow W)\;\frac{\Gamma,\,[\Delta]\Rightarrow\Sigma}{\Gamma,\,[\Delta]\Rightarrow\Sigma,\varphi}$$

Again, we can notice that in this framework it is hard to treat $\mathbf{L}_3$ (and other many-valued logics) as a kind of substructural logic. On the basis of Rousseau's and Takahashi's general results we can state

Theorem 5.13. *If* $\models_{L3}\varphi$, *then* $\vdash\Rightarrow\varphi$ *in the present calculus.*

Although this calculus is cut-free, cut may be expressed collectively by the following rules:

$$(Cut)\;\frac{\Gamma,\,[\Delta]\Rightarrow\Sigma,\varphi\quad\varphi,\Gamma,\,[\Delta]\Rightarrow\Sigma}{\Gamma,\,[\Delta]\Rightarrow\Sigma}\qquad(Cut[])\;\frac{\Gamma,\,[\Delta]\Rightarrow\Sigma,\varphi\quad\Gamma,\,[\varphi,\Delta]\Rightarrow\Sigma}{\Gamma,\,[\Delta]\Rightarrow\Sigma}$$

$$([]Cut)\;\frac{\Gamma,\,[\Delta,\varphi]\Rightarrow\Sigma\quad\varphi,\Gamma,\,[\Delta]\Rightarrow\Sigma}{\Gamma,\,[\Delta]\Rightarrow\Sigma}$$

Exercise 5.29. *Prove that structural rules (including cut) are validity-preserving (normal).*

Rousseau proves completeness semantically for cut-free version and point out that its admissibility is a consequence of completeness, but in Baaz, Fermüller and Zach [19], one may find a constructive and general proof of its admissibility for a class of n-sided SCs. Instead of presenting it here for the special case of $\mathbf{L_3}$ we rather consider if by means of admissible cut(s) one can prove that all theses of Hilbert system may be proved in this SC:

Theorem 5.14. *If $\vdash_{HL3} \varphi$, then $\vdash \Rightarrow \varphi$.*

PROOF: We left to the reader a tedious task of proving all axioms. As an appetiser let us prove the schema of an extremely simple thesis

$$(\Rightarrow\to) \frac{\dfrac{\varphi, \psi, [\varphi, \psi] \Rightarrow \varphi \qquad \varphi, \psi, [\varphi, \varphi] \Rightarrow \varphi}{(\Rightarrow\to) \dfrac{\varphi, [\varphi] \Rightarrow \psi \to \varphi}{\Rightarrow \varphi \to (\psi \to \varphi)}} \quad \mathcal{D}}{}$$

where $\mathcal{D}$ is:

$$([\to] \Rightarrow) \frac{\dfrac{\varphi, \psi, [\varphi, \psi, \psi] \Rightarrow \varphi \qquad \varphi, \varphi, \psi, [\psi] \Rightarrow \varphi, \psi}{(\Rightarrow\to) \dfrac{\varphi, \psi, [\psi, \psi \to \varphi] \Rightarrow \varphi}{\varphi, [\psi \to \varphi] \Rightarrow \psi \to \varphi}} \quad \varphi, \psi, [\varphi, \psi \to \varphi] \Rightarrow \varphi}{}$$

However, in order to complete the proof of this claim, we should be able to show in what way cuts simulate applications of MP in any proof. Suppose we have proved $\varphi \to \psi$ and φ, then by the induction hypothesis in SC we have proofs of $\Rightarrow \varphi \to \psi$ and $\Rightarrow \varphi$. How to get $\Rightarrow \psi$ on this basis? Let us consider the following proof:

$$(\to\Rightarrow) \frac{(W) \dfrac{\varphi, [\varphi] \Rightarrow \varphi}{\varphi, [\varphi, \varphi, \psi] \Rightarrow \varphi, \psi} \quad \dfrac{\psi, [\psi] \Rightarrow \psi}{\psi, \varphi, [\varphi, \varphi, \psi] \Rightarrow \psi} (W)}{\varphi \to \psi, \varphi, [\varphi, \varphi, \psi] \Rightarrow \psi} \quad ([\to] \Rightarrow) \dfrac{\dfrac{\varphi, [\varphi] \Rightarrow \psi}{\psi, \varphi \to \psi, \varphi, [\varphi] \Rightarrow \varphi, \psi} (W)}{}}{\varphi \to \psi, \varphi, [\varphi \to \psi, \varphi] \Rightarrow \psi}$$

It is easily seen that four applications of cut (more concretely, two applications of (Cut) and two of $(Cut[])$) using $\Rightarrow \varphi \to \psi$ and $\Rightarrow \varphi$ yields $\Rightarrow \psi$. □

Let us consider now the falsificationist interpretation for the same logic. This is usually realised in (labelled) tableaux framework but it is not difficult to provide a suitable SC defined on sequents of the form $\Sigma \Rightarrow [\Delta], \Gamma$[15]. Any such sequent is counted as axiom if either $\Gamma \cap \Delta$ or $\Gamma \cap \Sigma$ or $\Delta \cap \Sigma$ is nonempty. Logical rules are

---

[15]One may find such solution for example in Ripley [218], where such an interpretation is called negated conjunction. Note also that it is not a problem to use the former interpretation to obtain dual tableaux—see, e.g. Orłowska and Golińska-Pilarek [192].

$$(\neg \Rightarrow) \quad \frac{\Sigma \Rightarrow [\Delta], \Gamma, \varphi}{\neg\varphi, \Sigma \Rightarrow [\Delta], \Gamma} \qquad (\Rightarrow [\neg]) \quad \frac{\Sigma \Rightarrow [\varphi, \Delta], \Gamma}{\Sigma \Rightarrow [\neg\varphi, \Delta], \Gamma} \qquad (\Rightarrow \neg) \quad \frac{\varphi, \Sigma \Rightarrow [\Delta], \Gamma}{\Sigma \Rightarrow [\Delta], \Gamma, \neg\varphi}$$

$$(\wedge \Rightarrow) \quad \frac{\varphi, \psi, \Sigma \Rightarrow [\Delta], \Gamma}{\varphi \wedge \psi, \Sigma \Rightarrow [\Delta], \Gamma} \qquad (\Rightarrow \wedge) \quad \frac{\Sigma \Rightarrow [\Delta], \Gamma, \varphi \quad \Sigma \Rightarrow [\Delta], \Gamma, \psi}{\Sigma \Rightarrow [\Delta], \Gamma, \varphi \wedge \psi}$$

$$(\Rightarrow [\wedge]) \quad \frac{\varphi, \Sigma \Rightarrow [\Delta, \psi], \Gamma \quad \Sigma \Rightarrow [\Delta, \varphi, \psi], \Gamma \quad \psi, \Sigma \Rightarrow [\Delta, \varphi], \Gamma}{\Sigma \Rightarrow [\Delta, \varphi \wedge \psi], \Gamma}$$

$$(\vee \Rightarrow) \quad \frac{\varphi, \Sigma \Rightarrow [\Delta], \Gamma \quad \psi, \Sigma \Rightarrow [\Delta], \Gamma}{\varphi \vee \psi, \Sigma \Rightarrow [\Delta], \Gamma} \qquad (\Rightarrow \vee) \quad \frac{\Sigma \Rightarrow [\Delta], \Gamma, \varphi, \psi}{\Sigma \Rightarrow [\Delta], \Gamma, \varphi \vee \psi}$$

$$(\Rightarrow [\vee]) \quad \frac{\Sigma \Rightarrow [\Delta, \varphi], \Gamma, \psi \quad \Sigma \Rightarrow [\Delta, \varphi, \psi], \Gamma \quad \Sigma \Rightarrow [\Delta, \psi], \Gamma, \varphi}{\Sigma \Rightarrow [\Delta, \varphi \vee \psi], \Gamma}$$

$$(\Rightarrow\rightarrow) \quad \frac{\varphi, \Sigma \Rightarrow [\Delta], \Gamma, \psi}{\Sigma \Rightarrow [\Delta], \Gamma, \varphi \rightarrow \psi} \qquad (\Rightarrow [\rightarrow]) \quad \frac{\Sigma \Rightarrow [\Delta, \varphi], \Gamma, \psi \quad \varphi, \Sigma \Rightarrow [\Delta, \psi], \Gamma}{\Sigma \Rightarrow [\Delta, \varphi \rightarrow \psi], \Gamma}$$

$$(\rightarrow\Rightarrow) \quad \frac{\Sigma \Rightarrow [\Delta], \Gamma, \varphi \quad \Sigma \Rightarrow [\Delta, \varphi, \psi], \Gamma \quad \psi, \Sigma \Rightarrow [\Delta], \Gamma}{\varphi \rightarrow \psi, \Gamma, [\Delta] \Rightarrow \Sigma}$$

Note also that now if we want to prove the admissibility of cut there is only one rule, but of the form:

$$(3\text{-}Cut) \quad \frac{\Sigma \Rightarrow [\Delta], \Gamma, \varphi \quad \Sigma \Rightarrow [\Delta, \varphi], \Gamma \quad \varphi, \Sigma \Rightarrow [\Delta], \Gamma}{\Gamma, [\Delta] \Rightarrow \Sigma}$$

One can easily see that it is correct under falsificationist interpretation: if the conclusion is falsified, then at least one premiss must be falsified as well, since φ must take one of the values under any valuation.

As far as we know a constructive proof of cut admissibility was never provided for this kind of many-sided SCs, so its admissibility follows from the completeness theorem (see, e.g. Carnielli [44], for tableau systems of this sort). In what follows, we provide a sketch of such a proof but first, we take a look at soundness matters. One may easily prove

Lemma 5.4. *All rules are validity-preserving.*

PROOF: Now we prove it indirectly, i.e. by assuming that the conclusion is falsified and on this basis at least one premiss is not valid. Let us consider again all rules for $\wedge$. In case of $(\wedge \Rightarrow)$ iff all elements of the antecedent, including $\varphi \wedge \psi$ are true, then φ and ψ must be true and the premiss is falsified.

For $(\Rightarrow \wedge)$ if the conclusion is falsified, then $\varphi \wedge \psi$ is false so either φ or ψ is false. In each case, one premiss is falsified as well.

In case of $(\Rightarrow [\wedge])$, $h(\varphi \wedge \psi) = 1/2$, so either φ is false which falsifies the leftmost premiss, or ψ is true which falsifies the rightmost one. It remains the possibility that $h(\varphi) = h(\psi) = 1/2$ but then the middle premiss is falsified. $\square$

The definition of proof is the same. But now one proof tree is not enough in the positive case. To obtain a counterpart of theorem 5.14, one must prove

Theorem 5.15. *If $\models_{L3} \varphi$, then $\vdash \Rightarrow \varphi$ and $\vdash \Rightarrow [\varphi]$.*

General completeness theorems of this kind are in fact provided by Surma [251], Suchoń [247], Carnielli [44]. Again, instead of summarising them, we ask if it is possible to prove its syntactical counterpart, i.e. a claim that

Theorem 5.16. *If $\vdash_{HL3} \varphi$, then $\vdash \Rightarrow \varphi$ and $\vdash \Rightarrow [\varphi]$.*

PROOF: It is a tedious but possible task to find cut-free double proofs of all axioms. Let us examine again the same example of a thesis as in the verificationist SC. The first proof is

$$
\cfrac{\cfrac{\varphi, \psi \Rightarrow \varphi}{\varphi \Rightarrow \psi \to \varphi} \ (\Rightarrow\to)}{\Rightarrow \varphi \to (\psi \to \varphi)} \ (\Rightarrow\to)
$$

and the second

$$
(\Rightarrow\to) \ \cfrac{\cfrac{\psi \Rightarrow [\varphi], \varphi}{\Rightarrow [\varphi], \psi \to \varphi} \qquad \cfrac{\varphi \Rightarrow [\psi], \varphi \qquad \varphi, \psi \Rightarrow [\varphi]}{\varphi \Rightarrow [\psi \to \varphi]} \ (\Rightarrow [\to])}{\Rightarrow [\varphi \to (\psi \to \varphi)]} \ (\Rightarrow [\to])
$$

But what with simulation of MP in our SC? It is also possible to demonstrate by induction on the height of axiomatic proof. By the induction hypothesis, we have provable in our SC $\Rightarrow \varphi \to \psi$, $\Rightarrow [\varphi \to \psi]$, $\Rightarrow \varphi$ and $\Rightarrow [\varphi]$. We show that both $\Rightarrow \psi$ and $\Rightarrow [\psi]$ are provable. For the latter, first we prove

$$
\cfrac{\varphi \Rightarrow [\psi], \varphi \qquad \varphi \Rightarrow [\varphi, \psi, \psi] \qquad \varphi, \psi \Rightarrow [\psi]}{\varphi \to \psi, \varphi \Rightarrow [\psi]} \ (\to\Rightarrow)
$$

then we proceed

$$
\cfrac{\Rightarrow [\varphi] \qquad \Rightarrow \varphi \qquad \cfrac{\Rightarrow [\varphi \to \psi] \qquad \Rightarrow \varphi \to \psi \qquad \varphi \to \psi, \varphi \Rightarrow [\psi]}{\varphi, \Rightarrow [\psi]} \ (3\text{-}Cut)}{\Rightarrow [\psi]} \ (3\text{-}Cut)
$$

Exercise 5.30. *Prove $\Rightarrow \psi$ in the similar way.*

This is enough to establish our theorem with the help of (3-*Cut*). $\square$

Now the question is if we are able to provide a syntactic and constructive proof of cut admissibility in this case. In the most natural way, we can do it using Dragalin's strategy. Structural rules which are needed are exactly as for the verificationist version. Below we sketch the key points only, leaving to the reader the task of providing other details. First, we need

Lemma 5.5. *Axioms may be replaced by atomic versions.*

PROOF: As usual by induction on the complexity of active formulae. Take the case of implication

$$(\to\Rightarrow) \ \dfrac{\Sigma \Rightarrow [\Delta, \varphi], \Gamma, \psi, \varphi \qquad (\Rightarrow [\to]) \ \dfrac{\Sigma \Rightarrow [\Delta, \varphi, \varphi, \psi], \Gamma, \psi \qquad \psi, \Sigma \Rightarrow [\Delta, \varphi], \Gamma, \psi}{\varphi \to \psi, \Sigma \Rightarrow [\Delta, \varphi], \Gamma, \psi} \qquad \varphi \to \psi, \varphi, \Sigma \Rightarrow [\Delta, \psi], \Gamma}{\varphi \to \psi, \Sigma \Rightarrow [\Delta, \varphi \to \psi], \Gamma}$$

Provability of the rightmost premiss was demonstrated above, so we are done.

□

Exercise 5.31. *Prove the remaining (exactly eleven) cases of lemma 5.5.*
Prove h-p admissibility of W (three versions)
Prove h-p invertibility of all rules (with respect to all premisses)
Prove h-p admissibility of C (in three versions).

Having proved all auxiliary results stated in the exercise we are in the position to prove that:

Theorem 5.17. (3-*Cut*) *is admissible in 3-sided SC for* $\mathbf{L}_3$.

PROOF: We show only how to reduce the complexity of cut-formulae when both are principal on the example of conjunction. Consider the following application of (3-*Cut*):

$$\dfrac{\Sigma \Rightarrow [\Delta, \varphi \wedge \psi], \Gamma \qquad (\Rightarrow \wedge) \ \dfrac{\Pi \Rightarrow [\Lambda], \Theta, \varphi \qquad \Pi \Rightarrow [\Lambda], \Theta, \psi}{\Pi \Rightarrow [\Lambda], \Theta, \varphi \wedge \psi} \qquad (\wedge \Rightarrow) \ \dfrac{\varphi, \psi, \Xi \Rightarrow [\Upsilon], \Omega}{\varphi \wedge \psi, \Xi \Rightarrow [\Upsilon], \Omega}}{\Sigma, \Pi, \Xi \Rightarrow [\Delta, \Lambda, \Upsilon], \Gamma, \Theta, \Omega} \ (3\text{-}Cut)$$

where the leftmost premiss is deduced by

$$(\Rightarrow [\wedge]) \ \dfrac{\varphi, \Sigma \Rightarrow [\Delta, \psi], \Gamma \qquad \Sigma \Rightarrow [\Delta, \varphi, \psi], \Gamma \qquad \psi, \Sigma \Rightarrow [\Delta, \varphi], \Gamma}{\Sigma \Rightarrow [\Delta, \varphi \wedge \psi], \Gamma}$$

We can replace this with:

$$\dfrac{\Sigma, \Pi \Rightarrow [\Delta, \Lambda, \varphi], \Gamma, \Theta \qquad \dfrac{\Pi \Rightarrow [\Lambda], \Theta, \varphi \qquad \dfrac{\Pi \Rightarrow [\Lambda], \Theta, \psi \qquad \varphi, \psi, \Xi \Rightarrow [\Upsilon], \Omega \qquad \varphi, \Sigma \Rightarrow [\Delta, \psi], \Gamma}{\varphi, \Sigma, \Pi, \Xi \Rightarrow [\Delta, \Lambda, \Upsilon], \Gamma, \Theta, \Omega} \ (3\text{-}Cut)}{\dfrac{\Sigma, \Sigma, \Pi, \Pi, \Xi \Rightarrow [\Delta, \Delta, \Lambda, \Lambda, \Upsilon], \Gamma, \Gamma, \Theta, \Theta, \Omega}{\ }} \ (3\text{-}Cut)}{\Sigma, \Pi, \Xi \Rightarrow [\Delta, \Lambda, \Upsilon], \Gamma, \Theta, \Omega} \ (C)$$

where the leftmost premiss is deduced by

$$(3\text{-}Cut) \ \dfrac{\Pi \Rightarrow [\Lambda], \Theta, \psi \qquad \dfrac{\psi, \Sigma \Rightarrow [\Delta, \varphi], \Gamma \qquad \Sigma \Rightarrow [\Delta, \varphi, \psi], \Gamma}{\Sigma, \Sigma, \Pi \Rightarrow [\Delta, \Delta, \Lambda, \varphi], \Gamma, \Gamma, \Theta} \ (C)}{\Sigma, \Pi \Rightarrow [\Delta, \Lambda, \varphi], \Gamma, \Theta}$$

□

Exercise 5.32. *Prove the remaining cases of the principal cut-formulae.*

Check that due to context independency of the rules the reduction of height goes with no problems.

Prove the case with one premiss axiomatic.

We finish with a short comparison of these two approaches to n-sided SCs. On the basis of examples, one may think that the former approach produces more awkward proofs, whereas the latter is more artificial in demanding more proof trees. In fact, several examples may be provided for which proofs are simpler in the first system and still the second needs two proofs. Moreover, in case of more values with only one designated, we need more proof trees; one for each undesignated value (of course also more structure must be added to the succedent to express all these values). It seems to show that the falsificationist approach is in general worse but this is not necessarily so. Consider for example **LP** or any other logic with two designated values, then sequents in the system realising the falsificationist interpretation are of the form $\Sigma \, [\Delta] \Rightarrow \Gamma$ and we need only one proof tree for $\Rightarrow \varphi$. On the other hand, if we want to provide SC based on the verificationist reading, we need sequents of the form $\Gamma \Rightarrow [\Delta] \, \Sigma$ and now to prove that $\vdash \varphi$, we must provide a proof for $\Rightarrow \varphi$ or for $\Rightarrow [\varphi]$ (one is enough). We do not pursue this issues further, instead, we propose the reader

Exercise 5.33. *Construct logical rules for LP in both versions.*

There is one more question here. Using $\mathbf{L}_3$ as an example we restricted the application of such SC to proving theses. But what with logics without theses; how to define consequence relation for such SC[16]? Without going into details (see Ripley [218] for justification), we state only that in the verificationist framework $\Gamma \vdash \Delta$ corresponds in $\mathbf{K}_3$ to provability of $\Gamma, [\Gamma] \Rightarrow \Delta$, and in **LP** to $\Gamma \Rightarrow [\Delta], \Delta$. In case of the falsificationist framework it corresponds in $\mathbf{K}_3$ to $\Gamma \Rightarrow [\Delta_1], \Delta_2$ where $\Delta_1 \cup \Delta_2 = \Delta$, and in **LP** to $\Gamma_1, [\Gamma_2] \Rightarrow \Delta$, where $\Gamma_1 \cup \Gamma_2 = \Gamma$. One may notice again the serious computational complexities connected with proving validity of inferences under the falsificationist interpretation. For example, to prove that $\varphi, \psi \vdash_{LP} \chi$ we must provide proofs of four sequents: $\varphi, \psi \Rightarrow \chi$, $[\varphi, \psi] \Rightarrow \chi$, $\varphi, [\psi] \Rightarrow \chi$ and $\psi, [\varphi] \Rightarrow \chi$.

5.5.2   A More Uniform Approach to Structured Sequents

The approach based on n-sided sequents develops in two different ways the idea which is very natural for many-valued logics. But, as we have seen, despite the version it has serious drawbacks. First of all, it is not practical. Rules are often quite complicated and this leads (even in the case of this simple example-thesis) to involved proofs and, in the falsificationist version, to proliferation of their number. Moreover, in this approach a close relationship between $\mathbf{B}_4$ and $\mathbf{K}_3$ and **LP** is

---

[16]Even in the case of **LP** it is important since the set of theses coincide with **CPL** and we really need a system for demonstrating valid inferences.

obscured. In the framework of n-sided sequents, if we want to find a formalisation of $\mathbf{B}_4$ we must introduce sequents of the form $\Gamma, [\Delta] \Rightarrow [\Lambda], \Sigma$ where, depending on the kind of interpretation, either the antecedent or the succedent corresponds to both designated values. Further, for all connectives we must define four rules (rules for the verificationist interpretation may be found, e.g. in Baaz, Fermüller and Zach [18]), so this system is in many respects (structure of sequents, number of rules and proof trees which need to be constructed) significantly different and the obvious relation to its extensions is lost. We also noticed that the expression of consequence relations in this approach may be rather artificial even if we are using sequents apparently similar to standard ones.

One may think that this is a price for building SC which is directly based on semantics but this is by no means necessary. As we already remarked, in the framework of labelled tableaux a more efficient solution may be obtained when sets of values are introduced. Moreover, in this approach, the logics which are closely related may be formalised in a modular way by means of the same rules and exhibiting their relationships. This approach was developed greatly by Hähnle [106], in the setting of labelled tableaux but may be also transferred to SC setting. As promised above we present here a generalised SC for all many-valued logics we described so far. Again, it is a kind of SC operating on structured sequents formally identical to those considered so far but based on different motivations close to Hähnle's approach. To be more precise, these systems are different in many respects from Hähnle's original tableau systems but nevertheless, they realise essentially the same idea. Similar solutions were developed also by Degauquier [58]. In our treatment instead of (four) labels we will be using sequents of the form $\Gamma, [\Pi] \Rightarrow [\Sigma], \Delta$ where in each position we have a finite, possibly empty, set of formulae. Thus, we have almost classical sequents but where both the antecedent and the succedent is divided into two parts. Only the interpretation of parts of the sequent will vary for different logics. What is important is the fact that the basic intuition connected with the falsifying interpretation of a sequent is saved. Roughly, Γ contains accepted (designated values) and Δ rejected formulae; sets in [ ] are like their denials. In case Π and Σ are empty we just write $\Gamma \Rightarrow \Delta$ as in ordinary SC. Note that such systems may be developed also in terms of bisequents as we did for **S5** in section 4.8; instead of $\Gamma, [\Pi] \Rightarrow [\Sigma], \Delta$ we could use bisequents $\Gamma \Rightarrow \Delta \mid \Pi \Rightarrow \Sigma$ with suitably redefined rules. However, for our present purposes, it is better to choose this form of representation.

For all logics $\mathbf{L}$ under consideration we define:

$$\Gamma \vdash_L \Delta \text{ iff } \vdash \Gamma \Rightarrow \Delta$$

Hence, we start a proof search always with a *standard* sequent and apply suitable rules in an upside-down manner. Note that this solution is implicitly based on Suszko's thesis according to which there are only two 'real' truth values corresponding to sets of designated and nondesignated values in the characteristic matrix.

For $\mathbf{K}_3$ and $\mathbf{L}_3$ the informal interpretation of a sequent $\Gamma, [\Pi] \Rightarrow [\Sigma], \Delta$ is the following: Γ is the set of all true formulae, Π is the set of all formulae which are true or $1/2$, Σ—all false formulae and Δ—false or $1/2$. Such interpretation corresponds to the situation that at the first step we indirectly assume that a sequent $\Gamma \Rightarrow \Delta$ is not valid i.e. there is a valuation such that all elements of Γ have (the only) designated value but no element of Δ is designated (every one obtains 0 or $1/2$).

As axioms for both logics, we admit all sequents $\Gamma, [\Pi] \Rightarrow [\Sigma], \Delta$ such that either: $\Gamma \cap \Sigma$ or $\Gamma \cap \Delta$ or $\Pi \cap \Sigma$ is nonempty. Note that the case where $\Pi \cap \Delta \neq \varnothing$ is not treated as an axiom since in this case, it is possible that the formula in the intersection of both sets has the value $1/2$.

The rules for $\mathbf{K}_3$ are the following:

$$(\neg \Rightarrow) \quad \frac{\Gamma, [\Pi] \Rightarrow [\Sigma, \varphi], \Delta}{\neg \varphi, \Gamma, [\Pi] \Rightarrow [\Sigma], \Delta} \qquad\qquad (\Rightarrow \neg) \quad \frac{\Gamma, [\varphi, \Pi] \Rightarrow [\Sigma], \Delta}{\Gamma, [\Pi] \Rightarrow [\Sigma], \Delta, \neg \varphi}$$

$$([\neg] \Rightarrow) \quad \frac{\Gamma, [\Pi] \Rightarrow [\Sigma], \Delta, \varphi}{\Gamma, [\neg \varphi, \Pi] \Rightarrow [\Sigma], \Delta} \qquad\qquad (\Rightarrow [\neg]) \quad \frac{\varphi, \Gamma, [\Pi] \Rightarrow [\Sigma], \Delta}{\Gamma, [\Pi] \Rightarrow [\Sigma, \neg \varphi], \Delta}$$

$$(\wedge \Rightarrow) \quad \frac{\varphi, \psi, \Gamma, [\Pi] \Rightarrow [\Sigma], \Delta}{\varphi \wedge \psi, \Gamma, [\Pi] \Rightarrow [\Sigma], \Delta} \qquad (\Rightarrow \wedge) \quad \frac{\Gamma, [\Pi] \Rightarrow [\Sigma], \Delta, \varphi \quad \Gamma, [\Pi] \Rightarrow [\Sigma], \Delta, \psi}{\Gamma, [\Pi] \Rightarrow [\Sigma], \Delta, \varphi \wedge \psi}$$

$$([\wedge] \Rightarrow) \quad \frac{\Gamma, [\varphi, \psi, \Pi] \Rightarrow [\Sigma], \Delta}{\Gamma, [\varphi \wedge \psi, \Pi] \Rightarrow [\Sigma], \Delta} \qquad (\Rightarrow [\wedge]) \quad \frac{\Gamma, [\Pi] \Rightarrow [\Sigma, \varphi] \Delta \quad \Gamma, [\Pi] \Rightarrow [\Sigma, \psi], \Delta}{\Gamma, [\Pi] \Rightarrow [\Sigma, \varphi \wedge \psi], \Delta}$$

$$(\Rightarrow \vee) \quad \frac{\Gamma, [\Pi] \Rightarrow [\Sigma], \Delta, \varphi, \psi}{\Gamma, [\Pi] \Rightarrow [\Sigma], \Delta, \varphi \vee \psi} \qquad (\vee \Rightarrow) \quad \frac{\varphi, \Gamma, [\Pi] \Rightarrow [\Sigma] \Delta \quad \psi, \Gamma, [\Pi] \Rightarrow [\Sigma], \Delta}{\varphi \vee \psi, \Gamma, [\Pi] \Rightarrow [\Sigma], \Delta}$$

$$(\Rightarrow [\vee]) \quad \frac{\Gamma, [\Pi] \Rightarrow [\Sigma, \varphi, \psi], \Delta}{\Gamma, [\Pi] \Rightarrow [\Sigma, \varphi \vee \psi], \Delta} \qquad ([\vee] \Rightarrow) \quad \frac{\Gamma, [\varphi, \Pi] \Rightarrow [\Sigma], \Delta \quad \Gamma, [\psi, \Pi] \Rightarrow [\Sigma], \Delta}{\Gamma, [\varphi \vee \psi, \Pi] \Rightarrow [\Sigma], \Delta}$$

$$(\Rightarrow \rightarrow) \quad \frac{\Gamma, [\varphi, \Pi] \Rightarrow [\Sigma], \Delta, \psi}{\Gamma, [\Pi] \Rightarrow [\Sigma], \Delta, \varphi \rightarrow \psi} \qquad (\rightarrow \Rightarrow) \quad \frac{\Gamma, [\Pi] \Rightarrow [\Sigma, \varphi], \Delta \quad \psi, \Gamma, [\Pi] \Rightarrow [\Sigma], \Delta}{\varphi \rightarrow \psi, \Gamma, [\Pi] \Rightarrow [\Sigma], \Delta}$$

$$(\Rightarrow [\rightarrow]) \quad \frac{\varphi, \Gamma, [\Pi] \Rightarrow [\Sigma, \psi], \Delta}{\Gamma, [\Pi] \Rightarrow [\Sigma, \varphi \rightarrow \psi], \Delta} \qquad ([\rightarrow] \Rightarrow) \quad \frac{\Gamma, [\Pi] \Rightarrow [\Sigma], \Delta, \varphi \quad \Gamma, [\psi, \Pi] \Rightarrow [\Sigma], \Delta}{\Gamma, [\varphi \rightarrow \psi, \Pi] \Rightarrow [\Sigma], \Delta}$$

The proof of a sequent $\Gamma \Rightarrow \Delta$ is defined in the standard way. The calculus for $\mathbf{L}_3$ unsurprisingly is almost the same; only two rules for implication are different:

$$(\Rightarrow \rightarrow) \quad \frac{\varphi, \Gamma, [\Pi] \Rightarrow [\Sigma], \Delta, \psi \quad \Gamma, [\varphi, \Pi] \Rightarrow [\Sigma, \psi], \Delta}{\Gamma, [\Pi] \Rightarrow [\Sigma], \Delta, \varphi \rightarrow \psi}$$

$$(\rightarrow \Rightarrow) \quad \frac{\Gamma, [\psi, \Pi] \Rightarrow [\Sigma], \Delta, \varphi \quad \varphi, \psi, \Gamma, [\Pi] \Rightarrow [\Sigma], \Delta \quad \Gamma, [\Pi] \Rightarrow [\Sigma, \varphi, \psi], \Delta}{\varphi \rightarrow \psi, \Gamma, [\Pi] \Rightarrow [\Sigma], \Delta}$$

We can also provide rules for $\otimes$ and $\oplus$. $(\otimes \Rightarrow)$ and $(\Rightarrow \otimes)$ are identical as the respective rules for $\wedge$ but the remaining two rules are more involved:

$$([\otimes] \Rightarrow) \quad \frac{\varphi, \Gamma, [\psi, \Pi] \Rightarrow [\Sigma], \Delta \quad \psi, \Gamma, [\varphi, \Pi] \Rightarrow [\Sigma], \Delta}{\Gamma, [\varphi \otimes \psi, \Pi] \Rightarrow [\Sigma], \Delta}$$

$$(\Rightarrow [\otimes]) \quad \frac{\varphi, \Gamma, [\Pi] \Rightarrow [\Sigma, \psi], \Delta \quad \psi, \Gamma, [\Pi] \Rightarrow [\Sigma, \varphi], \Delta \quad \Gamma, [\Pi] \Rightarrow [\Sigma], \Delta, \varphi, \psi}{\Gamma, [\Pi] \Rightarrow [\Sigma, \varphi \otimes \psi], \Delta}$$

In case of $\oplus$, we have a dual situation.

Exercise 5.34. *Check that* $(\Rightarrow [\oplus])$ *and* $([\oplus] \Rightarrow)$ *are the same as the respective rules for* $\vee$.

Construct two-premiss rule for $(\Rightarrow \oplus)$ *and three-premiss rule for* $(\oplus \Rightarrow)$.

In case of **LP** and **RM**$_3$ we change a little an interpretation of a sequent $\Gamma, [\Pi] \Rightarrow [\Sigma], \Delta$: Γ is the set of all formulae true or $1/2$, Π is the set of all formulae which are true, Σ—all formulae false or $1/2$ and Δ—only false. Again, such an interpretation corresponds to the situation where at the first step we indirectly assume that a sequent $\Gamma \Rightarrow \Delta$ is not valid, i.e. there is an interpretation such that all elements of Γ have designated values (1 or $1/2$) but no element of Δ is designated.

As axioms for both logics, we admit all sequents $\Gamma, [\Pi] \Rightarrow [\Sigma], \Delta$ such that either: $\Gamma \cap \Delta$ or $\Pi \cap \Delta$ or $\Pi \cap \Sigma$ is nonempty. Note that now the case where $\Gamma \cap \Sigma \neq \varnothing$ is not treated as an axiom since in this case, it is possible that the formula in the intersection of both sets has the value $1/2$.

In case of **LP** we have all rules like for **K**$_3$, so the only difference between these two calculi is in the set of axioms. In case of **RM**$_3$ the rules for $\neg, \wedge, \vee$ are the same but for $\rightarrow$ there is a difference. $(\Rightarrow [\rightarrow])$ and $([\rightarrow] \Rightarrow)$ are like in **K**$_3$ and $(\Rightarrow \rightarrow)$ is like in **Ł**$_3$) but $(\rightarrow \Rightarrow)$ is in a sense dual to suitable rule from **Ł**$_3$)

$$(\rightarrow \Rightarrow) \quad \frac{\psi, \Gamma, [\Pi] \Rightarrow [\Sigma, \varphi], \Delta \quad \Gamma, [\varphi, \psi, \Pi] \Rightarrow [\Sigma], \Delta \quad \Gamma, [\Pi] \Rightarrow [\Sigma], \Delta, \varphi, \psi}{\varphi \rightarrow \psi, \Gamma, [\Pi] \Rightarrow [\Sigma], \Delta}$$

However, if we add $\otimes$ and $\oplus$ we again obtain a different set of rules. Now $([\oplus] \Rightarrow)$ and $(\Rightarrow [\otimes])$ are the same as for $\wedge$ but we have

$$(\otimes \Rightarrow) \quad \frac{\varphi, \Gamma, [\psi, \Pi] \Rightarrow [\Sigma], \Delta \quad \psi, \Gamma, [\varphi, \Pi] \Rightarrow [\Sigma], \Delta}{\varphi \otimes \psi, \Gamma, [\Pi] \Rightarrow [\Sigma], \Delta}$$

$$(\Rightarrow \otimes) \quad \frac{\Gamma, [\varphi, \Pi] \Rightarrow [\Sigma], \Delta, \psi \quad \Gamma, [\psi, \Pi] \Rightarrow [\Sigma], \Delta, \varphi \quad \Gamma, [\Pi] \Rightarrow [\Sigma, \varphi, \psi], \Delta}{\Gamma, [\Pi] \Rightarrow [\Sigma], \Delta, \varphi \otimes \psi}$$

Again in case of $\oplus$ we have a dual situation.

Exercise 5.35. *Check that* $(\Rightarrow \oplus)$ *and* $(\oplus \Rightarrow)$ *are the same as the respective rules for* $\vee$.

Construct two-premiss rule for $(\Rightarrow [\oplus])$ *and three-premiss rule for* $([\oplus] \Rightarrow)$.

One may also add rules for implication considered by Avron [12] (see previous section). Both rules for $[\varphi \rightarrow \psi]$ are the same as for Kleene's logic but for $\varphi \rightarrow \psi$

they are with side formulae also in outer position (similarly as in the respective rules for $\wedge$ and $\vee$).

Exercise 5.36. *Check that such rules for implication are correct for both Avron's logics* GM_3.

Finally, the case of $\mathbf{B}_4$ applies the following interpretation of a sequent $\Gamma, [\Pi] \Rightarrow [\Sigma], \Delta$: Γ is the set of all formulae which are evaluated as 1 or $\top$, Π—1 or $\bot$, Σ—0 or $\top$, and Δ—0 or $\bot$. Such an interpretation corresponds to the situation that at the first step we indirectly assume that a sequent $\Gamma \Rightarrow \Delta$ is not valid i.e. there is an interpretation such that all elements of Γ have designated values (1 or $\top$) but no element of Δ is designated (is 0 or $\bot$).

As axioms for $\mathbf{B}_4$ we admit all sequents $\Gamma, [\Pi] \Rightarrow [\Sigma], \Delta$ such that either: $\Gamma \cap \Delta$ or $\Pi \cap \Sigma$ is nonempty. Note that now, neither the case where $\Gamma \cap \Sigma \neq \varnothing$ nor where $\Pi \cap \Delta \neq \varnothing$ is treated as an axiom since in both cases it is possible that the formula in the intersection of both sets has the value $\top$ or $\bot$ respectively.

In case of $\mathbf{B}_4$, all rules are like for $\mathbf{K}_3$ with the exception of the two rules for implication:

$$(\Rightarrow\to) \ \frac{\varphi, \Gamma, [\Pi] \Rightarrow [\Sigma], \Delta, \psi}{\Gamma, [\Pi] \Rightarrow [\Sigma], \Delta, \varphi \to \psi} \qquad (\to\Rightarrow) \ \frac{\Gamma, [\Pi] \Rightarrow [\Sigma], \Delta, \varphi \quad \psi, \Gamma, [\Pi] \Rightarrow [\Sigma], \Delta}{\varphi \to \psi, \Gamma, [\Pi] \Rightarrow [\Sigma], \Delta}$$

Let us consider a few examples of proofs and disproofs.

$$\frac{\dfrac{[p] \Rightarrow p, q}{[p, \neg p] \Rightarrow q} \ ([\neg] \Rightarrow)}{\dfrac{[p \wedge \neg p] \Rightarrow q}{\Rightarrow p \wedge \neg p \to q} \ (\Rightarrow\to)} \ ([\wedge] \Rightarrow)$$

This is a legitimate proof tree in both **LP** and $\mathbf{K}_3$, however, only in the former, it provides a proof of the root sequent since the leaf is axiomatic.

$$\frac{(\neg \Rightarrow) \ \dfrac{p \Rightarrow [p], q}{p, \neg p \Rightarrow q} \qquad \dfrac{[p] \Rightarrow p, [q]}{[p, \neg p] \Rightarrow [q]} \ ([\neg] \Rightarrow)}{(\Rightarrow\to) \ \dfrac{(\wedge \Rightarrow) \ \dfrac{p \wedge \neg p \Rightarrow q}{}}{\Rightarrow p \wedge \neg p \to q} \qquad \dfrac{[p \wedge \neg p] \Rightarrow [q]}{} \ ([\wedge] \Rightarrow)}$$

This is a legitimate proof tree for both $\mathbf{L}_3$ and $\mathbf{RM}_3$ since $(\Rightarrow\to)$ is a rule sound for both logic. However, this is neither a proof in $\mathbf{L}_3$ nor in $\mathbf{RM}_3$; in the former because the rightmost leaf is not axiomatic and in the latter because of the leftmost leaf.

Let us consider the following two proofs in $\mathbf{L}_3$:

$$\cfrac{\cfrac{\cfrac{\cfrac{\cfrac{p \Rightarrow p, [q] \qquad p, [q] \Rightarrow [q]}{[p \to q], p \Rightarrow [q]}\ ([\to] \Rightarrow)}{[p \to q] \Rightarrow [q, \neg p]}\ (\Rightarrow [\neg])}{[p \to q], \neg q \Rightarrow [\neg p]}\ (\neg \Rightarrow)}{[p \to q] \Rightarrow [\neg q \to \neg p]}\ (\Rightarrow [\to])$$

and

$$\cfrac{\cfrac{\cfrac{[p, q] \Rightarrow p, [q] \qquad p, q, [p] \Rightarrow [q]}{p \to q, [p] \Rightarrow [q]}\ (\to \neg)}{p \to q \Rightarrow [q], \neg p}{p \to q, \neg q \Rightarrow \neg p}\ (\neg \to \neg)}{p \to q \Rightarrow \neg q \to \neg p}$$

$$\cfrac{[p] \Rightarrow [p, q]}{}$$

$$\cfrac{\cfrac{\cfrac{\cfrac{p, [q] \Rightarrow p, q \qquad p, q \Rightarrow q \qquad p \Rightarrow [p, q], q}{p \to q, p \Rightarrow q}\ (\to \Rightarrow)}{p \to q \Rightarrow q, [\neg p]}\ (\to [\neg])}{p \to q, [\neg q] \Rightarrow [\neg p]}\ (\neg \to \neg)}{}\ (\Rightarrow \to)$$

If we apply $(\Rightarrow \to)$ to them, we obtain an $\mathbf{L}_3$ proof of $\Rightarrow (p \to q) \to (\neg q \to \neg p)$.

Exercise 5.37. *Provide a* $\mathbf{RM}_3$ *proof of* $p \to q \Rightarrow \neg q \to \neg p$

Finally, an example of a disproof of the previous example in $\mathbf{B}_4$

$$\cfrac{\cfrac{\cfrac{\cfrac{\cfrac{[p] \Rightarrow [q], p \qquad q, [p] \Rightarrow [q]}{p \to q, [p] \Rightarrow [q]}\ (\to \Rightarrow)}{p \to q \Rightarrow [q], \neg p}\ (\Rightarrow \neg)}{p \to q, \neg q \Rightarrow \neg p}\ (\neg \Rightarrow)}{p \to q \Rightarrow \neg q \to \neg p}\ (\Rightarrow \to)}{\Rightarrow (p \to q) \to (\neg q \to \neg p)}\ (\Rightarrow \to)$$

Soundness of all these calculi is easy to establish. We will restrict the consideration to $\mathbf{L}_3$. Let us say that a sequent is verified by some interpretation in $\mathbf{L}_3$ (and in $\mathbf{K}_3$) if either at least one element of Π is assigned 0, or one of Δ—1, or one of Γ is not 1, or one of Σ is not 0. Otherwise, we will say that a sequent is falsified by this interpretation. It is sufficient to show for all rules that if the conclusion is falsified in some interpretation, then at least one premiss is also falsified. We will demonstrate the correctness (in the sense of verification-preservation) of $(\to \Rightarrow)$ in $\mathbf{L}_3$. Consider an interpretation falsifying the conclusion, i.e. assigning 1 to $\varphi \to \psi$ and to all elements of Γ, 1 or 1/2 to all formulae in Π, 0 to Σ and 0 or 1/2 to Δ. Easy calculation shows that in case of φ being 1/2 or 0, ψ must be 1 or 1/2. In any of these 4 cases, the leftmost premiss is falsified. It is also possible that both arguments of the implication are 1 or 0; the first case falsifies the central premiss and the second the rightmost one. Hence if all premisses are verified by some interpretation, then the conclusion must be also verified.

Exercise 5.38. *Check the correctness of all rules of* $\mathbf{K}_3$ *and* $\mathbf{L}_3$.
Define verifiability (falsifiability) of sequents in $\mathbf{LP}$ *and* $\mathbf{RM}_3$ *and check the correctness of all rules.*

Define verifiability (falsifiability) of sequents in **B**$_4$ *and check the correctness of all rules.*

For the moment, we will not demonstrate the completeness of defined calculi. However, let us conclude with one remark concerning proof search. All rules satisfy the subformula property so for any finite sequent the proof search must terminate.

5.5.3   Cut Admissibility

It is possible to provide a proof of admissibility of cut in two forms which will be called outer and inner (the same convention will be applied also to other rules or positions of formulae)

$$(Cut) \quad \frac{\Gamma, [\Pi] \Rightarrow [\Sigma], \Delta, \varphi \quad \varphi, \Lambda, [\Xi] \Rightarrow [\Theta], \Omega}{\Gamma, \Lambda, [\Pi, \Xi] \Rightarrow [\Sigma, \Theta], \Delta, \Omega}$$

$$([Cut]) \quad \frac{\Gamma, [\Pi] \Rightarrow [\Sigma, \varphi], \Delta \quad \Lambda, [\varphi, \Xi] \Rightarrow [\Theta], \Omega}{\Gamma, \Lambda, [\Pi, \Xi] \Rightarrow [\Sigma, \Theta], \Delta, \Omega}$$

We will provide a proof which follows Dragalin's strategy and in this setting requires some interesting modifications. First of all, we must provide proofs of several auxiliary results concerning the admissibility of structural rules and invertibility. In particular, we need to prove that in all considered systems

Lemma 5.6. *Axiomatic sequents may be replaced with atomic axioms.*

PROOF: as usual by induction on the complexity of active formula. For **B**$_4$ it is standard, the only difference is that we must examine an active formula in both forms: outer and inner, and in case of negation and inner implication, after reduction we obtain axiomatic sequents with the transition of some formulae (we obtain outer axioms from inner ones or vice versa). We will show only the latter case (of inner implication)

$$\frac{\varphi, \Gamma, [\Pi] \Rightarrow [\Sigma, \psi], \Delta, \varphi \quad \varphi, \Gamma, [\psi, \Pi] \Rightarrow [\Sigma, \psi], \Delta}{\cfrac{\varphi, \Gamma, [\varphi \to \psi, \Pi] \Rightarrow [\Sigma, \psi], \Delta}{\Gamma, [\varphi \to \psi, \Pi] \Rightarrow [\Sigma, \varphi \to \psi], \Delta} (\Rightarrow [\to])} ([\to] \Rightarrow)$$

For **K**$_3$, not only the case of outer implication is different (exactly dual to the previous case), but also we have additional axioms of the form: $\varphi, \Gamma, [\Pi] \Rightarrow [\Sigma, \varphi], \Delta$ and we must check all compound formulae for this case of being axiomatic. We show two cases:

$$\frac{\varphi, \psi, \Gamma, [\Pi] \Rightarrow [\Sigma, \varphi], \Delta \quad \varphi, \psi, \Gamma, [\Pi] \Rightarrow [\Sigma, \psi], \Delta}{\cfrac{\varphi, \psi, \Gamma, [\Pi] \Rightarrow [\Sigma, \varphi \wedge \psi], \Delta}{\varphi \wedge \psi, \Gamma, [\Pi] \Rightarrow [\Sigma, \varphi \wedge \psi], \Delta} (\wedge \Rightarrow)} (\Rightarrow [\wedge])$$

$$\frac{\dfrac{\varphi,\Gamma,[\Pi]\Rightarrow[\Sigma,\psi,\varphi],\Delta \qquad \psi,\varphi,\Gamma,[\Pi]\Rightarrow[\Sigma,\psi],\Delta}{\varphi,\varphi\rightarrow\psi,\Gamma,[\Pi]\Rightarrow[\Sigma,\psi],\Delta}\ (\rightarrow\Rightarrow)}{\varphi\rightarrow\psi,\Gamma,[\Pi]\Rightarrow[\Sigma,\varphi\rightarrow\psi],\Delta}\ (\Rightarrow[\rightarrow])$$

For $\mathbf{L_3}$, we must check additionally two cases with specific rules for implication

$$\frac{\dfrac{\varphi,\Gamma,[\psi,\Pi]\Rightarrow[\Sigma,\psi],\Delta,\varphi \qquad \varphi,\psi,\varphi,\Gamma,[\Pi]\Rightarrow[\Sigma,\psi],\Delta \qquad \varphi,\Gamma,[\Pi]\Rightarrow[\Sigma,\psi,\varphi,\psi],\Delta}{\varphi,\varphi\rightarrow\psi,\Gamma,[\Pi]\Rightarrow[\Sigma,\psi],\Delta}\ (\rightarrow\Rightarrow)}{\varphi\rightarrow\psi,\Gamma,[\Pi]\Rightarrow[\Sigma,\varphi\rightarrow\psi],\Delta}\ (\Rightarrow[\rightarrow])$$

and

$$\frac{\begin{array}{cc}\dfrac{\mathcal{D}_1}{\varphi,\varphi\rightarrow\psi,\Gamma,[\Pi]\Rightarrow[\Sigma],\Delta,\psi} & \dfrac{\mathcal{D}_2}{\varphi\rightarrow\psi,\Gamma,[\varphi,\Pi]\Rightarrow[\Sigma,\psi],\Delta}\end{array}}{\varphi\rightarrow\psi,\Gamma,[\Pi]\Rightarrow[\Sigma],\Delta,\varphi\rightarrow\psi}\ (\Rightarrow\rightarrow)$$

where $\mathcal{D}_1$ and $\mathcal{D}_2$ are, respectively

$$\frac{\varphi,\Gamma,[\psi,\Pi]\Rightarrow[\Sigma],\Delta,\psi,\varphi \qquad \varphi,\psi,\varphi,\Gamma,[\Pi]\Rightarrow[\Sigma],\Delta,\psi \qquad \varphi,\Gamma,[\Pi]\Rightarrow[\Sigma,\psi,\varphi],\Delta,\psi}{\varphi,\varphi\rightarrow\psi,\Gamma,[\Pi]\Rightarrow[\Sigma],\Delta,\psi}\ (\rightarrow\Rightarrow)$$

$$\frac{\Gamma,[\psi,\varphi,\Pi]\Rightarrow[\Sigma,\psi],\Delta,\varphi \qquad \varphi,\psi,\Gamma,[\varphi,\Pi]\Rightarrow[\Sigma,\psi],\Delta \qquad \Gamma,[\varphi,\Pi]\Rightarrow[\Sigma,\psi,\varphi,\psi],\Delta}{\varphi\rightarrow\psi,\Gamma,[\varphi,\Pi]\Rightarrow[\Sigma,\psi],\Delta}\ (\rightarrow\Rightarrow)$$

For $\mathbf{LP}$, the additional axioms have the form: $\Gamma,[\varphi,\Pi]\Rightarrow[\Sigma],\Delta,\varphi$. Again there is no problem with showing that we can always provide a reduction; we take only one case for illustration

$$\frac{\dfrac{\Gamma,[\varphi,\Pi]\Rightarrow[\Sigma],\Delta,\psi,\varphi \qquad \Gamma,[\psi,\varphi,\Pi]\Rightarrow[\Sigma],\Delta,\psi}{\Gamma,[\varphi,\varphi\rightarrow\psi,\Pi]\Rightarrow[\Sigma],\Delta,\psi}\ ([\rightarrow]\Rightarrow)}{\Gamma,[\varphi\rightarrow\psi,\Pi]\Rightarrow[\Sigma],\Delta,\varphi\rightarrow\psi}\ (\Rightarrow\rightarrow)$$

Finally, for $\mathbf{RM_3}$ $(\rightarrow\Rightarrow)$ is different. Since $(\Rightarrow\rightarrow)$ is the same as in $\mathbf{L_3}$, in case of active outer implication, we have the same root of suitable proof-figure but $\mathcal{D}_1$ and $\mathcal{D}_2$ are slightly different, now of the form

$$\frac{\psi,\varphi,\Gamma,[\Pi]\Rightarrow[\Sigma,\varphi],\Delta,\psi \qquad \varphi,\Gamma,[\varphi,\psi,\Pi]\Rightarrow[\Sigma],\Delta,\psi \qquad \varphi,\Gamma,[\Pi]\Rightarrow[\Sigma],\Delta,\psi,\varphi,\psi}{\varphi,\varphi\rightarrow\psi,\Gamma,[\Pi]\Rightarrow[\Sigma],\Delta,\psi}\ (\rightarrow\Rightarrow)$$

$$\frac{\psi,\Gamma,[\varphi,\Pi]\Rightarrow[\Sigma,\psi,\varphi],\Delta \qquad \Gamma,[\varphi,\psi,\varphi,\Pi]\Rightarrow[\Sigma,\psi],\Delta \qquad \Gamma,[\varphi,\Pi]\Rightarrow[\Sigma,\psi],\Delta,\varphi,\psi}{\varphi\rightarrow\psi,\Gamma,[\varphi,\Pi]\Rightarrow[\Sigma,\psi],\Delta}\ (\rightarrow\Rightarrow)$$

and (for the new axiom and the new rule):

$$([\to] \Rightarrow) \frac{\varphi, \Gamma, [\Pi] \Rightarrow [\Sigma], \Delta, \psi, \varphi \qquad \varphi, \Gamma, [\psi, \Pi] \Rightarrow [\Sigma], \Delta, \psi}{\varphi, \Gamma, [\varphi \to \psi, \Pi] \Rightarrow [\Sigma], \Delta, \psi}$$

$$\frac{\Gamma, [\varphi, \Pi] \Rightarrow [\Sigma, \psi], \Delta, \varphi \qquad \Gamma, [\varphi, \psi, \Pi] \Rightarrow [\Sigma, \psi], \Delta}{\Gamma, [\varphi, \varphi \to \psi, \Pi] \Rightarrow [\Sigma, \psi], \Delta} (\Rightarrow\to)$$

$$\frac{}{\Gamma, [\varphi \to \psi, \Pi] \Rightarrow [\Sigma], \Delta, \varphi \to \psi}$$

$\square$

Exercise 5.39. *Prove for each system at least four cases.*

Since then we will prove the next results for SC with atomic axioms. All the rules are context-independent, hence it is routine to demonstrate

Lemma 5.7. *Inner and outer weakening is h-p admissible.*

It is also not surprising that we can prove h-p invertibility of all rules

Lemma 5.8. *All rules are h-p invertible.*

Exercise 5.40. *Prove the lemma for $(\to\Rightarrow)$ in $\mathbf{L}_3$.*

On this basis we can prove h-p admissibility of outer and inner contraction.

Lemma 5.9. *Inner and outer contraction is h-p admissible*

A proof again follows the strategy of proofs from section 3.2 so we left it for the reader.

Exercise 5.41. *Prove the lemma 5.9.*

Now we are ready to prove the main result

Theorem 5.18. *Cut in both forms is admissible in all systems.*

PROOF: The structure of proof is like for G3 so we keep the overall structure but note that we carry the proof for both forms of cut simultaneously. It means in particular, that in the induction hypotheses we assume that the claim holds for both forms of cut of lesser height or complexity. Note also that in the case where one sequent is axiomatic, things are more complicated. In case of $\mathbf{B}_4$ there is no difference and we get rid of the application of cut as in G3 for **CPL**. However, for other logics, we have additional forms of axiomatic sequents, and in general, it is not possible to eliminate cut immediately. Consider the following situation in $\mathbf{K}_3$ and $\mathbf{L}_3$ respectively:

$$\frac{\Gamma, [\Pi] \Rightarrow [\Sigma], \Delta, p \qquad p, \Lambda, [\Xi] \Rightarrow [\Theta, p], \Omega}{\Gamma, \Lambda, [\Pi, \Xi] \Rightarrow [\Sigma, \Theta, p], \Delta, \Omega} (Cut)$$

Now, we cannot obtain the root from one of the premisses without cut. We must additionally perform induction on the height of the left premiss. In the basis we have two subcases. If $\Gamma, [\Pi] \Rightarrow [\Sigma], \Delta, p$ is an axiom with p parametric, then $\Gamma, \Lambda, [\Pi, \Xi] \Rightarrow [\Sigma, \Theta, p], \Delta, \Theta$ is also an axiom. Otherwise $p \in \Gamma$ and $\Gamma, \Lambda, [\Pi, \Xi] \Rightarrow [\Sigma, \Theta, p], \Delta, \Theta$ is an axiom as well. For the induction step just note that all rules are context independent hence permutable with cut.

Exercise 5.42. *Prove in the same way that cut is eliminable for* **LP** *and* **RM**$_3$

The cases where at least one premiss has parametric cut-formula are not troublesome, as we noticed above. So the only thing is to prove reduction when both cut-formulae are principal. The only new situations are connected with the outer cut on implications. We consider the case of **L**$_3$ and ask the reader to prove analogous case for **R**$_3$. The application of the cut is of the form

$$\frac{\Gamma, [\Pi] \Rightarrow [\Sigma], \Delta, \varphi \to \psi \qquad \varphi \to \psi, \Lambda, [\Xi] \Rightarrow [\Theta], \Omega}{\Gamma, \Lambda, [\Pi, \Xi] \Rightarrow [\Sigma, \Theta], \Delta, \Omega} \ (Cut)$$

where the left premiss is derived as follows:

$$\frac{\varphi, \Gamma, [\Pi] \Rightarrow [\Sigma], \Delta, \psi \qquad \Gamma, [\varphi, \Pi] \Rightarrow [\Sigma, \psi], \Delta}{\Gamma, [\Pi] \Rightarrow [\Sigma], \Delta, \varphi \to \psi} \ (\Rightarrow\to)$$

and the right

$$\frac{\Lambda, [\psi, \Xi] \Rightarrow [\Theta], \Omega, \varphi \qquad \varphi, \psi, \Lambda, [\Xi] \Rightarrow [\Theta], \Omega \qquad \Lambda, [\Xi] \Rightarrow [\Theta, \varphi, \psi], \Omega}{\varphi \to \psi, \Lambda, [\Xi] \Rightarrow [\Theta], \Omega} \ (\to\Rightarrow)$$

We construct the following proofs:

$$([Cut]) \ \frac{\Lambda, [\Xi] \Rightarrow [\Theta, \varphi, \psi], \Omega \qquad \Gamma, [\varphi, \Pi] \Rightarrow [\Sigma, \psi], \Delta}{([\Rightarrow C]) \ \dfrac{\Gamma, \Lambda, [\Pi, \Xi] \Rightarrow [\Sigma, \Theta, \psi, \psi], \Delta, \Omega}{([Cut]) \ \dfrac{\Gamma, \Lambda, [\Pi, \Xi] \Rightarrow [\Sigma, \Theta, \psi], \Delta, \Omega \qquad \Lambda, [\psi, \Xi] \Rightarrow [\Theta], \Omega, \varphi}{\Gamma, \Lambda, \Lambda, [\Pi, \Xi, \Xi] \Rightarrow [\Sigma, \Theta, \Theta], \Delta, \Omega, \Omega, \varphi}}}$$

and

$$(Cut) \ \frac{\varphi, \Gamma, [\Pi] \Rightarrow [\Sigma], \Delta, \psi \qquad \varphi, \psi, \Lambda, [\Xi] \Rightarrow [\Theta], \Omega}{(C \Rightarrow) \ \dfrac{\varphi, \varphi, \Gamma, \Lambda, [\Pi, \Xi] \Rightarrow [\Sigma, \Theta], \Delta, \Omega}{\varphi, \Gamma, \Lambda, [\Pi, \Xi] \Rightarrow [\Sigma, \Theta], \Delta, \Omega}}$$

The application of (Cut) on φ and several contractions yields the result, since all cuts are on cut-formulae of lower complexity. $\qquad\square$

Exercise 5.43. *Provide proofs of some cases, in particular, the last case for* **RM**$_3$

One may observe a close relationship of Avron's calculus to our generalised formalisation of **L**$_3$ which may be roughly specified like this: every nonstandard sequent $\Gamma, [\Pi] \Rightarrow [\Sigma], \Delta$ corresponds to a sequent $\Gamma, \neg\Sigma \Rightarrow \neg\Pi, \Delta$ in Avron's calculus and vice versa.

On the basis of this informal correspondence, we may provide Avron's characterisation of $\to$ for **K**$_3$ and **LP** (remember they have identical rules). On the other hand, we can provide suitable rules for generalised calculus on the basis of implication from GBS. We can state the correspondence between these two kinds of calculi formally by defining a translation function f:

For any sequent $\Gamma, \neg\Sigma \Rightarrow \neg\Pi, \Delta$ where Γ and Δ consists of unnegated formulae $f(\Gamma, \neg\Sigma \Rightarrow \neg\Pi, \Delta) = \Gamma, [\Pi] \Rightarrow [\Sigma], \Delta$.

Also we can define a reverse translation g for any $\Gamma, [\Pi] \Rightarrow [\Sigma], \Delta$,

$$g(\Gamma, [\Pi] \Rightarrow [\Sigma], \Delta) = \Gamma, \neg\Sigma \Rightarrow \neg\Pi, \Delta.$$

Now we can prove the following:

Theorem 5.19. *If $\vdash \Gamma \Rightarrow \Delta$ in cut-free GM_3 for any considered $\mathbf{L}$, then $\vdash f(\Gamma \Rightarrow \Delta)$ in a generalised calculus for $\mathbf{L}$.*

Exercise 5.44. *Prove the above theorem; proof goes by induction on the height of proof in respective GM_3 for $\mathbf{L}$.*

Since Cut-free versions of GM_3 are adequate for suitable logics, we obtain as a corrolary

Lemma 5.10. *Generalised calculi are complete.*

The reader could notice that in this case we informally applied the falsificationist interpretation. What happens if we appeal to the verificationist one? Nothing; the same rules provide a solution, in contrast to our formalisation of n-sided approach. Hence, when comparing these two approaches one may notice at least four advantages of the latter

- uniform rules independent of the kind of interpretation;

- rules more similar to classical ones (although in case of some notions of implication some complications are unavoidable);

- a straightforward expression of $\vdash$;

- simpler proofs.

These are the reasons that we paid more attention to this approach and treated it as more fundamental. Of course, other generalised kinds of SC may be also used to characterise many-valued logics. We finish our presentation with brief remarks concerning the application of hypersequent calculi.

All generalised SC for many-valued logics introduced so far were strongly based on semantical motivations, as such, they may be treated as so-called external systems. The kind of semantic commitment of these systems is specific for background matrix semantics. As such they have nothing to do with the phenomenon of substructurality which seems to show that this categorization is not absolute but rather relative to the kind of formal apparatus. Despite that, it was possible to apply to them general proof-theoretic techniques, including proofs of cut admissibility, in a similar way as in the case of those generalised SCs for **S5** which were also strongly based on the specific semantic intuitions (like, e.g. in Poggiolesi's HSC).

It may be interesting to consider if some kind of strictly syntactical (internal?) generalised systems was proposed for many-valued logics. Not surprisingly, the format of HSC is extensive enough for that aim. Moreover, in the context of HSC, the notion of substructurality may be reinvented also in the generalised form. Many studies present several HSCs for a variety of many-valued logics, including some infinitely-valued ones. In particular, one can find an extensive treatment of fuzzy logics in the framework of HSC in Metcalfe, Olivetti and Gabbay [176]. Since that book provides an excellent unified presentation of logics which are not dealt with in this volume, we restrict ourselves only to a brief illustration of how HSC may be used for the characterisation of $Ł_3$. In the formalisation due to Avron [10], we have the same set of rules which was applied for **RM**, but instead of IC rules now we have two IW rules. Moreover, the specific Splitting and Combining rules now are replaced with one rule of Mixing

$$\frac{G \mid \Gamma_1, \Gamma_2, \Gamma_3 \Rightarrow \Delta_1, \Delta_2, \Delta_3 \quad H \mid \Pi_1, \Pi_2, \Pi_3 \Rightarrow \Sigma_1, \Sigma_2, \Sigma_3}{G \mid H \mid \Gamma_1, \Pi_1 \Rightarrow \Delta_1, \Sigma_1 \mid \Gamma_2, \Pi_2 \Rightarrow \Delta_2, \Sigma_2 \mid \Gamma_3, \Pi_3 \Rightarrow \Delta_3, \Sigma_3}$$

In fact, simpler rule allows for the same effect. Ciabattoni, Gabbay and Olivetti [49] introduced

$$\frac{G \mid \Gamma, \Delta \Rightarrow \Pi, \Sigma \quad H \mid \Lambda, \Delta \Rightarrow \Theta, \Sigma}{G \mid H \mid \Gamma, \Lambda \Rightarrow \Pi, \Theta \mid \Delta \Rightarrow \Sigma}$$

Avron proves soundness on the basis of the same translation of hypersequents as the one provided for **RM**. Elimination of cut is proved also by means of the so called history method. We are not going into the details but rather stop at this point.

Appendix

We briefly survey the basic facts concerning set theory, induction and consequence relations which were often employed in the main text. They are collected here mainly for reference and to establish notation.

Sets, Relations, Functions

Let us recall that sets may be specified either by listing (enumeration) of their elements, e.g. $\{a, b, c, d\}$, or by means of set abstraction operator: $\{x : \varphi(x)\}$—the set of x satisfying condition φ. For example, the empty set: $\varnothing = \{x : x \neq x\}$.

The last way is also convenient for expression of operations on sets, like the ones applied in the book:

- the intersection of two sets: $A \cap B = \{x : x \in A \land x \in B\}$;
- the union (sum) of two sets: $A \cup B = \{x : x \in A \lor x \in B\}$;
- the difference of two sets: $A - B = \{x : x \in A \land x \notin B\}$;
- the power set: $\mathcal{P}(A) = \{B : B \subseteq A\}$

In the last case, a predicate of inclusion $\subseteq$ is used which admits the identity of both sets. For strict inclusion we use $\subset$.

Operations of intersection and union may be generalised for any (also infinite) collection of sets. It is convenient to introduce for that a notion of a family of sets on which we perform such an operation. Let $\mathcal{A}$ denote such nonempty family, then

the product is: $\bigcap \mathcal{A} = \{x : \forall A(A \in \mathcal{A} \to x \in A)\} = \{x : \forall A \in \mathcal{A}, x \in A\}$.
the union is: $\bigcup \mathcal{A} = \{x : \exists A(A \in \mathcal{A} \land x \in A)\} = \{x : \exists A \in \mathcal{A}, x \in A\}$.

In terms of sets, we can define the notions of

an ordered pair (tuple): $\langle a, b \rangle := \{\{a\}, \{a, b\}\}$ (often written simply as (a, b));
an ordered tripple: $\langle a, b, c \rangle := \langle \langle a, b \rangle, c \rangle$;
generally—ordered n-tuple (or sequence): $\langle a_1, ..., a_n \rangle := \langle \langle a_1, ..., a_{n-1} \rangle, a_n \rangle$;

In case of sets, the order and the repetition of elements does not matter, i.e. $\{a, b, c\} = \{b, c, a\} = \{a, b, c, a, c, c\}$, where a, b and c are any different objects.

© Springer Nature Switzerland AG 2021
A. Indrzejczak, *Sequents and Trees*, Studies in Universal Logic,
https://doi.org/10.1007/978-3-030-57145-0

In case of a sequence, both the order and the number of occurrences count, i.e. $\langle a, b \rangle \neq \langle b, a \rangle \neq \langle a, a, b \rangle$.

Multisets (sets with repetitions) are in between, in the sense that the order is inessential but the number of occurrences is important, for example

$$[a, b, c] \neq [a, b, a, c], \text{ but } [a, b, b, c] = [b, c, a, b].$$

Formally, multisets may be defined as tuples $\langle A, f_A \rangle$, where A is a set and f_A is a function which assigns to each element of A the number of its occurrences. We can define the operations of multiset intersection and union in at least two different ways. In this volume, we use an additive version

$$A \sqcap B := \langle A \cap B, f_{A,B} \rangle, \quad A \sqcup B := \langle A \cup B, f_{A,B} \rangle, \quad \text{where } f_{A,B}(x) = f_A(x) + f_B(x).$$

The notion of n-tuple allows for the introduction of the notion of n-ary relation and function as a subset of n-ary Cartesian product defined in the following way:

for binary relations $A \times B = \{\langle x, y \rangle : x \in A \wedge y \in B\}$;
in general: $A_1 \times ... \times A_n = \{\langle x_1, ..., x_n \rangle : x_1 \in A_1 \wedge ... \wedge x_n \in A_n\}$.

If $A_1 = ... = A_n$, then we write A^n, in particular A^2 instead of $A \times A$.

R is a binary relation on A, B iff $R \subseteq A \times B$.
R is an n-ary relation on $A_1, ..., A_n$ iff $R \subseteq A_1 \times ... \times A_n$.

If $R \subseteq A \times B$, then

- a domain of R: $D_l(R) = \{x \in A : \exists y \in B, \langle x, y \rangle \in R\}$,
- a codomain of R: $D_r(R) = \{x \in B : \exists y \in A, \langle y, x \rangle \in R\}$.

Instead of $\langle x, y \rangle \in R$ we usually write $R(xy)$ or Rxy.

Let $R \subseteq A^2$ and x, y, z denote arbitrary elements of A, then the following conditions denote some important conditions on relations

name	condition
seriality	$\forall x \exists y Rxy$
functionality	$\forall xyz (Rxy \wedge Rxz \to y = z)$
reflexivity	$\forall x Rxx$
irreflexivity	$\forall x \neg Rxx$
transitivity	$\forall xyz (Rxy \wedge Ryz \to Rxz)$
symmetry	$\forall xy (Rxy \to Ryx)$
assymetry	$\forall xy (Rxy \to \neg Ryx)$
antisymmetry	$\forall xy (Rxy \wedge x \neq y \to \neg Ryx)$
euclideaness	$\forall xyz (Rxy \wedge Rxz \to Ryz)$
(strong) connectedness	$\forall xy (Rxy \vee Ryx)$
(weak) connectedness	$\forall xy (Rxy \vee Ryx \vee x = y)$

A binary relation $R \subseteq A^2$ is an equivalence relation iff it is reflexive, symmetric and transitive. Let R be equivalence on A, then

1. $[x]_R = \{y : Rxy\}$ is an equivalence class of x wrt R;

2. $A_{/R} = \{[x]_R : x \in A\}$ is a collection of equivalence classes generated by R.

For equivalence relations the Abstraction Principle holds: If R is an equivalence on A, then $A_{/R}$ is a partition of A, i.e. it satisfies conditions of

- adequacy: $A = \bigcup A_{/R}$;

- disjointness: $\forall x, y \in A([x]_R \neq [y]_R \to [x]_R \cap [y]_R = \varnothing)$;

- nonemptyness: $\forall x \in A, [x]_R \neq \varnothing$.

This result allows for replacement of the relational semantics for **S5** with the simplified one.

Here is a list of some ordering relations:

- R is a quasi-order iff it is reflexive and transitive;

- R is a partial order iff it is an antisymmetric quasi-order;

- R is a (linear) order iff it is a strongly connected partial order;

- R is a strict partial order iff it is an asymmetric partial order;

- R is a strict (linear) order iff it is weakly connected strict partial order.

Both the relations and functions are sets so ordinary set operations may be performed on them but additionally, we have some specific operations

- transitive closure of $R \subseteq A^2$ is the least transitive relation $R^+ \subseteq A^2$ such that $R \subseteq R^+$;

- transitive reflexive closure of $R \subseteq A^2$ is the least transitive and reflexive relation $R^* \subseteq A^2$ such that $R \subseteq R^*$;

- converse of $R \subseteq A \times B$: $\breve{R}xy$ iff Ryx ($\breve{R} \subseteq B \times A$);

- composition (relative product) of $R \subseteq A \times B$ and $S \subseteq B \times C$:

$$R \circ Sxy \text{ iff } \exists z \in B(Rxz \wedge Szy) \ (R \circ S \subseteq A \times C)$$

The notion of a tree is characterized in several ways, not always equivalent, in mathematics, computer science, logic. We can define it as a relational structure $\mathfrak{T} = \langle \mathcal{T}, \mathcal{R} \rangle$, such that:

- there is a unique element $r \in \mathcal{T}$, called the *root* such that $\forall t \in \mathcal{T}, \mathcal{R}^* r t$;

- every element $t \in \mathcal{T}$ distinct from r has a unique predecessor i.e. there is only one $t' \in \mathcal{T}$, such that $\mathcal{R}t't$;

- $\mathcal{R}$ is acyclic, i.e. $\forall t \in \mathcal{T}$ it is not true that $\mathcal{R}^+ tt$.

All elements of $\mathcal{T}$ are called *nodes*, for every pair t, t' such that $\mathcal{R}tt'$, t is called *the parent* and t' – *a child*; if $\mathcal{R}^+tt'$, then t is *an ancestor* and t' is *a successor*. Every node t with no children is called *a leaf*. N-ary sequence $\langle t_1, \ldots, t_n \rangle$, where for each $i < n$ we have $\mathcal{R}t_i t_{i+1}$ is called *a path*; every maximal path is *a branch* (in finite case it is a sequence from the root to a leaf). The number of children of a node is *the branching factor* of this node; the branching factor of a tree is the biggest branching factor of its nodes. If the branching factor of a tree is a natural number, we have *finitely generated tree*. In particular, trees with nodes having at most two children are called *binary trees*.

The important result concerning such trees is

Lemma 5.11 (König). *Every finitely generated but infinite tree has at least one infinite branch.*

By the König lemma, to show that a tree is finite, it is sufficient to show that it is finitely generated and that every branch is finite.

An n-argument function (mapping) with the domain (the set of arguments) A^n and the range (the set of values) in B (denoted as $f : A^n \longrightarrow B$), is an $n+1$-argument relation $f \subseteq A^n \times B$ satisfying conditions

1. $\forall_{x_1, \ldots, x_n \in A} \exists_{y \in B} \langle x_1, \ldots, x_n, y \rangle \in f$

2. $\forall_{x_1, \ldots, x_n \in A} \forall_{y, z \in B} (\langle x_1, \ldots, x_n, y \rangle \in f \wedge \langle x_1, \ldots, x_n, z \rangle \in f \rightarrow y = z)$

If $B = A$, then f is an n-argument operation in A. Condition 2 enables an introduction of the functional notation which is more convenient than the relational one. For example, instead of $\langle x_1, \ldots, x_n, y \rangle \in f$ we can write $f(x_1, \ldots, x_n) = y$. In particular, an unary function defined on domain A and range B ($f : A \longrightarrow B$), is a binary relation $f \subseteq A \times B$ satisfying conditions of seriality and functionality

1. $\forall_{x \in A} \exists_{y \in B} \langle x, y \rangle \in f$

2. $\forall_{x \in A} \forall_{y, z \in B} (\langle x, y \rangle \in f \wedge \langle x, z \rangle \in f \rightarrow y = z)$

Kinds of mappings

Let $f : A^n \longrightarrow B$, then f is

- an injection (f. 1-1) iff $\forall_{x_1, \ldots, x_n, y_1, \ldots, y_n \in A} (f(x_1, \ldots, x_n) = f(y_1, \ldots, y_n) \rightarrow \langle x_1, \ldots, x_n \rangle = \langle y_1, \ldots, y_n \rangle)$

- a surjection (f. on) iff B is the range of f iff $\forall_{y \in B} \exists_{x_1, \ldots, x_n \in A} \, y = f(x_1, \ldots, x_n)$

- a bijection iff it is both an injection and a surjection.

If there is a bijection defined on A and B, then these sets are of the same cardinality, i.e. they have the same number of elements. Sets having bijection with the set of natural numbers or some of its subsets as a range are denumerable (and finite in the latter case). Their cardinality is denoted with $\aleph_0$ and they satisfy

Theorem 5.20 (enumeration). *Every denumerable set may be linearly ordered.*

This fact was applied many times in proofs of completeness since the set of formulae of considered languages is denumerable.

Mathematical Induction

This kind of reasoning was applied already in the XVII century by Pascal and Fermat. The name and definition comes from De Morgan. As one of the axioms of arithmetic, it appears in Peano and this is a paradigmatic case of the application of mathematical induction. Let us consider some formulations. Let N denote the set of natural numbers and sx the successor of x

1. $\varphi(0) \wedge \forall_{x \in N}(\varphi(x) \to \varphi(sx)) \to \forall_{x \in N}\varphi(x)$
2. $\varphi(0) \wedge \forall_{x \in N}(\varphi(x) \to \varphi(x+1)) \to \forall_{x \in N}\varphi(x)$
3. $0 \in A \wedge \forall_{x \in N}(x \in A \to sx \in A) \to A = N$

There is also a strong (or complete) induction:

1. $\varphi(0) \wedge \forall_{x \in N}(\forall_{y \in N}(0 \le y < x \to \varphi(y)) \to \varphi(x)) \to \forall_{x \in N}\varphi(x)$
2. $0 \in A \wedge \forall_{x \in N}(\forall_{y \in N}(0 \le y < x \to y \in A) \to x \in A) \to A = N$
3. $\forall_{x \in N}(\forall_{y \in N}(0 \le y < x \to \varphi(y)) \to \varphi(x)) \to \forall_{x \in N}\varphi(x)$
4. $\forall_{x \in N}(\forall_{y \in N}(0 \le y < x \to y \in A) \to x \in A) \to A = N$

The name 'strong induction' is misleading—both principles are interderivable (but we omit proofs here). Proofs by induction are based on this principle and contain two parts:

    a. the basis of induction—we prove that 0 satisfies φ

    b. the inductive step—we assume that x satisfies φ (the induction hypothesis) and we prove that sx also satisfies φ.

a. and b., by principle of induction imply that any number satisfies φ.

(in fact, in case of the complete induction, the first part is not necessary which is explicit in formulation 3 and 4 above).

An example

    We prove that: $\forall_{x \in N}\neg(sx = x)$ by induction on x

    a. basis: $\varphi(0) := \neg(s0 = 0)$

1.	$\forall_{x \in N}\neg(sx = 0)$	axiom of arithmetic
2.	$\neg(s0 = 0)$	1., $\forall E, x/0$

b. thesis: $\varphi(x) \rightarrow \varphi(sx) := \neg(sx = x) \rightarrow \neg(ssx = sx)$

1.	$\neg(sx = x)$	the induction hypothesis
2.	$\forall_{x,y \in N}(sx = sy \rightarrow x = y)$	axiom
3.	$ssx = sx \rightarrow sx = x$	2., $\forall E, x/sx, y/x$
4.	$\neg(ssx = sx)$	1., 3. MT

Why mathematical induction is a correct form of reasoning? The simplest answer is: because the set of natural numbers is an inductive (recursive) set. Moreover, it is not the only set which may be recursively defined and this is the reason that we can apply inductive proofs to other sets, i.e. we can apply induction to theses of the form

$$\forall_x (x \in A \rightarrow x \in B) \text{ (or } \forall_x(\varphi(x) \rightarrow \psi(x)))$$

on condition that A is inductive set (φ is a property defined recursively). Every inductive definition consists of three parts:

1. the basic condition specifies basic elements of A;

2. the inductive condition specifies operations by means of which the new elements are generated from the old ones (already belonging);

3. the closing condition is a claim that nothing more belongs to A except elements obtained by 1. or 2.

Alternatively, we can say that

A is inductive iff A is the least set satisfying conditions 1. and 2.

Formally, the schema of such definition may be described as follows:

A is inductive iff:

1. $B \subset A$,

2. $CL(A, O)$,

3. $\forall_C(B \subset C \wedge CL(C, O) \rightarrow A \subset C)$,

where B is the set of initial elements (generators), $CL(A, O)$ means that A is closed wrt every operation from O, i.e. $x_1, ..., x_n \in A \rightarrow o(x_1, ..., x_n) \in A$, for every n-argument operation (rules of construction) $o \in O$ ($n \geq 0$), and condition 3. states that A is the least set satisfying 1. and 2. The existence of such set is guaranteed by theorems of set theory—it is a product of all sets containing B and closed under all operations from O.

Example: the definition of the set of natural numbers N

N is the least set satisfying:

1. $0 \in N$,

2. if $x \in N$, then $sx \in N$.

It is an extremely simple example but in the text, we have other definitions of this kind. For example, a definition of the set of formulae FOR. Nowadays, a very popular way of expressing such definitions is by using Backus/Naur notation. A definition of FOR looks like that

$$\varphi \in FOR \text{ iff } \varphi := p \mid \neg\varphi \mid (\varphi \wedge \psi) \mid (\varphi \vee \psi) \mid (\varphi \rightarrow \psi)$$

it is equivalent to the standard recursive definition and inductive proofs apply as well.

Examples in Logic

We can distinguish two kinds of applications:

A. direct application of mathematical induction (usually in the strong form) to some chosen measure e.g.

1. the length of a formula (the number of symbols);

2. the complexity of a formula (the number of constants);

3. the length of a proof (the number of lines in linear proof).

B. introduction of structural induction—in the version suitable for FOR

If

1. every propositional symbol has property θ,

2a. if φ has θ, then $\neg\varphi$ has θ,

2b. if φ and ψ have θ, then $(\varphi \wedge \psi), (\varphi \vee \psi), (\varphi \rightarrow \psi)$ have θ,

then every formula has θ.

Let us consider three examples

1. the pairing of brackets;

2. the extensionality principle;

3. the deduction theorem.

The first one is very simple.

Theorem 5.21. *In every formula χ the number of left brackets $(nl(\chi))$ is equal to the number of right brackets $(nr(\chi))$.*

We perform structural induction:

a. the basis: formula χ is a propositional symbol, hence $nl(\chi) = nr(\chi) = 0$

b. the inductive hypothesis: the claim holds for φ and ψ

– if formula χ is $\neg\varphi$, then it has the same number of brackets as φ, so, by IH $nl(\chi) = nr(\chi)$

– if formula χ is $(\varphi \star \psi)$, where $\star$ is any binary connective, then the claim holds since $nl(\chi) = nl(\varphi) + nl(\psi) + 1$ and $nr(\chi) = nr(\varphi) + nr(\psi) + 1$, and by IH we have $nl(\varphi) = nr(\varphi)$ and $nl(\psi) = nr(\psi)$.

Therefore, it holds for any formula.   □

Example 2—the extensionality principle, by induction on the complexity of formula.

Theorem 5.22. *The extensionality principle is derivable in H-system for* **CPL**.

PROOF: We show that from $\varphi \leftrightarrow \psi$ it is derivable $\chi \leftrightarrow \chi[\varphi//\psi]$, where $\chi[\varphi//\psi]$ denotes a replacement of at least one occurrence of φ (as a subformula of χ) by ψ.

First we state the list of schemata of theses of **CPL** which are necessary for the proof

    a. $(\varphi \leftrightarrow \psi) \rightarrow (\neg\varphi \leftrightarrow \neg\psi)$
    b. $(\varphi \leftrightarrow \psi) \rightarrow (\varphi \wedge \chi \leftrightarrow \psi \wedge \chi)$
    c. $(\varphi \leftrightarrow \psi) \rightarrow (\chi \wedge \varphi \leftrightarrow \chi \wedge \psi)$
    d. $(\varphi \leftrightarrow \psi) \rightarrow (\varphi \vee \chi \leftrightarrow \psi \vee \chi)$
    e. $(\varphi \leftrightarrow \psi) \rightarrow (\chi \vee \varphi \leftrightarrow \chi \vee \psi)$
    f. $(\varphi \leftrightarrow \psi) \rightarrow (\varphi \rightarrow \chi \leftrightarrow \psi \rightarrow \chi)$
    g. $(\varphi \leftrightarrow \psi) \rightarrow (\chi \rightarrow \varphi \leftrightarrow \chi \rightarrow \psi)$
    h. $(\varphi \leftrightarrow \psi) \wedge (\gamma \leftrightarrow \delta) \rightarrow (\varphi \wedge \gamma \leftrightarrow \psi \wedge \delta)$
    i. $(\varphi \leftrightarrow \psi) \wedge (\gamma \leftrightarrow \delta) \rightarrow (\varphi \vee \gamma \leftrightarrow \psi \vee \delta)$
    j. $(\varphi \leftrightarrow \psi) \wedge (\gamma \leftrightarrow \delta) \rightarrow (\varphi \rightarrow \gamma \leftrightarrow \psi \rightarrow \delta)$

Now to prove our theorem, we apply complete induction on the length of χ and under the inductive hypothesis that the extensionality principle holds for any shorter formula. We consider 5 cases:

1. χ is a propositional symbol, so the operation is possible only for $\chi := \varphi$, but then it is trivial, i.e. $\chi[\varphi//\psi] := \psi$

2. $\chi := \neg\gamma$. By IH from $\varphi \leftrightarrow \psi$ we can derive $\gamma \leftrightarrow \gamma[\varphi//\psi]$, so by thesis a. we get $\neg\gamma \leftrightarrow \neg\gamma[\varphi//\psi]$.

3. $\chi := (\gamma \wedge \delta)$. We have three subcases:
    – the replacement of φ by ψ only in γ, i.e. $\chi[\varphi//\psi] := (\gamma[\varphi//\psi] \wedge \delta)$. By IH $\gamma \leftrightarrow \gamma[\varphi//\psi]$, by thesis b. we get $(\gamma \wedge \delta) \leftrightarrow (\gamma[\varphi//\psi] \wedge \delta)$.
    – the replacement of φ by ψ only in δ, i.e. $\chi[\varphi//\psi] := (\gamma \wedge \delta[\varphi//\psi])$. Analogous to the previous subcase but by c.
    – the replacement of φ by ψ in γ and in δ, i.e. $\chi[\varphi//\psi] := (\gamma[\varphi//\psi] \wedge \delta[\varphi//\psi])$. By IH $\gamma \leftrightarrow \gamma[\varphi//\psi]$ and $\delta \leftrightarrow \delta[\varphi//\psi]$. By h. we obtain $(\gamma \wedge \delta) \leftrightarrow (\gamma[\varphi//\psi] \wedge \delta[\varphi//\psi])$.

The cases 4. and 5. are proven similarly as 3. on the basis of suitable theses from the list for $\vee$ and $\rightarrow$. $\qquad$ $\square$

Example 3—the deduction theorem for H-system for **CPL**.

Theorem 5.23. *If* $\Gamma, \varphi \vdash \psi$, *then* $\Gamma \vdash \varphi \rightarrow \psi$.

PROOF: by complete induction on the lenght of a proof of $\Gamma, \varphi \vdash \psi$. By definition $\Gamma, \varphi \vdash \psi$ means that we have a finite sequence $\gamma_1, ..., \gamma_n$, with $\gamma_n := \psi$. We show that for any $1 \leq i \leq n$ the claim holds, i.e. that $\Gamma \vdash \varphi \rightarrow \gamma_i$ on the supposition that it holds for any $j < i$.

There are 4 cases to consider:

1. $\gamma_i \in \Gamma$: we add to the proof an instance of axiom 1.: $\gamma_i \rightarrow (\varphi \rightarrow \gamma_i)$ and apply MP to obtain $\varphi \rightarrow \gamma_i$.

2. $\gamma_i := \varphi$; we add to the proof an instance of axiom 1.: $\varphi \rightarrow (\varphi \rightarrow \varphi)$ and by MP we get $\varphi \rightarrow \varphi$ which is a thesis.

3. γ_i is an axiom—we proceed as in case 1.

4. γ_i was derived by MP from earlier lines of proof of the shape γ_j $(j < i)$ and $\gamma_j \rightarrow \gamma_i$. Since both premises satisfy IH, we have: $\varphi \rightarrow (\gamma_j \rightarrow \gamma_i)$ and $\varphi \rightarrow \gamma_j$. We add to the proof an instance of axiom 2: $(\varphi \rightarrow (\gamma_j \rightarrow \gamma_i)) \rightarrow ((\varphi \rightarrow \gamma_j) \rightarrow (\varphi \rightarrow \gamma_i))$. Two applications of MP yield $\varphi \rightarrow \gamma_i$.

Hence, in particular, this theorem holds for ψ (the case $i = n$), so $\Gamma \vdash \varphi \rightarrow \psi$ $\square$

Note that the definition of a tree we provided in the appendix was not inductive but it can be reformulated as an inductive definition in at least two ways. In one approach, the root is treated as the set of generators and inductive conditions specify how we build a tree upward. In the second, the set of generators is the set of leaves and inductive conditions specify how we build a tree downward. Depending on the way we defined inductively a tree, we can apply two different kinds of inductive proofs on trees. A detailed formal treatment may be found in Segerberg [235]; the careful reader noticed that both kinds of inductive definitions and respective inductive proofs were applied in the text.

Sequent Calculus and Consequence Relations

SC can lead to different ways of generating relations of consequence. Let us compare three such approaches:

1. Scott's consequence generated by $\Rightarrow$ in sequents built from finite sets.

2. Avron's consequence generated by $\Rightarrow$ in sequents built from multisets and with restricted set of structural conditions.

3. Consequence relations generated by sequent rules in SC $\vdash \subseteq \mathcal{P}(Sek) \times Sek$ where Sek is the set of all sequents.

Of course, other possibilities can be considered; the second approach may be defined on sequents built from other data structures, and the last one may be further generalised (see e.g. Zucker, Tragesser [277]).

Let us compare Scott's consequence with Tarski's well known definition. Originally, Tarski defined not a relation but an operation of consequence Cn : $\mathcal{P}(FOR) \longrightarrow \mathcal{P}(FOR)$ satisfying the following basic conditions:

- (REF) $\Gamma \subseteq Cn(\Gamma)$

- (MON) If $\Gamma \subseteq \Delta$, then $Cn(\Gamma) \subseteq Cn(\Delta)$

- (TR) $Cn(Cn(\Gamma)) \subseteq Cn(\Gamma)$

In relational approach they may be expressed as follows (via definition $\Gamma \models \varphi :=$ $\varphi \in Cn(\Gamma)$):

- (REF) $\varphi \models \varphi$

- (MON) If $\Gamma \models \varphi$, then $\Gamma, \psi \models \varphi$

- (TR) If $\Gamma \models \varphi$ and $\varphi, \Gamma \models \psi$, then $\Gamma \models \psi$

A concrete example of such relation is provided by syntactical deducibility relation for H-system defined in subsection 1.1.3. Additionally, relations of this kind are structural in the sense of being closed under substitution, and finitary iff for every φ and Γ, if $\varphi \in Cn(\Gamma)$, then there is a finite $\Delta \subseteq \Gamma$ which is sufficient for derivability of φ.

Boolean connectives may be characterised as follows:

- ($\neg$) $\varphi \in Cn(\Gamma)$ iff $Cn(\Gamma \cup \{\neg\varphi\}) = FOR$

- ($\wedge$) $Cn(\{\varphi, \psi\}) = Cn(\{\varphi \wedge \psi\})$

- ($\vee$) $Cn(\{\varphi\}) \cap Cn(\{\psi\}) = Cn(\{\varphi \vee \psi\})$

- ($\rightarrow$) $\psi \in Cn(\Gamma \cup \{\varphi\})$ iff $\varphi \rightarrow \psi \in Cn(\Gamma)$

  or equivalently

- ($\neg$) $\Gamma \models \varphi$ iff $\Gamma, \neg\varphi \models$

- ($\wedge$) $\varphi, \psi \models \gamma$ iff $\varphi \wedge \psi \models \gamma$

- ($\vee$) $\Gamma, \varphi \models \gamma$ and $\Gamma, \psi \models \gamma$ iff $\Gamma, \varphi \vee \psi \models \gamma$

- ($\rightarrow$) $\Gamma, \varphi \models \psi$ iff $\Gamma \models \varphi \rightarrow \psi$

Any relation of consequence in Scott's sense $\models \subseteq \mathcal{P}(FOR) \times \mathcal{P}(FOR)$ is characterised as follows:

- (REF) $\varphi \models \varphi$

- (MON) If $\Gamma \subseteq \Gamma'$ and $\Delta \subseteq \Delta'$ and $\Gamma \models \Delta$, then $\Gamma' \models \Delta'$

- (TR) If $\Gamma \models \Delta, \varphi$ and $\varphi, \Gamma \models \Delta$, then $\Gamma \models \Delta$

Boolean connectives may be characterised by means of these conditions

- ($\neg$) $\models \varphi, \neg\varphi$, and $\varphi, \neg\varphi \models$
- ($\wedge$) $\Gamma, \varphi, \psi \models \Delta$ iff $\Gamma, \varphi \wedge \psi \models \Delta$
- ($\vee$) $\Gamma \models \Delta, \varphi, \psi$ iff $\Gamma \models \Delta, \varphi \vee \psi$
- ($\rightarrow$) $\Gamma, \varphi \models \Delta, \psi$ iff $\Gamma \models \Delta, \varphi \rightarrow \psi$

Clearly, we can apply different conditions, for example

- ($\wedge'$) $\Gamma \models \Delta, \varphi$ and $\Gamma \models \Delta, \psi$ iff $\Gamma \models \Delta, \varphi \wedge \psi$
- ($\vee'$) $\varphi, \Gamma \models \Delta$ and $\psi, \Gamma \models \Delta$ iff $\varphi \vee \psi, \Gamma \models \Delta$
- ($\rightarrow'$) $\Gamma \models \Delta, \varphi$ and $\psi, \Gamma \models \Delta$ iff $\varphi \rightarrow \psi, \Gamma \models \Delta$

Both characterizations are equivalent under assumption that (REF), (MON) and (TR) hold. For example, we consider the case of conjunction

($\wedge$) implies ($\wedge'$)

$\Longrightarrow$ Suppose $\Gamma \models \Delta, \varphi$ and $\Gamma \models \Delta, \psi$. From $\varphi \wedge \psi \models \varphi \wedge \psi$ which is an instance of (REF), by ($\wedge$) we obtain $\varphi, \psi \models \varphi \wedge \psi$. By two applications of (TR) (and (MON)) to assumptions, we get $\Gamma \models \Delta, \varphi \wedge \psi$.

$\Longleftarrow$ Assume that $\Gamma \models \Delta, \varphi \wedge \psi$. Since (REF) and (MON) yield $\varphi, \psi, \Gamma \models \Delta, \varphi$ by ($\wedge$) we obtain $\varphi \wedge \psi, \Gamma \models \Delta, \varphi$. By (TR) we get $\Gamma \models \Delta, \varphi$. In a similar way, we obtain $\Gamma \models \Delta, \psi$.

($\wedge'$) implies ($\wedge$)

$\Longrightarrow$ Assume that $\Gamma, \varphi, \psi \models \Delta$. From $\varphi \wedge \psi \models \varphi \wedge \psi$ by ($\wedge'$) we derive $\varphi \wedge \psi \models \varphi$ and $\varphi \wedge \psi \models \psi$. Two applications of (TR) and (MON) to our assumption yield $\Gamma, \varphi \wedge \psi \models \Delta$.

$\Longleftarrow$ By (REF) and (MON) we have $\varphi, \psi, \Gamma \models \Delta, \varphi$ and $\varphi, \psi, \Gamma \models \Delta, \psi$ which by ($\wedge'$) yield $\varphi, \psi, \Gamma \models \Delta, \varphi \wedge \psi$. The latter together with the assumption $\Gamma, \varphi \wedge \psi \models \Delta$ by (TR) lead to $\Gamma, \varphi, \psi \models \Delta$. $\square$

Scott was comparing his approach to Tarski's notion and observed that the relationship defined by

$$\Gamma \vdash \varphi \text{ iff } \varphi \in Cn(\Gamma)$$

is not necessarily unique. In fact, every Scott's consequence uniquely determines Tarki's one, but the converse, in general, does not hold. Every Cn determines rather a class of Scott's relations $\vdash$. We can put these observations more precisely

1. Every $\vdash \subseteq \mathcal{P}(FOR) \times \mathcal{P}(FOR)$ satisfying (REF), (MON), (TR) determines $Cn_{\vdash}(\Gamma) = \{\varphi : \Delta \vdash \varphi\}$, where Δ is any finite subset of Γ.

2. For every Cn, we can define two relations

$\Gamma \vdash_{min} \Delta$ iff $Cn(\Gamma) \cap \Delta \neq \varnothing$ and

$\Gamma \vdash_{max} \Delta$ iff $\bigcap_{\varphi \in \Delta} Cn(\Gamma' \cup \{\varphi\}) \subseteq Cn(\Gamma')$, for any $\Gamma' \supseteq \Gamma$.

Both relations are Scott's consequence relations, and moreover, they determine a class of consequence relations C such that for every $\vdash \in C$

- $\vdash_{min} \subseteq \vdash \subseteq \vdash_{max}$

- $Cn = Cn_{\vdash}$ iff $\vdash_{min} \subseteq \vdash \subseteq \vdash_{max}$

This result seems to suggest that the notion of Scott's consequence may be more subtle research tool than Tarski's one. Of course, in finite case both approaches are equivalent; directly, if we have a disjunction in the language, or in the more complex way otherwise (details may be found in Wójcicki [273]). However, in case of infinite sets a situation is different. Let us consider a set $H_{\vdash}$ of admissible valuations (homomorphisms) satisfying some (Tarski's or Scott's) $\vdash$ (i.e. $h \in H_{\vdash}$ iff for no $\Gamma \vdash \Delta$ holds $h\Gamma \subseteq \{1\}$ and $h\Delta \subseteq \{0\}$).

Since every semantics generates a relation of consequence, we can consider a relation generated by $H_{\vdash}$, i.e. $\vdash_{H_{\vdash}}$, and also a set of admissible valuations $H_{\vdash_H}$ satisfying $\vdash_H$. It can be easily shown that the following Galois properties hold:

1. $\vdash_1 \subseteq \vdash_2$ implies $H_{\vdash_2} \subseteq H_{\vdash_1}$

2. $H_1 \subseteq H_2$ implies $\vdash_{H_2} \subseteq \vdash_{H_1}$

3. $H \subseteq H_{\vdash_H}$

4. $\vdash \subseteq \vdash_{H_{\vdash}}$

One may find proofs of points 1 and 2 for Tarski's case in Wójcicki [273], and generalize it for Scott's relation. More interesting are points 3 and 4. The latter may be strengthened to equality on the basis of some abstract version of Lindenbaum lemma. This result holds also for Scott's relation but a proof requires different construction called Scott's Atlas (see Dunn and Hardegree [69]). The important difference between Tarski's and Scott's relations is expressed in point 3. Every Tarski's relation determines a semantics but not necessarily unique. On the other hand, for Scott's relation, this property may be strengthened to equality which means that every relation determines exactly one semantics. This property is dual to completeness; Dunn and Hardegree [69] call it absoluteness.

Generalizing Scott's approach to SCs based on multisets, we can obtain the following hierarchy of consequence relations (see Avron [8]):

- Any $\vdash$ on finite multisets satisfying (REF) and (TR) is a simple consequence relation.

- Simple consequence relation satisfying condition (C) (of contraction) and its converse (so defined on sets indeed) is regular.

- Regular consequence relation satisfying (MON) is a Scott's consequence relation.

- Regular consequence relation satisfying (MON) restricted to at most one formula in the succedent is Tarski's consequence relation.

In case of simple consequence relations, even satisfying (MON), we obtain a possibility of defining substructural logics, which is not possible on the basis of Scott's notion of consequence. For example, conditions $(\wedge), (\vee), (\rightarrow)$ characterise multiplicative constants whereas conditions $(\wedge'), (\vee'), (\rightarrow')$ define additive ones. The reader is invited to derive suitable rules for multiplicative and additive constants from respective conditions.

The last kind of relation, generated by sequent rules, i.e. in which $\vdash$ corresponds not to $\Rightarrow$ but to the horizontal line between premisses and conclusion, was investigated by Font, Jansana (see for example Font [88]). We do not discuss it here.

Bibliography

[1] D'AGOSTINO, M. 1999. Tableau Methods for Classical Propositional Logic. In *Handbook of Tableau Methods*, ed. M. D'Agostino, 45–123. Dordrecht: Kluwer Academic Publishers.

[2] AGUZZOLI, S., and A. CIABATTONI. 2000. Finiteness of Infinite-valued Lukasiewicz Logics. *Journal of Logic, Language and Information* 9: 5–29.

[3] AHO, A.V., and J.D. ULLMAN. 1995. *Foundations of Computer Science in C*. New York: W. H. Freeman and CO.

[4] ANDERSON, A.R., and N.D. BELNAP. 1975. *Entailment: The Logic of Relevance and Necessity*, vol I. Princeton: Princeton University Press.

[5] ANDREWS, P.B. 1986. *An Introduction to Mathematical Logic and Type Theory: To Truth Through Proof*. Orlando: Harcourt Academic Press.

[6] ASSER, G. 1959. *Einführung in die Mathematische Logik*, Leipzig 1959 (Teil I), 1972 (Teil II).

[7] AVRON, A. 1987. A Constructive Analysis of RM. *Journal of Symbolic Logic* 52: 939–951.

[8] AVRON, A. 1991. Simple Consequence Relations. *Information and Computation* 92: 105–139.

[9] AVRON, A. 1993. Gentzen-Type Systems, Resolution and Tableaux. *Journal of Automated Reasoning* 10 (2): 265–281.

[10] AVRON, A. 1996. The Method of Hypersequents in the Proof Theory of Propositional Non-Classical Logics. In *Logic: From Foundations to Applications*, ed. W. Hodges, et al., 1–32. Oxford: Oxford Science Publication.

[11] AVRON, A. 1998. Two Types of Multiple-Conclusion Systems. *Logic Journal of the IGPL* 6 (5): 695–717.

[12] AVRON, A. 2001. Classical Gentzen-Type Methods in Propositional Many-Valued Logics. In *Proceedings of IJCAR'01*, vol. 2083, 529–543. LNCS.

[13] AVRON, A., F. HONSELL, M. MICULAN, and C. PARAVANO. 1998. Encoding Modal Logics in Logical Frameworks. *Studia Logica* 60: 161–202.

© Springer Nature Switzerland AG 2021
A. Indrzejczak, *Sequents and Trees*, Studies in Universal Logic,
https://doi.org/10.1007/978-3-030-57145-0

[14] AVRON, A., and I. LEV. 2001. Canonical Propositional Gentzen-Type Systems. In *Proceedings of IJCAR'01*, vol. 2083, 529–543. LNCS.

[15] AVRON, A., O. ARIELI, and A. ZAMANSKY. 2018. *Theory of Effective Propositional Paraconsistent Logics*. College Publications.

[16] BAAZ, M., and A. CIABATTONI. 2002. A Schütte-Tait style cut-elimination proof for first-order Gödel Logic. In *Automated Reasoning with Tableaux and Related Methods, Tableaux '02*, vol. 2381 of LNAI, 24–38.

[17] BAAZ, M., A. CIABATTONI, and C.G. FERMÜLLER. 2003. Hypersequent Calculi for Gödel Logics – a Survey. *Journal of Logic and Computation* 13: 1–27.

[18] BAAZ, M., C.G. FERMÜLLER, and R. ZACH. 1992. Dual Systems of Sequents and Tableaux for Many-valued Logics. Technical Report TUW-E185.2-BFZ,2-92.

[19] BAAZ, M., C.G. FERMÜLLER, and R. ZACH. 1994. Elimination of cuts in first-order finite-valued logics. *Journal of Information Processing and Cybernetics* 29 (6): 333–355.

[20] BAELDE, D., A. LICK, and S. SCHMITZ. 2018. A Hypersequent Calculus with Clusters for Linear Frames. In *Advances in Modal Logic 12*, ed. G. Bezhanishvili et al., 43–62. College Publications.

[21] BASIN, D., S. MATHEWS, and L. VIGANO. 1997. Labelled Propositional Modal Logics: Theory and Practice. *Journal of Logic and Computation* 7 (6): 685–717.

[22] BASIN, D., S. MATHEWS, and L. VIGANO. 1998. Natural Deduction for Non-Classical Logics. *Studia Logica* 60: 119–160.

[23] BEALL, J.C., and B.C. VAN FRAASSEN. 2010. *Possibilities and Paradox. An Introduction to Modal and Many-Valued Logic*. Oxford.

[24] BEDNARSKA, K., and A. INDRZEJCZAK. 2015. Hypersequent Calculi for S5 - the Methods of Cut-elimination. *Logic and Logical Philosophy* 24 (3): 277–311.

[25] BELNAP, N.D. 1982. Display Logic. *Journal of Philosophical Logic* 11: 375–417.

[26] BERNAYS, P. 1965. Betrachtungen zum Sequenzen-Kalkul. In *Contributions to Logic and Methodology in honor of J. M. Bocheński*, ed. A.T. Tymieniecka, 1–44. North-Holland, Amsterdam.

[27] BETH, E. 1955. *Semantic Entailment and Formal Derivability*, Mededelingen der Kon. Ned. Akad. v. Wet. 18 13.

[28] BETH, E.W. 1959. *The Foundations of Mathematics*. North Holland, Amsterdam.

[29] BEZIAU, J-Y. 2001. Sequents and Bivaluations. *Logique et Analyse* 176: 373–394.

[30] BILKOVA, M. 2007. The Uniform Interpolation and Propositional Quantifiers in Modal Logics. *Studia Logica* 85 (1): 1–31.

[31] BIMBO, K. 2015. *Proof Theory. Sequent Calculi and Related Formalisma.* CRC Press.

[32] BLACKBURN, P. 2000. Internalizing Labelled Deduction. *Journal of Logic and Computation* 10 (1): 137–168.

[33] BLACKBURN, P., M. DERIJKE, and Y. VENEMA. 2001. *Modal Logic.* Cambridge: Cambridge University Press.

[34] BLAMEY, S., and L. HUMBERSTONE. 1991. A Perspective on Modal Sequent Logic. *Publications of the Research Institute for Mathematical Sciences, Kyoto University* 27: 763–782.

[35] BOLOTOV, A., A. BASUKOSKI, O. GRIGORIEV, and V. SHANGIN. 2006. Natural Deduction Calculus for Linear-Time Temporal Logic. In *Proceedings of Jelia 2006*, LNAI, Springer 4160.

[36] BOLOTOV A., O. GRIGORIEV, and V. SHANGIN. 2006. Natural Deduction Calculus for CTL. In *IEEE John Vincent Atanasoff Symposium on Modern Computing*, 175–183.

[37] BOOLOS, G. 1984. Don't Eliminate Cut. *Journal of Philosophical Logic* 7: 373–378.

[38] BOOLOS, G., T. BURGESS, and R.C. JEFFREY. 2007. *Computability and Logic.* Cambridge.

[39] BRAÜNER, T. 2009. *Hybrid Logic and its Proof-Theory*, Roskilde.

[40] BRIGHTON, J. 2015. Cut elimination for GLS using the terminability of its regress process. *Journal of Philosophical Logic* 45: 147–153.

[41] BRÜNNLER, T. 2000. A cut-free Gentzen formulation of the modal logic S5. *Logic Journal of the IGPL* 8: 629–643.

[42] BULL, R.A. 1992. Cut Elimination for Propositional Dynamic Logic without Star. *Zeitschrift für Mathematische Logik und Grundlagen der Mathematik* 38: 85–100.

[43] BUSS, S.R. 1998. An Introduction to Proof Theory. In *Handbook of Proof Theory*, ed. S. Buss. Elsevier.

[44] CARNIELLI, W.A. 1991. On Sequents and Tableaux for Many-valued Logics. *Journal of Non-Classical Logic* 8 (1): 59–76.

[45] CASARI, E. 1997. *Introduzione alla Logica.* Torino: UTET.

[46] CASTELLINI, C. 2005. *Automated Reasoning in Quantified Modal and Temporal Logic*, Ph.D. thesis, University of Edinburgh.

[47] CASTELLINI, C., and A. SMAILL. 2000. A systematic presentation of quantified modal logics. *Logic Journal of the IGPL* 10: 571–599.

[48] CHAGROV, A., and M. ZAKHARYASCHEV. 1997. *Modal Logic*. Oxford: Oxford University Press.

[49] CIABATTONI, A., D. GABBAY, and N. OLIVETTI. 1998. Cut-free proof systems for logics of weak excluded middle. *Soft Computing* 2 (4): 147–156.

[50] CIABATTONI. 2004. Automated Generation of Analytic Calculi for Logics with Linearity. In *Proceedings of CSL'04*, vol. 3210, 503–517. LNCS. *Studia Logica* 82: 95–119 (2006).

[51] CIABATTONI, A., and K. TERUI. 2006. Towards a Semantic Characterization of Cut-elimination. *Studia Logica* 82: 95–119.

[52] CIABATTONI, A., G. METCALFE, and F. MONTAGNA. 2010. Algebraic and proof-theoretic characterizations of truth stressers for MTL and its extensions. *Fuzzy Sets and Systems* 161 (3): 369–389.

[53] CIABATTONI, A., R. RAMANAYAKE, and H. WANSING. 2014. Hypersequent and Display Calculi – A Unified Perspective. *Studia Logica* 102 (6): 1245–1294.

[54] COPELAND, J.B. 2002. The Genesis of Possible World Semantics. *Journal of Philosophical Logic* 31: 99–137.

[55] CURRY, H.B. 1950. *A Theory of Formal Deducibility*. Notre Dame: University of Notre Dame Press.

[56] CURRY, H.B. 1963. *Foundations of Mathematical Logic*. New York: McGraw-Hill.

[57] DAVIS, M., and H. PUTNAM. 1960. A Computing Procedure for Quantification Theory. *Journal of the Association for Computing Machinery* 7: 201–215.

[58] DEGAUQUIER, V. 2016. Partial and Paraconsistent Three-Valued Logics. *Logic and Logical Philosophy* 25 (2): 143–171.

[59] DOHERTY, P. 1991. A constraint-based approach to proof-procedures for multi-valued logics. In *WOCFAI-91*, Paris.

[60] DOŠEN, K. 1985. Sequent-Systems for Modal Logic. *Journal of Symbolic Logic* 50: 149–159.

[61] DOŠEN, K. 1989. Logical constants as punctuation marks. *Notre Dame Journal of Formal Logic* 30: 362–381.

[62] Došen K., and P. Schroeder-Heister (ed.). 1994. *Substructural Logics.* Oxford: Oxford University Press.

[63] Dowek, G. 2011. *Proofs and Algorithms. An Introduction to Logic and Computability.* Springer.

[64] Dragalin, A.G. 1988. *Mathematical Intuitionism. Introduction to Proof Theory.* Providence: American Mathematical Society.

[65] Dummett, M. 2000. *Elements of Intuitionism.* Oxford.

[66] Dummett, M. 1993. *The Logical Basis of Metaphysics.* Harvard.

[67] Dunn, J.M. 1973. A "Gentzen system" for positive relevant implication. *Journal of Symbolic Logic* 38: 356–357.

[68] Dunn, J.M. 1986. Relevance Logic and Entailment. In *Handbook of Philosophical Logic*, vol. III, ed. D. Gabbay, F. Guenthner, 117–224. Dordrecht: Reidel Publishing Company.

[69] Dunn, J.M., and G.M. Hardegree. 2001. *Algebraic Methods in Philosophical Logic.* Oxford: Clarendon.

[70] Dyckhoff, R. 1992. Contraction-free sequent calculi for intuitionistic logic. *Journal of Symbolic Logic* 57 (3): 795–807.

[71] Dyckhoff, R. 1997. Dragalin's proof of cut-admissibility for the intuitionistic sequent calculi G3i and G3i. Research Report CS/97/8, St Andrews.

[72] Dyckhoff, R., and L. Pinto. 1996. A Permutation-free Sequent Calculus for Intuitionistic Logic. Research Report CS/96/9, St Andrews.

[73] Ebbinghaus, H.D., J. Flum, and W. Thomas. 1984. *Mathematical Logic.* Berlin: Springer.

[74] Ershow, Y.L., and E.A. Palyutin. 1984. *Mathematical Logic.* Moscow: MIR.

[75] Ferrari, M., C. Fiorentini, and G. Fiorino. 2013. Contraction-Free Linear Depth Sequent Calculi for Intuitionistic Propositional Logic with the Subformula Property and Minimal Depth Counter-Models. *Journal of Automated Reasoning* 51: 129–149.

[76] Feys, R. 1950. Les systemes formalises des modalites. *Revue Philosophicue de Louvain* 48: 478–509.

[77] Feys, R., and J. Ladriere. 1955. Supplementary notes in: *Recherches sur la deduction logique*, french translation of Gentzen, Press Univ. de France, Paris.

[78] Fine, K. 1985. *Reasoning with Arbitrary Objects.* Oxford: Blackwell.

[79] Fitch, F. 1952. *Symbolic Logic.* New York: Ronald Press Co.

[80] FITCH, F. 1966. Tree Proofs in Modal Logic (abstract). *The Journal of Symbolic Logic* 31: 152.

[81] FITTING, M. 1968. *Intuitionistic Model Theory*. Berlin: Springer.

[82] FITTING, M. 1972. Tableau methods of proof for Modal Logics. *Notre Dame Journal of Formal Logic* 13 (2): 237–247.

[83] FITTING, M. 1983. *Proof Methods for Modal and Intuitionistic Logics*. Dordrecht: Reidel.

[84] FITTING, M. 1995. Tableaus for many-valued modal logic. *Studia Logica* 55: 63–87.

[85] FITTING, M. 1996. *First-Order Logic and Automated Theorem Proving*. Berlin: Springer.

[86] FITTING, M. 1999. Simple propositional S5 tableau system. *Annals of Pure and Applied Logic* 96: 101–115.

[87] FITTING, M. 2007. Modal Proof Theory. In *Handbook of Modal Logic*, ed. P. Blackburn et al., 84–138. Elsevier.

[88] FONT, J.M., and R. Jansana. 1996. *A General Algebraic Semantics for Sentential Logics*. Berlin: Springer.

[89] FORBES, G. 2001. *Modern Logic*. New York.

[90] GAO, F., and G. TOURLAKIS. 2015. A short and readable proof of cut elimination for two first-order modal logics. *Bulletin of the Section of Logic* 44.

[91] GABBAY, D. 1996. *LDS - Labelled Deductive Systems*. Oxford: Clarendon Press.

[92] GARSON, J.W. 2006. *Modal Logic for Philosophers*. Cambridge: Cambridge University Press.

[93] GALLIER, J.H. 1986. *Logic for Computer Science*. New York: Harper and Row.

[94] GENTZEN, G. 1932. Über die Existenz unabhängiger Axiomensysteme zu unendlichen Satzsystemen. *Mathematische Annalen* 107: 329–350.

[95] GENTZEN, G. 1934. Untersuchungen über das Logische Schliessen. *Mathematische Zeitschrift* 39: 176–210 and 39: 405–431.

[96] GENTZEN, G. 1936. Die Widerspruchsfreiheit der reinen Zahlentheorie. *Mathematische Annalen* 112: 493–565.

[97] GENTZEN, G. 1938. Neue Fassung des Widerspruchsfreiheitsbeweises für die reine Zahlentheorie. *Forschungen zur Logik und zur Grundlegung der exakten Wissenschaften*, New Series 4, Leipzig 19–44.

[98] GIRARD, J.Y. 1987. Linear Logic. *Theoretical Computer Science* 50: 1–101.

[99] GIRARD, J.Y. 1987. *Proof Theory and Logical Complexity.* Napoli: Bibliopolis.

[100] GIRLE, R. 2000. *Elementary Modal Logic.* London: Acumen.

[101] GOLDBLATT, R.I. 1987. *Logics of Time and Computation,* CSLI Lecture Notes, Stanford.

[102] GORÉ, R. 1992. *Cut-free Sequent and Tableau Systems for Propositional Normal Modal Logics,* Ph.D. thesis, University of Cambridge.

[103] GORÉ, R. 1999. Tableau Methods for Modal and Temporal Logics. In *Handbook of Tableau Methods,* ed. M. D'Agostino, 297–396. Dordrecht: Kluwer Academic Publishers.

[104] GOUBAULT-LARRECQ, J., and I. MACKIE. 2001. *Proof Theory and Automated Deduction.* Kluwer.

[105] GÖDEL, K. 1930. Die Vollständigkeit der Axiome des Logischen Funktionenkalküls. *Monatschefte für Mathematik und Physik* 37: 349–360.

[106] HÄHNLE, R. 1994. *Automated Deduction in Multiple-Valued Logics.* Oxford University Press.

[107] HÄHNLE, R. 2001. Tableaux and Related Methods. In *Handbook of Automated Reasoning,* ed. A. Robinson, and A. Voronkov, 101–177. Amsterdam: Elsevier.

[108] D'AGOSTINO, M. (ed.). 1999. *Handbook of Tableau Methods.* Dordrecht: Kluwer Academic Publishers.

[109] HASENJAEGER, G. 1972. *Introduction to the Basic Concepts and Problems of Modern Logic.* Dordrecht: Reidel.

[110] HERBELIN, H. 1995. A lambda-calculus Structure Isomorphic to Gentzen-style Sequent Calculus Structure. In *Proceedings of LNCS 533,* 156–173. Springer.

[111] HERBRAND J. 1928. Abstract in: *Comptes Rendus des Seances de l'Academie des Sciences,* vol. 186, 1275 Paris.

[112] HERBRAND J. 1930. Recherches sur la theorie de la demonstration. In *Travaux de la Societe des Sciences et des Lettres de Varsovie, Classe III, Sciences Mathematiques et Physiques, Warsovie.*

[113] HERMES, H. 1963. *Einführung in die Mathematische Logik.* Stuttgart: Teubner.

[114] HERTZ, P. 1929. Über Axiomensysteme für beliebige Satzsysteme. *Mathematische Annalen* 101: 457–514.

[115] HEUERDING, A., M. SEYFRIED, and H. ZIMMERMANN. 1996. Efficient loop-check for backward proof search in some non-classical propositional logics. In *Tableaux 96, LNCS 1071*, ed. P. Miglioli, et al., 210–225. Berlin: Springer.

[116] HINTIKKA, J. 1955. Form and Content in Quantification Theory. *Acta Philosophica Fennica* 8: 8–55.

[117] HINTIKKA, J. 1957. Quantifiers in Deontic Logic. *Societas Scientiarum Fennica, Commentationes Humanarum Literarum XXIII.*

[118] HODGES, W. 1983. Elementary Predicate Logic. In *Handbook of Philosophical Logic*, vol. I, ed. D. Gabbay, and F. Guenther, 1–132. Dordrecht: Kluwer.

[119] HODGES, W. 1994. Logical Features of Horn Clauses. In *Handbook of Logic in AI and Logic Programming*, vol. I, ed. by D. Gabbay, C.J. Hogger, and J.A. Robinson, 449–503. Oxford: Clarendon.

[120] HORROCKS, I. 1997. *Optimising Tableaux Decision Procedures for Description Logics*, Ph.D. Thesis, University of Manchester.

[121] HORROCKS, I., U. SATLER, and S. TOBIES. 2000. Practical Reasoning for Very Expressive Description Logics. *Logic Journal of the IGPL* 8 (3): 239–263.

[122] HOWE, J.M. 1997. Two Loop Detection Mechanisms. A Comparison. In *TABLEAUX, LNiCS 1071*, ed. D. Galmiche, 210–225.

[123] HUDELMAIER, J. 1992. Bounds for Cut-elimination in Intuitionistic Propositional Logic. *Archive for Mathematical Logic* 31: 331–354.

[124] HUGHES, G.E., and M.J. CRESSWELL. 1996. *A New Introduction to Modal Logic.* London: Routledge.

[125] INDRZEJCZAK, A. 1996. Cut-free Sequent Calculus for S5. *Bulletin of the Section of Logic* 25 (2): 95–102.

[126] INDRZEJCZAK, A. 1997. Generalised Sequent Calculus for Propositional Modal Logics. *Logica Trianguli* 1: 15–31.

[127] INDRZEJCZAK, A. 2000. Multiple Sequent Calculus for Tense Logics. *Abstracts of AiML and ICTL* 2000: 93–104, Leipzig.

[128] INDRZEJCZAK, A. 2005. Sequent Calculi for Monotonic Modal Logics. *Bulletin of the Section of logic* 34 (3): 151–164.

[129] INDRZEJCZAK, A. 2009. Suszko's Contribution to the Theory of Nonaxiomatic Proof Systems. *Bulletin of the Section of Logic* 38 (3-4): 151–162.

[130] INDRZEJCZAK, A. 2010. *Natural Deduction, Hybrid Systems and Modal Logics.* Springer.

[131] INDRZEJCZAK, A. 2011. Admissibility of Cut in Congruent Modal Logics. *Logic and Logical Philosophy* 20 (3): 189–203.

[132] INDRZEJCZAK, A. 2012. Cut-free Hypersequent Calculus for S4.3. *Bulletin of the Section of Logic* 41 (1/2): 89–104.

[133] INDRZEJCZAK, A. 2014. A Survey of Nonstandard Sequent Calculi. *Studia Logica* 102 (6): 1295–1322.

[134] INDRZEJCZAK, A. 2014. Contraction Contracted. *Bulletin of the Section of logic* 43 (3-4): 139–153.

[135] INDRZEJCZAK, A. 2015. Eliminability of Cut in Hypersequent Calculi for some Modal Logics of Linear Frames. *Information Processing Letters* 115 (2): 75–81.

[136] INDRZEJCZAK, A. 2016. Linear Time in Hypersequent Framework. *The Bulletin of Symbolic Logic* 22 (1): 121–144.

[137] INDRZEJCZAK, A. 2016. Simple Cut Elimination Proof for Hybrid Logic. *Logic and Logical Philosophy* 25 (2): 129–141.

[138] INDRZEJCZAK, A. 2016. Simple Decision Procedure for S5 in Standard Cut-free Sequent Calculus. *The Bulletin of the Section of Logic* 45 (2): 125–140.

[139] INDRZEJCZAK, A. 2017. Tautology Elimination, Cut Elimination and S5. *Logic and Logical Philosophy* 26 (4): 461–471.

[140] INDRZEJCZAK, A. 2017. Cut Elimination Theorem for Non-Commutative Hypersequent Calculus. *The Bulletin of the Section of Logic* 46 (1-2): 135–149.

[141] INDRZEJCZAK, A., Fregean Description Theory in Proof-Theoretical Setting. *Logic and Logical Philosophy* 28 (1): 137–155.

[142] INDRZEJCZAK, A. 2018. Cut-Free Modal Theory of Definite Descriptions. In *Advances in Modal Logic 12*, ed. G. Bezhanishvili et al., 387–406. College Publications.

[143] INDRZEJCZAK, A. 2019. Cut Elimination in Hypersequent Calculus for Some Logics of Linear Time. *The Review of Symbolic Logic* 12 (4): 806–822.

[144] JAŚKOWSKI, S. 1934. On the Rules of Suppositions in Formal Logic. *Studia Logica* 1: 5–32.

[145] KAMIDE, N., and H. WANSING. 2015. *Proof Theory of N4-Related Paraconsistent Logics*. College Publications.

[146] KANGER, S. 1957. *Provability in Logic*. Stockholm: Almqvist & Wiksell.

[147] KASHIMA, R. 1994. Cut-free sequent calculi for some tense logics. *Studia Logica* 53: 119–135.

[148] KETONEN, O. 1944. *Untersuchungen zum Prädikatenkalkül*, Annalea Acad. Sci. Fenn. Ser. A. I. 32, Helsinki.

[149] KLEENE, S.C. 1952. *Introduction to Metamathematics*. North Holland, Amsterdam.

[150] KLEENE, S.C. 1967. *Mathematical Logic*. New York: Willey.

[151] KLEENE, S.C. 1952. Permutability of inferences in Gentzen's calculi LK and LJ. *Memoirs of the American Mathematical Society* 10: 1–26.

[152] KRIPKE, S. 1959. A Completeness Theorem in Modal Logic. *Journal of Symbolic Logic* 24: 1–14.

[153] KRIPKE, S. 1963. Semantical Analysis of Modal Logic I. *Zeitschrift für Mathematische Logik und Grundlegen der Mathematik* 9: 67–96.

[154] KUROKAWA, H. 2014. Hypersequent Calculi for Modal Logics Extending S4. In *New Frontiers in Artificial Intelligence (2013)*, 51–68. Springer.

[155] KUZNETS, R., and B. LELLMANN. 2018. Interpolation for Intermediate Logics via Hyper- and Linear Nested Sequents. In *Advances in Modal Logic 12*, ed. G. Bezhanishvili et al., 473–492. College Publications.

[156] LAHAV, O. 2013. From Frame Properties to Hypersequent Rules in Modal Logics. In *Proceedings of LICS*, 408–417. Springer.

[157] LEBLANC, H. 1963. Proof routines for the propositional calculus. *Notre Dame Journal of Formal Logic* 4 (2): 81–104.

[158] LEBLANC, H. 1966. Two separation theorems for natural deduction. *Notre Dame Journal of Formal Logic* 7 (2): 81–104.

[159] LELLMANN, B., and D. PATTINSON. 2013. Correspondence between modal Hilbert axioms and sequent rules with an application to S5. In *TABLEAUX 2013, LNCS 8123*, 219–233. Springer.

[160] LELLMANN, B. 2014. Axioms vs hypersequent rules with context restrictions. In *Proceedings of IJCAR*, 307–321. Springer.

[161] LELLMANN, B. 2015. Linear nested sequents, 2-sequents and hypersequents. In *TABLEAUX 2015, LNAI 9323*, 135–150. Springer.

[162] LEMMON, E.J. 1965. *Beginning Logic*. London: Nelson.

[163] LESZCZYŃSKA-JASION, D., M. URBAŃSKI, and A. WIŚNIEWSKI. 2013. Socratic Trees. *Studia Logica* 101 (5): 959–986.

[164] LESZCZYŃSKA-JASION, D. 2018. *From Questions to Proofs. Between Logic of Questions and Proof Theory*. Poznań: Facutly of Social Sciences Publishers.

[165] LEŚNIEWSKI, S. 1929. Gründzuge eines Neuen Systems der Grundlagen der Mathematik. *Fundamenta Mathematicae* 14: 1–81.

[166] LYALETSKI, A.V. 2011. A note on the cut rule. In *Abstracts of the International Conference "Maltsev Meeting"*, vol. 137. Novosibirsk.

[167] LAWROW, I.A., and Ł.L. MAKSIMOWA. 2004. *Zadania z teorii mnogości, logiki matematycznej i teorii algorytmów.* Warszawa: PWN.

[168] MAEHARA, S. 1954. Eine Darstellung der Intuitionistiche Logik in der Klassischen. *Nagoya Mathematical Journal* 7: 45–64.

[169] MAEHARA, S. 1970. A general theory of completeness proofs. *Annals of the Japan Association for Philosophy of Science* 3 (5): 242–256.

[170] MALINOWSKI, G. 1993. *Many-Valued Logics.* Oxford.

[171] MARX, M., S. MIKULAS, and M. REYNOLDS. 2000. The Mosaic Method for Temporal Logics. In *Automated Reasoning with Analytic Tableaux and Related Methods, Proceedings of International Conference TABLEAUX 2000,* Saint Andrews, Scotland, LNAI 1847, ed. R. Dyckhoff, 324–340. Springer.

[172] MARTIN-LÖF, P. 1968. *Notes on Constructive Mathematics.* Almqvist and Wiksell.

[173] MASSACCI, F. 1994. Strongly Analytic Tableaux for Normal Modal Logics. In *Proc. CADE-12,* ed. A. Bundy. LNAI 814: 723–737. Springer.

[174] MASSACCI, F. 1998. Single Step Tableaux for Modal Logics: methodology, computations, algorithms. *Technical Report TR-04,* Dipartimento di Informatica e Sistemistica, Universita di Roma "La Sapienza".

[175] MENDELSON, E. 1964. *Introduction to Mathematical Logic.* Chapman and Hall.

[176] METCALFE, G., N. OLIVETTI, and D. GABBAY. 2008. *Proof Theory for Fuzzy Logics.* Springer.

[177] MINTS, G. 1968. Some calculi of modal logic. [in Russian]. *Trudy Mat. Inst. Steklov* 98: 88–111.

[178] MINTS, G. 1972. Cut elimination theorem in relevant logics. [in Russian]. In *Essays of Constructive Mathematics and Mathematical Logic,* ed. J.V. Matijasevitch, and O.A. Silenko, 90–97. Nauka.

[179] MINTS, G. 1974. Systems of Lewis and system T. [in Russian]. Supplement to Russian edition of Feys R., *Modal Logic,* 422–509. Nauka.

[180] MINTS, G. 1970. Cut-free calculi of the S5 type. *Studies in Constructive Mathematics and Mathematical Logic* II: 79–82.

[181] MINTS, G. 1992. *Selected Papers in Proof Theory.* North-Holland.

[182] MINTS, G. 1997. Indexed systems of sequents and cut elimination. *Journal of Philosophical Logic* 26: 671–696.

[183] MINTS, G. 2002. *A Short Introduction to Intuitionistic Logic.* Kluwer.

[184] MONDADORI, M. 1988. Classical analytical deduction. *Annali dell'Universita di Ferrara*.

[185] NEGRI, S., and J. von PLATO. 2001. *Structural Proof Theory*. Cambridge: Cambridge University Press.

[186] NEGRI, S., and J. von PLATO. 2011. *Proof Analysis. A Contribution to Hilbert's Last Problem*. Cambridge: Cambridge University Press.

[187] NEGRI, S. 2005. Proof Analysis in Modal Logic. *Journal of Philosophical Logic* 34: 507–544.

[188] NISHIMURA, H. 1980. A Study of Some Tense Logics by Gentzen's Sequential Method. *Publications of the Research Institute for Mathematical Sciences, Kyoto University* 16: 343–353.

[189] OHNISHI, M., and K. MATSUMOTO. 1957. Gentzen Method in Modal Calculi I. *Osaka Mathematical Journal* 9: 113–130.

[190] OHNISHI, M., and K. MATSUMOTO. 1959. Gentzen Method in Modal Calculi II. *Osaka Mathematical Journal* 11: 115–120.

[191] ONO, H. 1998. Proof-Theoretic Methods in Nonclassical Logic – An Introduction. In *Theories of Types and Proofs*, MSJ-Memoir 2, ed. M. Takahashi, 207–254. Mathematical Society of Japan.

[192] ORŁOWSKA, E., and J. GOLIŃSKA-PILAREK. 2011. *Dual Tableaux: Foundations, Methodology, Case Studies*. Springer.

[193] PAOLI, F. 2002. *Substructural Logics: A Primer*. Dordrecht: Kluwer.

[194] PERZANOWSKI, J. 1973. The Deduction Theorems for the modal propositional calculi formalised after the manner of Lemmon I. *Reports on Mathematical Logic* 1: 1–12.

[195] PERZANOWSKI, J. 1976. The Deduction Theorems for the modal propositional calculi formalised after the manner of Lemmon II. *Reports on Mathematical Logic* 1: 1–12.

[196] PFENNING, F. 2000. Structural cut Elimination. *Information and Computation* 157: 84–141.

[197] VON PLATO, J. 2008. Gentzen's proof of normalization for ND. *The Bulletin of Symbolic Logic* 14 (2): 240–257.

[198] POGGIOLESI, F. 2011. *Gentzen Calculi for Modal Propositional Logic*. Springer.

[199] POGORZELSKI, W.A. 1973. *Klasyczny rachunek zdań*. Warszawa: PWN.

[200] POPPER, K. 1947. Logic without assumptions. *Proceedings of the Aristotelian Society* 47: 251–292.

[201] POPPER, K. 1947. New Foundations for Logic. *Mind* 56.

[202] POTTINGER, G. 1983. Uniform cut-free formulations of T, S4 and S5 (abstract). *Journal of Symbolic Logic* 48: 900.

[203] PRAWITZ, D. 1965. *Natural Deduction*. Stockholm: Almqvist and Wiksell.

[204] PŘENOSIL, A. 2017. Cut Elimination, Identity Elimination and Interpolation in Super-Belnap Logics. *Studia Logica* 105 (6): 1255–1290.

[205] PRIEST, G. 2001. *An Introduction to Non-classical Logic*. Cambridge: Cambridge University Press.

[206] PRIOR, A.N. 1960. A runabout inference ticket. *Analysis* 21: 38–39.

[207] PRIJATELJ, A. 1996. Contraction and Gentzen style formulation of Lukasiewicz logics. *Studia Logica* 57: 437–456.

[208] VAN QUINE, W.O. 1950. *Methods of Logic*. New York: Colt.

[209] RAMANAYAKE, R. 2015. Embedding the Hypersequent Calculus in the Display Calculus. *Journal of Logic and Computation* 25 (3): 921–942.

[210] RASIOWA, H., and R. SIKORSKI. 1963. *The Mathematics of Metamathematics*. Warszawa: PWN.

[211] RENTERIA, C., and E. HAUSLER. 2002. Natural Deduction for CTL. *Bulletin of the Section of Logic* 31 (4): 231–240.

[212] RESCHER, N. 1969. *Many-Valued Logic*. New York: McGraw-Hill.

[213] RESCHER, N., and A. URQUHART. 1971. *Temporal Logic*. New York: Springer.

[214] RESTALL, G. 2000. *Substructural Logics*. Routledge.

[215] RESTALL, G. 2007. Proofnets for S5: sequents and circuits for modal logic. *Lecture Notes in Logic* 28: 151–172.

[216] RESTALL, G., *Proof Theory and Philosophy*. http://consequently.org/writing/ptp.

[217] RIEGER, L. 1967. *Algebraic Methods of Mathematical Logic*. Prague: Academia.

[218] RIPLEY, D. 2012. Conservatively Extending Classical Logic with Transparent truth. *The Review of Symbolic Logic* 5 (2): 354–378.

[219] ROBINSON, J.A. 1965. A Machine Oriented Logic based on the Resolution Principle. *Journal of the Association for Computing Machinery* 12: 23–41.

[220] ROORDA, D. 1991. *Resource Logic*. Amsterdam.

[221] ROUSSEAU, G. 1967. Sequents in Many Valued Logic. *Fundamenta Mathematicae*, LX 1: 22–23.

[222] RUSSO, A. 1995. Modal Labelled Deductive Systems. Department of Computing, Imperial College, London, Technical Report 95/7.

[223] SAMBIN, G., and S. VALENTINI. 1980. A Modal Sequent Calculus for a Fragment of Arithmetic. *Studia Logica* 39: 245–256.

[224] SATO, M. 1977. A Study of Kripke-type Models for Some Modal Logics by Gentzen's Sequential Method. *Publications of the Research Institute for Mathematical Sciences, Kyoto University* 13: 381–468.

[225] SATO, M. 1980. A Cut-Free Gentzen-Type System for the Modal Logic S5. *The Journal of Symbolic Logic* 45 (1): 67–84.

[226] SCHROEDER-HEISTER, P. 1984. Popper's theory of deductive inference and the concept of a logical constant. *History and Philosophy of Logic* 5: 79–110.

[227] SCHROEDER-HEISTER, P. 2002. Resolution and the origins of structural reasoning: early proof-theoretic ideas of Hertz and Gentzen. *The Bulletin of Symbolic Logic* 8 (2): 246–265.

[228] SCHROEDER-HEISTER, P. 2006. Popper's structuralist theory of logic. In *Karl Popper: A Centenary Assesment, vol III: Science*, ed. I. Jarvie, K. Milford, and D. Miller, 17–36. Aldershot: Ashgate Publishing.

[229] SCHROEDER-HEISTER, P., Proof-theoretic Semantics. In *The Stanford Encyclopedia of Philosophy*, ed. E.N. Zalta. https://plato.stanford.edu

[230] SCHROEDER-HEISTER, P. 2014. The Calculus of Higher-Level Rules, Propositional Quantification and the Foundational Approach to Proof-Theoretic Harmony. *Studia Logica* 102 (6): 1185–1216.

[231] SCHRÖTER, K. 1955. Methoden zur Axiomatisierung beliebiger Aussagen- und Praedikaten-kalküle. *Zeitschrift für Mathematische Logik und Grundlagen der Mathematik* 1: 241–251.

[232] SCHÜTTE, K. 1977. *Proof Theory.* Berlin: Springer.

[233] SCHWICHTENBERG, H. 1977. Proof Theory. In *Handbook of Mathematical Logic*, vol. 1, ed. J. Barwise. North-Holland, Amsterdam.

[234] SCOTT, D. 1974. Rules and derived rules. In *Logical Theory and Semantical Analysis*, ed. S. Stenlund, 147–161.

[235] SEGERBERG, K. 1982. *Classical Propositional Operators.* Oxford: Clarendon Press.

[236] SEREBRIANNIKOV, O. 1982. Gentzen's Hauptsatz for Modal Logic with Quantifiers. In *Intensional Logic: Theory and Applications; Acta Philosophica Fennica 35*, ed. I. Niniluoto, and E. Saarinen, 79–88.

[237] SHIMURA, T. 1991. Cut-free systems for the modal logic S4.3 and S4.3GRZ. *Reports on Mathematical Logic* 25: 57–73.

[238] SHOESMITH, D.J., and T.J. SMILEY. 1978. *Multiple-Conclusion Logic.* Cambridge.

[239] SHRAMKO, Y., and H. WANSING. 2008. Suszko's Thesis, Inferential Many-Valuedness, and a Notion of a Logical System. *Studia Logica* 88: 405–429.

[240] SHVARTS, G.F. 1989. Gentzen style systems for K45 and K45D. *Logic at Botic '89*, 245–256. Berlin: Springer.

[241] SIMPSON, A. 1994. *The Proof Theory and Semantics of Intuitionistic Modal Logic*, Ph.D. thesis, University of Edinburgh.

[242] SMULLYAN, R. 1965. Analytic Natural Deduction. *The Journal of Symbolic Logic* 30 (2): 123–139.

[243] SMULLYAN, R. 1966. Trees and Nest Structures. *The Journal of Symbolic Logic* 31 (3): 303–321.

[244] SMULLYAN, R. 1968. Analytic Cut. *The Journal of Symbolic Logic* 33 (4): 560–564.

[245] SMULLYAN, R. 1968. *First-Order Logic.* Berlin: Springer.

[246] STOUPPA, P. 2007. A deep inference system for the modal logic S5. *Studia Logica* 85: 199–214.

[247] SUCHOŃ, W. 1974. La methode de Smullyan de construire le calcul n-valent de Lukasiewicz avec implication and negation. *Reports on Mathematical Logic* 2: 37–42.

[248] SUNDHOLM, G. 1984. Systems of Deduction. In *Handbook of Philosophical Logic*, vol. I, ed. D. Gabbay, and F. Guenthner, 133–188. Dordrecht: Reidel Publishing Company.

[249] SUPPES, P. 1957. *Introduction to Logic.* Princeton: Van Nostrand.

[250] SURMA, S.J. 1965. *Wprowadzenie do metamatematyki*, vol. I. Kraków.

[251] SURMA, S.J. 1974. A method of the construction of finite Łukasiewiczian algebras and its application to a Gentzen-style characterisation of finite logics. *Reports on Mathematical Logic* 2: 49–54.

[252] SUSZKO, R. 1948. W sprawie logiki bez aksjomatów. *Kwartalnik Filozoficzny* 17 (3/4): 199–205.

[253] SUSZKO, R. 1949. *O analitycznych aksjomatach i logicznych regułach wnioskowania*, Poznańskie Towarzystwo Przyjaciół Nauk, Prace Komisji Filozoficznej 7/5.

[254] SUSZKO, R. 1957. Formalna teoria wartości logicznych. *Studia Logica* 6: 145–320.

[255] SUSZKO, R. 1977. The Fregean Axiom and Polish Mathematical Logic in the 1920's. *Studia Logica* 36 (4): 377–380.

[256] SZABO, M.E. 1969. *The Collected Papers of Gerhard Gentzen.* North-Holland, Amsterdam.

[257] TAIT, W.W. 1968. Normal Derivability in Classical Logic. In *The Sintax and Semantics of Infinitary Languages*, LNM 72, 204–236.

[258] TAKAHASHI, M. 1967. Many-valued logics of extended Gentzen style I. *Science Reports of the Tokyo Kyoiku Daigaku* 9 (231): 95–116.

[259] TAKANO, M. 1992. Subformula Property as a substitute for Cut-Elimination in Modal Propositional Logics. *Mathematica Japonica* 37 (6): 1129–1145.

[260] TAKANO, M. 2001. A Modified Subformula Property for the Modal Logics K5 and K5D. *Bulletin of the Section of Logic* 30 (2): 115–123.

[261] TAKEUTI, G. 1987. *Proof Theory.* North-Holland, Amsterdam.

[262] TARSKI, A. 1930. Fundamentale Begriffe der Methodologie der Deduktiven Wissenschaften. *Monatschefte für Mathematik und Physik* 37: 361–404.

[263] TROELSTRA, A.S. 1992. *Lectures on Linear Logic.* Stanford: CSLI Publications.

[264] TROELSTRA, A.S., and H. Schwichtenberg. 1996. *Basic Proof Theory.* Oxford: Oxford University Press.

[265] URQUHART, A. 1984. The Undecidability of Entailment and Relevant Implication. *The Journal of Symbolic Logic* 49: 1059–1073.

[266] URQUHART, A. 1986. Many-valued Logics. In *Handbook of Philosophical Logic*, vol. III, ed. D. Gabbay, and F. Guenthner, 71–116. Dordrecht: Reidel Publishing Company.

[267] URQUHART, A. 1995. The Complexity of Propositional Proofs. *The Bulletin of Symbolic Logic* 1: 425–467.

[268] WAALER, A., and L. WALLEN. 1999. Tableaux for Intuitionistic Logics. In *Handbook of Tableau Methods*, ed. M. D'Agostino, 255–296. Dordrecht: Kluwer Academic Publishers.

[269] WANSING, H. 1999. *Displaying Modal Logics.* Dordrecht: Kluwer Academic Publishers.

[270] WANSING, H. 2002. Sequent Systems for Modal Logics. In *Handbook of Philosophical Logic*, vol. IV, ed. D. Gabbay, and F. Guenthner, 89–133. Dordrecht: Reidel Publishing Company.

[271] WINTEIN, S., and R. MUSKENS. 2017. Interpolation Methods for Dunn Logics and Their Extensions. *Studia Logica* 105 (6): 1319–1348.

[272] WIŚNIEWSKI, A. 2004. Socratic Proofs. *Journal of Philosophical Logic* 33 (3): 299–326.

[273] WÓJCICKI, R. 1988. *Theory of Logical Calculi.* Dordrecht: Kluwer.

[274] VIGANO, L. 2000. *Labelled Non-Classical Logics.* Kluwer.

[275] VOROBEV, N.N. 1952. The derivability problem in the constructive propositional calculus with strong negation. [in Russian]. *Doklady Akademii Nauk SSSR* 85: 689–692.

[276] ZEMAN, J.J. 1973. *Modal Logic.* Oxford: Oxford University Press.

[277] ZUCKER, J., and R. TRAGESSER. 1987. The adequacy problem for inferential logic. *Journal of Philosophical Logic* 7: 501–516.

Index

Symbols

© Springer Nature Switzerland AG 2021
A. Indrzejczak, *Sequents and Trees*, Studies in Universal Logic,
https://doi.org/10.1007/978-3-030-57145-0